W9-CHX-571

Computational Methods in Stochastic Dynamics

Computational Methods in Applied Sciences

Volume 26

Series Editor

E. Oñate
International Center for Numerical Methods in Engineering (CIMNE)
Technical University of Catalonia (UPC)
Edificio C-1, Campus Norte UPC
Gran Capitán, s/n
08034 Barcelona Spain
onate@cimne.upc.edu
www.cimne.com

For further volumes:
www.springer.com/series/6899

Manolis Papadrakakis • George Stefanou •
Vissarion Papadopoulos
Editors

Computational Methods in Stochastic Dynamics

Volume 2

Editors
Manolis Papadrakakis
Institute of Structural Analysis &
Antiseismic Research
National Technical University of Athens
Athens, Greece

Vissarion Papadopoulos
Institute of Structural Analysis &
Antiseismic Research
National Technical University of Athens
Athens, Greece

George Stefanou
Institute of Structural Analysis &
Antiseismic Research
National Technical University of Athens
Athens, Greece

ISSN 1871-3033 Computational Methods in Applied Sciences
ISBN 978-94-007-5133-0 ISBN 978-94-007-5134-7 (eBook)
DOI 10.1007/978-94-007-5134-7
Springer Dordrecht Heidelberg New York London

Library of Congress Control Number: 2012949417

Printed on acid-free paper

Springer is part of Springer Science+Business Media (www.springer.com)

This book is dedicated to the memory of Prof. Gerhart I. Schuëller, a pioneer in the field of Computational Stochastic Dynamics

Preface

The considerable influence of inherent uncertainties on structural behavior has led the engineering community to recognize the importance of a stochastic approach to structural problems. Issues related to uncertainty quantification and its influence on the reliability of the computational models, are continuously gaining in significance. In particular, the problems of dynamic response analysis and reliability assessment of structures with uncertain system and excitation parameters have been the subject of continuous research over the last two decades as a result of the increasing availability of powerful computing resources and technology. This book is a follow up of a previous book with the same subject and focuses on advanced computational methods and software tools which can highly assist in tackling complex problems in stochastic dynamic/seismic analysis and design of structures. The selected chapters are authored by some of the most active scholars in their respective areas and represent some of the most recent developments in this field.

This edited book is primarily intended for researchers and post-graduate students who are familiar with the fundamentals and wish to study or to advance the state of the art on a particular topic in the field of computational stochastic structural dynamics. Nevertheless, practicing engineers could benefit as well from it as most code provisions tend to incorporate probabilistic concepts in the analysis and design of structures. The book consists of 21 chapters which are extended versions of papers presented at the recent COMPDYN 2011 Conference. The chapters can be grouped into several thematic topics including dynamic analysis of stochastic systems, reliability-based design, structural control and health monitoring, model updating, system identification, wave propagation in random media, seismic fragility analysis and damage assessment.

In Chap. 1, A. Batou and C. Soize examine the random dynamic response of a multibody system with uncertain rigid bodies. A stochastic model of an uncertain rigid body is constructed by modeling the mass, the center of mass and the tensor of inertia by random variables. The prior probability distributions of these random variables are computed using the maximum entropy principle under the constraints defined by the available information. Several uncertain rigid bodies are linked to each other in order to calculate the random response of a multibody dynamic system.

A numerical application consisting of five rigid bodies is proposed to illustrate the theoretical developments.

In Chap. 2, V. Papadopoulos and O. Kokkinos extend the concept of Variability Response Functions (VRFs) to linear stochastic systems under dynamic excitation. An integral form for the variance of the dynamic response of stochastic systems is considered, involving a Dynamic VRF (DVRF) and the spectral density function of the stochastic field modeling the uncertain system properties. The uncertain property considered is the flexibility of the system. The same integral expression can be used to calculate the mean response of a dynamic system using a Dynamic Mean Response Function (DMRF) which is a function similar to the DVRF. These integral forms are used to efficiently compute the mean and variance of the transient system response along with time dependent spectral-distribution-free upper bounds.

A. Kundu and S. Adhikari provide the theoretical development and simulation results of a novel Galerkin subspace projection scheme for damped linear dynamic systems with stochastic coefficients and homogeneous Dirichlet boundary conditions (Chap. 3). The fundamental idea is to solve the stochastic dynamic system in the frequency domain by projecting the solution into a reduced finite dimensional spatio-random vector basis spanning the stochastic Krylov subspace to approximate the response. Galerkin weighting coefficients are employed to minimize the error induced by the use of the reduced basis. The statistical moments of the solution are evaluated at all frequencies to illustrate and compare the stochastic system response with the deterministic case. The results are validated with direct Monte Carlo simulation for different correlation lengths and variability of randomness.

An efficient approach for modeling nonlinear systems subjected to general non-Gaussian excitations is developed by X.F. Xu and G. Stefanou in Chap. 4. This chapter describes the formulation of an n-th order convolved orthogonal expansion (COE) method. For linear vibration systems, the statistics of the output are directly obtained as the first-order COE about the underlying Gaussian process. The COE method is next verified by its application on a weakly nonlinear oscillator. In dealing with strongly nonlinear dynamics problems, a variational method is presented by formulating a convolution-type action and using the COE representation as trial functions.

In Chap. 5 by L. Pichler et al., various finite difference (FD) and finite element methods (FEM) are discussed for the numerical solution of the Fokker–Planck equation allowing the investigation of the evolution of the probability density function of linear and nonlinear systems. The results are compared using various numerical examples. Despite the greater numerical effort, the FEM is preferable over FD, because it yields more accurate results. However, at this moment the FEM is only suitable for dimension less or equal to 3. In the case of 3D and 4D problems, a stabilized multi-scale FEM provides a tool with a high order of accuracy, preserving numerical efficiency due to the fact that a coarser mesh size can be used.

There are various approaches to deal with uncertainty propagation in stochastic dynamics. In Chap. 6, M. Corradi et al. examine some classical structural problems in order to investigate which probabilistic approach better propagates the uncertainty from input to output, in terms of accuracy and computational cost. The examined methods are: Univariate Dimension Reduction methods, Polynomial Chaos Expansion, First-Order Second Moment method, and algorithms based on the Evidence Theory for epistemic uncertainty. The performances of these methods are compared in terms of moment estimations and probability density function construction corresponding to several scenarios of reliability-based design and robust design. The structural problems examined are: (i) the static, dynamic and buckling behavior of a composite plate, (ii) the reconstruction of the deformed shape of a beam from measured surface strains.

Chapter 7 by F. Steinigen et al. is devoted to enhanced computational algorithms to simulate the load-bearing behavior of reinforced concrete structures under dynamic loading. In order to take into account uncertain data of reinforced concrete, fuzzy and fuzzy stochastic analyses are presented. The capability of the fuzzy dynamic analysis is demonstrated by an example in which a steel bracing system and viscous damping connectors are designed to enhance the structural resistance of a reinforced concrete structure under seismic loading.

W. Verhaeghe et al. use the concept of interval fields to deal with uncertainties of spatial character arising in the context of groundwater transport models needed to predict the flow of contaminants (Chap. 8). The main focus of the chapter is on the application of interval fields to a geo-hydrological problem. The uncertainty taken into account is the material layers' hydraulic conductivity. The results presented are the uncertainties on the contaminant's concentration near a river. Another objective of the chapter is to define an input uncertainty elasticity of the output, i.e. to identify the locations in the model, whose uncertainties mostly influence the uncertainty on the output. Such a quantity indicates where to perform additional in situ point measurements to reduce the uncertainty on the output the most.

Although reliability analysis methods have matured in recent years, the problem of reliability-based structural design still poses a challenge in stochastic dynamics. In Chap. 9, A. Naess et al. extend their recently developed enhanced Monte Carlo approach to the problem of reliability-based design. The objective is to optimize a design parameter α so that the system, represented by a set of failure modes or limit states, achieves a target reliability. Monte Carlo sampling occurs at a range of values for α that result in failure probabilities larger than the target and thus the design problem essentially amounts to a statistical estimation of a high quantile. Several examples of the approach are provided in the chapter.

Chapter 10 by H. Jensen et al. presents a general framework for reliability-based design of base-isolated structural systems under uncertain conditions. The uncertainties about the structural parameters as well as the variability of future excitations are characterized in a probabilistic manner. Nonlinear elements composed by hysteretic devices are used for the isolation system. The optimal design problem is formulated as a constrained minimization problem which is solved by a sequential approximate optimization scheme. First excursion probabilities that account for the uncertainties in the system parameters as well as in the excitation are used to characterize the system reliability. The approach explicitly takes into account all nonlinear characteristics of the combined structural system (superstructure-isolation system) during the design process. Numerical results highlight the beneficial effects of isolation systems in reducing the superstructure response.

The influence of structural uncertainties on actively controlled smart beams is investigated in Chap. 11 by A. Moutsopoulou et al. The dynamical problem of a model smart composite beam is treated using a simplified modeling of the actuators and sensors, both being realized by means of piezoelectric layers. In particular, a practical robust controller design methodology is developed, which is based on recent theoretical results on H_∞ control theory and μ-analysis. Numerical examples demonstrate the vibration-suppression property of the proposed smart beams under stochastic loading.

The field of Structural Health Monitoring (SHM) has significantly evolved in the last years due to the technological advances and the evolution of advanced smart systems for damage detection and signal processing. In Chap. 12, G. Saad and R. Ghanem present a robust data assimilation approach based on a stochastic variation of the Kalman Filter where polynomial functions of random variables are used to represent the uncertainties inherent to the SHM process. The presented methodology is combined with a non-parametric modeling technique to tackle SHM of a four-story shear building subjected to a base motion consistent with the El-Centro earthquake and undergoing a preset damage in the first floor. The purpose of the problem is localizing the damage in both space and time, and tracking the state of the system throughout and subsequent to the damage time. The application of the introduced data assimilation technique to SHM enhances its applicability to a wide range of structural problems with strongly nonlinear dynamic behavior and with uncertain and complex governing laws.

The accurate prediction of the response of spacecraft systems during launch and ascent phase is a crucial aspect in design and verification stages which requires accurate numerical models. The enhancement of numerical models based on experimental data is denoted model updating and focuses on the improvement of the correlation between finite element (FE) model and test structure. In aerospace industry, the examination of the agreement between model and real structure involves the comparison of the modal properties of the structure. Chapter 13 by B. Goller et al. is devoted to the efficient model updating of a satellite in a Bayesian setting based on experimental modal data. A detailed FE model of the satellite is used for demonstrating the applicability of the employed updating procedure to large-scale complex aerospace structures.

In Chap. 14, B. Rosič and H. Matthies deal with the identification of properties of stochastic elastoplastic systems in a Bayesian setting. The inverse problem is formulated in a probabilistic framework where the unknown uncertain quantities are embedded in the form of their probability distributions. With the help of stochastic functional analysis, a new update procedure is introduced as a direct, purely algebraic way of computing the posterior, which is comparatively inexpensive to evaluate. Such description requires the solution of the convex minimization problem in a stochastic setting for which the extension of the classical optimization algorithm in predictor-corrector form is proposed as the solution procedure. The identification method is finally validated through a series of virtual experiments taking into account the influence of the measurement error and the order of the approximation on the posterior estimate.

Chapter 15 deals with the study of SH surface waves in a half space with random heterogeneities. C. Du and X. Su prove both theoretically and numerically that surface waves exist in a half space which has small, random density, but the mean value of the density is homogeneous. Historically, this type of half space is often treated as homogeneous using deterministic methods. In this investigation, a closed-form dispersion equation is derived stochastically and the frequency spectrum, dispersion equation and phase/group velocity are computed numerically to study how the random inhomogeneities will affect the dispersion properties of the half space with random density. The results of this research may find their application in various fields, such as in seismology and in non-destructive test/evaluation of structures with randomly distributed micro-cracks or heterogeneities.

The following six chapters are devoted to earthquake engineering applications. P. Jehel et al. (Chap. 16) investigate the seismic fragility of a moment-resisting reinforced concrete frame structure in the area of the Cascadia subduction zone situated in the South-West of Canada and the North-West of the USA. According to shaking table tests, the authors first validate the capability of an inelastic fiber beam/column element, using a recently developed concrete constitutive law, for representing the seismic behavior of the tested frame coupled to either a commonly used Rayleigh damping model or a proposed new model. Then, for each of the two damping models, they perform a structural fragility analysis and investigate the amount of uncertainty to be induced by damping models.

In Chap. 17 by Y. Vargas et al., a detailed study of the seismic response of a reinforced concrete building is conducted using a probabilistic approach in the framework of Monte Carlo simulation. The building is representative for office buildings in Spain but the procedures used and the results obtained can be extended to other types of buildings. The purpose of the work is twofold: (i) to analyze the differences when static and dynamic analysis techniques are used and (ii) to obtain a measure of the uncertainties involved in the assessment of structural vulnerability. The results show that static procedures are somehow conservative and that uncertainties increase with the severity of the seismic actions and with the damage. Low damage state fragility curves have little uncertainty while high damage state fragility curves show great scattering.

Seismic pounding can induce severe damage and losses in buildings. The corresponding risk is particularly relevant in densely inhabited metropolitan areas, due to the inadequate clearance between buildings. Chapter 18 by E. Tubaldi and M. Barbato proposes a reliability-based procedure for assessing the level of safety corresponding to a given value of the separation distance between adjacent buildings exhibiting linear elastic behavior. The seismic input is modeled as a non-stationary random process and the first-passage reliability problem corresponding to the pounding event is solved employing analytical techniques involving the determination of specific statistics of the response processes. The proposed procedure is applied to estimate the probability of pounding between linear single-degree-of-freedom systems and to evaluate the reliability of simplified design code formulae used to determine building separation distances. Furthermore, the capability of the proposed method to deal with complex systems is demonstrated by assessing the effectiveness of the use of viscous dampers in reducing the probability of pounding between adjacent buildings modeled as multi-degree-of-freedom systems.

In Chap. 19, A. Elenas provides a methodology to quantify the relationship between seismic intensity parameters and structural damage. First, a computer-supported elaboration of ground motion records provides several peak, spectral and energy seismic parameters. After that, nonlinear dynamic analyses are carried out to provide the structural response for a set of seismic excitations. Among the several response characteristics, the overall structure damage indices after Park/Ang and the maximum inter-storey drift ratio are selected to represent the structural response. Correlation coefficients are evaluated to express the grade of interrelation between seismic acceleration parameters and structural damage. The presented methodology is applied to a reinforced concrete frame building, designed according to the rules of the recent Eurocodes, and the numerical results show that the spectral and energy parameters provide strong correlation to the damage indices.

As demonstrated in the previous chapter, there is interdependence between seismic intensity parameters and structural damage. In Chap. 20, A. Elenas et al. proceed to the classification of seismic damage in buildings using an adaptive neuro-fuzzy inference system. The seismic excitations are simulated by artificial accelerograms and their intensity is described by seismic parameters. The proposed system is trained using a number of seismic events and tested on a reinforced concrete structure. The results show that the proposed fuzzy technique contributes to the development of an efficient blind prediction of seismic damage. The recognition scheme achieves correct classification rates over 90%.

The book closes with a study on damage identification of historical masonry structures under seismic excitation by G. De Matteis et al. (Chap. 21). The seismic behavior of a physical 1:5.5 scaled model of the church of the Fossanova Abbey (Italy) is examined by means of numerical and experimental analyses. As it mostly influences the seismic vulnerability of the Abbey, the central transversal three-bay complex of the church was investigated in detail by means of a shaking table test on a 1:5.5 scaled physical model in the Laboratory of the Institute for Earthquake Engineering and Engineering Seismology in Skopje. In this chapter, a brief review of the numerical activity related to the prediction of the shaking table test response of the model is first proposed. Then, the identification of frequency decay during collapse is performed through decomposition of the measured power spectral density matrix. Finally, the localization and evolution of damage in the structure is analyzed and the obtained numerical results show a very good agreement with the experimental data.

The book editors would like to express their deep appreciation to all contributors for their active participation in the COMPDYN 2011 Conference and for the time and effort devoted to the completion of their contributions to this volume. Special thanks are also due to the reviewers for their constructive comments and suggestions which enhanced the quality of the book. Finally, the editors would like to thank the personnel of Springer for their most valuable support during the publication process.

Athens, Greece
April 2012

Manolis Papadrakakis
George Stefanou
Vissarion Papadopoulos

Contents

Chapter 1
Random Dynamical Response of a Multibody System with Uncertain Rigid Bodies

Anas Batou and Christian Soize

Abstract This work is devoted to the construction of the random dynamical response of a multibody system with uncertain rigid bodies. We construct a stochastic model of an uncertain rigid body by modeling the mass, the center of mass and the tensor of inertia by random variables. The prior probability distributions of these random variables are constructed using the maximum entropy principle under the constraints defined by the available information. A generator of independent realizations are then developed. Several uncertain rigid bodies can be linked each to the others in order to calculate the random response of a multibody dynamical system. An application is proposed to illustrate the theoretical development.

1 Introduction

This work is devoted to the construction of a probabilistic model of uncertainties for a rigid multibody dynamical system made up of uncertain rigid bodies. In some cases, the mass distribution inside a rigid body is not perfectly known and must be considered as random (for example, the distribution of passengers inside a vehicle) and therefore, this unknown mass distribution inside the rigid body induces uncertainties in the model of this rigid body. Here, we propose a new probabilistic modeling for uncertain rigid bodies in the context of the multibody dynamics. Concerning the modeling of uncertainties in multibody dynamical system, a very few previous researches have been carried out. These researches concerned parameters which describe the joints linking each rigid body to the others and the external sources (see [3, 8, 12, 13, 16]), but not rigid bodies themselves. In the field of uncertain rigid bodies, a first work has been proposed in [9, 10], in which the authors take into account uncertain rigid bodies for rotor dynamical systems using the nonparamet-

A. Batou (✉) · C. Soize
Laboratoire Modélisation et Simulation Multi Echelle, MSME UMR 8208 CNRS, Université Paris-Est, 5 bd Descartes, 77454 Marne-la-Vallee, France
e-mail: anas.batou@univ-paris-est.fr

C. Soize
e-mail: christian.soize@univ-paris-est.fr

M. Papadrakakis et al. (eds.), *Computational Methods in Stochastic Dynamics*, Computational Methods in Applied Sciences 26,
DOI 10.1007/978-94-007-5134-7_1,

ric probabilistic approach [19, 20] consisting in replacing the mass and gyroscopic matrices by random matrices.

In this paper, a general and complete stochastic model is constructed for an uncertain rigid body. The mass, the center of mass and the tensor of inertia which describe the rigid body are modeled by random variables. The prior probability distributions of the random variables are constructed using the maximum entropy principle [6, 7] from Information Theory [17, 18]. The generator of independent realizations corresponding to the prior probability distributions of these random quantities are developed and presented. Then, several uncertain rigid bodies can be linked each to the others in order to calculate the random response of an uncertain multibody dynamical system. The stochastic multibody dynamical equations are solved using the Monte Carlo simulation method.

Section 2 is devoted to the construction of the nominal model for the rigid multibody dynamical system by using the classical method. In Sect. 3, we propose a general probability model for an unconstrained uncertain rigid body and then, the uncertain rigid multibody dynamical system is obtained by joining this unconstrained uncertain rigid body to the other rigid bodies. The last section is devoted to an application which illustrates the proposed theory.

2 Nominal Model for the Rigid Multibody Dynamical System

In this paper, the usual model of a rigid multibody dynamical system for which all the mechanical properties are known will be called the nominal model. This section is devoted to the construction of the nominal model for a rigid multibody dynamical system. This nominal model is constructed as in [14, 15] and is summarized below.

2.1 Dynamical Equations for a Rigid Body of the Multibody System

Let RB_i be the rigid body occupying a bounded domain Ω_i with a given geometry. Let ξ be the generic point of the three dimensional space (see Fig. 1.1). Let $\mathbf{x} = (x_1, x_2, x_3)$ be the position vector of point ξ defined in a fixed inertial frame $(O, x_{0,1}, x_{0,2}, x_{0,3})$, such that $\mathbf{x} = \overrightarrow{O\xi}$. A rigid body is classically defined by three quantities.

1. The first one is the mass m_i of RB_i which is such that

$$m_i = \int_{\Omega_i} \rho(\mathbf{x})\, d\mathbf{x}, \tag{1.1}$$

   where $\rho(\mathbf{x})$ is the mass density.

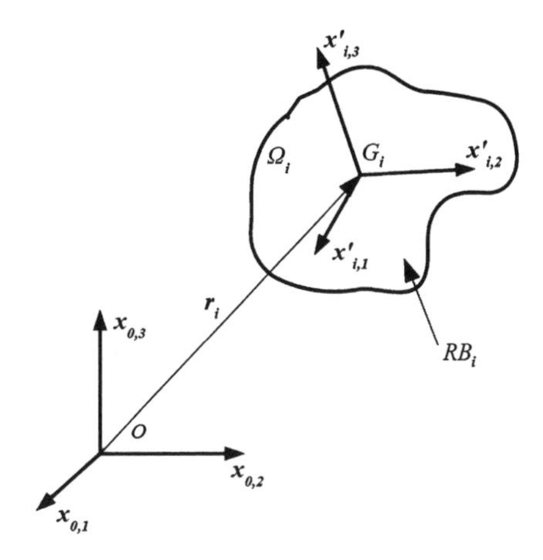

Fig. 1.1 Rigid body RB_i

2. The second quantity is the position vector $\mathbf{r}_i$ of the center of mass G_i, defined in the fixed inertial frame, by

$$\mathbf{r}_i = \frac{1}{m_i} \int_{\Omega_i} \mathbf{x} \rho(\mathbf{x})\, d\mathbf{x}. \tag{1.2}$$

3. Let $(G_i, x'_{i,1}, x'_{i,2}, x'_{i,3})$ be the local frame for which the origin is G_i and which is deduced from the fixed frame $(O, x_{0,1}, x_{0,2}, x_{0,3})$ by the translation $\overrightarrow{OG_i}$ and a rotation defined by the three Euler angles α_i, β_i and γ_i. The third quantity is the positive-definite matrix $[J_i]$ of the tensor of inertia in the local frame such that

$$[J_i]\mathbf{u} = -\int_{\Omega_i} \mathbf{x}' \times \mathbf{x}' \times \mathbf{u} \rho\big(\mathbf{x}'\big)\, d\mathbf{x}', \quad \forall \mathbf{u} \in \mathbb{R}^3, \tag{1.3}$$

in which the vector $\mathbf{x}' = (x'_1, x'_2, x'_3)$ of the components of vector $\overrightarrow{G_i \xi}$ are given in $(G_i, x'_{i,1}, x'_{i,2}, x'_{i,3})$. In the above equation, $\mathbf{u} \times \mathbf{v}$ denotes the cross product between the vectors $\mathbf{u}$ and $\mathbf{v}$.

2.2 Matrix Model for the Rigid Multibody Dynamical System

The rigid multibody dynamical system is made up of n_b rigid bodies and ideal joints including rigid joints, joints with given motion (rheonomic constraints) and vanishing joints (free motion). The interactions between the rigid bodies are realized by these ideal joints but also by springs, dampers or actuators which produce forces between the bodies. In this paper, only n_c holonomic constraints are considered. Let $\mathbf{u}$ be the vector in $\mathbb{R}^{6n_b}$ such that $\mathbf{u} = (\mathbf{r}_1, \dots, \mathbf{r}_{n_b}, \mathbf{s}_1, \dots, \mathbf{s}_{n_b})$ in which $\mathbf{s}_i = (\alpha_i, \beta_i, \gamma_i)$ is the rotation vector. The n_c constraints are given by n_c implicit equations which are globally written as $\boldsymbol{\varphi}(\mathbf{u}, t) = 0$. The $(6n_b \times 6n_b)$ mass matrix $[M]$ is defined by

$$[M] = \begin{bmatrix} [M^r] & 0 \\ 0 & [M^s] \end{bmatrix}, \tag{1.4}$$

where the $(3n_b \times 3n_b)$ matrices $[M^r]$ and $[M^s]$ are defined by

$$\left[M^r\right] = \begin{bmatrix} m_1[I_3] & \cdots & 0 \\ \vdots & \ddots & \vdots \\ 0 & \cdots & m_{n_b}[I_3] \end{bmatrix}, \qquad \left[M^s\right] = \begin{bmatrix} [J_1] & \cdots & 0 \\ \vdots & \ddots & \vdots \\ 0 & \cdots & [J_{n_b}] \end{bmatrix}, \tag{1.5}$$

in which $[I_3]$ is the (3×3) identity matrix. The function $\{\mathbf{u}(t) \in [0, T]\}$ is then the solution of the following differential equation (see [15])

$$\begin{bmatrix} [M] & [\boldsymbol{\varphi}_{\mathbf{u}}]^T \\ [\boldsymbol{\varphi}_{\mathbf{u}}] & [0] \end{bmatrix} \begin{bmatrix} \ddot{\mathbf{u}} \\ \boldsymbol{\lambda} \end{bmatrix} = \begin{bmatrix} \mathbf{q} - \mathbf{k} \\ -\frac{d}{dt}\boldsymbol{\varphi}_t - [\frac{d}{dt}\boldsymbol{\varphi}_{\mathbf{u}}]\dot{\mathbf{u}} \end{bmatrix}, \tag{1.6}$$

with the initial conditions

$$\mathbf{u}(0) = \mathbf{u}_0, \qquad \dot{\mathbf{u}}(0) = \mathbf{v}_0, \tag{1.7}$$

in which $\mathbf{k}(\dot{\mathbf{u}})$ is the vector of the Coriolis forces and where $[\boldsymbol{\varphi}_{\mathbf{u}}(\mathbf{u}(t), t)]_{ij} = \partial\varphi_i(\mathbf{u}(t), t)/\partial u_j(t)$ and $\boldsymbol{\varphi}_t = \partial\boldsymbol{\varphi}/\partial t$. The vector $\mathbf{q}(\mathbf{u}, \dot{\mathbf{u}}, t)$ is constituted of the applied forces and torques induced by springs, dampers and actuators. The vector $\boldsymbol{\lambda}(t)$ is the vector of the Lagrange multipliers. Equation (1.6) can be solved using an adapted integration algorithm (see for instance [2]).

3 Stochastic Model for a Multibody Dynamical System with Uncertain Rigid Bodies

Firstly, a stochastic model for an uncertain rigid body of the multibody dynamical system is proposed and secondly, the stochastic model for the multibody dynamical system with uncertain rigid bodies is constructed joining the stochastic model of the uncertain rigid bodies.

3.1 Stochastic Model for an Uncertain Rigid Body of the Multibody Dynamical System

The properties of the nominal model of the rigid body RB_i are defined by its mass $\underline{m}_i$, the position vector $\underline{\mathbf{r}}_{0,i}$ of its center of mass $\underline{G}_i$ at initial time $t = 0$ and the matrix $[\underline{J}_i]$ of its tensor of inertia with respect to the local frame $(\underline{G}_i, \underline{x}'_{i,1}, \underline{x}'_{i,2}, \underline{x}'_{i,3})$. The probabilistic model of uncertainties for this rigid body is constructed by replacing these three parameters by the following three random variables: the random mass M_i, the random position vector $\mathbf{R}_{0,i}$ of its random center of mass $\mathbf{G}_i$ at initial time $t = 0$ and the random matrix $[\mathbf{J}_i]$ of its random tensor of inertia with respect to the random local frame $(\mathbf{G}_i, \underline{x}'_{i,1}, \underline{x}'_{i,2}, \underline{x}'_{i,3})$. The probability density functions

(PDF) of these three random variables are constructed using the maximum entropy principle (see [6, 7, 17]), that is to say, in maximizing the uncertainties in the model under the constraints defined by the available information.

3.1.1 Construction of the PDF for the Random Mass

(i) *Available information.*
Let $E\{\cdot\}$ be the mathematical expectation. The available information for the random mass M_i is defined as follows. Firstly, the random variable M_i must be positive almost surely. Secondly, the mean value of the random mass M_i must be equal to the value $\underline{m}_i$ of the nominal model. Thirdly, as it is proven in [20], the random mass must verify the inequality $E\{M_i^{-2}\} < +\infty$ in order that a second-order solution exists for the stochastic dynamical system. In addition, it is also proven that this constraint can be replaced by $|E\{\log M_i\}| < +\infty$.

(ii) *Maximum entropy principle.*
The probability density function $\mu \mapsto p_{M_i}(\mu)$ of the random variable M_i is constructed by maximizing the entropy under the constraints defined above. The solution of this optimization problem is the PDF of a gamma random variable defined on $]0, +\infty[$. This PDF depends on two parameters which are the nominal value $\underline{m}_i$ and the coefficient of variation δ_{M_i} of the random variable M_i such that $\delta_{M_i} = \sigma_{M_i}/\underline{m}_i$ where σ_{M_i} is the standard deviation of the random variable M_i. Therefore, the PDF of the random mass is completely defined by the mean value $\underline{m}_i$ and by the dispersion parameter δ_{M_i}.

3.1.2 Construction of the PDF for the Random Position Vector $\mathbf{R}_{0,i}$

In this subsection, the PDF of the random initial position vector $\mathbf{R}_{0,i}$ of the center of mass of RB_i at initial time $t = 0$ is constructed.

(i) *Available information.*
The position vector $\underline{\mathbf{r}}_{0,i}$ of the center of mass $\underline{G}_i$ at initial time $t = 0$ of the nominal model is given. However, the real position is not exactly known and $\underline{\mathbf{r}}_{0,i}$ only corresponds to a mean position. Consequently, there is an uncertainty about the real position and this is the reason why this position is modeled by the random vector $\mathbf{R}_{0,i}$. Some geometrical and mechanical considerations lead us to introduce an admissible domain D_i of random vector $\mathbf{R}_{0,i}$. We introduce the vector $\mathbf{h}$ of the parameters describing the geometry of domain D_i. In addition, the mean value of the random vector $\mathbf{R}_{0,i}$ must be equal to the value $\underline{\mathbf{r}}_{0,i}$ of the nominal model. Therefore, the available information for random variable $\mathbf{R}_{0,i}$ can be written as

$$\mathbf{R}_{0,i} \in D_i(\mathbf{h}) \quad \text{a.s.}, \tag{1.8a}$$

$$E\{\mathbf{R}_{0,i}\} = \underline{\mathbf{r}}_{0,i} \in D_i(\mathbf{h}). \tag{1.8b}$$

(ii) *Maximum entropy principle.*
The probability density function $\mathbf{a} \mapsto p_{\mathbf{R}_{0,i}}(\mathbf{a})$ of random variable $\mathbf{R}_{0,i}$ is then constructed by maximizing the entropy with the constraints defined by the available information in Eqs. (1.8a) and (1.8b). The solution of this optimization problem depends on two parameters which are $\underline{\mathbf{r}}_{0,i}$ and vector-valued parameter $\mathbf{h}$, and is such that

$$p_{\mathbf{R}_{0,i}}(\mathbf{a};\mathbf{h}) = \mathbb{1}_{D_i(\mathbf{h})}(\mathbf{a}) C_0 e^{-\langle \boldsymbol{\lambda}, \mathbf{a} \rangle}. \tag{1.9}$$

The positive valued parameter C_0 and vector $\boldsymbol{\lambda}$ are the unique solution of the equations

$$C_0 \int_{D_i(\mathbf{h})} e^{-\langle \boldsymbol{\lambda}, \mathbf{a} \rangle} \, d\mathbf{a} = 1, \tag{1.10a}$$

$$C_0 \int_{D_i(\mathbf{h})} \mathbf{a} e^{-\langle \boldsymbol{\lambda}, \mathbf{a} \rangle} \, d\mathbf{a} = \underline{\mathbf{r}}_{0,i}. \tag{1.10b}$$

(iii) *Generator of independent realizations.*
The independent realizations of random variable $\mathbf{R}_{0,i}$ must be generated using the constructed PDF $p_{\mathbf{R}_{0,i}}$. Such a generator can be obtained using the Monte Carlo Markov Chain (MCMC) method (Metropolis–Hastings algorithm [5]).

3.1.3 Random Matrix $[\mathbf{J}_i]$ of the Random Tensor of Inertia

In this subsection, the random matrix $[\mathbf{J}_i]$ of the random tensor of inertia with respect to $(\mathbf{G}_i, \underline{x}'_{i,1}, \underline{x}'_{i,2}, \underline{x}'_{i,3})$ is defined and an algebraic representation of this random matrix is constructed. The mass distribution around the random center of mass $\mathbf{G}_i$ is uncertain and consequently, the tensor of inertia is also uncertain. This is the reason why the matrix $[\underline{J}_i]$ of the tensor of inertia of the nominal model with respect to $(\underline{G}_i, \underline{x}'_{i,1}, \underline{x}'_{i,2}, \underline{x}'_{i,3})$ is replaced by a random matrix $[\mathbf{J}_i]$ which is constructed by using the maximum entropy principle. We introduce the positive-definite matrix $[Z_i]$ independent of m_i such that

$$[Z_i] = \frac{1}{m_i} \left\{ \frac{\mathrm{tr}([J_i])}{2} [I_3] - [J_i] \right\}. \tag{1.11}$$

Then $[J_i]$ can be calculated as a function of $[Z_i]$,

$$[J_i] = m_i \left\{ \mathrm{tr}\big([Z_i]\big)[I_3] - [Z_i] \right\}. \tag{1.12}$$

It can be proven that $[Z_i]$ is positive definite and that each positive definite matrix $[J_i]$ constructed using Eq. (1.12), where $[Z_i]$ is a given positive definite matrix, can be interpreted as the matrix of a tensor of inertia of a physical rigid body (see [1]). In the literature, the matrix $m_i[Z_i]$ is referred as to the Euler tensor. The probabilistic modeling $[\mathbf{J}_i]$ of $[J_i]$ consists in introducing the random matrix $[\mathbf{Z}_i]$ and in using Eq. (1.12) in which m_i is replaced by the random variable M_i and where $[Z_i]$ is replaced by $[\mathbf{Z}_i]$. We then obtain

$$[\mathbf{Z}_i] = \frac{1}{M_i}\left\{\frac{\mathrm{tr}([\mathbf{J}_i])}{2}[I_3] - [\mathbf{J}_i]\right\}, \tag{1.13}$$

$$[\mathbf{J}_i] = M_i\left\{\mathrm{tr}\left([\mathbf{Z}_i]\right)[I_3] - [\mathbf{Z}_i]\right\}. \tag{1.14}$$

(i) *Available information concerning random matrix* $[\mathbf{Z}_i]$.
Let us introduce (1) the nominal value $[\underline{Z}_i]$ of deterministic matrix $[Z_i]$ such that $[\underline{Z}_i] = (1/\underline{m}_i)\{\mathrm{tr}([\underline{J}_i])/2[I_3] - [\underline{J}_i]\}$ and (2) the upper bound $[Z_i^{\max}]$ of random matrix $[\mathbf{Z}_i]$. Then, the available information for $[\mathbf{Z}_i]$ can be summarized as follows,

$$\begin{aligned}
[\mathbf{Z}_i] &\in \mathbb{M}_3^+(\mathbb{R}) \quad \text{a.s.},\\
\left\{[Z_i^{\max}] - [\mathbf{Z}_i]\right\} &\in \mathbb{M}_3^+(\mathbb{R}) \quad \text{a.s.},\\
E\left\{[\mathbf{Z}_i]\right\} &= [\underline{Z}_i],\\
E\left\{\log\left(\det[\mathbf{Z}_i]\right)\right\} &= C_i^l, \quad |C_i^l| < +\infty,\\
E\left\{\log\left(\det\left([Z_i^{\max}] - [\mathbf{Z}_i]\right)\right)\right\} &= C_i^u, \quad |C_i^u| < +\infty.
\end{aligned} \tag{1.15}$$

For more convenience, random matrix $[\mathbf{Z}_i]$ is normalized as follow. Matrix $[\underline{Z}_i]$ being positive definite, its Cholesky decomposition yields $[\underline{Z}_i] = [\underline{L}_{Z_i}]^T[\underline{L}_{Z_i}]$ in which $[\underline{L}_{Z_i}]$ is an upper triangular matrix in the set $\mathbb{M}_3(\mathbb{R})$ of all the (3×3) real matrices. Then, random matrix $[\mathbf{Z}_i]$ can be rewritten as

$$[\mathbf{Z}_i] = [\underline{L}_{Z_i}]^T[\mathbf{G}_i][\underline{L}_{Z_i}], \tag{1.16}$$

in which the matrix $[\mathbf{G}_i]$ is a random matrix for which the available information is

$$\begin{aligned}
[\mathbf{G}_i] &\in \mathbb{M}_3^+(\mathbb{R}) \quad \text{a.s.},\\
\left\{[G_i^{\max}] - [\mathbf{G}_i]\right\} &\in \mathbb{M}_3^+(\mathbb{R}) \quad \text{a.s.},\\
E\left\{[\mathbf{G}_i]\right\} &= [I_3],\\
E\left\{\log\left(\det[\mathbf{G}_i]\right)\right\} &= C_i^{l'}, \quad |C_i^{l'}| < +\infty,\\
E\left\{\log\left(\det\left([G_i^{\max}] - [\mathbf{G}_i]\right)\right)\right\} &= C_i^{u'}, \quad |C_i^{u'}| < +\infty,
\end{aligned} \tag{1.17}$$

in which $C_i^{l'} = C_i^l - \log(\det[\underline{Z}_i])$, $C_i^{u'} = C_i^u - \log(\det[\underline{Z}_i])$ and where the matrix $[G_i^{\max}]$ is an upper bound for random matrix $[\mathbf{G}_i]$ and is defined by $[G_i^{\max}] = ([\underline{L}_{Z_i}]^T)^{-1}[Z_i^{\max}][\underline{L}_{Z_i}]^{-1}$.

(ii) *Maximum entropy principle.*
The probability distribution of random matrix $[\mathbf{G}_i]$ is constructed using the maximum entropy principle under the constraints defined by the available information given by Eq. (1.17). The probability density function $p_{[\mathbf{G}_i]}([G])$ with respect to the volume element $\widetilde{d}G$ of random matrix $[\mathbf{G}_i]$ is then written as

$$\begin{aligned}
p_{[\mathbf{G}_i]}\left([G]\right) = \mathbb{1}_{\mathbb{M}_3^+(\mathbb{R})}\left([G]\right) \times \mathbb{1}_{\mathbb{M}_3^+(\mathbb{R})}\left([G_i^{\max}] - [G]\right) \times C_{G_i} \times \left(\det[G]\right)^{-\lambda_l}\\
\times \left(\det\left([G_i^{\max}] - [G]\right)\right)^{-\lambda_u} \times e^{-\mathrm{tr}([\mu][G])},
\end{aligned} \tag{1.18}$$

in which the positive valued parameter C_{G_i} is a normalization constant, the real parameters $\lambda_l < 1$ and $\lambda_u < 1$ are Lagrange multipliers relative to the two last constraints defined by Eq. (1.17) and the symmetric real matrix $[\mu]$ is a Lagrange multiplier relative to the third constraint defined by Eq. (1.17). This probability density function is a particular case of the Kummer-Beta matrix variate distribution (see [4, 11]) for which the lower bound is a zero matrix. Parameters C_{G_i}, λ_l, λ_u and matrix $[\mu]$ are the unique solution of the equations

$$\begin{aligned} E\{\mathbb{1}_{\mathbb{M}_3^S(\mathbb{R})}([\mathbf{G}_i])\} &= 1, \\ E\{[\mathbf{G}_i]\} &= [I_3], \\ E\{\log(\det[\mathbf{G}_i])\} &= C_i^{l'}, \\ E\{\log(\det([G_i^{\max}] - [\mathbf{G}_i]))\} &= C_i^{u'}. \end{aligned} \tag{1.19}$$

For fixed values of λ_l and λ_u, parameters C_{G_i} and $[\mu]$ can be estimated using Eq. (1.19). In Eq. (1.19), since the parameters $C_i^{l'}$ and $C_i^{u'}$ have no real physical meaning, the parameters λ_l and λ_u are kept as parameters which then allows the "shape" of the PDF to be controlled. If experimental data are available for the responses of the dynamical system, then the two parameters λ_l and λ_u can be identified solving an inverse problem. If experimental data are not available, these two parameters allow a sensitivity analysis of the solution to be carried out with respect to the level of uncertainties.

(iii) *Properties for random matrix* $[\mathbf{J}_i]$.
It is proven in [1] that using Eq. (1.14) and the available information defined by Eq. (1.15), the following important properties for random matrix $[\mathbf{J}_i]$ can be deduced,

$$\left\{\frac{1}{2}\operatorname{tr}([\mathbf{J}_i])[I_3] - [\mathbf{J}_i]\right\} \in \mathbb{M}_3^+(\mathbb{R}) \quad \text{a.s.,} \tag{1.20a}$$

$$\{[\mathbf{J}_i^{\max}] - [\mathbf{J}_i]\} \in \mathbb{M}_3^+(\mathbb{R}) \quad \text{a.s.,} \tag{1.20b}$$

$$E\{[\mathbf{J}_i]\} = [\underline{J}_i], \tag{1.20c}$$

$$\{\lambda_l < -2\} \Rightarrow E\{\|[\mathbf{J}_i]^{-1}\|^2\} < +\infty, \tag{1.20d}$$

in which the random matrix $[\mathbf{J}_i^{\max}]$, which represents a random upper bound for random matrix $[\mathbf{J}_i]$, is defined by

$$[\mathbf{J}_i^{\max}] = M_i\{\operatorname{tr}([Z_i^{\max}])[I_3] - [Z_i^{\max}]\}. \tag{1.21}$$

It should be noted that Eq. (1.20a) implies that each realization of random matrix $[\mathbf{J}_i]$ corresponds to the matrix of a tensor of inertia of a physical rigid body. In addition, this equation implies that random matrix $[\mathbf{J}_i]$ is almost surely positive definite. Equation (1.20b) provides a random upper bound for random matrix $[\mathbf{J}_i]$. Equation (1.20c) corresponds to a construction for which the mean value of random matrix $[\mathbf{J}_i]$ is equal to the nominal value $[\underline{J}_i]$. Finally, Eq. (1.20d) is necessary for that the random solution of the nonlinear dynamical system be a second-order stochastic process.

(iv) *Generator of independent realizations for random matrix* $[\mathbf{J}_i]$.
The generator of independent realizations of random matrix $[\mathbf{G}_i]$ is based on the Monte Carlo Markov Chain (MCMC) (Metropolis–Hastings algorithm [5] with the PDF defined by Eq. (1.18). Then, independent realizations of random matrix $[\mathbf{Z}_i]$ are obtained using Eq. (1.16). Finally, independent realizations of random matrix $[\mathbf{J}_i]$ are obtained using Eq. (1.14) and independent realizations of random mass M_i.

3.2 Stochastic Matrix Model for a Multibody Dynamical System with Uncertain Rigid Bodies and Its Random Response

In order to limit the developments, it is assumed that only one of the n_b rigid bodies denoted by RB_i of the rigid multibody system is uncertain. The extension to several uncertain rigid bodies is straightforward. Let the $6n_b$ random coordinates be represented by the $\mathbb{R}^{6n_b}$-valued stochastic process $\mathbf{U} = (\mathbf{R}_1, \ldots, \mathbf{R}_{n_b}, \mathbf{S}_1, \ldots, \mathbf{S}_{n_b})$ indexed by $[0, T]$ and let the n_c random Lagrange multipliers be represented by the $\mathbb{R}^{n_c}$-valued stochastic process $\boldsymbol{\Lambda}$ indexed by $[0, T]$. The deterministic Eq. (1.6) becomes the following stochastic equation

$$\begin{bmatrix} [\mathbf{M}] & [\boldsymbol{\varphi}_{\mathbf{u}}]^T \\ [\boldsymbol{\varphi}_{\mathbf{u}}] & [0] \end{bmatrix} \begin{bmatrix} \ddot{\mathbf{U}} \\ \boldsymbol{\Lambda} \end{bmatrix} = \begin{bmatrix} \mathbf{q} - \mathbf{K} \\ -\frac{d}{dt}\boldsymbol{\varphi}_t - [\frac{d}{dt}\boldsymbol{\varphi}_{\mathbf{u}}]\dot{\mathbf{U}} \end{bmatrix}, \tag{1.22}$$

$$\mathbf{U}(0) = \mathbf{U}_0, \qquad \dot{\mathbf{U}}(0) = \mathbf{v}_0, \quad \text{a.s.} \tag{1.23}$$

in which the vector $\mathbf{U}_0 = (\mathbf{r}_{0,1}, \ldots, \mathbf{R}_{0,i}, \ldots, \mathbf{r}_{0,n_b}, \mathbf{s}_{0,1}, \ldots, \mathbf{s}_{0,n_b})$ is random due to the random vector $\mathbf{R}_{0,i}$. For all given real vector $\dot{\mathbf{u}}$, the vector $\mathbf{K}(\dot{\mathbf{u}})$ of the Coriolis forces is random due to the random matrix $[\mathbf{J}_i]$. The random mass matrix $[\mathbf{M}]$ is defined by

$$[\mathbf{M}] = \begin{bmatrix} [\mathbf{M}^r] & 0 \\ 0 & [\mathbf{M}^s] \end{bmatrix}, \tag{1.24}$$

in which the $(3n_b \times 3n_b)$ random matrices $[\mathbf{M}^r]$ and $[\mathbf{M}]^s$ are defined by

$$[\mathbf{M}^r] = \begin{bmatrix} m_1[I_3] & & \cdots & & 0 \\ & \ddots & & & \\ \vdots & & M_i[I_3] & & \vdots \\ & & & \ddots & \\ 0 & & \cdots & & m_{n_b}[I_3] \end{bmatrix}, \tag{1.25}$$

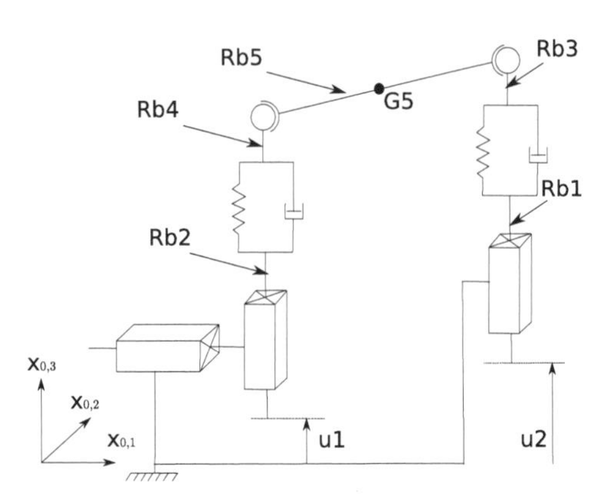

Fig. 1.2 Rigid multibody system

$$[\mathbf{M}^s] = \begin{bmatrix} [J_1] & & \cdots & & 0 \\ & \ddots & & & \\ \vdots & & [\mathbf{J}_i] & & \vdots \\ & & & \ddots & \\ 0 & & \cdots & & [J_{n_b}] \end{bmatrix}. \tag{1.26}$$

Random Eqs. (1.22) and (1.23) are solved using the Monte Carlo simulation method.

4 Application

In this section, we present a numerical application which validates the methodology presented in this paper.

4.1 Description of the Nominal Model

The rigid multibody model is made up of five rigid bodies and six joints which are described in the fixed frame $(O, x_{0,1}, x_{0,2}, x_{0,3})$ (see Fig. 1.2). The plan defined by $(O, x_{0,1}, x_{0,2})$ is identified below as the "ground". The gravity forces in the $x_{0,3}$-direction are taken into account.

(i) *Rigid bodies.*
In the initial configuration, the rigid bodies $Rb1$, $Rb2$, $Rb3$ and $Rb4$ are cylinders for which the axes follow the $x_{0,3}$-direction. In the initial configuration, the rigid body $Rb5$ is supposed to be symmetric with respect to the planes $(G5, x_{0,1}, x_{0,2})$ and $(G5, x_{0,1}, x_{0,3})$ in which $G5$ is the center of mass of $Rb5$.

(ii) *Joints.*
- The joint *Ground*-$Rb1$ is made up of a prismatic joint following $x_{0,3}$-direction. The displacement following $x_{0,3}$-direction (see Fig. 1.2), denoted

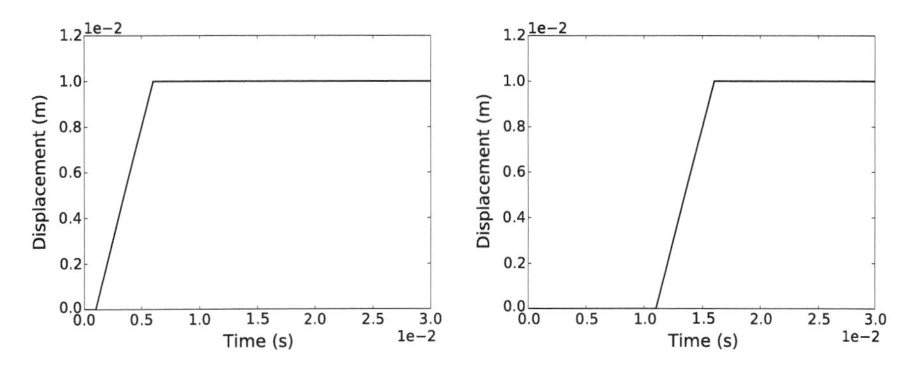

Fig. 1.3 Imposed displacement $u1(t)$ (*left figure*) and $u2(t)$ (*right figure*)

by $u1(t)$, is imposed. The joint *Ground*-*Rb*2 is a prismatic joint following $x_{0,3}$-direction. The displacement following $x_{0,3}$-direction (see Fig. 1.2), denoted by $u2(t)$, is imposed. The displacement following $x_{0,1}$-direction is unconstrained. Imposed displacements $u1(t)$ and $u2(t)$ are plotted in Fig. 1.3 for t in $[0, 0.03]$ s.

- The joints *Rb*1-*Rb*3 and *Rb*2-*Rb*4 are constituted of 6D spring-dampers.
- Finally, the joints *Rb*3-*Rb*5 and *Rb*4-*Rb*5 are $x_{0,2}$-direction revolute joints.

4.2 Random Response of the Stochastic Model

Rigid body *Rb5a* is considered as uncertain and is therefore modeled by a random rigid body. As explained in Sect. 3, the elements of inertia of the uncertain rigid Body *Rb*5 are replaced by random quantities. The fluctuation of the response is controlled by four parameters δ_{M_5}, $\mathbf{h}$, λ_l and λ_u. A sensitivity analysis is carried out with respect to these four parameters. Statistics on the transient response are estimated using the Monte Carlo simulation method with 500 independent realizations. The initial velocities and angular velocities are zero. The observation point P_{obs} belongs to *Rb*5. Four different cases are analyzed:

1. *Case 1*: M_5 is random, $\mathbf{R}_{0,5}$ is deterministic and $[\mathbf{Z}_5]$ is deterministic.
   We choose $\delta_{M_5} = 0.5$. The confidence region, with a probability level $P_c = 0.90$, of the random acceleration of point P_{obs} is plotted in Fig. 1.4. It can be noted that the acceleration is sensitive to the mass uncertainties.
2. *Case 2*: M_i is deterministic, $\mathbf{R}_{0,5}$ is random and $[\mathbf{Z}_5]$ is deterministic.
   The domain of $\mathbf{R}_{0,5}$ is supposed to be a parallelepiped which is centered at point $(0, 0, 0.55)$ for which its edges are parallel to the directions $x_{0,1}$, $x_{0,2}$ and $x_{0,3}$ and for which the lengths following these three directions are respectively 0.5, 0.2 and 0.02. The confidence region, with a probability level $P_c = 0.90$, of the random acceleration of point P_{obs} is plotted in Fig. 1.5. We can remark that the angular acceleration is sensitive to uncertainties on initial center of mass of *Rb*5.

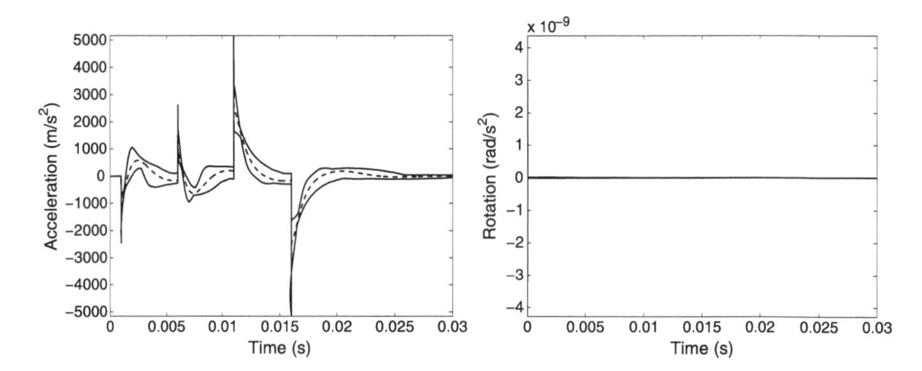

Fig. 1.4 Random transient acceleration of point P_{obs}, Case 1: confidence region (*upper* and *lower solid lines*) and mean response (*dashed line*); $x_{0,3}$-acceleration (*left figure*) and $x_{0,1}$-angular acceleration (*right figure*)

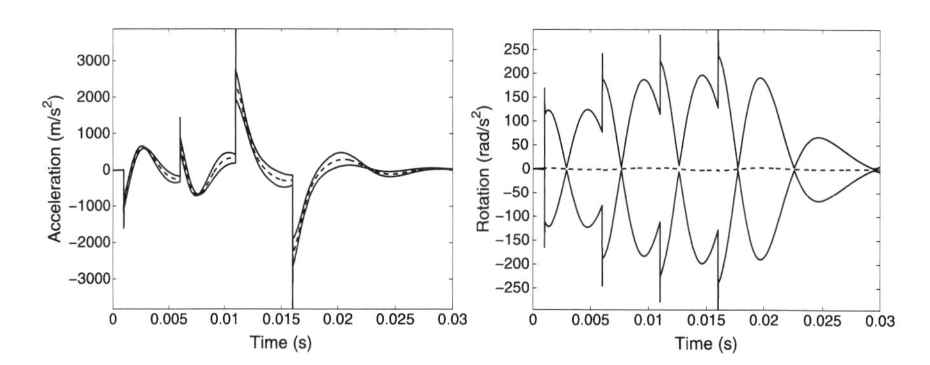

Fig. 1.5 Random transient acceleration of point P_{obs}, Case 2: confidence region (*upper* and *lower solid lines*) and mean response (*dashed line*); $x_{0,3}$-acceleration (*left figure*) and $x_{0,1}$-angular acceleration (*right figure*)

3. *Case 3*: M_5 is deterministic, $\mathbf{R}_{0,5}$ is deterministic and $[\mathbf{Z}_5]$ is random.
   We choose $\lambda_l = -5$ and $\lambda_u = -5$ for random matrix $[\mathbf{Z}_5]$. The confidence region, with a probability level $P_c = 0.90$, of the random acceleration of point P_{obs} is plotted in Fig. 1.6. We can remark, as it was expected, that the angular acceleration is very sensitive to uncertainties on the tensor of inertia. We can also remark a high sensitivity of the acceleration.
4. *Case 4*: M_5, $\mathbf{R}_{0,5}$ and $[\mathbf{Z}_5]$ are random.
   The values of the parameters of the PDF are those fixed in the three previous cases. The confidence region, with a probability level $P_c = 0.90$, of the random acceleration of point P_{obs} is plotted in Fig. 1.7. It can be viewed that (1) the randomness on the acceleration is mainly due to the randomness of mass M_5, (2) the randomness on the angular acceleration is mainly due to the randomness of the initial position $\mathbf{R}_{0,5}$ of the center of mass and the random tensor $[\mathbf{Z}_5]$.

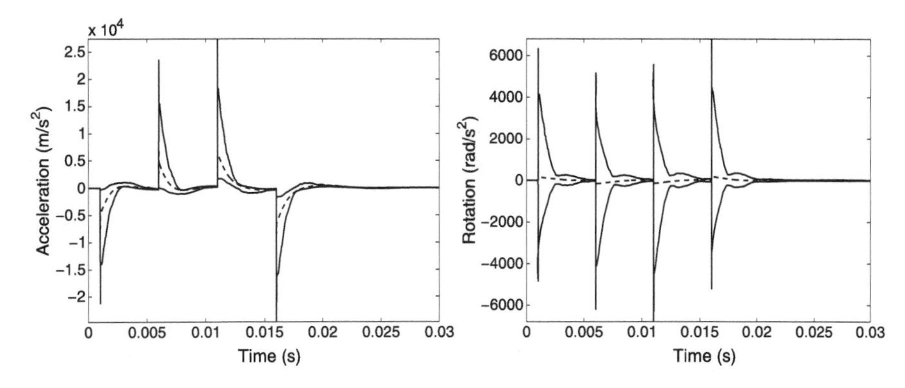

Fig. 1.6 Random transient acceleration of point P_{obs}, Case 3: confidence region (*upper* and *lower solid lines*) and mean response (*dashed line*); $x_{0,3}$-acceleration (*left figure*) and $x_{0,1}$-angular acceleration (*right figure*)

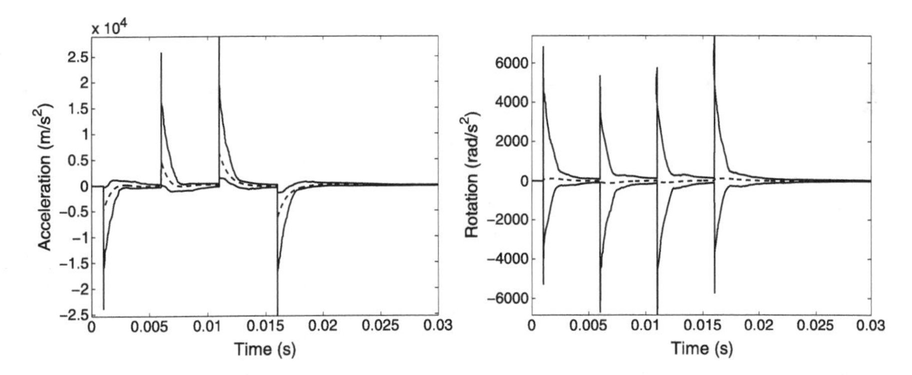

Fig. 1.7 Random transient acceleration of point P_{obs}, Case 4: confidence region (*upper* and *lower solid lines*) and mean response (*dashed line*); $x_{0,3}$-acceleration (*left figure*) and $x_{0,1}$-angular acceleration (*right figure*)

5 Conclusion

We have presented a complete and general probabilistic modeling of uncertain rigid bodies taking into account all the known mechanical and mathematical properties defining a rigid body. This probabilistic model of uncertainties is used to construct the stochastic equations of uncertain multibody dynamical systems. The random dynamical responses can then be calculated. In the proposed probabilistic model, the mass, the center of mass and the tensor of inertia are modeled by random variables for which the prior probability density functions are constructed using the maximum entropy principle under the constraints defined by all the available mathematical, mechanical and design properties. Several uncertain rigid bodies can be linked each to the others in order to obtain the stochastic dynamical model of the uncertain multibody dynamical system. The theory proposed has been illustrated analyzing a simple example. The results obtained clearly show the role played by uncertainties and the sensitivity of the responses due to uncertainties on (1) the mass (2) the center of mass and (3) the tensor of inertia. Such a prior stochastic model allows the robustness of the responses to be analyzed with respect to uncertainties. If experimental data were available on the responses, then the parameters which control the level of uncertainties could be estimated by solving an inverse stochastic problem.

References

1. Batou, A., Soize, C.: Rigid multibody system dynamics with uncertain rigid bodies. Multibody Syst. Dyn. **27**, 285–319 (2012)
2. Baumgarte, J.: Stabilization of constraints and integrals of motion in dynamical systems. Comput. Methods Appl. Mech. Eng. **1**(1), 1–16 (1972)
3. Carrarini, A.: Reliability based analysis of the crosswind stability of railway vehicles. J. Wind Eng. Ind. Aerodyn. **95**, 493–509 (2007)
4. Das, S., Ghanem, R.: A bounded random matrix approach for stochastic upscaling. Multiscale Model. Simul. **8**(1), 296–325 (2009)
5. Hastings, W.K.: Monte Carlo sampling methods using Markov chains and their applications. Biometrika **109**, 57–97 (1970)
6. Jaynes, E.T.: Information theory and statistical mechanics. Phys. Rev. **106**(4), 620–630 (1957)
7. Jaynes, E.T.: Information theory and statistical mechanics. II. Phys. Rev. **108**(2), 171–190 (1957)
8. Li, L., Corina Sandu, C.: On the impact of cargo weight, vehicle parameters, and terrain characteristics on the prediction of traction for off-road vehicles. J. Terramech. **44**, 221–238 (2007)
9. Murthy, R., Mignolet, M.P., El-Shafei, A.: Nonparametric stochastic modeling of uncertainty in rotordynamics—part I: formulation. J. Eng. Gas Turbines Power **132**(9), 092501 (2010)
10. Murthy, R., Mignolet, M.P., El-Shafei, A.: Nonparametric stochastic modeling of uncertainty in rotordynamics—part II: applications. J. Eng. Gas Turbines Power **132**(9), 092502 (2010)
11. Nagar, D.K., Gupta, A.K.: Matrix-variate Kummer-Beta distribution. J. Aust. Math. Soc. A **73**, 11–25 (2002)
12. Negrut, D., Datar, M., Gorsich, D., Lamb, D.: A framework for uncertainty quantification in nonlinear multi-body system dynamics. In: Proceedings of the 26th Army Science Conference, Orlando, FL (2008)
13. Sandu, A., Sandu, C., Ahmadian, M.: Modeling multibody dynamic systems with uncertainties. Part I: theoretical and computational aspects. Multibody Syst. Dyn. **23**, 375–395 (2006)
14. Schiehlen, W.: Multibody Systems Handbook. Springer, Berlin (1990)
15. Schiehlen, W.: Multibody system dynamics: roots and perspectives. Multibody Syst. Dyn. **1**, 149–188 (1997)
16. Schmitt, K.P., Anitescu, M., Negrut, D.: Efficient sampling for spatial uncertainty quantification in multibody system dynamics applications. Int. J. Numer. Methods Eng. **80**, 537–564 (2009)
17. Shannon, C.E.: A mathematical theory of communication. Bell Syst. Tech. J. **27**, 379–423 (1948)
18. Shannon, C.E.: A mathematical theory of communication. Bell Syst. Tech. J. **27**, 623–656 (1948)
19. Soize, C.: A nonparametric model of random uncertainties on reduced matrix model in structural dynamics. Probab. Eng. Mech. **15**(3), 277–294 (2000)
20. Soize, C.: Maximum entropy approach for modeling random uncertainties in transient elastodynamics. J. Acoust. Soc. Am. **109**(5), 1979–1996 (2001)

Chapter 2
Dynamic Variability Response for Stochastic Systems

Vissarion Papadopoulos and Odysseas Kokkinos

Abstract In this study we implement the concept of Variability Response Functions (VRFs) in dynamic systems. The variance of the system response can be readily estimated by an integral involving the Dynamic VRF (DVRF) and the uncertain system parameter power spectrum. With the proposed methodology spectral and probability distribution-free upper bounds can be easily derived. Also an insight is provided with respect to the mechanisms controlling the system's response. The necessarily asserted conjecture of independence of the DVRF to the spectral density and the marginal probability density is validated numerically through brute-force Monte Carlo simulations.

Keywords Dynamic Variability Response Functions · Stochastic finite element analysis · Upper bounds · Stochastic dynamic systems

1 Introduction

Over the past two decades a lot of research has been dedicated to the stochastic analysis of structural systems involving uncertain parameters in terms of material or geometry with the implementation of stochastic finite element methodologies. Although these methods have proven to be highly accurate and computationally efficient for a variety of problems, there is still a wide range of problems in stochastic mechanics involving combinations of strong non-linearities and/or large variations of system properties as well as non-Gaussian system properties that can be solved with reasonable accuracy only through a computationally expensive Monte Carlo simulation approach [3–5, 12].

In all aforementioned cases, the spectral/correlation characteristics and the marginal probability distribution function (pdf) of the stochastic fields describing

V. Papadopoulos (✉) · O. Kokkinos
Institute of Structural Analysis and Antiseismic Research, National Technical University of Athens, 9 Iroon Polytechneiou, Zografou Campus, Athens 15780, Greece
e-mail: vpapado@central.ntua.gr

O. Kokkinos
e-mail: okokki@central.ntua.gr

M. Papadrakakis et al. (eds.), *Computational Methods in Stochastic Dynamics*,
Computational Methods in Applied Sciences 26,
DOI 10.1007/978-94-007-5134-7_2,

the uncertain system parameters are required in order to estimate the response variability of a stochastic static or dynamic system. As there is usually a lack of experimental data for the quantification of such probabilistic quantities, a sensitivity analysis with respect to various stochastic parameters is often implemented. In this case, however, the problems that arise are the increased computational effort, the lack of insight on how these parameters control the response variability of the system and the inability to determine bounds of the response variability.

In this framework and to tackle the aforementioned issues, the concept of the variability response function (*VRF*) has been proposed in the late 1980s [10], along with different aspects and applications of the *VRF* [1, 13]. A development of this approach was presented in a series of papers [7–9], where the existence of closed-form integral expressions for the variance of the response displacement of the form

$$\mathrm{Var}[u] = \int_{-\infty}^{\infty} VRF(\kappa, \sigma_{ff}) S_{ff}(\kappa)\, d\kappa \tag{2.1}$$

was demonstrated for linear stochastic systems under static loads using a flexibility-based formulation. It was shown that the *VRF* depends on standard deviation σ_{ff} but appears to be independent of the functional form of the spectral density function $S_{ff}(\kappa)$ modeling the inverse of the elastic modulus. The existence however of this integral expression had to be conjectured for statically indeterminate as well as for general stochastic finite element systems. A rigorous proof of such existence is available only for statically determinate systems for which *VRF* is independent of σ_{ff} as well [8]. Further investigations [6] verified the aforementioned results but showed that *VRF* has a slight dependence on the marginal pdf of the stochastic field modeling the flexibility.

The present paper extends the aforementioned approach to linear statically determinate stochastic systems under dynamic excitations. Although the derivation of an analytic expression for the variability response function of the dynamic system (*DVRF*), if possible at all, is extremely cumbersome, a numerical computation of the *DVRF* can be easily achieved to provide results for the variance time history of the dynamic system response. As in previous works [7–9], the existence of the *DVRF* and a similar to Eq. (2.1) integral form expression has to be conjectured. This assumption is numerically validated by comparing the results from Eq. (2.1) with brute force Monte Carlo simulations. It is demonstrated that the *DVRF* is highly dependent on the standard deviation σ_{ff} of the inverse of the elastic modulus and, based on numerical evidence further presented but, to this point, not to a full proof verification technique, appears to be almost independent of the functional form of $S_{ff}(\kappa)$ as well as of the marginal *pdf* of the flexibility. In addition, an integral expression similar to that of Eq. (2.1) is proposed for the mean system response involving a Dynamic Mean Response Function (*DMRF*), which is a function similar to the *DVRF*.

Both integral forms for the mean and variance can be used to efficiently compute the first and second order statistics of the transient system response with reasonable accuracy, together with time dependant spectral-distribution-free upper bounds.

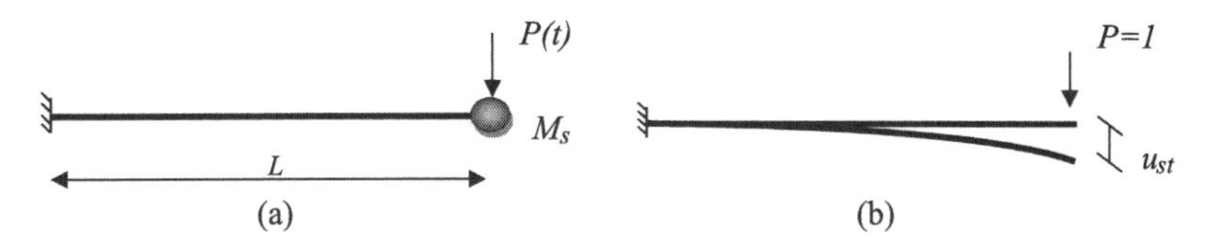

Fig. 2.1 One degree of freedom oscillator: (**a**) Geometry and loading (**b**) Static displacement for unit load

They also provide an insight into the mechanisms controlling the uncertainty propagation with respect to both space and time and in particular the mean and variability time histories of the stochastic system dynamic response.

2 Dynamic Analysis of a Stochastic Single Degree of Freedom Oscillator

For the single degree of freedom statically determinate stochastic oscillator of length L and mass M_s in Fig. 2.1(a), loaded with a dynamic deterministic load $P(t)$, the inverse of the elastic modulus is considered to vary randomly along the length of the beam according to the following expression:

$$\frac{1}{E(x)} = F_0\bigl(1 + f(x)\bigr) \tag{2.2}$$

is the elastic modulus, F_0 is the mean value of the inverse of $E(x)$, and $f(x)$ is a zero-mean homogeneous stochastic field modeling the variation of $1/E(x)$ around its mean value F_0.

The displacement time history $u(t)$ of the oscillator can be derived from the solution of Duhamel's integral:

$$u(t) = \frac{1}{\omega_D} \int_0^t P(\tau)e^{-\xi\omega(t-\tau)} \sin\bigl(\omega_D(t-\tau)\bigr)\, d\tau \tag{2.3}$$

where ξ is the damping ratio and $\omega_D = \omega\sqrt{1-\xi^2}$ with ω being the circular frequency of the system. Due to the system uncertainty in Eq. (2.2), the circular frequency ω is a random variable given by the following relation:

$$\omega = \sqrt{k/M_s} \tag{2.4}$$

where k is the stiffness of the oscillator which can be derived from the static displacement of the oscillator for a unit static deterministic load at the end of the beam (Fig. 2.1(b)) as follows:

$$k = \frac{1}{u_{st}} = \left[-\frac{F_0}{I} \int_0^L (x-\alpha)M(\alpha)\bigl(1 + f(\alpha)\bigr)\, d\alpha\right]^{-1} \tag{2.5}$$

where I is the moment of inertia of the beam and $M(\alpha)$ is the moment at position α.

In the general case where the load is arbitrary and the system is initially at rest, the deterministic displacement at the right end of the beam can be derived by numerically solving the Duhamel's integral. In the special case of a sinusoidal $P(t) = P_0 \sin(\bar{\omega}t)$ the solution of Eq. (2.3) leads to the following expression for $u(t)$:

$$u(t) = u_0(t) + u_p(t) \tag{2.6}$$

where

$$u_0(t) = e^{-\xi\omega t}(A\sin\omega_D t + B\cos\omega_D t) \tag{2.7a}$$

$$u_p(t) = C_1 \sin\bar{\omega}t + C_2\cos\bar{\omega}t \tag{2.7b}$$

$$A = \frac{P_0}{K} * \frac{1}{(1-\beta^2)^2 + (2\xi\beta)^2} * \frac{2\beta\xi^2 - (1-\beta^2)\beta}{\sqrt{1-\xi^2}} \tag{2.7c}$$

$$\mathrm{B} = -\frac{P_0}{K} * \frac{2\xi\beta}{(1-\beta^2)^2 + (2\xi\beta)^2} \tag{2.7d}$$

$$C_1 = \frac{P_0}{K} * \frac{1}{(1-\beta^2)^2 + (2\xi\beta)^2}(1-\beta^2) \tag{2.7e}$$

$$C_2 = -\frac{P_0}{K} * \frac{1}{(1-\beta^2)^2 + (2\xi\beta)^2}(2\xi\beta) \tag{2.7f}$$

$$\beta = \bar{\omega}/\omega \tag{2.7g}$$

In the trivial case in which a static load $P(t) = P_0$ is suddenly applied, the response displacement is given by

$$u(t) = \frac{P_0}{k}\left[1 - \left(\cos\omega_D t + \frac{\xi}{\sqrt{1-\xi^2}}\sin\omega_D t\right)e^{-\xi\omega t}\right] \tag{2.7h}$$

3 Response Variance and Mean Value of the Dynamic Response

Following a procedure similar to the one presented in [8] for linear stochastic systems under static loading, it is possible to express the variance of the dynamic response of the stochastic system in the following integral form expression:

$$\mathrm{Var}\big[u(t)\big] = \int_{-\infty}^{\infty} DVRF(t,\kappa,\sigma_{ff})S_{ff}(\kappa)\,d\kappa \tag{2.8a}$$

where *DVRF* is the dynamic version of a *VRF*, assumed to be a function of deterministic parameters of the problem related to geometry, loads and (mean) material properties and the standard deviation σ_{ff} of the stochastic field that models the system flexibility. A similar integral expression can provide an estimate for the mean value of the dynamic response of the system using the Dynamic Mean Response Function (*DMRF*) [9]:

$$\varepsilon\big[u(t)\big] = \int_{-\infty}^{\infty} DMRF(t,\kappa,\sigma_{ff})S_{ff}(\kappa)\,d\kappa \tag{2.8b}$$

DMRF is assumed to be a function similar to the *DVRF* in the sense that it also depends on deterministic parameters of the problem as well as σ_{ff}. It is extremely difficult however, to prove that the *DVRF* (same counts for *DMRF*) is independent (or even approximately independent) of the marginal pdf and the functional form of the power spectral density of the stochastic field $f(x)$. As in [7–9], the aforementioned assumptions are considered to form a conjecture which is numerically validated here by comparing the results from Eqs. (2.8a) and (2.8b) with brute force MCS.

The derivation of an analytic expression for the *DVRF* and *DMRF*, if possible at all, is an extremely cumbersome task. A numerical computation, however can be easily achieved, as described in the following section and then fed into the Eqs. (2.8a) and (2.8b) to provide estimates of the mean and variance of the dynamic system response.

3.1 Numerical Estimation of the DVRF and the DMRF Using Fast Monte Carlo Simulation

The numerical estimation of DVRF and DMRF involves a fast Monte Carlo simulation (FMCS) whose basic idea is to consider the random field $f(x)$ as a random sinusoid [7, 8] and plug its monochromatic power spectrum into Eqs. (2.8a) and (2.8b), in order to compute the respective mean and variance response at various wave numbers. The steps of the FMCS approach are the following:

(i) Generate N (10–20) sample functions of the below random sinusoid with standard deviation σ_{ff} and wave number $\bar{\kappa}$ modeling the variation of the inverse of the elastic modulus $1/E$ around its mean F_0:

$$f_j(x) = \sqrt{2}\sigma_{ff}\cos(\bar{\kappa}x + \varphi_j) \tag{2.9}$$

where $j = 1, 2, \ldots, N$ and φ_j varies randomly under uniform distribution in the range $[0, 2\pi]$.

(ii) Using these N generated sample functions it is straightforward to compute their respective dynamic mean and response variance, $\varepsilon[u(t)]_{\bar{\kappa}}$ and $\mathrm{Var}[u(t)]_{\bar{\kappa}}$, respectively for a given time step t.

(iii) The value of the *DMRF* at wave number $\bar{\kappa}$ can then be computed as follows

$$DMRF(t, \bar{\kappa}, \sigma_{ff}) = \frac{\varepsilon[u(t)]_{\bar{\kappa}}}{\sigma_{ff}^2} \tag{2.10a}$$

and likewise the value of the *DVRF* at wave number $\bar{\kappa}$

$$DVRF(t, \bar{\kappa}, \sigma_{ff}) = \frac{\mathrm{Var}[u(t)]_{\bar{\kappa}}}{\sigma_{ff}^2} \tag{2.10b}$$

Both previous equations are direct consequences of the integral expressions in Eqs. (2.8a) and (2.8b) in the case that the stochastic field becomes a random sinusoid.

(iv) Get *DMRF* and *DVRF* as a function of both time t and wave number κ by repeating previous steps for various wave numbers and different time steps. The entire procedure can be repeated for different values of the standard deviation σ_{ff} of the random sinusoid.

3.2 Bounds of the Mean and Variance of the Dynamic Response

Upper bounds on the mean and variance of the dynamic response of the stochastic system can be established directly from Eqs. (2.8a) and (2.8b), as follows:

$$\varepsilon\left[u(t)\right] = \int_{-\infty}^{\infty} DMRF(t,\kappa,\sigma_{ff})S_{ff}(\kappa)\,d\kappa \leq DMRF\left(t,\kappa^{\max}(t),\sigma_{ff}\right)\sigma_{ff}^{2} \tag{2.11a}$$

$$\mathrm{Var}\left[u(t)\right] = \int_{-\infty}^{\infty} DVRF(t,\kappa,\sigma_{ff})S_{ff}(\kappa)\,d\kappa \leq DVRF\left(t,\kappa^{\max}(t),\sigma_{ff}\right)\sigma_{ff}^{2} \tag{2.11b}$$

where $\kappa^{\max}(t)$ is the wave number at which *DMRF* and *DVRF*, corresponding to a given time step t and value of σ_{ff}, reach their maximum value. An envelope of time evolving upper bounds on the mean and variance of the dynamic system response can be extracted from Eqs. (2.11a) and (2.11b). As in the case of linear stochastic systems under static loads [7–9], this envelope is physically realizable since the form of the stochastic field that produces it is the random sinusoid of Eq. (2.9) with $\bar{\kappa} = \kappa^{\max}(t)$.

4 Numerical Example

For the cantilever beam shown in Fig. 2.1 with length $L = 4$ m, the inverse of the modulus of elasticity is assumed to vary randomly along its length according to Eq. (2.2) with $F_0 = (1.25 \times 10^8\ \mathrm{kN/m})^{-1}$ and $I = 0.1\ \mathrm{m}^4$. A concentrated mass $M_s = 3.715 \times 10^3$ kg is assumed at the right end of the beam. The damping ratio is taken as $\xi = 5\%$ and the mean eigenperiod of this one d.o.f. oscillator is calculated at $T_0 = 0.5$ s.

Two load cases are considered: LC1 consisting of a concentrated dynamic periodic load $P(t) = 100\sin(\bar{\omega}t)$ and LC2 consisting of $P(t) = -M_s\ddot{U}_g(t)$ where $\ddot{U}_g(t)$ is the acceleration time history of the 1940 El Centro earthquake.

The spectral density function (*SDF*) of Fig. 2.2 was used for the modeling of the inverse of the elastic modulus stochastic field, given by:

$$S_{ff}(\kappa) = \frac{1}{4}\sigma^2 b^3 \kappa^2 e^{-b|\kappa|} \tag{2.12}$$

with $b = 10$ being a correlation length parameter.

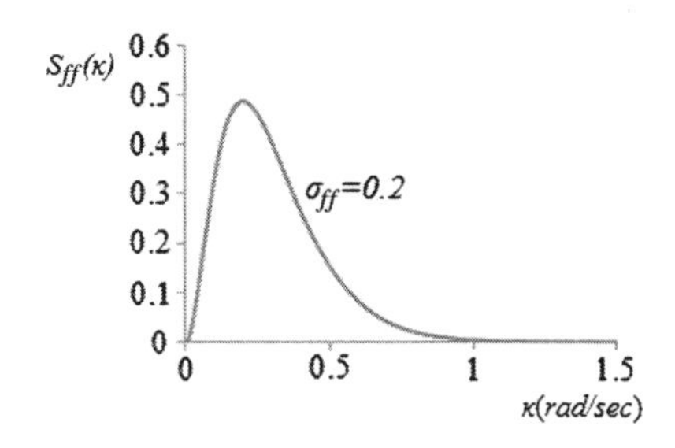

Fig. 2.2 Spectral density function for stochastic field $f(x)$ standard deviation $\sigma_{ff} = 0.2$

In order to demonstrate the validity of the proposed methodology, a truncated Gaussian and a lognormal pdf were used to model $f(x)$. For this purpose, an underlying Gaussian stochastic field denoted by $g(x)$ is generated using the spectral representation method [11] and the power spectrum of Eq. (2.12). The truncated Gaussian field $f_{\mathrm{TG}}(x)$ is obtained by simply truncating $g(x)$ in the following way: $-0.9 \leq g(x) \leq 0.9$, while the lognormal $f_{\mathrm{L}}(x)$ is obtained from the following transformation as a translation field [2]:

$$f_{\mathrm{L}}(x) = F_{\mathrm{L}}^{-1}\{G[g(x)]\} \tag{2.13}$$

The *SDF* of the underlying Gaussian field in Eq. (2.12) and the corresponding spectral densities of the truncated Gaussian and the Lognormal fields denoted $S_{f_{\mathrm{TG}} f_{\mathrm{TG}}}(\kappa)$ and $S_{f_{\mathrm{L}} f_{\mathrm{L}}}(\kappa)$, respectively, will be different. These are computed from the following formula

$$S_{f_i f_i}(\kappa) = \frac{1}{2\pi L_x} \left| \int_0^{L_x} f_i(x) e^{-i\kappa x} \, dx \right|^2; \quad i = \mathrm{TG}, \mathrm{L} \tag{2.14}$$

where L_x is the length of the sample functions of the non-Gaussian fields modeling flexibility. As the sample functions of the non-Gaussian fields are non-ergodic, the estimation of power spectra in Eq. (2.14) is performed in an ensemble average sense [2].

LC1: Dynamic Periodic Load at the End of the Beam Figures 2.3 and 2.4 present *DMRF* and *DVRF*, respectively, computed with FMCS for a periodic load with frequency $\bar{\omega} = 2$ and three different values of the standard deviation $\sigma_{ff} = 0.2$, $\sigma_{ff} = 0.4$ and $\sigma_{ff} = 0.6$. From these figures it can be observed that *DVRF* do not follow any particular pattern with respect to any increase or decrease of σ_{ff} in contrast to *DMRF* and to what has been observed in Papadopoulos and Deodatis [7] for the corresponding static problem, albeit the mean and variability response increases as σ_{ff} increases, as shown in Fig. 2.4. Figures 2.5(a) and (b) present plots of *DMRF* and *DVRF* as a function of t for a fixed wave number $\kappa = 2$ and $\sigma_{ff} = 0.2$. From the above Figs. 2.3, 2.4 and 2.5 it appears that *DMRF* and *DVRF* have a significant variation along the wave number κ axis and the time axis t. Both functions and especially *DVRF* have an initial transient phase and then appear to be periodic. It is reminded here that *DVRF* and *DMRF* are functions of the imposed dynamic loading. This explains the fact that they do not approach zero with t increasing, since the applied dynamic load is periodic with constant amplitude which does not decay.

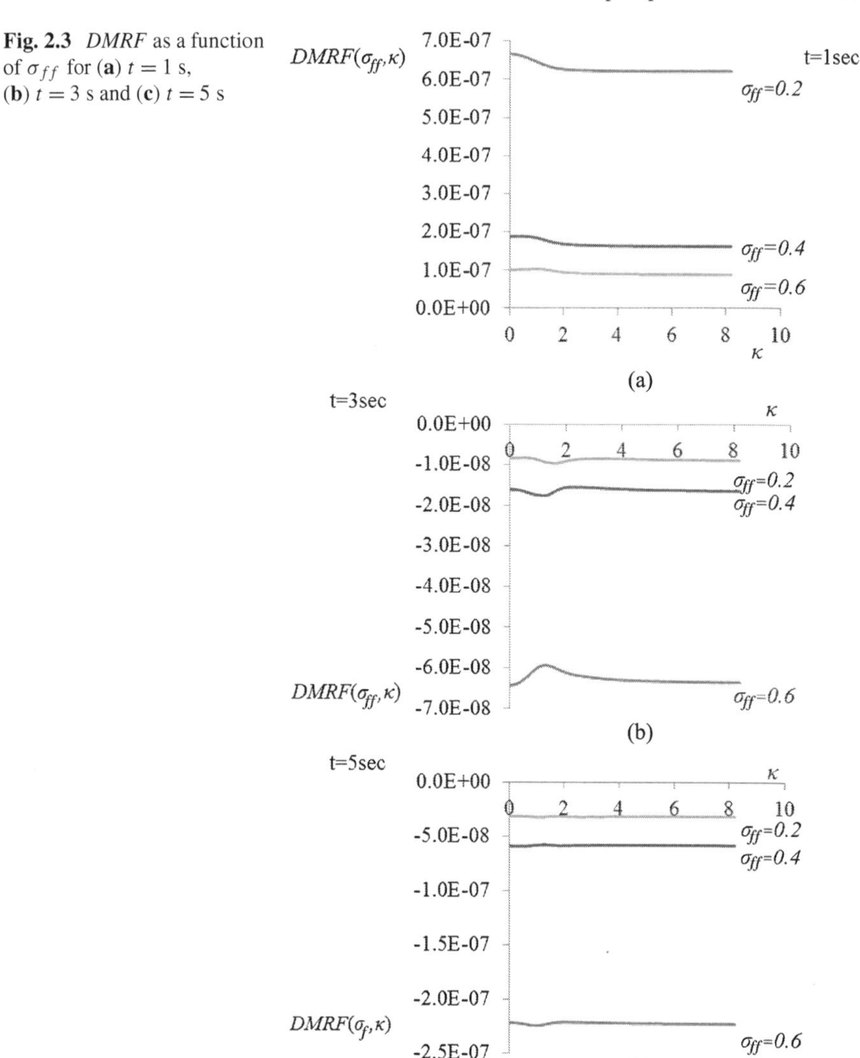

Fig. 2.3 *DMRF* as a function of σ_{ff} for (**a**) $t = 1$ s, (**b**) $t = 3$ s and (**c**) $t = 5$ s

Figures 2.6(a) and (b) present comparatively the results of the computed response variance time histories using the integral expression of Eq. (2.8a) and MCS, for two different standard deviations of a truncated Gaussian stochastic field used for the modeling of flexibility. The underlying Gaussian field is modeled with the power spectral density of Eq. (2.12) and two different standard deviations $\sigma_{gg} = 0.4$ and $\sigma_{gg} = 0.6$. The corresponding standard deviations of the truncated Gaussian field $f(x)$ are computed as $\sigma_{ff} = 0.3912$ and $\sigma_{ff} = 0.5286$, respectively. Figures 2.7(a) and (b) present the same results with Fig. 2.8 but for the mean response of the oscillator. The deterministic displacement time history is also plotted in Fig. 2.7(c) for comparison purposes. From these figures it can be observed that the mean and

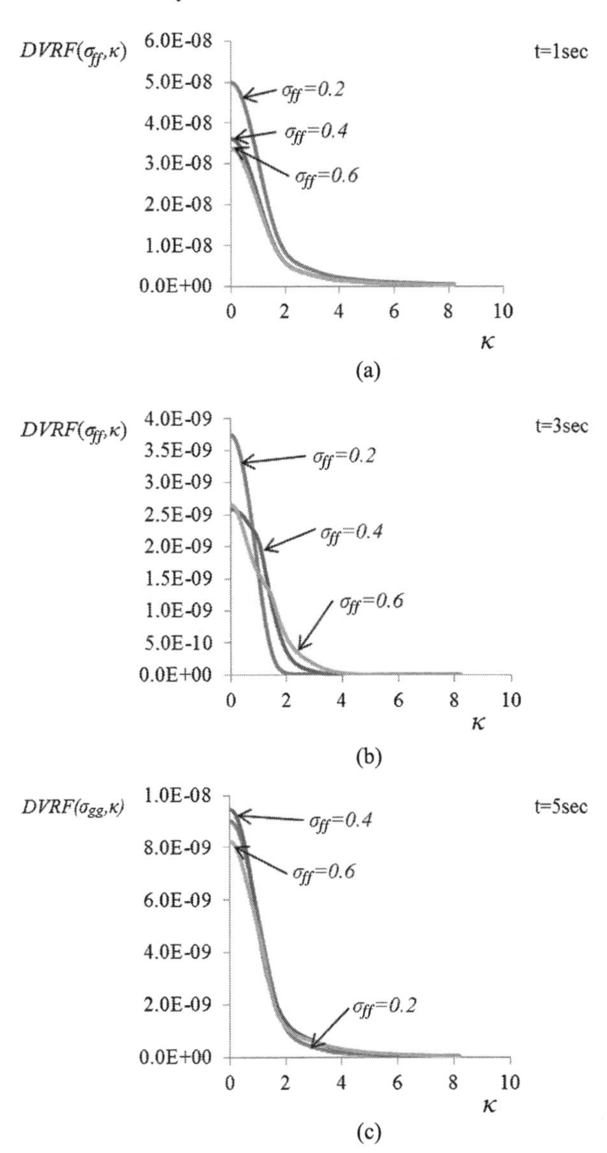

Fig. 2.4 *DVRF* as a function of the σ_{ff} for (**a**) $t = 1$ s, (**b**) $t = 3$ s and (**c**) $t = 5$ s

variability response time histories obtained with the integral expressions of Eqs. (2.8a) and (2.8b) are in close agreement with the corresponding MCS estimates. In all cases examined the maximum error in the computed $\text{Var}[u(t)]$, observed at the peak values of the variance, is less than 25%, while in all other time steps this error is less than 3–4%. In the case of $\varepsilon[u(t)]$, the predictions of Eq. (2.8b) are almost identical to the ones obtained with MCS, with an error of less than 3% in all cases. From Figs. 2.7(a)–(c), it can be observed that in all cases, the mean response time

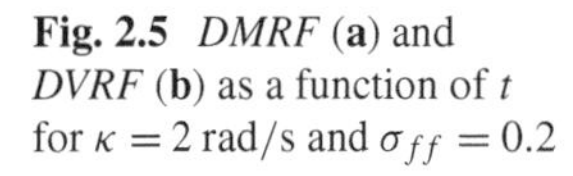

Fig. 2.5 *DMRF* (**a**) and *DVRF* (**b**) as a function of t for $\kappa = 2$ rad/s and $\sigma_{ff} = 0.2$

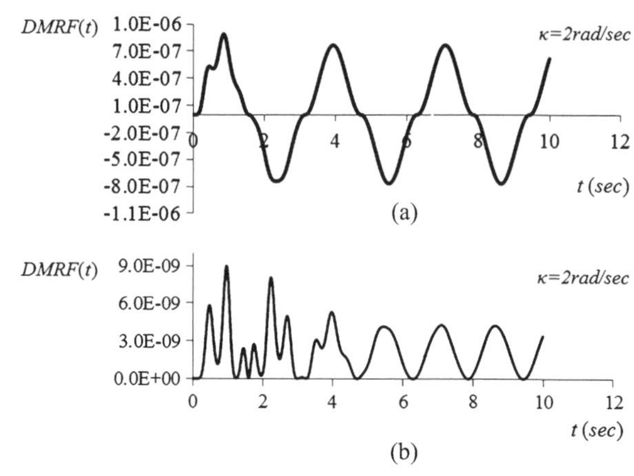

history for all cases examined is almost identical to the deterministic one, with the exception of the first cycle where slight differences in the peak values are observed.

Figures 2.8(a) and (b) repeat the same comparisons with the previous Figs. 2.6 and 2.7 but for the case of a lognormal stochastic field used for the modeling of flexibility with $\sigma_{ff} = 0.2$ and lower bound $l_b = -0.8$. The conclusions extracted previously for the case of truncated Gaussian fields also apply here.

LC2: El Centro Earthquake Figures 2.9(a) and (b) present 3D plots of the *DMRF* and *DVRF* as a function of frequency κ and time t (s) for $\sigma_{ff} = 0.2$. From these figures it can be observed that *DMRF* and *DVRF* have a significant variation in both κ and t axis, without being periodic in contrast to what has been observed in LC1. In addition, both *DMRF* and *DVRF* approach a zero value with time increasing due to the fact that ground accelerations decay and vanish after some time.

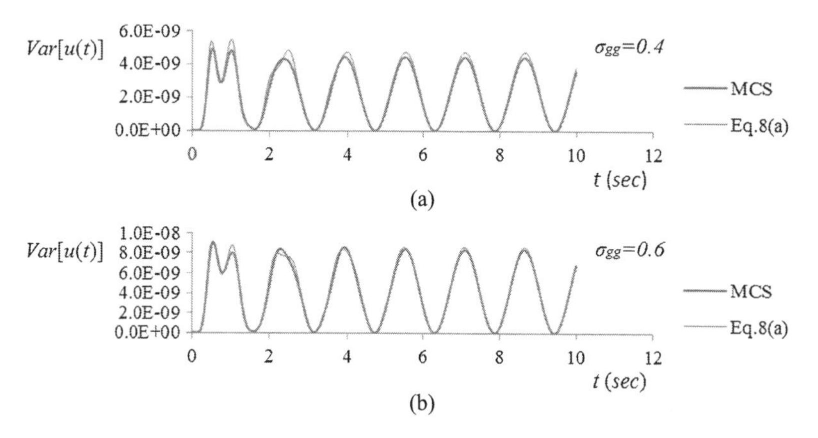

Fig. 2.6 Time histories of the variance of the response displacement for a truncated Gaussian field with (**a**) $\sigma_{gg} = 0.4$, and (**b**) $\sigma_{gg} = 0.6$. Comparison of results obtained from Eq. (2.11a) and MCS

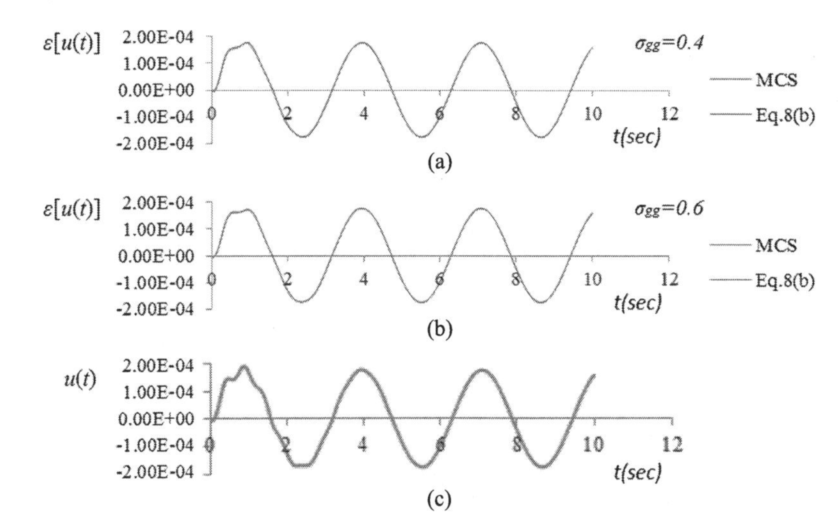

Fig. 2.7 Time histories of: (**a**) mean response displacement for a truncated Gaussian field with $\sigma_{gg} = 0.4$, (**b**) $\sigma_{gg} = 0.6$ and (**c**) the deterministic displacement. Comparison of results obtained from Eq. (2.11b) and MCS

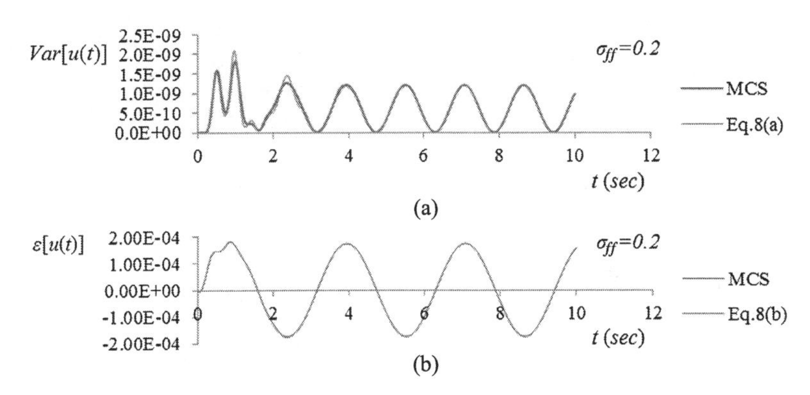

Fig. 2.8 Comparative results from Eqs. (2.11a) and (2.11b) and MCS for a lognormal field with $\sigma_{ff} = 0.2$ for (**a**) the variance and (**b**) the mean of the response displacement time history

Figures 2.10(a) and (b) present a comparison of the response variance computed with Eq. (2.8a) and MCS, in the case of a truncated Gaussian stochastic field modeling flexibility with $\sigma_{gg} = 0.4$ and 0.6, while Figs. 2.11(a) and (b) present the same results for the mean dynamic response of the stochastic oscillator along with the corresponding deterministic displacement time history (Fig. 2.11(c)). Figures 2.12(a) and (b) repeat the same comparisons for the case of a lognormal stochastic field used for the modeling of flexibility and $\sigma_{ff} = 0.3$ and lower bound $l_b = -0.8$.

From the above figures it can be observed that, as in LC2, the mean and variability response time histories obtained with the integral expressions of Eqs. (2.8a)

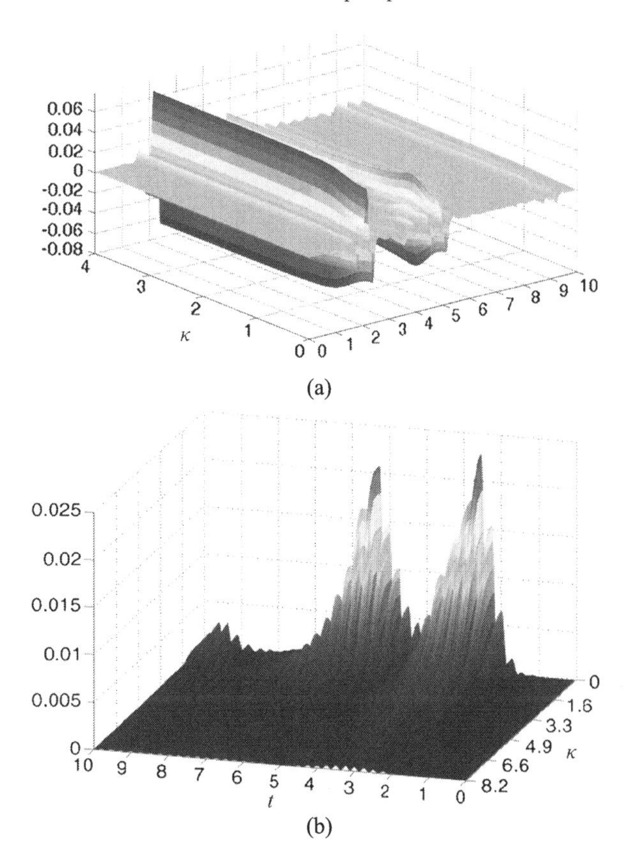

Fig. 2.9 3D plots of (**a**) *DMRF* and (**b**) *DVRF*, as a function of frequency κ (rad/m) and time t (s) for LC2 and $\sigma_{ff} = 0.2$

and (2.8b) are in close agreement with the corresponding MCS estimates, in all cases. Again, the maximum error in the computed $\text{Var}[u(t)]$ was observed at the peak values of the variance and is less than 25%, while in all other time steps this error is less than 3–4%. In the case of $\varepsilon[u(t)]$, the predictions of Eq. (2.8b) are very close to the ones obtained with MCS, with a error of less than 3% in all cases. From Figs. 2.9(a)–(c), it can be observed that, in contrast to what was observed in LC2, the mean response time history differs significantly from the corresponding deterministic one, in terms of both frequencies and amplitudes.

Upper Bounds on the Mean and Variance of the Response of LC3 Spectral-distribution-free upper bounds on both the mean and variance of the response are obtained via Eqs. (2.11a) and (2.11b), respectively. Results of this calculation are presented in Figs. 2.13(a) and (b), in which the time dependant upper bounds on the mean and variance of the response displacement are plotted against time for a standard deviation $\sigma_{ff} = 0.4$.

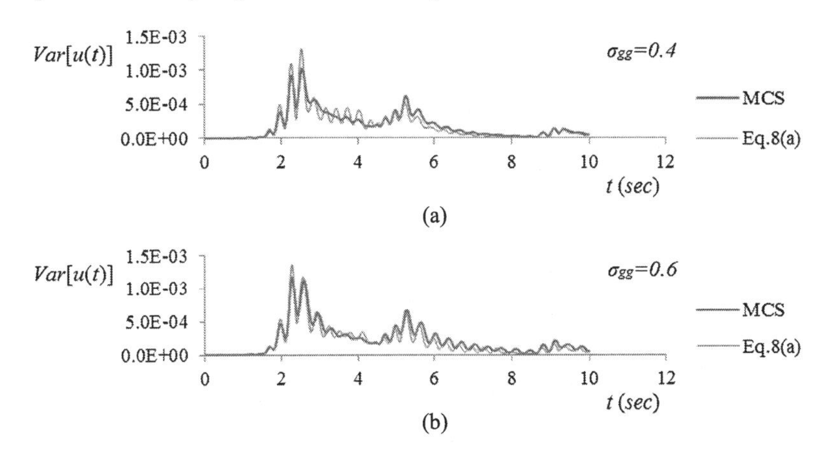

Fig. 2.10 Time histories of the variance of the response displacement for a truncated Gaussian field for (**a**) $\sigma_{gg} = 0.4$ and (**b**) $\sigma_{gg} = 0.6$. Comparison of results obtained from Eq. (2.8a) and MCS

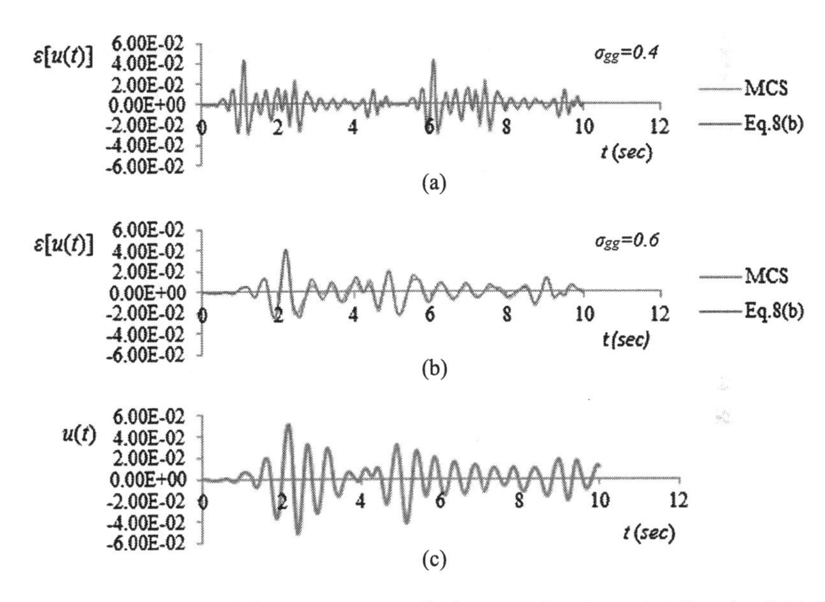

Fig. 2.11 Time histories of the mean response displacement for a truncated Gaussian field with (**a**) $\sigma_{gg} = 0.4$, (**b**) $\sigma_{gg} = 0.6$ and (**c**) of the deterministic response displacement. Comparison of results obtained from Eq. (2.8b) and MCS

Sensitivity Analysis for LC3 Using the Integral Expressions in Eqs. (2.8a) and (2.8b) Finally, a sensitivity analysis is performed using Eqs. (2.8a) and (2.8b) at minimum computational cost, with respect to three different values of the correlation length parameter of the *SDF* in Eq. (2.12) and $\sigma_{ff} = 0.2$. Respective results are shown in Figs. 2.14(a) and (b).

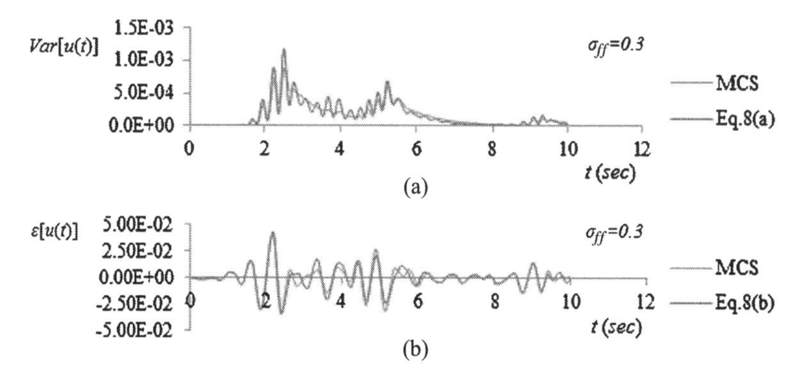

Fig. 2.12 Comparative results from Eqs. (2.11a) and (2.11b) and MCS for a lognormal field with $\sigma_{ff} = 0.2$ for (**a**) the variance and (**b**) the mean of the response displacement time history

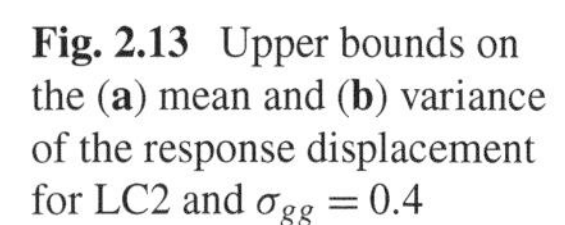
Fig. 2.13 Upper bounds on the (**a**) mean and (**b**) variance of the response displacement for LC2 and $\sigma_{gg} = 0.4$

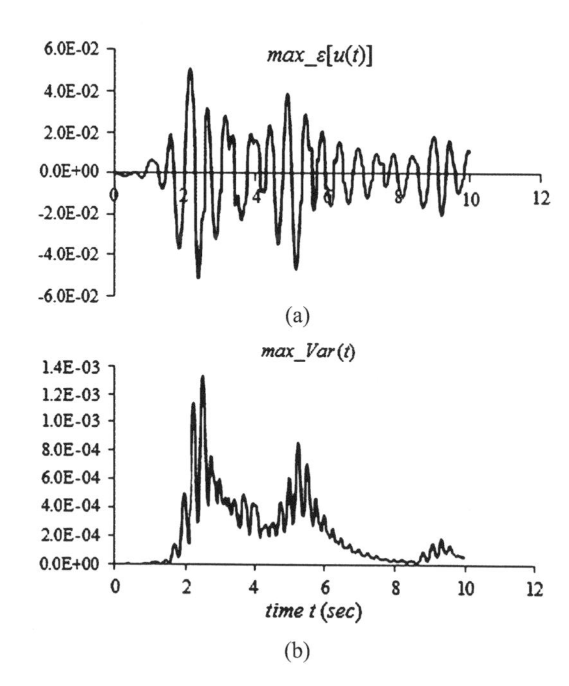

5 Concluding Remarks

In the present work, Dynamic Variability Response Functions and Dynamic Mean Response Functions are obtained for a linear stochastic single d.o.f. oscillator with random material properties under dynamic excitation. The inverse of the modulus of elasticity was considered as the uncertain system parameter.

It is demonstrated that, as in the case of stochastic systems under static loading, *DVRF* and *DMRF* depend on the standard deviation of the stochastic field modeling the uncertain parameter but appear to be almost independent of its power spectral

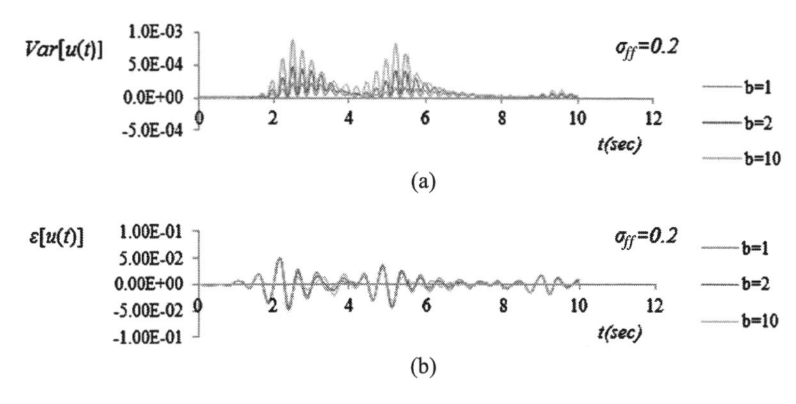

Fig. 2.14 (**a**) Mean and (**b**) variance time histories of the response displacement computed from Eqs. (2.8b) and (2.8a), respectively for three different values of the correlation length parameter b of the *SDF* in Eq. (2.12)

density and marginal *pdf*. The results obtained from the integral expressions are close to those obtained with MCS reaching a maximum error of the order of 20–25%.

As in the case of stochastic systems under static loading, the *DVRF* and *DMRF* provide with an insight of the dynamic system sensitivity to the stochastic parameters and the mechanisms controlling the response mean and variability and their evolution in time.

References

1. Graham, L., Deodatis, G.: Weighted integral method and variability response functions for stochastic plate bending problems. Struct. Saf. **20**, 167–188 (1998)
2. Grigoriu, M.: Applied Non-Gaussian Processes: Examples, Theory, Simulation, Linear Random Vibration, and MATLAB Solutions. Prentice Hall, New York (1995)
3. Grigoriu, M.: Evaluation of Karhunen–Loève, spectral and sampling representations for stochastic processes. J. Eng. Mech. **132**, 179–189 (2006)
4. Liu, W.K., Belytschko, T., Mani, A.: Probabilistic finite elements for nonlinear structural dynamics. Comput. Methods Appl. Mech. Eng. **56**, 61–86 (1986)
5. Matthies, H.G., Brenner, C.E., Bucher, C.G., Guedes Soares, C.: Uncertainties in probabilistic numerical analysis of structures and solids—stochastic finite elements. Struct. Saf. **19**, 283–336 (1997)
6. Miranda, M.: On the response variability of beam structures with stochastic parameters. Ph.D. thesis, Columbia University (2008)
7. Papadopoulos, V., Deodatis, G.: Response variability of stochastic frame structures using evolutionary field theory. Comput. Methods Appl. Mech. Eng. **195**(9–12), 1050–1074 (2006)
8. Papadopoulos, V., Deodatis, G., Papadrakakis, M.: Flexibility-based upper bounds on the response variability of simple beams. Comput. Methods Appl. Mech. Eng. **194**(12–16), 1385–1404 (2005)
9. Papadopoulos, V., Papadrakakis, M., Deodatis, G.: Analysis of mean response and response variability of stochastic finite element systems. Comput. Methods Appl. Mech. Eng. **195**(41–43), 5454–5471 (2006)

10. Shinozuka, M.: Structural response variability. J. Eng. Mech. **113**(6), 825–842 (1987)
11. Shinozuka, M., Deodatis, G.: Simulation of stochastic processes by spectral representation. Appl. Mech. Rev. **44**(4), 191–203 (1991)
12. Stefanou, G.: The stochastic finite element method: past, present and future. Comput. Methods Appl. Mech. Eng. **198**, 1031–1051 (2009)
13. Wall, F.J., Deodatis, G.: Variability response functions of stochastic plane stress/strain problems. J. Eng. Mech. **120**(9), 1963–1982 (1994)

Chapter 3
A Novel Reduced Spectral Function Approach for Finite Element Analysis of Stochastic Dynamical Systems

Abhishek Kundu and Sondipon Adhikari

Abstract This work provides the theoretical development and simulation results of a novel Galerkin subspace projection scheme for damped dynamic systems with stochastic coefficients and homogeneous Dirichlet boundary conditions. The fundamental idea involved here is to solve the stochastic dynamic system in the frequency domain by projecting the solution into a reduced finite dimensional spatio-random vector basis spanning the stochastic Krylov subspace to approximate the response. Subsequently, Galerkin weighting coefficients have been employed to minimize the error induced due to the use of the reduced basis and a finite order of the spectral functions and hence to explicitly evaluate the stochastic system response. The statistical moments of the solution have been evaluated at all frequencies to illustrate and compare the stochastic system response with the deterministic case. The results have been validated with direct Monte-Carlo simulation for different correlation lengths and variability of randomness.

1 Introduction

Due to the significant development in computational hardware it is now possible to solve very high resolution models in various computational physics problems, ranging from fluid mechanics to nano-bio mechanics. However, the spatial resolution is not enough to determine the credibility of the numerical model, the physical model as well its parameters are also crucial. Since neither of these may not be exactly known, there has been increasing research activities over the past three decades to model the governing partial differential equations within the framework of stochastic equations. We refer to few recent review papers [5, 25, 32]. Consider a bounded domain $\mathcal{D} \in \mathbb{R}^d$ with piecewise Lipschitz boundary $\partial\mathcal{D}$, where $d \leq 3$ is the spatial dimension and $t \in \mathbb{R}^+$ is the time. Further, consider that $(\Theta, \mathcal{F}, P)$ is a probability

A. Kundu (✉) · S. Adhikari
Swansea University, Singleton Park, Swansea, SA2 8PP, UK
e-mail: a.kundu.577613@swansea.ac.uk

S. Adhikari
e-mail: s.adhikari@swansea.ac.uk

M. Papadrakakis et al. (eds.), *Computational Methods in Stochastic Dynamics*, Computational Methods in Applied Sciences 26,
DOI 10.1007/978-94-007-5134-7_3,

space where $\theta \in \Theta$ is a sample point from the sampling space Θ, $\mathcal{F}$ is the complete σ-algebra over the subsets of Θ and P is the probability measure. We consider a linear stochastic partial differential equation (PDE) of the form

$$\mathfrak{F}_\alpha\big[u(\mathbf{r},t,\theta)\big] + \mathfrak{L}_\beta\big[u(\mathbf{r},t,\theta)\big] = p(\mathbf{r},t); \quad \mathbf{r} \text{ in } \mathcal{D} \tag{3.1}$$

with the associated Dirichlet condition

$$u(\mathbf{r},t,\theta) = 0; \quad \mathbf{r} \text{ on } \partial\mathcal{D}. \tag{3.2}$$

These second order time varying equations typically arise in case of structural vibration problems in mechanical engineering. Here $\mathfrak{F}_\alpha$ and $\mathfrak{L}_\beta$ denote the linear stochastic differential operators with coefficients $\alpha(\mathbf{r},\theta)$ and $\beta(\mathbf{r},\theta)$ as the second order random fields such that $\alpha, \beta : \mathbb{R}^d \times \Theta \to \mathbb{R}$. We assume the random fields to be stationary and square integrable. Depending on the physical problem the random fields $\alpha[u(\mathbf{r},\theta)]$ and $\beta[u(\mathbf{r},\theta)]$ can be used to model different physical quantities. $p(\mathbf{r},t)$ denotes the deterministic excitation field for which the solution $u(\mathbf{r},t,\theta)$ is sought. The purpose of this paper is to investigate a new solution approach for Eq. (3.1) after the discretization of the spatio-random fields using the well established techniques of stochastic finite element method (SFEM) as can be found in references [10, 14, 21].

The random fields in Eq. (3.1) can be discretized to represent them as spectral functions using a finite number of random variables using one of the established techniques available in literatures [10, 16]. Hence the stochastic PDE along with the boundary conditions would result in an equation of the form

$$\mathbf{M}(\theta)\ddot{\mathbf{u}}(\theta,t) + \mathbf{C}(\theta)\dot{\mathbf{u}}(\theta,t) + \mathbf{K}(\theta)\mathbf{u}(\theta,t) = \mathbf{f}_0(t) \tag{3.3}$$

where $\mathbf{M}(\theta) = \mathbf{M}_0 + \sum_{i=1}^{p_1} \mu_i(\theta)\mathbf{M}_i \in \mathbb{R}^{n\times n}$ is the random mass matrix, $\mathbf{K}(\theta) = \mathbf{K}_0 + \sum_{i=1}^{p_2} \nu_i(\theta)\mathbf{K}_i \in \mathbb{R}^{n\times n}$ is the random stiffness matrix along with $\mathbf{C}(\theta) \in \mathbb{R}^{n\times n}$ as the random damping matrix. Here the mass and stiffness matrices have been expressed in terms of their deterministic components ($\mathbf{M}_0$ and $\mathbf{K}_0$) and the corresponding random contributions ($\mathbf{M}_i$ and $\mathbf{K}_i$) obtained from discretizing the stochastic field with a finite number of random variables ($\mu_i(\theta)$ and $\nu_i(\theta)$) and their corresponding spatial basis functions. This has been elaborated in Sect. 2.1. In the present work proportional damping is considered for which $\mathbf{C}(\theta) = \zeta_1\mathbf{M}(\theta) + \zeta_2\mathbf{K}(\theta)$, where ζ_1 and ζ_2 are deterministic scalars. For the harmonic analysis of the structural system considered in Eq. (3.3), it is represented in the frequency domain as

$$\big[-\omega^2\mathbf{M}(\theta) + i\omega\mathbf{C}(\theta) + \mathbf{K}(\theta)\big]\widetilde{\mathbf{u}}(\theta,\omega) = \widetilde{\mathbf{f}}_0(\omega) \tag{3.4}$$

where $\widetilde{\mathbf{u}}(\theta,\omega)$ is the complex frequency domain system response amplitude and $\widetilde{\mathbf{f}}_0(\omega)$ is the amplitude of the harmonic force. Now we group the random variables associated with the mass and damping matrices of Eq. (3.3) as

$$\xi_i(\theta) = \mu_i(\theta) \quad \text{for } i = 1, 2, \ldots, p_1$$

and

$$\xi_{i+p_1}(\theta) = \nu_i(\theta) \quad \text{for } i = 1, 2, \dots, p_2.$$

Thus the total number of random variables used to represent the mass and the stiffness matrices becomes $p = p_1 + p_2$. Following this, the expression for the linear structural system in Eq. (3.4) can be expressed as

$$\left(\mathbf{A}_0(\omega) + \sum_{i=1}^{p} \xi_i(\theta)\mathbf{A}_i(\omega)\right)\widetilde{\mathbf{u}}(\omega,\theta) = \widetilde{\mathbf{f}}_0(\omega) \tag{3.5}$$

where $\mathbf{A}_0$ and $\mathbf{A}_i \in \mathbb{C}^{n\times n}$ represent the complex deterministic and stochastic parts respectively of the mass, the stiffness and the damping matrices ensemble. For the case of proportional damping the matrices $\mathbf{A}_0$ and $\mathbf{A}_i$ can be written as

$$\mathbf{A}_0(\omega) = \left[-\omega^2 + i\omega\zeta_1\right]\mathbf{M}_0 + [i\omega\zeta_2 + 1]\mathbf{K}_0 \tag{3.6}$$

and

$$\begin{aligned} \mathbf{A}_i(\omega) &= \left[-\omega^2 + i\omega\zeta_1\right]\mathbf{M}_i \quad \text{for } i = 1, 2, \dots, p_1 \\ \mathbf{A}_{i+p_1}(\omega) &= [i\omega\zeta_2 + 1]\mathbf{K}_i \quad \text{for } i = 1, 2, \dots, p_2. \end{aligned} \tag{3.7}$$

The paper has been arranged as follows. In Sect. 2 a brief overview of spectral stochastic finite element method is presented. The projection theory in the vector space is developed in Sect. 3. In Sect. 4 an error minimization approach in the Hilbert space is proposed. The idea of the reduced orthonormal vector basis is introduced in Sect. 5. The post processing of the results to obtain the response moments are discussed in Sect. 6. Based on the theoretical results, a simple computational approach is shown in Sect. 7 where the proposed method of reduced orthonormal basis is applied to the stochastic mechanics of an Euler–Bernoulli beam. From the theoretical developments and numerical results, some conclusions are drawn in Sect. 8.

2 Overview of the Spectral Stochastic Finite Element Method

2.1 Discretization of the Stochastic PDE

First consider $a(\mathbf{r},\theta)$ is a Gaussian random field with a covariance function $C_a : \mathbb{R}^d \times \mathbb{R}^d \to \mathbb{R}$ defined in the domain $\mathcal{D}$. Since the covariance function is square bounded, symmetric and positive definite, it can be represented by a spectral decomposition in an infinite dimensional Hilbert space. Using this spectral decomposition, the random process $a(\mathbf{r},\theta)$ can be expressed [see for example, [10, 27]] in a generalized Fourier type of series known as the Karhunen–Loève (KL) expansion

$$a(\mathbf{r},\theta) = a_0(\mathbf{r}) + \sum_{i=1}^{\infty} \sqrt{\nu_i}\widetilde{\xi}_i(\theta)\varphi_i(\mathbf{r}) \tag{3.8}$$

Here $a_0(\mathbf{r})$ is the mean function, $\widetilde{\xi}_i(\theta)$ are uncorrelated standard Gaussian random variables, ν_i and $\varphi_i(\mathbf{r})$ are eigenvalues and eigenfunctions satisfying the integral equation

$$\int_{\mathcal{D}} C_a(\mathbf{r}_1, \mathbf{r}_2)\varphi_j(\mathbf{r}_1)\,\mathrm{d}\mathbf{r}_1 = \nu_j \varphi_j(\mathbf{r}_2), \quad \forall j = 1, 2, \ldots \tag{3.9}$$

The Gaussian random field model is not applicable for strictly positive quantities arising in many practical problems. Equation (3.8) could also represent the Karhunen–Loève expansion of a non-Gaussian random field, which is also well defined. Alternatively, when $a(\mathbf{r}, \theta)$ is a general non-Gaussian random field, it can be expressed in a mean-square convergent series in random variables and spatial functions using the polynomial chaos expansion. For example Ghanem [9] expanded log-normal random fields in a polynomial chaos expansion. In general, non Gaussian random fields can be expressed in a series like

$$a(\mathbf{r}, \theta) = a_0(\mathbf{r}) + \sum_{i=1}^{\infty} \xi_i(\theta) a_i(\mathbf{r}) \tag{3.10}$$

using Wiener–Askey chaos expansion [33–35]. Here $\xi_i(\theta)$ are in general non-Gaussian and correlated random variables and $a_i(\mathbf{r})$ are deterministic functions. In this paper we use this general form of the decomposition of the random field.

Truncating the series in Eq. (3.10) upto the M-th term and using the same approach for the governing PDE (3.1) with boundary conditions, the discretized system equation in the frequency domain (3.3) can be represented by Eq. (3.5), with $M = p$. It is given here once again for convenience.

$$\left[\mathbf{A}_0(\omega) + \sum_{i=1}^{M} \xi_i(\theta)\mathbf{A}_i(\omega)\right]\mathbf{u}(\theta, \omega) = \mathbf{f}_0(\omega). \tag{3.11}$$

The 'tilde' sign has been omitted from the notations of the frequency domain quantities of $\mathbf{u}(\theta, \omega)$ and $\mathbf{f}_0(\omega)$ for the sake of notational convenience and this shall be followed henceforth. The global matrices in Eq. (3.11) can be expressed as

$$\mathbf{A}_i = \sum_e \mathbf{A}_i^{(e)}; \quad i = 0, 1, 2, \ldots, M \tag{3.12}$$

The element matrices $\mathbf{A}_i^{(e)}$ are defined over an element domain $\mathcal{D}_e \in \mathcal{D}$ such that $\mathcal{D}_e \cap \mathcal{D}_{e'} = \varnothing$ for $e \neq e'$ and $\bigcup_{\forall e} \mathcal{D}_e = \mathcal{D}$ and can be given by

$$\mathbf{A}_0^{(e)} = \int_{\mathcal{D}_e} a_0(\mathbf{r})\mathbf{B}^{(e)^T}(\mathbf{r})\mathbf{B}^{(e)}(\mathbf{r})\,\mathrm{d}\mathbf{r} \tag{3.13}$$

and

$$\mathbf{A}_i^{(e)} = \sqrt{\nu_i}\int_{\mathcal{D}_e} a_i(\mathbf{r})\mathbf{B}^{(e)^T}(\mathbf{r})\mathbf{B}^{(e)}(\mathbf{r})\,\mathrm{d}\mathbf{r}; \quad i = 1, 2, \ldots, M \tag{3.14}$$

In the above equations the $\mathbf{B}^{(e)}(\mathbf{r})$ is a deterministic matrix related to the shape function used to interpolate the solution within the element e. For the elliptic problem it can be shown [37] that $\mathbf{B}^{(e)}(\mathbf{r}) = \nabla \mathbf{N}^{(e)}(\mathbf{r})$. The necessary technical details to obtain the discrete stochastic algebraic equations from the stochastic partial differential equation (3.1) have become standard in the literature. Excellent references, for example [2, 3, 10, 20] are available on this topic. In Eq. (3.11), $\mathbf{A}_0(\omega) \in \mathbb{C}^{n\times n}$ and $\mathbf{A}_i(\omega) \in \mathbb{C}^{n\times n}$; $i = 1, 2, \ldots, M$ are symmetric matrices which are deterministic in nature, $\mathbf{u}(\omega, \theta) \in \mathbb{C}^n$ is the solution vector and $\mathbf{f}_0 \in \mathbb{C}^n$ in the input vector. We assume that the eigenvalues of the generalized eigenvalue problem with the deterministic mass ($\mathbf{M}_0$) and stiffness matrices $\mathbf{K}_0$ are distinct. The number of terms M in Eq. (3.11) can be selected based on the accuracy desired for the representation of the underlying random field. One of the main aim of a stochastic finite element analysis is to obtain $\mathbf{u}(\omega, \theta)$ for $\theta \in \Theta$ from Eq. (3.11) in an efficient manner and is the main topic of this paper. We propose a solution technique for Eq. (3.11) when $\xi_i(\theta)$ are in general non-Gaussian and correlated random variables.

2.2 *Brief Review of the Solution Techniques*

The solution of the set of stochastic linear algebraic equations (3.11) is a key step in the stochastic finite element analysis. As a result, several methods have been proposed. These methods include, first- and second-order perturbation methods [14, 18], Neumann expansion method [1, 36], Galerkin approach [11], linear algebra based methods [6, 7, 17] and simulation methods [26]. More recently efficient collocation methods have been proposed [8, 19]. Another class of methods which have been used widely in the literature is known as the spectral methods (see [25] for a recent review). These methods include the polynomial chaos (PC) expansion [10], stochastic reduced basis method [22, 29, 30] and Wiener–Askey chaos expansion [33–35]. According to the polynomial chaos expansion, second-order random variables $u_j(\theta)$ can be represented by the mean-square convergent expansion

$$\begin{aligned} u_j(\theta) &= u_{i_0} h_0 + \sum_{i_1=1}^{\infty} u_{i_1} h_1\big(\xi_{i_1}(\theta)\big) \\ &\quad + \sum_{i_1=1}^{\infty} \sum_{i_2=1}^{i_1} u_{i_1,i_2} h_2\big(\xi_{i_1}(\theta), \xi_{i_2}(\theta)\big) \\ &\quad + \sum_{i_1=1}^{\infty} \sum_{i_2=1}^{i_1} \sum_{i_3=1}^{i_2} u_{i_1 i_2 i_3} h_3\big(\xi_{i_1}(\theta), \xi_{i_2}(\theta), \xi_{i_3}(\theta)\big) \end{aligned}$$

$$+ \sum_{i_1=1}^{\infty} \sum_{i_2=1}^{i_1} \sum_{i_3=1}^{i_2} \sum_{i_4=1}^{i_3} u_{i_1 i_2 i_3 i_4} h_4(\xi_{i_1}(\theta), \xi_{i_2}(\theta), \xi_{i_3}(\theta), \xi_{i_4}(\theta))$$

$$+ \cdots, \tag{3.15}$$

where $u_{i_1,\dots,i_r}$ are deterministic constants to be determined and $h_r(\xi_{i_1}(\theta), \dots, \xi_{i_r}(\theta))$ is the rth order homogeneous Chaos. When $\xi_i(\theta)$ are Gaussian random variables, the functions $h_r(\xi_{i_1}(\theta), \dots, \xi_{i_r}(\theta))$ are the rth order Hermite polynomial so that it becomes orthonormal with respect to the Gaussian probability density function. The same idea can be extended to non-Gaussian random variables, provided more generalized functional basis are used [33–35] so that the orthonormality with respect to the probability density functions can be retained. When we have a random vector, as in the case of the solution of Eq. (3.11), then it is natural to replace the constants $u_{i_1,\dots,i_r}$ by vectors $\mathbf{u}_{i_1,\dots,i_r} \in \mathbb{R}^n$. Suppose the series is truncated after P number of terms. The value of P depends on the number of basic random variables M and the order of the PC expansion r as

$$P = \sum_{j=0}^{r} \frac{(M+j-1)!}{j!(M-1)!} = \binom{M+r}{r} \tag{3.16}$$

After the truncation, there are P number of unknown vectors of dimension n. Then a mean-square error minimization approach can be applied and the unknown vectors can be solved using the Galerkin approach [10]. Since P increases very rapidly with the order of the chaos r and the number of random variables M, the final number of unknown constants Pn becomes very large. As a result several methods have been developed (see for example [4, 22, 29–31]) to reduce the computational cost. In the polynomial chaos based solution approach, the *only* information used to construct the basis is the probability density function of the random variables. In the context of the discretized Eq. (3.11), more information such as the matrices $\mathbf{A}_i$, $i = 0, 1, 2, \dots, M$ are available. It may be possible to construct alternative basis using these matrices. Here we investigate such an approach, where instead of projecting the solution in the space of orthonormal polynomials, the solution is projected in an orthonormal vector basis generated from the coefficient matrices.

3 Spectral Decomposition in the Vector Space

3.1 Derivation of the Spectral Functions

Following the spectral stochastic finite element method, or otherwise, an approximation to the solution of Eq. (3.11) can be expressed as a linear combination of functions of random variables and deterministic vectors. Recently Nouy [23, 24] discussed the possibility of an optimal spectral decomposition. The aim is to use small number of terms to reduce the computation without loosing the accuracy. Here an orthonormal vector basis is considered. Fixing a value of θ, say $\theta = \theta_1$, the solution of Eq. (3.11) $\mathbf{u}(\theta_1)$ can be expanded in a complete basis as $\mathbf{u}(\theta_1) = \alpha_1^{(1)}\boldsymbol{\phi}_1 + \alpha_2^{(1)}\boldsymbol{\phi}_2 + \cdots + \alpha_n^{(1)}\boldsymbol{\phi}_n$. Repeating this for $\theta_1, \theta_2, \ldots$ eventually the whole sample-space can be covered and it would be possible to expand $\mathbf{u}(\theta)$, $\forall\theta \in \Theta$ as a linear combination of $\boldsymbol{\phi}_1, \boldsymbol{\phi}_2, \ldots, \boldsymbol{\phi}_n$.

We use the eigenvectors $\boldsymbol{\phi}_k \in \mathbb{R}^n$ of the generalized eigenvalue problem

$$\mathbf{K}_0\boldsymbol{\phi}_k = \lambda_k\mathbf{M}_0\boldsymbol{\phi}_k; \quad k = 1, 2, \ldots, n \tag{3.17}$$

Since the matrices $\mathbf{K}_0$ and $\mathbf{M}_0$ are symmetric and generally non-negative definite, the eigenvectors $\boldsymbol{\phi}_k$ for $k = 1, 2, \ldots, n$ form an orthonormal basis. Note that in principle any orthonormal basis can be used. This choice is selected due to the analytical simplicity as will be seen later. For notational convenience, define the matrix of eigenvalues and eigenvectors

$$\boldsymbol{\lambda}_0 = \text{diag}[\lambda_1, \lambda_2, \ldots, \lambda_n] \in \mathbb{R}^{n\times n} \quad \text{and} \quad \boldsymbol{\Phi} = [\boldsymbol{\phi}_1, \boldsymbol{\phi}_2, \ldots, \boldsymbol{\phi}_n] \in \mathbb{R}^{n\times n} \tag{3.18}$$

Eigenvalues are ordered in the ascending order so that $\lambda_1 < \lambda_2 < \cdots < \lambda_n$. Since $\boldsymbol{\Phi}$ is an orthonormal matrix we have $\boldsymbol{\Phi}^{-1} = \boldsymbol{\Phi}^T$ so that the following identities can easily be established

$$\begin{aligned}\boldsymbol{\Phi}^T\mathbf{A}_0\boldsymbol{\Phi} &= \boldsymbol{\Phi}^T\left([-\omega^2 + i\omega\zeta_1]\mathbf{M}_0 + [i\omega\zeta_2 + 1]\mathbf{K}_0\right)\boldsymbol{\Phi}\\ &= \left(-\omega^2 + i\omega\zeta_1\right)\mathbf{I} + (i\omega\zeta_2 + 1)\boldsymbol{\lambda}_0 \quad \text{which gives,}\end{aligned}$$

$$\boldsymbol{\Phi}^T\mathbf{A}_0\boldsymbol{\Phi} = \boldsymbol{\Lambda}_0; \qquad \mathbf{A}_0 = \boldsymbol{\Phi}^{-T}\boldsymbol{\Lambda}_0\boldsymbol{\Phi}^{-1} \quad \text{and} \quad \mathbf{A}_0^{-1} = \boldsymbol{\Phi}\boldsymbol{\Lambda}_0^{-1}\boldsymbol{\Phi}^T \tag{3.19}$$

where $\boldsymbol{\Lambda}_0 = (-\omega^2 + i\omega\zeta_1)\mathbf{I} + (i\omega\zeta_2 + 1)\boldsymbol{\lambda}_0$ and $\mathbf{I}$ is the identity matrix. Hence, $\boldsymbol{\Lambda}_0$ can also be written as

$$\boldsymbol{\Lambda}_0 = \text{diag}[\lambda_{0_1}, \lambda_{0_2}, \ldots, \lambda_{0_n}] \in \mathbb{C}^{n\times n} \tag{3.20}$$

where $\lambda_{0_j} = (-\omega^2 + i\omega\zeta_1) + (i\omega\zeta_2 + 1)\lambda_j$ and λ_j is as defined in Eq. (3.18). We also introduce the transformations

$$\widetilde{\mathbf{A}}_i = \boldsymbol{\Phi}^T\mathbf{A}_i\boldsymbol{\Phi} \in \mathbb{C}^{n\times n}; \quad i = 0, 1, 2, \ldots, M. \tag{3.21}$$

Note that $\widetilde{\mathbf{A}}_0 = \boldsymbol{\Lambda}_0$ is a diagonal matrix and

$$\mathbf{A}_i = \boldsymbol{\Phi}^{-T}\widetilde{\mathbf{A}}_i\boldsymbol{\Phi}^{-1} \in \mathbb{C}^{n\times n}; \quad i = 1, 2, \ldots, M. \tag{3.22}$$

Suppose the solution of Eq. (3.11) is given by

$$\hat{\mathbf{u}}(\omega, \theta) = \left[\mathbf{A}_0(\omega) + \sum_{i=1}^{M}\xi_i(\theta)\mathbf{A}_i(\omega)\right]^{-1}\mathbf{f}_0(\omega) \tag{3.23}$$

Using Eqs. (3.18)–(3.22) and the orthonormality of $\boldsymbol{\Phi}$ one has

$$\begin{aligned}\hat{\mathbf{u}}(\omega,\theta) &= \left[\boldsymbol{\Phi}^{-T}\boldsymbol{\Lambda}_0(\omega)\boldsymbol{\Phi}^{-1} + \sum_{i=1}^{M}\xi_i(\theta)\boldsymbol{\Phi}^{-T}\widetilde{\mathbf{A}}_i\boldsymbol{\Phi}^{-1}\right]^{-1}\mathbf{f}_0(\omega)\\ &= \boldsymbol{\Phi}\boldsymbol{\Psi}\big(\omega,\boldsymbol{\xi}(\theta)\big)\boldsymbol{\Phi}^T\mathbf{f}_0(\omega)\end{aligned} \tag{3.24}$$

where

$$\boldsymbol{\Psi}\big(\omega,\boldsymbol{\xi}(\theta)\big) = \left[\boldsymbol{\Lambda}_0(\omega) + \sum_{i=1}^{M}\xi_i(\theta)\widetilde{\mathbf{A}}_i(\omega)\right]^{-1} \tag{3.25}$$

and the M-dimensional random vector

$$\boldsymbol{\xi}(\theta) = \left\{\xi_1(\theta),\xi_2(\theta),\ldots,\xi_M(\theta)\right\}^T \tag{3.26}$$

Now we separate the diagonal and off-diagonal terms of the $\widetilde{\mathbf{A}}_i$ matrices as

$$\widetilde{\mathbf{A}}_i = \boldsymbol{\Lambda}_i + \boldsymbol{\Delta}_i, \quad i = 1,2,\ldots,M \tag{3.27}$$

Here the diagonal matrix

$$\boldsymbol{\Lambda}_i = \mathrm{diag}\,[\widetilde{\mathbf{A}}_i] = \mathrm{diag}\,[\lambda_{i_1},\lambda_{i_2},\ldots,\lambda_{i_n}] \in \mathbb{C}^{n\times n} \tag{3.28}$$

and the matrix containing only the off-diagonal elements $\boldsymbol{\Delta}_i = \widetilde{\mathbf{A}}_i - \boldsymbol{\Lambda}_i$ is such that $\mathrm{Trace}\,(\boldsymbol{\Delta}_i) = 0$. Using these, from Eq. (3.25) one has

$$\boldsymbol{\Psi}\big(\omega,\boldsymbol{\xi}(\theta)\big) = \Bigg[\underbrace{\boldsymbol{\Lambda}_0(\omega) + \sum_{i=1}^{M}\xi_i(\theta)\boldsymbol{\Lambda}_i(\omega)}_{\boldsymbol{\Lambda}(\omega,\boldsymbol{\xi}(\theta))} + \underbrace{\sum_{i=1}^{M}\xi_i(\theta)\boldsymbol{\Delta}_i(\omega)}_{\boldsymbol{\Delta}(\omega,\boldsymbol{\xi}(\theta))}\Bigg]^{-1} \tag{3.29}$$

where $\boldsymbol{\Lambda}(\omega,\boldsymbol{\xi}(\theta)) \in \mathbb{C}^{n\times n}$ is a diagonal matrix and $\boldsymbol{\Delta}(\omega,\boldsymbol{\xi}(\theta))$ is an off-diagonal only matrix. In the subsequent expressions we choose to omit the inclusion of frequency dependence of the individual matrices for the sake of notational simplicity, so that $\boldsymbol{\Psi}(\omega,\boldsymbol{\xi}(\theta)) \equiv \boldsymbol{\Psi}(\boldsymbol{\xi}(\theta))$ and so on. Hence,we rewrite Eq. (3.29) as

$$\boldsymbol{\Psi}\big(\boldsymbol{\xi}(\theta)\big) = \left[\boldsymbol{\Lambda}\big(\boldsymbol{\xi}(\theta)\big)\left[\mathbf{I}_n + \boldsymbol{\Lambda}^{-1}\big(\boldsymbol{\xi}(\theta)\big)\boldsymbol{\Delta}\big(\boldsymbol{\xi}(\theta)\big)\right]\right]^{-1} \tag{3.30}$$

The above expression can be represented using a Neumann type of matrix series [36] as

$$\boldsymbol{\Psi}\big(\boldsymbol{\xi}(\theta)\big) = \sum_{s=0}^{\infty}(-1)^s\left[\boldsymbol{\Lambda}^{-1}\big(\boldsymbol{\xi}(\theta)\big)\boldsymbol{\Delta}\big(\boldsymbol{\xi}(\theta)\big)\right]^s\boldsymbol{\Lambda}^{-1}\big(\boldsymbol{\xi}(\theta)\big) \tag{3.31}$$

Taking an arbitrary r-th element of $\hat{\mathbf{u}}(\theta)$, Eq. (3.24) can be rearranged to have

$$\hat{u}_r(\theta) = \sum_{k=1}^{n} \Phi_{rk} \left(\sum_{j=1}^{n} \Psi_{kj}\big(\boldsymbol{\xi}(\theta)\big)\big(\boldsymbol{\phi}_j^T \mathbf{f}_0\big) \right) \tag{3.32}$$

Defining

$$\Gamma_k\big(\boldsymbol{\xi}(\theta)\big) = \sum_{j=1}^{n} \Psi_{kj}\big(\boldsymbol{\xi}(\theta)\big)\big(\boldsymbol{\phi}_j^T \mathbf{f}_0\big) \tag{3.33}$$

and collecting all the elements in Eq. (3.32) for $r = 1, 2, \ldots, n$ one has

$$\hat{\mathbf{u}}(\theta) = \sum_{k=1}^{n} \Gamma_k\big(\boldsymbol{\xi}(\theta)\big)\boldsymbol{\phi}_k \tag{3.34}$$

This shows that the solution vector $\hat{\mathbf{u}}(\theta)$ can be projected in the space spanned by $\boldsymbol{\phi}_k$.

Now assume the series in Eq. (3.31) is truncated after m-th term. We define the truncated function

$$\boldsymbol{\Psi}^{(m)}\big(\boldsymbol{\xi}(\theta)\big) = \sum_{s=0}^{m} (-1)^s \big[\boldsymbol{\Lambda}^{-1}\big(\boldsymbol{\xi}(\theta)\big)\boldsymbol{\Delta}\big(\boldsymbol{\xi}(\theta)\big)\big]^s \boldsymbol{\Lambda}^{-1}\big(\boldsymbol{\xi}(\theta)\big) \tag{3.35}$$

From this one can obtain a sequence for different m

$$\hat{\mathbf{u}}^{(m)}(\theta) = \sum_{k=1}^{n} \Gamma_k^{(m)}\big(\boldsymbol{\xi}(\theta)\big)\boldsymbol{\phi}_k; \quad m = 1, 2, 3, \ldots \tag{3.36}$$

Since $\theta \in \Theta$ is arbitrary, comparing (3.11) and (3.23) we observe that $\hat{\mathbf{u}}^{(m)}(\theta)$ is the solution of Eq. (3.11) for every θ when $m \to \infty$. This implies that

$$\operatorname{Prob}\left\{\theta \in \Theta : \lim_{m\to\infty} \hat{\mathbf{u}}^{(m)}(\theta) = \hat{\mathbf{u}}(\theta)\right\} = 1 \tag{3.37}$$

Therefore, $\hat{\mathbf{u}}(\theta)$ is the solution of Eq. (3.11) in probability. In this derivation, the probability density function of the random variables has not been used. Therefore, the random variables can be general as long as the solution exists.

Remark 1 The matrix power series in (3.31) is different from the classical Neumann series [36]. The classical Neumann series is a power series in $\mathbf{A}_0^{-1}[\boldsymbol{\Delta}\mathbf{A}(\boldsymbol{\xi}(\theta))]$, where the first term is deterministic and the second term is random. The elements of this matrix series are polynomials in $\xi_i(\theta)$. In contrast, the series in (3.31) is in terms of $[\boldsymbol{\Lambda}^{-1}(\boldsymbol{\xi}(\theta))][\boldsymbol{\Delta}(\boldsymbol{\xi}(\theta))]$, where both terms are random. The elements of this matrix series are not simple polynomials in $\xi_i(\theta)$, but are in terms of a ratio of polynomials as seen in Eq. (3.39). The convergence of this series depends on the spectral radius of

$$\mathbf{R} = \boldsymbol{\Lambda}^{-1}\big(\boldsymbol{\xi}(\theta)\big)\boldsymbol{\Delta}\big(\boldsymbol{\xi}(\theta)\big) \tag{3.38}$$

A generic term of this matrix can be obtained as

$$R_{rs} = \frac{\Delta_{rs}}{\Lambda_{rr}} = \frac{\sum_{i=1}^{M} \xi_i(\theta)\Delta_{i_{rs}}}{\lambda_{0_r} + \sum_{i=1}^{M} \xi_i(\theta)\lambda_{i_r}} = \frac{\sum_{i=1}^{M} \xi_i(\theta)\widetilde{A}_{i_{rs}}}{\lambda_{0_r} + \sum_{i=1}^{M} \xi_i(\theta)\widetilde{A}_{i_{rr}}}; \quad r \neq s \tag{3.39}$$

Since $\mathbf{A}_0$ is positive definite, $\lambda_{0_r} > 0$ for all r. It can be seen from Eq. (3.39) that the spectral radius of $\mathbf{R}$ is also controlled by the diagonal dominance of the $\widetilde{\mathbf{A}}_i$ matrices. If the diagonal terms are relatively larger than the off-diagonal terms, the series will converge faster even if the relative magnitude of λ_{0_r} is not large.

The series in (3.36) approaches to the exact solution of the governing Eq. (3.11) for every $\theta \in \Theta$ for $m \to \infty$. For this reason it converges in probability 1.

Definition 1 The functions $\Gamma_k(\boldsymbol{\xi}(\theta))$, $k = 1, 2, \ldots, n$ are called the spectral functions as they are expressed in terms of the spectral properties of the coefficient matrix $\mathbf{A}_0$ arising in the discretized equation.

For certain class of problems the series in Eq. (3.34) can give useful physical insights into the uncertainty propagation. For structural mechanics problems, the matrix $\mathbf{A}_0$ is the stiffness matrix and its eigenvectors $\boldsymbol{\phi}_k$ are proportional to vibrational mode with a lumped mass assumption [28]. Equation (3.34) says that the response of a stochastic system is a linear combination of fundamental deformation modes weighted by the random variables Γ_k.

3.2 Properties of the Spectral Functions

In this section we discuss some important properties of these functions. From the series expansion in Eq. (3.31) we have

$$\begin{aligned}\boldsymbol{\Psi}\big(\boldsymbol{\xi}(\theta)\big) &= \boldsymbol{\Lambda}^{-1}\big(\boldsymbol{\xi}(\theta)\big) - \boldsymbol{\Lambda}^{-1}\big(\boldsymbol{\xi}(\theta)\big)\boldsymbol{\Delta}\big(\boldsymbol{\xi}(\theta)\big)\boldsymbol{\Lambda}^{-1}\big(\boldsymbol{\xi}(\theta)\big) \\ &\quad + \boldsymbol{\Lambda}^{-1}\big(\boldsymbol{\xi}(\theta)\big)\boldsymbol{\Delta}\big(\boldsymbol{\xi}(\theta)\big)\boldsymbol{\Lambda}^{-1}\big(\boldsymbol{\xi}(\theta)\big)\boldsymbol{\Delta}\big(\boldsymbol{\xi}(\theta)\big)\boldsymbol{\Lambda}^{-1}\big(\boldsymbol{\xi}(\theta)\big) \\ &\quad + \cdots \end{aligned} \tag{3.40}$$

Since $\boldsymbol{\Lambda}(\boldsymbol{\xi}(\theta))$ is a diagonal matrix, its inverse is simply a diagonal matrix containing the inverse of each of the diagonal elements. Also recall that the diagonal of $\boldsymbol{\Delta}(\boldsymbol{\xi}(\theta))$ contains only zeros. Different terms of the series in (3.40) can be obtained using a simple recursive relationship [36]. The numerical computation of the series is therefore computationally very efficient. For further analytical results, truncating the series upto different terms, we define spectral functions of different order.

Definition 2 The first-order spectral functions $\Gamma_k^{(1)}(\boldsymbol{\xi}(\theta))$, $k = 1, 2, \ldots, n$ are obtained by retaining one term in the series (3.40).

Retaining one term in (3.40) we have

$$\boldsymbol{\Psi}^{(1)}(\boldsymbol{\xi}(\theta)) = \boldsymbol{\Lambda}^{-1}(\boldsymbol{\xi}(\theta)) \quad \text{or} \quad \Psi_{kj}^{(1)}(\boldsymbol{\xi}(\theta)) = \frac{\delta_{kj}}{\lambda_{0_k} + \sum_{i=1}^{M} \xi_i(\theta)\lambda_{i_k}} \tag{3.41}$$

Using the definition of the spectral function in Eq. (3.33), the first-order spectral functions can be explicitly obtained as

$$\Gamma_k^{(1)}(\boldsymbol{\xi}(\theta)) = \sum_{j=1}^{n} \Psi_{kj}^{(1)}(\boldsymbol{\xi}(\theta))(\boldsymbol{\phi}_j^T \mathbf{f}_0) = \frac{\boldsymbol{\phi}_k^T \mathbf{f}_0}{\lambda_{0_k} + \sum_{i=1}^{M} \xi_i(\theta)\lambda_{i_k}} \tag{3.42}$$

From this expression it is clear that $\Gamma_k^{(1)}(\boldsymbol{\xi}(\theta))$ are correlated non-Gaussian random variables. Since we assumed that all eigenvalues λ_{0_k} are distinct, every $\Gamma_k^{(1)}(\boldsymbol{\xi}(\theta))$ in Eq. (3.42) are different for different values of k.

Definition 3 The second-order spectral functions $\Gamma_k^{(2)}(\boldsymbol{\xi}(\theta))$, $k = 1, 2, \ldots, n$ are obtained by retaining two terms in the series (3.40).

Retaining two terms in (3.40) we have

$$\boldsymbol{\Psi}^{(2)}(\boldsymbol{\xi}(\theta)) = \boldsymbol{\Lambda}^{-1}(\boldsymbol{\xi}(\theta)) - \boldsymbol{\Lambda}^{-1}(\boldsymbol{\xi}(\theta))\boldsymbol{\Delta}(\boldsymbol{\xi}(\theta))\boldsymbol{\Lambda}^{-1}(\boldsymbol{\xi}(\theta)) \tag{3.43}$$

or

$$\begin{aligned}\Psi_{kj}^{(2)}(\boldsymbol{\xi}(\theta)) = {} & \frac{\delta_{kj}}{\lambda_{0_k} + \sum_{i=1}^{M} \xi_i(\theta)\lambda_{i_k}} \\ & - \frac{\sum_{i=1}^{M} \xi_i(\theta)\Delta_{i_{kj}}}{(\lambda_{0_k} + \sum_{i=1}^{M} \xi_i(\theta)\lambda_{i_k})(\lambda_{0_j} + \sum_{i=1}^{M} \xi_i(\theta)\lambda_{i_j})}\end{aligned} \tag{3.44}$$

Using the definition of the spectral function in Eq. (3.33), the second-order spectral functions can be obtained in closed-form as

$$\begin{aligned}\Gamma_k^{(2)}(\boldsymbol{\xi}(\theta)) = {} & \frac{\boldsymbol{\phi}_k^T \mathbf{f}_0}{\lambda_{0_k} + \sum_{i=1}^{M} \xi_i(\theta)\lambda_{i_k}} \\ & - \sum_{\substack{j=1 \\ j \neq k}}^{n} \frac{(\boldsymbol{\phi}_j^T \mathbf{f}_0)\sum_{i=1}^{M} \xi_i(\theta)\Delta_{i_{kj}}}{(\lambda_{0_k} + \sum_{i=1}^{M} \xi_i(\theta)\lambda_{i_k})(\lambda_{0_j} + \sum_{i=1}^{M} \xi_i(\theta)\lambda_{i_j})}\end{aligned} \tag{3.45}$$

The second-order function can be viewed as adding corrections to the first-order expression derived in Eq. (3.42).

Definition 4 The vector of spectral functions of order s can be obtained by retaining s terms in the series (3.40) and can be expressed as

$$\boldsymbol{\Gamma}^{(s)}(\boldsymbol{\xi}(\theta)) = \left[\mathbf{I}_n - \mathbf{R}(\boldsymbol{\xi}(\theta)) + \mathbf{R}(\boldsymbol{\xi}(\theta))^2 - \mathbf{R}(\boldsymbol{\xi}(\theta))^3 \ldots s^{\text{th}}\text{ term}\right]\boldsymbol{\Gamma}^{(1)}(\boldsymbol{\xi}(\theta)) \quad (3.46)$$

where $\mathbf{I}_n$ is the n-dimensional identity matrix and $\mathbf{R}$ is defined in Eq. (3.38) as $\mathbf{R}(\boldsymbol{\xi}(\theta)) = [\boldsymbol{\Lambda}^{-1}(\boldsymbol{\xi}(\theta))][\boldsymbol{\Delta}(\boldsymbol{\xi}(\theta))]$. Different terms of this series can be obtained recursively from the previous term [36].

4 Error Minimization Using the Galerkin Approach

In Section 3.1 we derived the spectral functions such that a projection in an orthonormal basis converges to the exact solution in probability 1. The spectral functions are expressed in terms of a convergent infinite series. First, second and higher order spectral functions obtained by truncating the infinite series have been derived. We have also showed that they have the same functional form as the exact solution of Eq. (3.11). This motivates us to use these functions as 'trial functions' to construct the solution. The idea is to minimize the error arising due to the truncation. A Galerkin approach is proposed where the error is made orthogonal to the spectral functions.

We express the solution vector by the series representation

$$\hat{\mathbf{u}}(\theta) = \sum_{k=1}^{n} c_k \widehat{\Gamma}_k(\boldsymbol{\xi}(\theta))\boldsymbol{\phi}_k \quad (3.47)$$

Here the functions $\widehat{\Gamma}_k : \mathbb{C}^M \to \mathbb{C}$ are the spectral functions and the constants $c_k \in \mathbb{C}$ need to be obtained using the Galerkin approach. The functions $\widehat{\Gamma}_k(\boldsymbol{\xi}(\theta))$ can be the first-order (3.42), second-order (3.45) or any higher-order spectral functions (3.46) and $\boldsymbol{\phi}_k$ are the eigenvectors introduced earlier in Eq. (3.17). Substituting the expansion of $\hat{\mathbf{u}}(\theta)$ in the governing equation (3.11), the error vector can be obtained as

$$\boldsymbol{\varepsilon}(\theta) = \left(\sum_{i=0}^{M} \mathbf{A}_i \xi_i(\theta)\right)\left(\sum_{k=1}^{n} c_k \widehat{\Gamma}_k(\boldsymbol{\xi}(\theta))\boldsymbol{\phi}_k\right) - \mathbf{f}_0 \in \mathbb{C}^n \quad (3.48)$$

where $\xi_0 = 1$ is used to simplify the first summation expression. The expression (3.47) is viewed as a projection where $\{\widehat{\Gamma}_k(\boldsymbol{\xi}(\theta))\boldsymbol{\phi}_k\} \in \mathbb{C}^n$ are the basis functions and c_k are the unknown constants to be determined. We wish to obtain the coefficients c_k using the Galerkin approach so that the error is made orthogonal to the basis functions, that is, mathematically

$$\boldsymbol{\varepsilon}(\theta) \perp \left(\widehat{\Gamma}_j(\boldsymbol{\xi}(\theta))\boldsymbol{\phi}_j\right) \quad \text{or} \quad \left\langle \widehat{\Gamma}_j(\boldsymbol{\xi}(\theta))\boldsymbol{\phi}_j, \boldsymbol{\varepsilon}(\theta)\right\rangle = 0 \quad \forall j = 1, 2, \ldots, n \quad (3.49)$$

Here $\langle \mathbf{u}(\theta), \mathbf{v}(\theta) \rangle = \int_\Theta P(d\theta)\mathbf{u}(\theta)\mathbf{v}(\theta)$ defines the inner product norm. Imposing this condition and using the expression of $\boldsymbol{\varepsilon}(\theta)$ from Eq. (3.48) one has

$$\mathrm{E}\left[\left(\widehat{\Gamma}_j\big(\boldsymbol{\xi}(\theta)\big)\boldsymbol{\phi}_j\right)^T \left(\sum_{i=0}^{M} \mathbf{A}_i \xi_i(\theta)\right) \left(\sum_{k=1}^{n} c_k \widehat{\Gamma}_k\big(\boldsymbol{\xi}(\theta)\big)\boldsymbol{\phi}_k\right) - \left(\widehat{\Gamma}_j\big(\boldsymbol{\xi}(\theta)\big)\boldsymbol{\phi}_j\right)^T \mathbf{f}_0 \right] = 0 \tag{3.50}$$

Interchanging the E[·] and summation operations, this can be simplified to

$$\sum_{k=1}^{n} \left(\sum_{i=0}^{M} (\boldsymbol{\phi}_j^T \mathbf{A}_i \boldsymbol{\phi}_k) \mathrm{E}\left[\xi_i(\theta) \widehat{\Gamma}_j^T\big(\boldsymbol{\xi}(\theta)\big) \widehat{\Gamma}_k\big(\boldsymbol{\xi}(\theta)\big)\right] \right) c_k = \mathrm{E}\left[\widehat{\Gamma}_j^T\big(\boldsymbol{\xi}(\theta)\big)\right] \left(\boldsymbol{\phi}_j^T \mathbf{f}_0\right) \tag{3.51}$$

or

$$\sum_{k=1}^{n} \left(\sum_{i=0}^{M} \widetilde{A}_{i_{jk}} D_{ijk} \right) c_k = b_j \tag{3.52}$$

Defining the vector $\mathbf{c} = \{c_1, c_2, \ldots, c_n\}^T$, these equations can be expressed in a matrix form as

$$\mathbf{S}\mathbf{c} = \mathbf{b} \tag{3.53}$$

with

$$S_{jk} = \sum_{i=0}^{M} \widetilde{A}_{i_{jk}} D_{ijk}; \quad \forall j, k = 1, 2, \ldots, n \tag{3.54}$$

where

$$\widetilde{A}_{i_{jk}} = \boldsymbol{\phi}_j^T \mathbf{A}_i \boldsymbol{\phi}_k, \tag{3.55}$$

$$D_{ijk} = \mathrm{E}\left[\xi_i(\theta) \widehat{\Gamma}_j^T\big(\boldsymbol{\xi}(\theta)\big) \widehat{\Gamma}_k\big(\boldsymbol{\xi}(\theta)\big)\right] \tag{3.56}$$

and

$$b_j = \mathrm{E}\left[\widehat{\Gamma}_j^T\big(\boldsymbol{\xi}(\theta)\big)\right] \left(\boldsymbol{\phi}_j^T \mathbf{f}_0\right). \tag{3.57}$$

Higher order spectral functions can be used to improve the accuracy and convergence of the series (3.47). This will be demonstrated in the numerical examples later in the paper.

Remark 2 (Comparison with the classical spectral SFEM) We compare this Galerkin approach with the classical spectral stochastic finite element approach for further insight. The number of equations to be solved for the unknown coefficients in Eq. (3.53) is n, the same dimension as the original governing equation

(3.11). There are only n unknown constants, as opposed to nP unknown constants arising in the polynomial chaos expansion. The coefficient matrix $\mathbf{S}$ and the vector $\mathbf{b}$ in Eq. (3.53) should be obtained numerically using the Monte Carlo simulation or other numerical integration technique. In the classical PC expansion, however, the coefficient matrix and the associated vector are obtained exactly in closed-form. In addition, the coefficient matrix is a sparse matrix whereas the matrix $\mathbf{S}$ in Eq. (3.53) is in general a fully populated matrix.

The series in Eq. (3.47) can also be viewed as an enhanced Neumann expansion method where the approximating functions have been generated using a Neumann type expansion. It can be observed that the matrix $\mathbf{S}$ in Eq. (3.53) is symmetric. Therefore, one need to determine $n(n+1)/2$ number of coefficients by numerical methods. Any numerical integration method, such as the Gaussian quadrature method, can be used to obtain the elements of D_{ijk} and b_j in Eq. (3.55). In this paper Monte Carlo simulation is used. The samples of the spectral functions $\widehat{\Gamma}_k(\boldsymbol{\xi}(\theta))$ can be simulated from Eqs. (3.42), (3.45) or (3.46) depending on the order. These can be used to compute D_{ijk} and b_j from Eq. (3.55). The simulated spectral functions can also be 'recycled' to obtain the statistics and probability density function (pdf) of the solution. In summary, compared to the classical spectral stochastic finite element method, the proposed Galerkin approach results in a smaller size matrix but requires numerical integration techniques to obtain its entries. The numerical method proposed here therefore can be considered as a hybrid analytical-simulation approach.

5 Model Reduction Using a Reduced Number of Basis

The Galerkin approach proposed in the previous section requires the solution of $n \times n$ algebraic equations. Although in general this is smaller compared to the polynomial chaos approach, the computational cost can still be high for large n as the coefficient matrix is in general a dense matrix. The aim of this section is to reduce it further so that, in addition to large number of random variables, problems with large degrees of freedom can also be solved efficiently.

Suppose the eigenvalues of $\mathbf{A}_0$ are arranged in an increasing order such that

$$\lambda_{0_1} < \lambda_{0_2} < \cdots < \lambda_{0_n} \tag{3.58}$$

From the expression of the spectral functions observe that the eigenvalues appear in the denominator:

$$\Gamma_k^{(1)}\big(\boldsymbol{\xi}(\theta)\big) = \frac{\boldsymbol{\phi}_k^T \mathbf{f}_0}{\lambda_{0_k} + \sum_{i=1}^{M} \xi_i(\theta)\lambda_{i_k}} \tag{3.59}$$

The numerator $(\boldsymbol{\phi}_k^T \mathbf{f}_0)$ is the projection of the force on the deformation mode. Since the eigenvalues are arranged in an increasing order, the denominator of $|\Gamma_{k+r}^{(1)}(\boldsymbol{\xi}(\theta))|$

is larger than the denominator of $|\Gamma_k^{(1)}(\boldsymbol{\xi}(\theta))|$ according a suitable measure. The numerator $(\boldsymbol{\phi}_k^T \mathbf{f}_0)$ depends on the nature of forcing and the eigenvectors. Although this quantity is deterministic, in general an ordering cannot be easily established for different values of k. Because all the eigenvectors are normalized to unity, it is reasonable to consider that $(\boldsymbol{\phi}_k^T \mathbf{f}_0)$ does not vary significantly for different values of k. Using the ordering of the eigenvalues, one can select a small number ε such that $\lambda_1/\lambda_q < \varepsilon$ for some value of q, where λ_j is the eigenvalue of the generalized eigenvalue problem defined in Eq. (3.17). Based on this, we can approximate the solution using a truncated series as

$$\hat{\mathbf{u}}(\theta) \approx \sum_{k=1}^{q} c_k \widehat{\Gamma}_k\big(\boldsymbol{\xi}(\theta)\big)\boldsymbol{\phi}_k \tag{3.60}$$

where c_k, $\widehat{\Gamma}_k(\boldsymbol{\xi}(\theta))$ and $\boldsymbol{\phi}_k$ are obtained following the procedure described in the previous section by letting the indices j, k only upto q in Eqs. (3.54) and (3.55). The accuracy of the series (3.60) can be improved in two ways, namely, (a) by increasing the number of terms q, or (b) by increasing the order of the spectral functions $\widehat{\Gamma}_k(\boldsymbol{\xi}(\theta))$.

Model reduction techniques have been widely used within the scope of proper orthogonal decomposition (POD) method [12, 13, 15]. Here the eigenvalues of a symmetric positive definite matrix (the covariance matrix of a snapshot the system response) are used for model reduction. In spite of this similarity, the reduction method proposed here is different from a POD since it only considers the operator and not the solution itself. Reduction based on eigen-solution is of classical nature in various areas of applied mathematics, engineering and physics and extensive studies exist on this topic. It should be noted that the truncation in series (3.60) introduces errors. A rigorous mathematical quantification of error arising due to this truncation is beyond the scope of this article. The ratio of the eigenvalues λ_1/λ_q gives a good indication, but the projection of the force on the eigenvector $(\boldsymbol{\phi}_k^T \mathbf{f}_0)$ is also of importance. Since this quantity is problem dependent, care should be taken while applying this reduction method.

Remark 3 The reduction of the original problem by a projection on the set of dominant eigenvectors of a part of the operator is rather classical in model reduction techniques. It relies on the strong hypothesis that the solution can be well represented on this set of vectors. The impact of this truncation on the solution or a quantity of interest is not estimated in the article. The truncation criteria is based on the spectral decay of a part of the operator but not on the solution itself. By introducing this reduction, some essential features of the solution may not be always captured. The proposed method will only capture the projection of the solution $\mathbf{u}$ on the reduced basis $\boldsymbol{\phi}_k$, $k = 1, 2, \ldots, q$, which could be unadapted to the complete representation of $\mathbf{u}$.

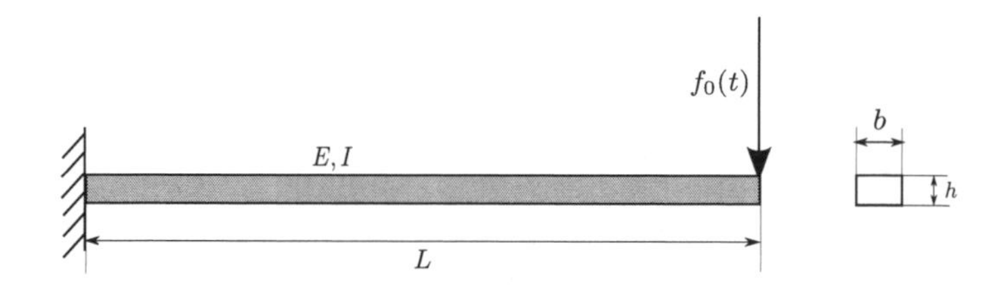

Fig. 3.1 Schematic diagram of the Euler–Bernoulli beam with a point load at the free end

6 Post Processing: Moments of the Solution

For the practical application of the method developed here, the efficient computation of the response moments and pdf is of crucial importance. A simulation based algorithm is proposed in this section. The coefficients c_k in Eq. (3.51) can be calculated from a reduced set of equations given by (3.53). The reduced equations can be obtained by letting the indices j, k upto $q < n$ in Eqs. (3.54) and (3.55). After obtaining the coefficient vector $\mathbf{c} \in \mathbb{C}^q$, the statistical moments of the solution can be obtained from Eqs. (3.61) and (3.62) using the Monte Carlo simulation. The spectral functions used to obtain the vector $\mathbf{c}$ itself, can be reused to obtain the statistics and pdf of the solution. The mean vector can be obtained as

$$\bar{\mathbf{u}} = \mathrm{E}\big[\big|\hat{\mathbf{u}}(\theta)\big|\big] = \sum_{k=1}^{q} |c_k| \mathrm{E}\big[\big|\widehat{\Gamma}_k\big(\boldsymbol{\xi}(\theta)\big)\big|\big]\boldsymbol{\phi}_k \tag{3.61}$$

where $|\cdot|$ is the absolute value of the complex quantities. The covariance of the solution vector can be expressed as

$$\boldsymbol{\Sigma}_u = \mathrm{E}\big[\big(\big|\hat{\mathbf{u}}(\theta)\big| - \bar{\mathbf{u}}\big)\big(\big|\hat{\mathbf{u}}(\theta)\big| - \bar{\mathbf{u}}\big)\big] = \sum_{k=1}^{q}\sum_{j=1}^{q} |c_k c_j| \Sigma_{\Gamma_{kj}} \boldsymbol{\phi}_k \boldsymbol{\phi}_j \tag{3.62}$$

where the elements of the covariance matrix of the spectral functions are given by

$$\Sigma_{\Gamma_{kj}} = \mathrm{E}\big[\big(\big|\widehat{\Gamma}_k\big(\boldsymbol{\xi}(\theta)\big)\big| - \mathrm{E}\big[\big|\widehat{\Gamma}_k\big(\boldsymbol{\xi}(\theta)\big)\big|\big]\big)\big(\big|\widehat{\Gamma}_k\big(\boldsymbol{\xi}(\theta)\big)\big| - \mathrm{E}\big[\big|\widehat{\Gamma}_k\big(\boldsymbol{\xi}(\theta)\big)\big|\big]\big)\big] \tag{3.63}$$

Based on the results derived in the paper, a hybrid reduced simulation-analytical approach can thus be realized in practice. The method is applicable to general structural dynamics problems with general non-Gaussian random fields. In the following section this approach has been applied to a physical problem.

7 Illustrative Application: The Stochastic Mechanics of an Euler–Bernoulli Beam

In this section we apply the computational method to a cantilever beam with stochastic bending modulus. Figure 3.1 shows the configuration of the cantilever beam with a harmonic point load at its free end. We assume that the bending modulus is a homogeneous stationary Gaussian random field of the form

$$EI(x,\theta) = EI_0\big(1 + a(x,\theta)\big) \tag{3.64}$$

where x is the coordinate along the length of the beam, EI_0 is the estimate of the mean bending modulus, $a(x,\theta)$ is a zero mean stationary Gaussian random field. The autocorrelation function of this random field is assumed to be

$$C_a(x_1,x_2) = \sigma_a^2 e^{-(|x_1-x_2|)/\mu_a} \tag{3.65}$$

where μ_a is the correlation length and σ_a is the standard deviation. We use the base-line parameters as the length $L = 1$ m, cross-section $(b \times h)$ 39×5.93 mm^2 and Young's modulus $E = 2 \times 10^{11}$ Pa. In study we consider deflection of the tip of the beam under harmonic loads of amplitude $\tilde{f}_0 = 1.0$ N. The correlation length considered in this numerical study is $\mu_a = L/2$. The number of terms retained (M) in the Karhunen–Loève expansion (3.8) is selected such that $\nu_M/\nu_1 = 0.01$ in order to retain 90% of the variability. For this correlation length the number of terms M comes to 18. For the finite element discretization, the beam is divided into 40 elements. Standard four degrees of freedom Euler–Bernoulli beam model is used [37]. After applying the fixed boundary condition at one edge, we obtain the number of degrees of freedom of the model to be $n = 80$.

7.1 Results

The proposed method has been compared with a direct Monte Carlo Simulation (MCS), where both have been performed with 10,000 samples. For the direct MCS, Eq. (3.23) is solved for each sample and the mean and standard deviation is derived by assembling the responses. The calculations have been performed for all the four values of σ_a to simulate increasing uncertainty. This is done to check the accuracy of the proposed method against the direct MCS results for varying degrees of uncertainty.

Figure 3.2(a) presents the ratio of the eigenvalues of the generalized eigenvalue problem (3.17) for which the ratio of the eigenvalues is taken with the first eigenvalue. We choose the reduced basis of the problem based on $\lambda_1/\lambda_q < \varepsilon$, where $\varepsilon = 0.01$, and they are highlighted in the figure. Figure 3.2(b) shows the frequency domain response of the deterministic system for both damped and undamped conditions. We have applied a constant modal damping matrix with the damping coefficient $\alpha = 0.02$ (which comes to 1% damping). It is also to be noted that the mass and damping matrices are assumed to be deterministic in nature, while it has to be emphasized that the approach is equally valid for random mass, stiffness and damping matrices. The frequency range of interest for the present study is 0–600 Hz with an interval of 2 Hz. In Fig. 3.2(b), the tip deflection is shown on a log scale for a unit

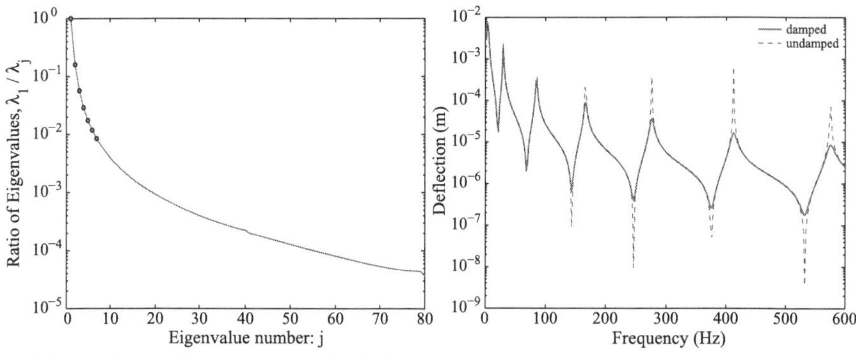

(a) Ratio of eigenvalues of the generalized eigenvalue problem.

(b) Frequency domain reponse of the tip of the beam under point load for the undamped and damped conditions (constant modal damping)

Fig. 3.2 The eigenvalues of the generalized eigenvalue problem involving the mass and stiffness matrices given in Eq. (3.17). For $\varepsilon = 0.01$, the number of reduced eigenvectors $q = 7$ such that $\lambda_1/\lambda_j < \varepsilon$

amplitude harmonic force input. The resonance peak amplitudes of the response of the undamped system definitely depends on the frequency resolution of the plot.

The frequency response of the mean deflection of the tip of the beam is shown in Fig. 3.3 for the cases for cases of $\sigma_a = \{0.05, 0.10, 0.15, 0.20\}$. The figures show a comparison of the direct MCS simulation results with different orders of the solution following Eq. (3.31), where the orders $s = 2, 3, 4$. A very good agreement between the MCS simulation and the proposed spectral approach can be observed in the figures. All the results have been compared with the response of the deterministic system which shows that the uncertainty has an added damping effect at the resonance peaks. This can be explained by the fact that the parametric variation of the beam, results in its peak response for the different samples to get distributed around the resonance frequency zones instead of being concentrated at a particular frequency, and when the subsequent averaging is applied, it smooths out the response peaks to a fair degree. The same explanation holds for the anti-resonance frequencies. It can also be observed that increased variability of the parametric uncertainties (as is represented by the increasing value of σ_a) results in an increase of this added damping effect which is consistent with the previous explanation.

The standard deviation of the frequency domain response of the tip deflection for different spectral order of solution of the reduced basis approach is compared with the direct MCS and is shown in Fig. 3.4, for different values of σ_a. We find that the standard deviation is maximum at the resonance frequencies which is expected due to the differences in the resonance peak of each sample. It is again observed that the direct MCS solution and the reduced order approach give almost identical results, which demonstrate the effectiveness of the proposed approach.

Figure 3.5 shows the standard deviation of the response of the beam at two frequencies 154 Hz and 412 Hz, which correspond to the anti-resonance and reso-

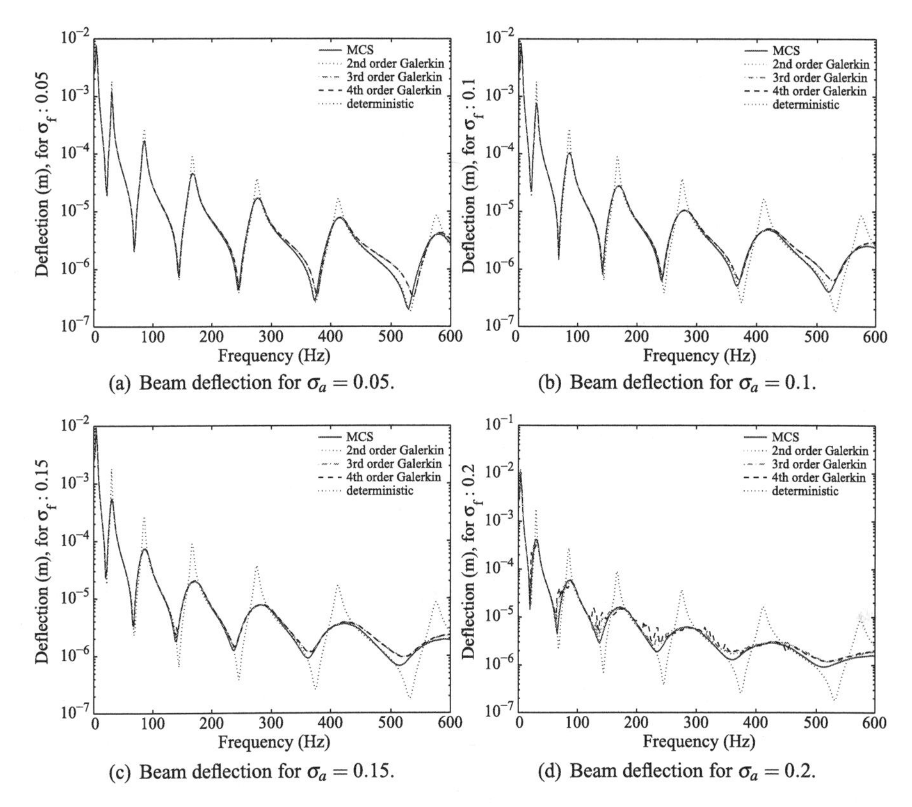

(a) Beam deflection for $\sigma_a = 0.05$.

(b) Beam deflection for $\sigma_a = 0.1$.

(c) Beam deflection for $\sigma_a = 0.15$.

(d) Beam deflection for $\sigma_a = 0.2$.

Fig. 3.3 The frequency domain response of the deflection of the tip of the Euler–Bernoulli beam under unit amplitude harmonic point load at the free end. The response is obtained with 10,000 sample MCS and for $\sigma_a = \{0.05, 0.10, 0.15, 0.20\}$. The proposed Galerkin approach requires the solution of a 7×7 linear system of equation only

nance frequencies of the cantilever beam respectively. The standard deviation values have been obtained for a set of 4 values of σ_a, which represents the different degrees of variability of the system uncertainty. The results obtained with the Galerkin approach for the different order of spectral functions have been compared to the direct MCS, and a good agreement between the two results have been obtained. It is interesting to point out here that the standard deviation decreases with the values of σ_a for the anti-resonance frequency while it increases for the resonance frequencies. This is consistent with the results shown in Fig. 3.4 which shows that an increased value of the variance of the random field has the effect of an increasing added damping on the system, when an averaging is done over the sample space. Thus the resonance response is expected to reduce with the increased variability of the random field while the anti-resonance response will increase.

The probability density function of the deflection of the tip of the cantilever beam for different degrees of variability of the random field is shown in Fig. 3.6.

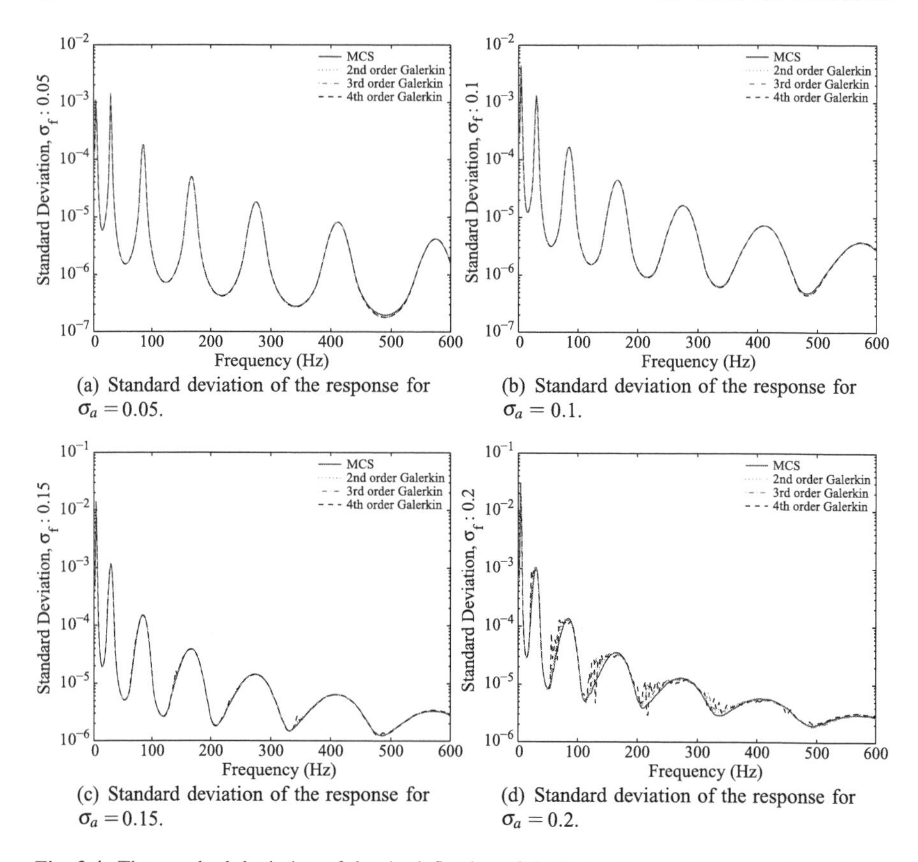

(a) Standard deviation of the response for $\sigma_a = 0.05$.

(b) Standard deviation of the response for $\sigma_a = 0.1$.

(c) Standard deviation of the response for $\sigma_a = 0.15$.

(d) Standard deviation of the response for $\sigma_a = 0.2$.

Fig. 3.4 The standard deviation of the tip deflection of the Euler–Bernoulli beam under unit amplitude harmonic point load at the free end. The response is obtained with 10,000 sample MCS and for $\sigma_a = \{0.05, 0.10, 0.15, 0.20\}$

The probability density functions have been calculated at the frequency of 412 Hz, which is a resonance frequency of the beam. The results indicate that with the increase in the degree of uncertainty (variance) of the system, the lower values of deflection has a higher probability which is absolutely consistent with the standard deviation curve shown in Fig. 3.5(a) and the comparison of the mean deflection of the stochastic system with the deterministic response in Fig. 3.3. This shows that the increase in the variability of the stochastic system has a damping effect on the response.

The results establish the applicability of this spectral reduced basis method with Galerkin error minimization technique as a satisfactory working model for providing solution of the stochastic structural systems. The method is found to be consistent with the direct MCS approach, while being much more computationally efficient than the latter.

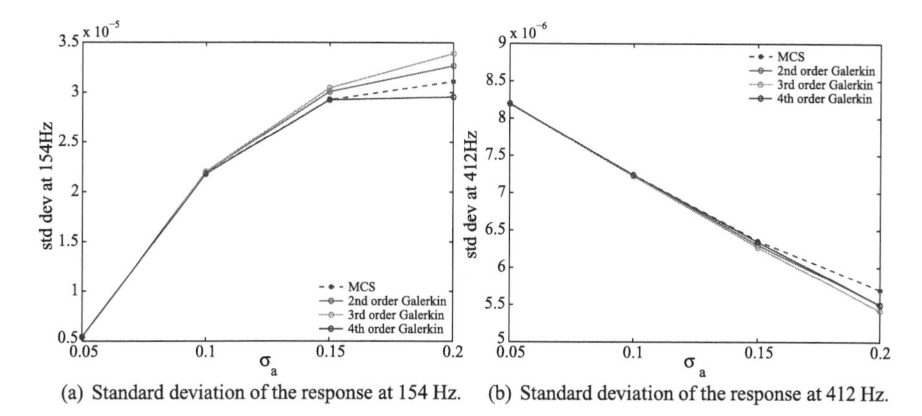

(a) Standard deviation of the response at 154 Hz. (b) Standard deviation of the response at 412 Hz.

Fig. 3.5 The standard deviation of the deflection of the tip versus the variability (σ_a) of the random field of the Euler–Bernoulli beam under unit amplitude harmonic point load at the free end at two frequencies 154 Hz and 412 Hz, which correspond to off-resonance and resonance frequencies respectively. The plots are shown for 4 different values of $\sigma_a = \{0.05, 0.10, 0.15, 0.20\}$ and calculated with 10,000 random samples

8 Conclusions

Here we have considered the discretized stochastic partial differential equation for structural systems with generally non-Gaussian random fields. In the classical spectral stochastic finite element approach, the solution is projected into an infinite dimensional orthonormal basis functions and the associated constant vectors are obtained using the Galerkin type of error minimization approach. Here an alternative approach is proposed. The solution is projected into a finite dimensional reduced vector basis and the associated coefficient functions are obtained. The coefficient functions, called as the *spectral functions*, are expressed in terms of the spectral properties of the matrices appearing in the discretized governing equation. It is shown that then the resulting series converges to the exact solution in probability 1. This is a stronger convergence compared to the classical polynomial chaos which converges in the mean-square sense in the Hilbert space. Using an analytical approach, it is shown that the proposed spectral decomposition has the same functional form as the exact solution, which is not a polynomial, but a ratio of polynomials where the denominator has a higher degree than the numerator.

Using the spectral functions, a Galerkin error minimization approach has been developed. It is shown that the number of unknown constants can be obtained by solving a system of linear equations which have a dimension much smaller than the dimension of the original discretized equation. A simple numerical approach to obtain the reduced dimension has been suggested based on the ratio of the eigenvalues of the generalized eigenvalue problem involving the deterministic mass and stiffness matrices of the baseline model. A numerical approach using a general-

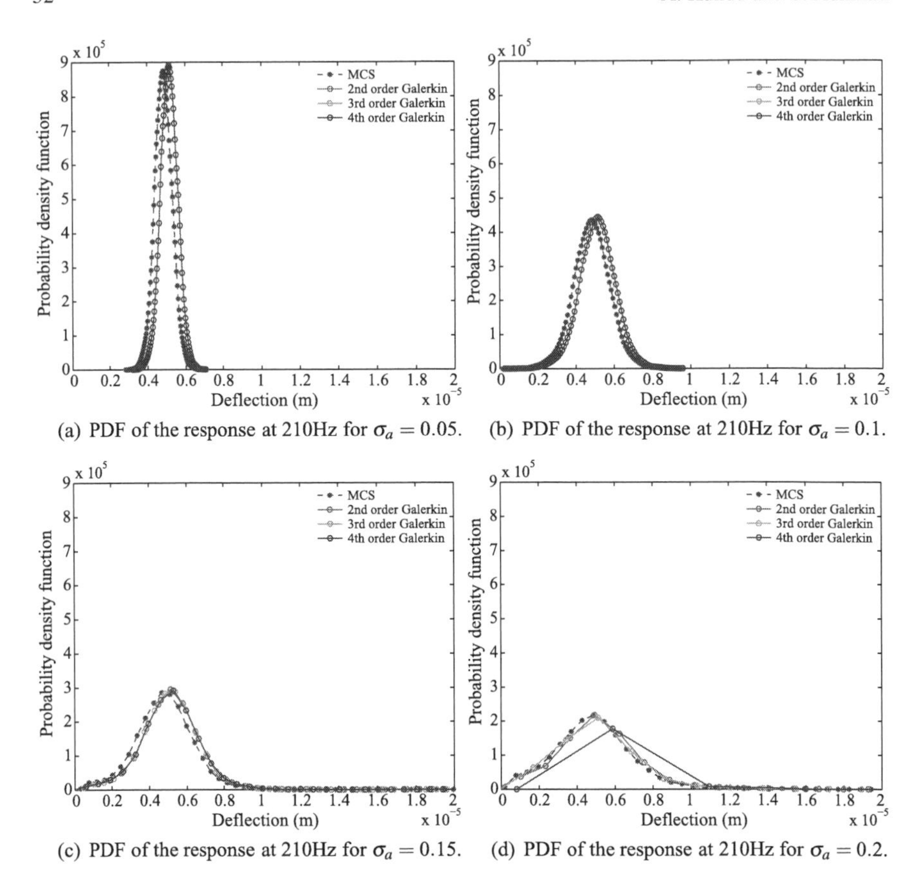

(a) PDF of the response at 210Hz for $\sigma_a = 0.05$. (b) PDF of the response at 210Hz for $\sigma_a = 0.1$.

(c) PDF of the response at 210Hz for $\sigma_a = 0.15$. (d) PDF of the response at 210Hz for $\sigma_a = 0.2$.

Fig. 3.6 The probability density function (PDF) of the tip deflection of the Euler–Bernoulli beam at 210 Hz under unit amplitude harmonic point load at the free end. The response is obtained with 10,000 samples and for $\sigma_a = \{0.05, 0.10, 0.15, 0.20\}$

order spectral function has been developed. Based on these, a hybrid analytical-simulation approach is proposed to obtain the statistical properties of the solution.

The computational efficiency of the proposed reduced spectral approach has been demonstrated for large linear systems with non-Gaussian random variables. It may be possible to extend the underlying idea to the class of non-linear problems. For example, the proposed spectral approach can be used for every linearisation step or every time step. Further research is necessary in this direction.

Acknowledgements AK acknowledges the financial support from the Swansea University through the award for Zienkiewicz scholarship. SA acknowledges the financial support from The Royal Society of London through the Wolfson Research Merit Award.

References

1. Adhikari, S., Manohar, C.S.: Dynamic analysis of framed structures with statistical uncertainties. Int. J. Numer. Methods Eng. **44**(8), 1157–1178 (1999)
2. Babuska, I., Tempone, R., Zouraris, G.E.: Galerkin finite element approximations of stochastic elliptic partial differential equations. SIAM J. Numer. Anal. **42**(2), 800–825 (2004)
3. Babuska, I., Tempone, R., Zouraris, G.E.: Solving elliptic boundary value problems with uncertain coefficients by the finite element method: the stochastic formulation. Comput. Methods Appl. Mech. Eng. **194**(12–16), 1251–1294 (2005)
4. Blatman, G., Sudret, B.: An adaptive algorithm to build up sparse polynomial chaos expansions for stochastic finite element analysis. Probab. Eng. Mech. **25**(2), 183–197 (2010)
5. Charmpis, D.C., Schuëller, G.I., Pellissetti, M.F.: The need for linking micromechanics of materials with stochastic finite elements: a challenge for materials science. Comput. Mater. Sci. **41**(1), 27–37 (2007)
6. Falsone, G., Impollonia, N.: A new approach for the stochastic analysis of finite element modelled structures with uncertain parameters. Comput. Methods Appl. Mech. Eng. **191**(44), 5067–5085 (2002)
7. Feng, Y.T.: Adaptive preconditioning of linear stochastic algebraic systems of equations. Commun. Numer. Methods Eng. **23**(11), 1023–1034 (2007)
8. Ganapathysubramanian, B., Zabaras, N.: Sparse grid collocation schemes for stochastic natural convection problems. J. Comput. Phys. **225**(1), 652–685 (2007)
9. Ghanem, R.: The nonlinear Gaussian spectrum of log-normal stochastic processes and variables. J. Appl. Mech. **66**, 964–973 (1989)
10. Ghanem, R., Spanos, P.D.: Stochastic Finite Elements: A Spectral Approach. Springer, New York (1991)
11. Grigoriu, M.: Galerkin solution for linear stochastic algebraic equations. J. Eng. Mech. **132**(12), 1277–1289 (2006)
12. Kerfriden, P., Gosselet, P., Adhikari, S., Bordas, S.: Bridging the proper orthogonal decomposition methods and augmented Newton–Krylov algorithms: an adaptive model order reduction for highly nonlinear mechanical problems. Comput. Methods Appl. Mech. Eng. **200**(5–8), 850–866 (2011)
13. Khalil, M., Adhikari, S., Sarkar, A.: Linear system identification using proper orthogonal decomposition. Mech. Syst. Signal Process. **21**(8), 3123–3145 (2007)
14. Kleiber, M., Hien, T.D.: The Stochastic Finite Element Method. Wiley, Chichester (1992)
15. Lenaerts, V., Kerschen, G., Golinval, J.C.: Physical interpretation of the proper orthogonal modes using the singular value decomposition. J. Sound Vib. **249**(5), 849–865 (2002)
16. Li, C.C., Kiureghian, A.D.: Optimal discretization of random fields. J. Eng. Mech. **119**(6), 1136–1154 (1993)
17. Li, C.F., Feng, Y.T., Owen, D.R.J.: Explicit solution to the stochastic system of linear algebraic equations $(\alpha_1 A_1 + \alpha_2 A_2 + \cdots + \alpha_m A_m)x = b$. Comput. Methods Appl. Mech. Eng. **195**(44–47), 6560–6576 (2006)
18. Liu, W.K., Belytschko, T., Mani, A.: Random field finite-elements. Int. J. Numer. Methods Eng. **23**(10), 1831–1845 (1986)
19. Ma, X., Zabaras, N.: An adaptive hierarchical sparse grid collocation algorithm for the solution of stochastic differential equations. J. Comput. Phys. **228**(8), 3084–3113 (2009)
20. Matthies, H.G., Keese, A.: Galerkin methods for linear and nonlinear elliptic stochastic partial differential equations. Comput. Methods Appl. Mech. Eng. **194**(12–16), 1295–1331 (2005)
21. Matthies, H.G., Brenner, C.E., Bucher, C.G., Soares, C.G.: Uncertainties in probabilistic numerical analysis of structures and solids—stochastic finite elements. Struct. Saf. **19**(3), 283–336 (1997)
22. Nair, P.B., Keane, A.J.: Stochastic reduced basis methods. AIAA J. **40**(8), 1653–1664 (2002)
23. Nouy, A.: A generalized spectral decomposition technique to solve a class of linear stochastic partial differential equations. Comput. Methods Appl. Mech. Eng. **196**(45–48), 4521–4537 (2007)

24. Nouy, A.: Generalized spectral decomposition method for solving stochastic finite element equations: invariant subspace problem and dedicated algorithms. Comput. Methods Appl. Mech. Eng. **197**(51–52), 4718–4736 (2008)
25. Nouy, A.: Recent developments in spectral stochastic methods for the numerical solution of stochastic partial differential equations. Arch. Comput. Methods Eng. **16**, 251–285 (2009). doi:10.1007/s11831-009-9034-5
26. Papadrakakis, M., Papadopoulos, V.: Robust and efficient methods for stochastic finite element analysis using Monte Carlo simulation. Comput. Methods Appl. Mech. Eng. **134**(3–4), 325–340 (1996)
27. Papoulis, A., Pillai, S.U.: Probability, Random Variables and Stochastic Processes, 4th edn. McGraw-Hill, Boston (2002)
28. Petyt, M.: Introduction to Finite Element Vibration Analysis. Cambridge University Press, Cambridge (1998)
29. Sachdeva, S.K., Nair, P.B., Keane, A.J.: Comparative study of projection schemes for stochastic finite element analysis. Comput. Methods Appl. Mech. Eng. **195**(19–22), 2371–2392 (2006)
30. Sachdeva, S.K., Nair, P.B., Keane, A.J.: Hybridization of stochastic reduced basis methods with polynomial chaos expansions. Probab. Eng. Mech. **21**(2), 182–192 (2006)
31. Sarkar, A., Benabbou, N., Ghanem, R.: Domain decomposition of stochastic PDEs: theoretical formulations. Int. J. Numer. Methods Eng. **77**(5), 689–701 (2009)
32. Stefanou, G.: The stochastic finite element method: past, present and future. Comput. Methods Appl. Mech. Eng. **198**(9–12), 1031–1051 (2009)
33. Wan, X.L., Karniadakis, G.E.: Beyond Wiener–Askey expansions: handling arbitrary pdfs. J. Sci. Comput. **27**(1–3), 455–464 (2006)
34. Xiu, D.B., Karniadakis, G.E.: The Wiener–Askey polynomial chaos for stochastic differential equations. SIAM J. Sci. Comput. **24**(2), 619–644 (2002)
35. Xiu, D.B., Karniadakis, G.E.: Modeling uncertainty in flow simulations via generalized polynomial chaos. J. Comput. Phys. **187**(1), 137–167 (2003)
36. Yamazaki, F., Shinozuka, M., Dasgupta, G.: Neumann expansion for stochastic finite element analysis. J. Eng. Mech. **114**(8), 1335–1354 (1988)
37. Zienkiewicz, O.C., Taylor, R.L.: The Finite Element Method, 4th edn. McGraw-Hill, London (1991)

Chapter 4
Computational Stochastic Dynamics Based on Orthogonal Expansion of Random Excitations

X. Frank Xu and George Stefanou

Abstract A major challenge in stochastic dynamics is to model nonlinear systems subject to general non-Gaussian excitations which are prevalent in realistic engineering problems. In this work, an n-th order convolved orthogonal expansion (COE) method is proposed. For linear vibration systems, the statistics of the output can be directly obtained as the first-order COE about the underlying Gaussian process. The COE method is next verified by its application on a weakly nonlinear oscillator. In dealing with strongly nonlinear dynamics problems, a variational method is presented by formulating a convolution-type action and using the COE representation as trial functions.

1 Introduction

To evaluate the probabilistic response of a structural dynamic system subject to parametric and external excitations, there are generally two approaches [6]. The first approach uses Fokker–Planck–Kolmogorov (FPK) equation to directly find the probability density function (PDF) by assuming a white noise excitation. To solve FPK equation especially for nonlinear systems, various techniques have been proposed, including weighted residual, path integral, etc., which however are all limited to systems of low dimension. The second approach includes perturbation method, moment closure method, and statistical equivalent techniques. While the perturbation method is limited to weak nonlinearity, the accuracy of moment closure method and statistical equivalent techniques on highly nonlinear problems remains an open question. Solutions of nonlinear random oscillators subject to stochastic forcing

X.F. Xu (✉)
Department of Civil, Environmental and Ocean Engineering, Stevens Institute of Technology, Hoboken, NJ 07030, USA
e-mail: xxu1@stevens.edu

G. Stefanou
Institute of Structural Analysis & Antiseismic Research, National Technical University of Athens, 15780 Athens, Greece
e-mail: stegesa@mail.ntua.gr

M. Papadrakakis et al. (eds.), *Computational Methods in Stochastic Dynamics*, Computational Methods in Applied Sciences 26,
DOI 10.1007/978-94-007-5134-7_4,

have also been obtained by means of the generalized polynomial chaos expansion, as described in [5].

A major deficiency of the existing approaches is their incapability in dealing with general non-Gaussian excitations which are prevalent in realistic engineering problems [8]. The marginal PDF and power spectral density (PSD) of a loading process play a major role in determining the response of systems, e.g. seismic wave in earthquake engineering. Therefore, a new approach to model dynamic systems subject to non-Gaussian excitations is highly desired.

A novel stochastic computation method based on orthogonal expansion of random fields has been recently proposed [10]. In this study, the idea of orthogonal expansion is extended to the so-called n-th order convolved orthogonal expansions (COE) especially in dealing with nonlinear dynamics. The COE is first verified by its application on a weakly nonlinear oscillator. Next in dealing with strongly nonlinear dynamics problems, a variational method is presented by formulating the convolution-type action and using the COE representation as trial functions [12]. Theoretically, substitution of the trial response function into the stochastic action will lead to the optimal solution. The effect of using different trial functions (COE of different orders) on the accuracy and efficiency of the proposed approach will be the subject of future investigation.

2 Convolved Orthogonal Expansions

2.1 *The Zero-th Order Convolved Orthogonal Expansion*

An underlying stationary Gaussian excitation $h_1(t, \vartheta)$ is characterized with the autocorrelation function $\rho(t)$ and unit variance, where $\vartheta \in \Theta$ indicates a sample point in random space. Based on the so-called diagonal class of random processes [1], the zero-th order convolved (or memoryless) orthogonal expansion of $h_1(t, \vartheta)$ is proposed as [10]

$$u(t, \vartheta) = \sum_{i=0} u_i(t) h_i(t, \vartheta) \tag{4.1}$$

where the random basis function h_i corresponds to the i-th degree Hermite polynomial with $h_0 = 1$. According to the generalized Mehler's formula [7] the correlations among the random basis functions are given as

$$\begin{aligned}
&R_{s_1 s_2 \cdots s_n}(t_1, t_2, \ldots, t_n) \\
&\quad = \overline{h_{s_1}(t_1, \vartheta) \cdots h_{s_n}(t_n, \vartheta)} \\
&\quad = \sum_{\nu_{12}=0}^{\infty} \cdots \sum_{\nu_{n-1,n}=0}^{\infty} \delta_{s_1 r_1} \cdots \delta_{s_n r_n} \prod_{j<k} \frac{\rho^{\nu_{jk}}(t_j - t_k)}{\nu_{jk}!} s_1! \cdots s_n!
\end{aligned} \tag{4.2}$$

where

$$r_k = \sum_{j \neq k} \nu_{jk}, \quad \nu_{jk} = \nu_{kj}, \quad \delta_{s_k r_k} = \begin{cases} 1, & s_k = r_k \\ 0, & s_k \neq r_k \end{cases}$$

and the overbar denotes ensemble average. Following Eq. (4.2), the two-point and three-point correlation functions are specifically obtained as

$$R_{ij}(t_1 - t_2) = \overline{h_i(t_1, \vartheta) h_j(t_2, \vartheta)} = \delta_{ij} i! \rho^i(t_1 - t_2) \tag{4.3}$$

$$R_{ijk}(t_1 - t_2, t_1 - t_3, t_2 - t_3) = \overline{h_i(t_1, \vartheta) h_j(t_2, \vartheta) h_k(t_3, \vartheta)}$$
$$= \frac{i!j!k!}{i'!j'!k'!} \rho^{k'}(t_1 - t_2) \rho^{j'}(t_1 - t_3) \rho^{i'}(t_2 - t_3) \tag{4.4}$$

$$i' = \frac{j+k-i}{2}, \qquad j' = \frac{i+k-j}{2}, \qquad k' = \frac{i+j-k}{2}$$

where i', j', k' must be non-negative integers, otherwise $R_{ijk} = 0$.

The correlation relations can be extended to the derivatives of the random basis functions, e.g.

$$R_{ij,pq}(t_1 - t_2) = \overline{h^{(0)}_{i,p}(t_1, \vartheta) h^{(0)}_{j,q}(t_2, \vartheta)} = \delta_{ij} i! \frac{\partial^{p+q}}{\partial t_1^p \, \partial t_2^q} \rho^i(t_1 - t_2)$$
$$= \delta_{ij} (-1)^q i! \frac{\partial^{p+q}}{\partial \tau^{p+q}} \rho^i(\tau) \tag{4.5}$$

where $\tau = t_1 - t_2$, and the subscripts p, q denote p-th and q-th derivatives. Similarly, the derivations can be made for the convolution of the random basis functions, e.g.

$$C_{ij} = \overline{h_i(t_1, \vartheta) * h_j(t_2, \vartheta)} = \delta_{ij} i! \int_{-\infty}^{\infty} \rho^i(t_1 - 2t_2) \, dt_2 = \delta_{ij} i! \tau_i$$
$$C_{ij,11} = \overline{h_{i,1}(t_1, \vartheta) * h_{j,1}(t_2, \vartheta)} = \delta_{ij} i! \int_{-\infty}^{\infty} \frac{\partial^2}{\partial t_1 \, \partial t_2} \rho^i(t_1 - 2t_2) \, dt_2 = 0 \tag{4.6}$$

where $\tau_i = \int_0^\infty \rho^i(t) \, dt$ is the correlation time.

2.2 n-th Order Convolved Orthogonal Expansion

The idea of the memoryless orthogonal expansion presented above can be generalized to an n-th order convolved orthogonal expansion (COE) for representation of nonlinear output processes

$$u(t, \vartheta) = \sum_{n=0} \sum_{i=0} u_i^{(n)}(t) h_i^{(n)}(t, \vartheta) \tag{4.7}$$

$$h_i^{(n)}(t,\vartheta) = \overbrace{g * g * \cdots * g}^{n} * h_i = g^{*n} * h_i \tag{4.8}$$

where g is a given kernel, and the symbol * denotes the convolution operator. For notational simplicity, the superscript (0) of the zero-th order COE is usually omitted throughout the paper. The memoryless orthogonal expansion thus corresponds to the zero-th order COE with $n = 0$ in (4.7). The correlation functions of the n-th order basis functions are therefore obtained as

$$\begin{aligned}
&R_{s_1\cdots s_n}^{m_1\cdots m_n}(t_1, t_2, \ldots, t_n) \\
&= \overline{h_{s_1}^{(m_1)}(t_1,\vartheta)\cdots h_{s_n}^{(m_n)}(t_n,\vartheta)} \\
&= \sum_{\nu_{12}=0}^{\infty}\cdots\sum_{\nu_{n-1,n}=0}^{\infty} \delta_{s_1 r_1}\cdots\delta_{s_n r_n}\int_{-\infty}^{\infty}\cdots\int_{-\infty}^{\infty} g^{*m_1}(t_1,\tau_1)\cdots g^{*m_n}(t_n,\tau_n) \\
&\quad\times \prod_{j<k}\frac{\rho^{\nu_{jk}}(t_j - t_k)}{\nu_{jk}!} s_1!\cdots s_n!\, d\tau_1\cdots d\tau_n
\end{aligned} \tag{4.9}$$

with the two-point correlations

$$\begin{aligned}
R_{ij}^{mn}(t_1,t_2) &= \overline{h_i^{(m)}(t_1,\vartheta) h_j^{(n)}(t_2,\vartheta)} \\
&= \delta_{ij} i! \int_{-\infty}^{\infty}\int_{-\infty}^{\infty} g^{*m}(t_1,\tau_1) g^{*n}(t_2,\tau_2)\rho^i(\tau_1-\tau_2)\, d\tau_1\, d\tau_2
\end{aligned} \tag{4.10}$$

The derivatives of the n-th order basis functions can be similarly obtained, e.g.

$$\begin{aligned}
&R_{ij,pq}^{mn}(t_1,t_2) \\
&= \overline{h_{i,p}^{(m)}(t_1,\vartheta) h_{j,q}^{(n)}(t_2,\vartheta)} \\
&= \delta_{ij} i! \frac{\partial^{p+q}}{\partial\tau_1^p\,\partial\tau_2^q}\int_{-\infty}^{\infty}\int_{-\infty}^{\infty} g^{*m}(t_1,\tau_1) g^{*n}(t_2,\tau_2)\rho^i(\tau_1-\tau_2)\, d\tau_1\, d\tau_2
\end{aligned} \tag{4.11}$$

When the kernel g is stationary, by letting $U = \mathcal{F}(u)$, $H = \mathcal{F}(h)$, $S = \mathcal{F}(R)$, $G^n = \mathcal{F}(g^{*n})$ and $\hat{\rho}^{*i} = \mathcal{F}(\rho^i)$, with $\mathcal{F}$ being the Fourier transform operator, we specially rewrite the two-point correlation functions of the COE basis functions in frequency domain

$$S_{ij}^{mn}(\omega) = \overline{H_i^{(m)}(\omega,\vartheta) H_j^{(n)}(\omega,\vartheta)} = \delta_{ij} i! G^m(\omega)\tilde{G}^n(\omega)\hat{\rho}^{*i}(\omega) \tag{4.12}$$

$$\begin{aligned}
S_{ij,pq}^{mn}(\omega) &= (\omega\sqrt{-1})^{p+q}\,\overline{H_i^{(m)}(\omega,\vartheta)\tilde{H}_j^{(n)}(\omega,\vartheta)} \\
&= \delta_{ij} i! (\omega\sqrt{-1})^{p+q} G^m(\omega)\tilde{G}^n(\omega)\hat{\rho}^{*i}(\omega)
\end{aligned} \tag{4.13}$$

where the tilde denotes complex conjugate.

Remark The advantage of the n-th order COE (4.7) can be demonstrated by comparing it with the classical Volterra series expansion [9]

$$u(t,\vartheta)=\sum_{n=0}\frac{1}{n!}\int_{-\infty}^{\infty}\cdots\int_{-\infty}^{\infty}k^{(n)}(t_1,t_2,\ldots,t_n)$$
$$\times h(t-t_1,\vartheta)h(t-t_2,\vartheta)\cdots h(t-t_n,\vartheta)\,dt_1\,dt_2\cdots dt_n \quad (4.14)$$

and the random variable-based polynomial chaos expansion (PCE)

$$u(t,\vartheta)=\sum_{n=0}u_n(t)\xi_n(\vartheta) \quad (4.15)$$

The Volterra representation typically suffers from severe difficulties in solving for the unknown kernels $k^{(n)}$. In the COE representation, the kernels are all explicitly given, and the problem is significantly reduced to solving of the unknown coefficients $u_i^{(n)}$. The random variable-based PCE, on the other hand, suffers from curse of dimensionality in using random variables $\xi_n(\vartheta)$ to represent random processes. By using random process-based expansions, the COE (4.7) circumvents much of this deficiency.

3 The COE Method in Random Vibration

3.1 Linear Oscillators

Suppose the linear oscillator

$$\begin{aligned}&\ddot{u}+2\zeta\omega_n\dot{u}+\omega_n^2u=f/m\\&u(0)=\dot{u}(0)=0\end{aligned} \quad (4.16)$$

is subjected to a non-stationary non-Gaussian translation input, i.e.

$$f(t,\vartheta)=\sum_{i=0}f_i(t)h_i(t,\vartheta) \quad (4.17)$$

By using the Green function

$$\begin{aligned}g(t)&=\frac{1}{\omega_n\sqrt{1-\zeta^2}}e^{-\zeta\omega_n t}\sin\left(\omega_n\sqrt{1-\zeta^2}t\right)\\G(\omega)&=\frac{1}{\omega_n^2-\omega^2+\sqrt{-1}2\zeta\omega\omega_n}\end{aligned} \quad (4.18)$$

the first three correlations of the non-stationary output u can be directly calculated from

$$\bar{u}(t) = \int_0^t g(t-\tau) f_0(\tau)\, d\tau \tag{4.19}$$

$$\begin{aligned} R_{uu}(t_1, t_2) = & \int_0^{t_2} \int_0^{t_1} g(t_1 - \tau_1) g(t_2 - \tau_2) \\ & \times \sum_{i=0} i! \rho^i (\tau_1 - \tau_2) f_i(\tau_1) f_i(\tau_2)\, d\tau_1\, d\tau_2 \end{aligned} \tag{4.20}$$

$$\begin{aligned} R_{uuu}(t_1, t_2, t_3) = & \int_0^{t_3} \int_0^{t_2} \int_0^{t_1} g(t_1 - \tau_1) g(t_2 - \tau_2) g(t_3 - \tau_3) \\ & \times \sum_{i,j,k=0} R_{ijk}(\tau_1 - \tau_2, \tau_1 - \tau_3, \tau_2 - \tau_3)\, d\tau_1\, d\tau_2\, d\tau_3 \end{aligned} \tag{4.21}$$

where R_{ijk} is given in Eq. (4.4).

When the excitation in Eq. (4.16) is stationary, each term f_i of Eq. (4.17) becomes constant and the output can be directly given as

$$u(t, \vartheta) = \sum_{i=0} f_i h_i^{(1)}(t, \vartheta) \tag{4.22}$$

which is a special case of the COE representation (4.7). Note that, with the Green function g and the underlying Gaussian process being given, the stationary PDF of the output in Eq. (4.22) can be rapidly estimated by using Monte Carlo method in the frequency domain.

A numerical example of application of the COE on linear oscillator is given in [10]. Hereby we extend this example to provide a parametric study demonstrating the effect of the PSD of the excitation on the output statistics. The excitation force f is assumed to be a stationary lognormal random process $f(t, \vartheta) = \exp(\sigma_z Z(t, \vartheta))$, with the underlying process Z being Gaussian–Markov with the correlation function

$$\rho(\tau) = \exp\left(-\frac{|\tau|}{t_c}\right) \tag{4.23}$$

The lognormal excitation can be represented as a zeroth-order COE

$$f(t, \vartheta) = \sum_{i=0}^{N} f_i h_i(t, \vartheta) \tag{4.24}$$

with

$$f_n = \frac{1}{\sqrt{2\pi n!}} \int_{-\infty}^{+\infty} \exp(\sigma_z z) \Phi_n(z) e^{-z^2/2}\, dz = \frac{\sigma_z^n}{n!} \exp\left(\frac{\sigma_z^2}{2}\right) \tag{4.25}$$

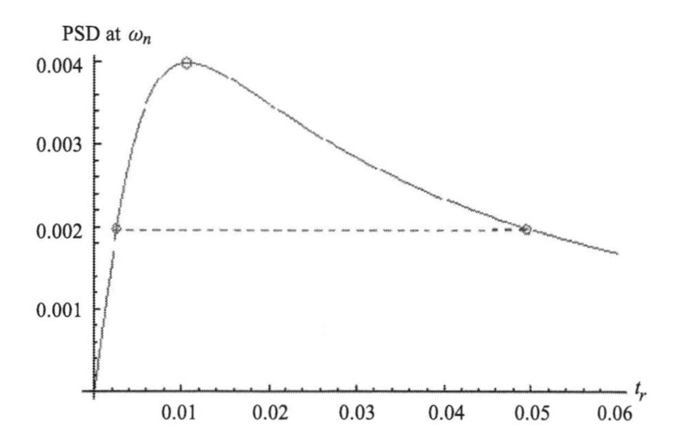

Fig. 4.1 The three points selected corresponding to a fast mode, probabilistic resonance and a slow mode, respectively (from left to right)

The correlation function of the lognormal process can be derived from Eqs. (4.24)–(4.25) as $\rho_f(\tau) = [\exp(\sigma_z^2\rho(\tau)) - 1]/[\exp(\sigma_z^2) - 1]$ with the corresponding PSD expressed in terms of Gamma function and incomplete Gamma function

$$\begin{aligned}\hat{\rho}_f(\omega) = &-\frac{\delta(\omega)}{e^{\sigma_z^2}-1} \\ &+\frac{t_c}{(e^{\sigma_z^2}-1)\pi}\Big[(-\sigma_z^2)^{\sqrt{-1}\omega t_c}\big(\Gamma(-\sqrt{-1}\omega t_c) - \Gamma(-\sqrt{-1}\omega t_c, -\sigma_z^2)\big) \\ &+(-\sigma_z^2)^{-\sqrt{-1}\omega t_c}\big(\Gamma(\sqrt{-1}\omega t_c) - \Gamma(\sqrt{-1}\omega t_c, -\sigma_z^2)\big)\Big], \quad \omega \geq 0 \end{aligned} \tag{4.26}$$

Let the mass be normalized as $m = 1/\omega_n$, $\sigma_z = 1$, and choose the natural frequency $\omega_n = 40\pi$. To study the effect of PSD of the non-Gaussian lognormal excitation on the response, we modulate the PSD of the lognormal process by using three values of t_c, i.e. 0.0108, 0.05 and 0.00254 (Fig. 4.1). The first value corresponds to the maximum of the PSD at $\omega_n = 40\pi$, i.e. to "probabilistic resonance". The excitation with $t_c = 0.00254$ the smallest correlation time is highly fluctuating, and is termed as a fast mode. Accordingly the excitation with $t_c = 0.05$ is termed as a slow mode. Note the above t_c values are selected based on the undamped frequency, which approximate well the damped cases in most of engineering applications.

As shown in Fig. 4.2, the mean displacement of Eq. (4.19) is independent of PSD. Figure 4.3 shows that the variance of probabilistic resonance is the largest for lightly damped cases. The fast mode has the smallest variance, which is similar to the amplitude of the harmonic motion. It is interesting to note that in the moderately damped case (when ζ is larger than approximately 0.2), the variance of probabilistic resonance is smaller than that of the slow mode. In Fig. 4.4, the coefficient of variation consistently decreases from probabilistic resonance to slow mode, and to fast mode, respectively.

The third centered moment (Fig. 4.5) decreases with the frequency mode, i.e. the fast mode has very small values of the third centered moment. As expected, this trend is also observed for the skewness in Fig. 4.6, where it is shown that, when the

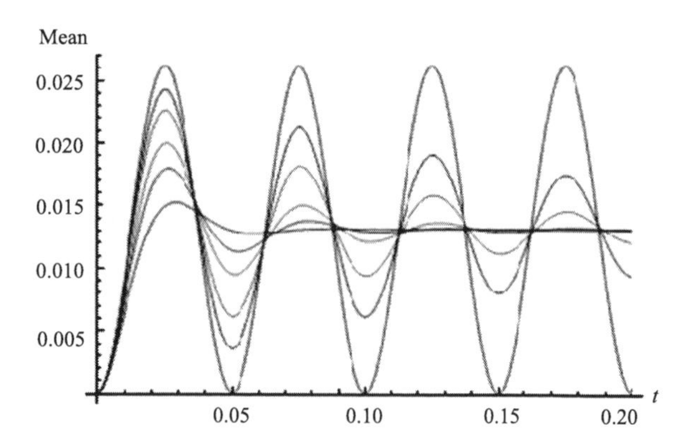

Fig. 4.2 Mean of displacement with the *curves* from exterior to interior corresponding to damping ratio $\zeta = 0, 0.05, 0.1, 0.2, 0.3, 0.5$, respectively

excitation mode becomes faster, the displacement output tends to the Gaussian case in terms of skewness.

With regard to the multi-degree-of-freedom linear systems, the oscillator equations given above can be directly applied by using the modal decomposition as shown in [11].

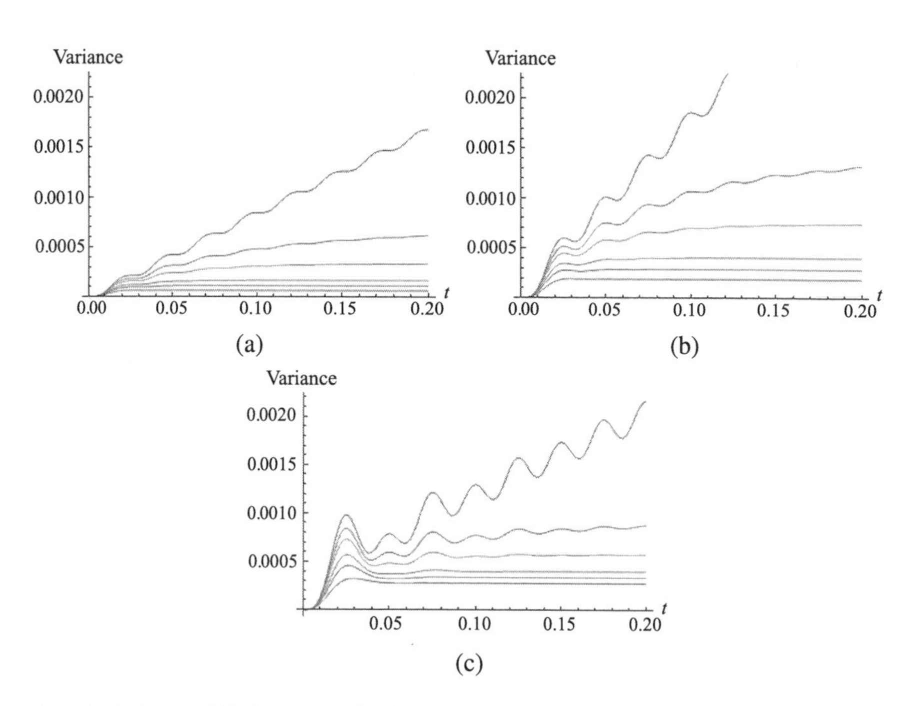

Fig. 4.3 Variance of displacement with the *curves* from top to bottom corresponding to damping ratio $\zeta = 0, 0.05, 0.1, 0.2, 0.3, 0.5$, respectively. (**a**) fast mode, (**b**) probabilistic resonance, (**c**) slow mode

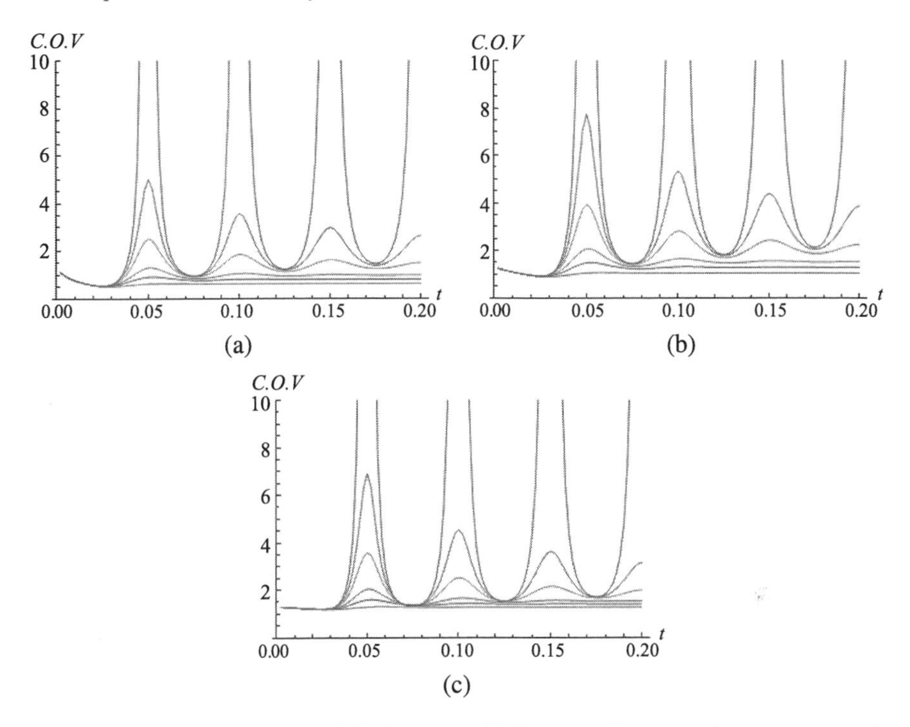

Fig. 4.4 Coefficient of variation of displacement with the *curves* from top to bottom corresponding to damping ratio $\zeta = 0, 0.05, 0.1, 0.2, 0.3, 0.5$, respectively. (**a**) fast mode, (**b**) probabilistic resonance, (**c**) slow mode

3.2 Weakly Nonlinear Oscillators

The accurate computation of the response of nonlinear single-degree-of-freedom (SDOF) oscillators under stochastic loading is important in earthquake engineering where equivalent nonlinear SDOF systems are often used in order to avoid the computationally intensive nonlinear response history analysis of MDOF systems, see e.g. [3].

In this section, a Duffing oscillator subjected to a Gaussian white noise excitation with intensity D is considered

$$\ddot{u} + 2\zeta\omega_n\dot{u} + \omega_n^2\left(u + \alpha u^3\right) = W \tag{4.27}$$

The Gaussian response of the linear filter can be given as $u_0 = \sigma_0 h_1$ where h_1 is characterized by unit variance and PSD [2]

$$S = \frac{D}{\sigma_0^2}\left|G(\omega)\right|^2 \tag{4.28}$$

$$\sigma_0^2 = \frac{1}{2\pi}D\int_{-\infty}^{\infty}\left|\frac{1}{\omega_n^2 - \omega^2 + \sqrt{-1}2\zeta\omega\omega_n}\right|^2 d\omega = \frac{D}{4\zeta\omega_n^3} \tag{4.29}$$

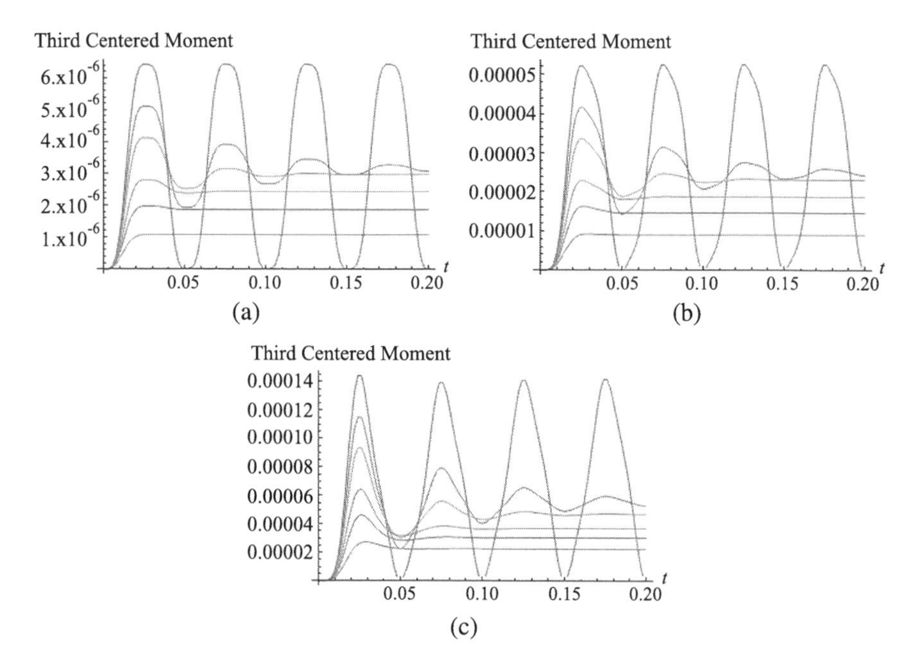

Fig. 4.5 Third centered moment of displacement with the *curves* from top to bottom corresponding to damping ratio $\zeta = 0, 0.05, 0.1, 0.2, 0.3, 0.5$, respectively. (**a**) fast mode, (**b**) probabilistic resonance, (**c**) slow mode

For small α, the nonlinear output of Eq. (4.27) can be approximated as

$$u = \sigma_0 h_1 - \alpha\omega_n^2\sigma_0^3 g * h_1^3 + 3\alpha^2\omega_n^4\sigma_0^5 g * \left(h_1^2 g * h_1^3\right) + O\left(\alpha^3\right) \tag{4.30}$$

By noting $h_1^3 = h_3 + 3h_1$ and $h_1^2 = h_2 + 1$, Eq. (4.30) can be rewritten in terms of the random basis functions

$$\begin{aligned} u = {} & \sigma_0 h_1 - \alpha\omega_n^2\sigma_0^3 g * (h_3 + 3h_1) \\ & + 3\alpha^2\omega_n^4\sigma_0^5 g * \left[(h_2 + 1) g * (h_3 + 3h_1)\right] + O\left(\alpha^3\right) \end{aligned} \tag{4.31}$$

$$\begin{aligned} U = {} & \sigma_0 H_1 - \alpha\omega_n^2\sigma_0^3 G(H_3 + 3H_1) \\ & + 3\alpha^2\omega_n^4\sigma_0^5 G\left[\left(H_2 + \delta(0)\right) * G(H_3 + 3H_1)\right] + O\left(\alpha^3\right) \end{aligned} \tag{4.32}$$

By using the correlations of Eqs. (4.3)–(4.4) and (4.12), it follows that the stationary mean

$$\bar{u} = O\left(\alpha^3\right) \tag{4.33}$$

and the stationary PSD

$$S_{UU} = U\tilde{U} = \sigma_0^2 S - 3\alpha\omega_n^2\sigma_0^4 (G + \tilde{G}) S + O\left(\alpha^2\right) \tag{4.34}$$

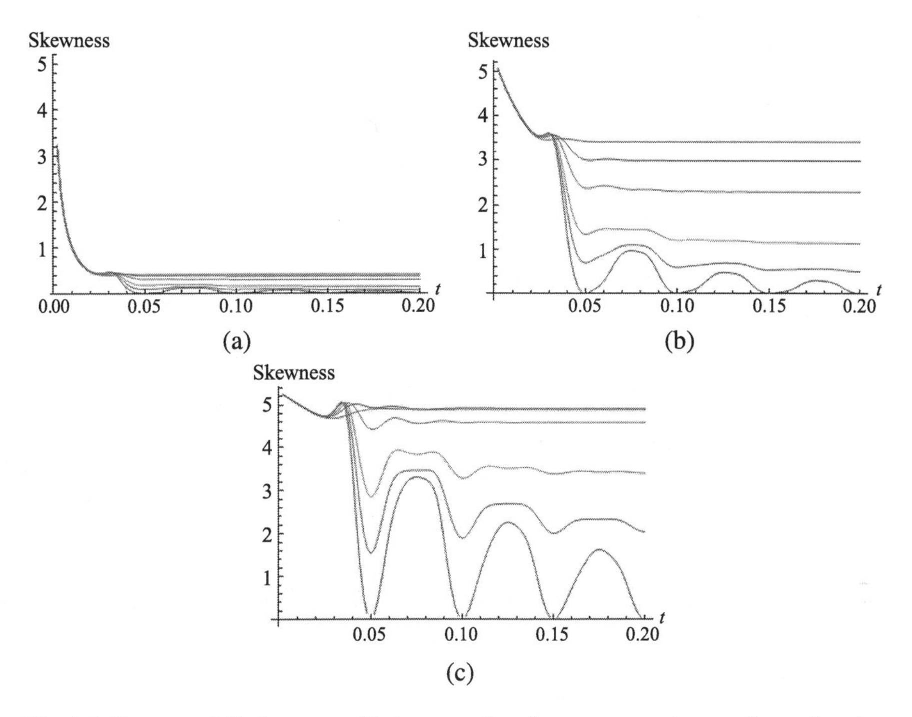

Fig. 4.6 Skewness of displacement with the *curves* from bottom to top corresponding to damping ratio $\zeta = 0, 0.05, 0.1, 0.2, 0.3, 0.5$, respectively. (**a**) fast mode, (**b**) probabilistic resonance, (**c**) slow mode

Since

$$\int_{-\infty}^{\infty} \frac{2(\omega_n^2 - \omega^2)}{(\omega_n^2 - \omega^2)^2 + (2\zeta\omega\omega_n)^2} \left| \frac{1}{\omega_n^2 - \omega^2 + \sqrt{-1}2\zeta\omega\omega_n} \right|^2 d\omega = \frac{\pi}{2\zeta\omega_n^5} \tag{4.35}$$

(see e.g. [2]) the variance calculated from the first two terms of Eq. (4.34) is simply obtained as

$$\sigma^2 = \sigma_0^2 \left(1 - 3\alpha\sigma_0^2\right) \tag{4.36}$$

which is identical to the result obtained using other approaches e.g. [4, 6]. In addition to serving as verification to the COE method, this example shows simplicity and efficiency of the orthogonal expansions in nonlinear problems.

3.3 Strongly Nonlinear Oscillators

For strongly nonlinear systems, the perturbation method is inapplicable. In this part, a variational method will be presented following the variational principles formulated for random media elastodynamics [12]. The variational functional, or action, of a nonlinear oscillator

$$\ddot{u} + 2\zeta\omega_n\dot{u} + \omega_n^2\big(u + g(u,\dot{u})\big) = f \tag{4.37}$$

can be formulated by using the convolution form

$$\delta\ell = \delta u * \big[\ddot{u} + 2\zeta\omega_n\dot{u} + \omega_n^2\big(u + g(u,\dot{u})\big) - f\big] = 0 \tag{4.38}$$

For a convolved nonlinear term $g(u,\dot{u}) = \alpha u^{*3}$, the action is derived from Eq. (4.38) as

$$\begin{aligned}\ell(u) &= \frac{1}{2}\dot{u} * \dot{u} + \zeta u * \dot{u} + \frac{1}{2}\omega_n^2 u * u \\ &\quad + \frac{1}{4}\alpha\omega_n^2(u * u * u * u) - f * u + \dot{u}(0)u\end{aligned} \tag{4.39}$$

where any trial function u satisfies the specified initial condition $u(0)$. To the authors' knowledge, the action (4.39) is the first convolution-type variational form formulated for nonlinear dissipative systems. It is especially noted that the classical point-wise Lagrangian form does not work on the dissipative term.

For nonlinear random vibrations, the stochastic action is directly obtained by taking ensemble average of Eq. (4.39), i.e.

$$\delta\bar{\ell} = 0 \tag{4.40}$$

with the trial function u based on the COE representation (4.7).

For stationary solutions, Eq. (4.40) can be rewritten in frequency domain as

$$\begin{aligned}&\delta\bar{L}(U) = 0 \\ &\bar{L}(U) = \left(-\frac{1}{2}\omega^2 + \sqrt{-1}\omega\zeta + \frac{1}{2}\omega_n^2\right)\overline{U^2} + \frac{1}{4}\alpha\omega_n^2\overline{U^4} - \overline{FU}\end{aligned} \tag{4.41}$$

Suppose the excitation is stationary

$$f(t,\vartheta) = \sum_{i=0} f_i h_i(t,\vartheta) \tag{4.42}$$

and choose the zeroth-order COE

$$u(t,\vartheta) = \sum_{i=0} u_i h_i(t,\vartheta) \tag{4.43}$$

as the trial function for the stationary solution. By substituting Eq. (4.43) into Eqs. (4.39)–(4.40) and taking derivative with respect to u_i, it leads to a series of equations to solve for u_i

$$\frac{\partial\bar{\ell}}{\partial u_i} = 0 \tag{4.44}$$

Similarly the first- or higher-order COE can be chosen as the trial function. The detail of numerical examples and investigation of accuracy and computational efficiency of the different trial functions will be the subject of future research.

4 Conclusion

By developing a diagonal class of random fields/stochastic processes to represent high-dimensional uncertainty, the proposed convolved orthogonal expansion method opens a new direction to deal with nonlinear stochastic dynamics. The advantages of the proposed method over the classical Volterra series representation and the random variable-based polynomial chaos expansions have been highlighted in the preceding sections. Future work will be devoted to the investigation of its efficiency in computing of large and strongly nonlinear dynamical systems.

References

1. Barrett, J.F., Lampard, D.G.: An expansion for some second-order probability distributions and its application to noise problems. IRE Trans. Inf. Theory **IT-I**, 10–15 (1955)
2. Elishakoff, I.: Probabilistic Theory of Structures, 2nd edn. Dover, New York (1999)
3. Han, S.W., Wen, Y.K.: Method of reliability-based seismic design. I: equivalent nonlinear systems. J. Struct. Eng. **123**(3), 256–263 (1997)
4. Lin, Y.K.: Probabilistic Theory of Structural Dynamics. Krieger, Melbourne (1976)
5. Lucor, D., Karniadakis, G.E.: Adaptive generalized polynomial chaos for nonlinear random oscillators. SIAM J. Sci. Comput. **26**(2), 720–735 (2005)
6. Manolis, G.D., Koliopoulos, P.K.: Stochastic Structural Dynamics in Earthquake Engineering. WIT Press, Ashurst (2001)
7. Slepian, D.: On the symmetrized Kronecker power of a matrix and extensions of Mehler's formula for Hermite polynomials. SIAM J. Math. Anal. **3**, 606–616 (1972)
8. Stefanou, G.: The stochastic finite element method: past, present and future. Comput. Methods Appl. Mech. Eng. **198**, 1031–1051 (2009)
9. Volterra, V.: Theory of Functionals and of Integral and Integro-Differential Equations. Dover, New York (1959)
10. Xu, X.F.: Stochastic computation based on orthogonal expansion of random fields. Comput. Methods Appl. Mech. Eng. **200**(41–44), 2871–2881 (2011)
11. Xu, X.F.: Quasi-weak and weak formulations of stochastic finite element methods. Probab. Eng. Mech. **28**(Special issue on CSM 6), 103–109 (2012)
12. Xu, X.F.: Variational principles of random media elastodynamics. In preparation

Chapter 5
Numerical Solution of the Fokker–Planck Equation by Finite Difference and Finite Element Methods—A Comparative Study

L. Pichler, A. Masud, and L.A. Bergman

Abstract Finite element and finite difference methods have been widely used, among other methods, to numerically solve the Fokker–Planck equation for investigating the time history of the probability density function of linear and nonlinear 2d and 3d problems; also the application to 4d problems has been addressed. However, due to the enormous increase in computational costs, different strategies are required for efficient application to problems of dimension ≥ 3. Recently, a stabilized multi-scale finite element method has been effectively applied to the Fokker–Planck equation. Also, the alternating directions implicit method shows good performance in terms of efficiency and accuracy. In this paper various finite difference and finite element methods are discussed, and the results are compared using various numerical examples.

1 Introduction

The response of linear systems subjected to additive Gaussian white noise or linearly filtered Gaussian white noise is Gaussian. The derivation for an N-dimensional system can be found, e.g. in [11]. For the case of nonlinear systems subjected to additive Gaussian white noise, analytical solutions are restricted to certain scalar systems. It has been shown (see [2]), that the response of a multi-dimensional memoryless nonlinear system subjected to additive Gaussian white noise forms a vector Markov process, with transition probability density function satisfying both the forward (Fokker–Planck) and backward Kolmogorov equations for which numerical

L. Pichler · L.A. Bergman (✉)
Department of Aerospace Engineering, University of Illinois at Urbana-Champaign, 104 S. Wright Street, Urbana, IL 61801, USA
e-mail: lbergman@illinois.edu

L. Pichler
e-mail: Lukas.Pichler@ymail.com

A. Masud
Department of Civil and Environmental Engineering, University of Illinois at Urbana-Champaign, 205 N. Mathews Avenue, Urbana, IL 61801, USA
e-mail: amasud@illinois.edu

M. Papadrakakis et al. (eds.), *Computational Methods in Stochastic Dynamics*,
Computational Methods in Applied Sciences 26,
DOI 10.1007/978-94-007-5134-7_5,

approximations in terms of finite element and finite difference methods can be pursued.

1.1 Scope of This Work

A number of numerical methods have been introduced over the past five decades to obtain approximate results for the solution of the Fokker–Planck equation (FPE). Many of these approximations can be shown to be accurate. This work deals with a review of several finite element and finite difference methods. A comparison and assessment of different methods is carried out by means of various numerical examples including a 2d linear, 2d unimodal and bimodal Duffing oscillators, 3d linear and 3d Duffing oscillators.

The goal is to evaluate the transient solution for the probability density function (PDF) of the oscillator due to stochastic (white noise) excitation. Thus, the forward Kolmogorov or Fokker–Planck equation is of interest and will be approximated within the numerical methods.

1.2 Background

The finite element method was first applied by [1] to determine the reliability of the linear, single degree-of-freedom oscillator subjected to stationary Gaussian white noise. The initial-boundary value problem associated with the backward Kolmogorov equation was solved numerically using a Petrov–Galerkin finite element method.

Reference [10] solved the stationary form of the FPE adopting the finite element method (FEM) to calculate the stationary probability density function of response. The weighted residual statement for the Fokker–Planck equation was first integrated by parts to yield the weak form of the equation.

The transient form of the FPE was analyzed in [16] using a Bubnov–Galerkin FEM. It was shown that the initial-boundary value problem can be modified to evaluate the first passage problem. A comparison for the reliability was carried out with the results obtained from the backward Kolmogorov equation.

A drawback of the FEM is the quickly increasing computational cost with increasing dimension. Thus, while 2 and 3 dimensional systems have been analyzed in the literature, the analysis of 4d or 5d problems reaches the limits of today's computational capabilities and are not yet feasible.

Computationally more economical—in terms of memory requirements, and when considering the effort spent for the assembly of the mass and stiffness matrices—are finite difference methods. The application of central differences is, as expected, only feasible for the case of 2d linear systems because of stability issues. The stability is a function of the nonlinearity and the dimension (ratio Δt and $\prod_{i=1}^{n} \Delta x_i$), thus being unfavorable for the use of this simple method.

A successful approach to overcome the limitations of simple finite differences was achieved by [20] in terms of higher order finite differences. The solution of a 4d system using higher order finite differences is reported in [20]. A comparison of various higher order FD formulations is presented in [9].

A viable approach to achieve higher accuracy is the application of operator splitting methods. Their capabilities with respect to the numerical solution of the FPE has so far received little attention. Operator splitting methods provide a tool to reduce the computational costs by the reduction of the solution to a series of problems of dimension one order less than the original problem. Thus, more efficiency, required for the solution of problems of larger dimension (≥ 3), can be achieved. An operator splitting method for the solution of the 2d Duffing oscillator is presented in [23]. An operator splitting scheme for 3d oscillators subjected to additive and multiplicative white noise is given by [22]. The method consists of a series of consecutive difference equations for the three fluxes and is numerically stable. The alternating directions implicit method (ADI) [15] is adopted in this paper for a series of problems, and acceptably accurate results are achieved at low cost. The implementation of the method is straightforward and can be readily extended to higher dimensions.

Recent work by Masud et al. introduced a stabilized multi-scale finite element method which allows for a reduction of the number of elements for given accuracy and, thus, the efficiency of the computation can be increased by an order of magnitude when solving a 3d problem.

Several four-state dynamical systems were studied in [17, 18] in which the Fokker–Planck equation was solved using a global weighted residual method and extended orthogonal functions.

Meshless methods were proposed in [6, 7] to solve the transient FPE and [8] for the stationary FPE. Considerable reduction of the memory storage requirements is expected due to coarse meshes employed, and thus a standard desktop PC suffices to carry out the numerical analysis.

In addition, many numerical packages now provide the capability to solve partial differential equations by means of finite element and finite difference methods. However, in most cases these tools are limited to 2d and can only solve special forms of elliptic, parabolic or hyperbolic partial differential equations (PDE). The implementation of FD and FEM into computational software is shown for the cases of COMSOL (2d linear) and FEAP (general 3d).

2 The Fokker–Planck Equation

The Fokker–Planck equation for a n-dimensional system subjected to external Gaussian white noise excitation is given by

$$\frac{\partial p}{\partial t} = -\sum_{j=1}^{n} \frac{\partial}{\partial x_j}(z_j p) + \frac{1}{2}\left(\sum_{i=1}^{n}\sum_{j=1}^{n} \frac{\partial^2}{\partial x_i\,\partial x_j}(H_{ij} p)\right) \tag{5.1}$$

where p denotes the transition probability density function, $\mathbf{x}$ the n-dimensional state space vector and $\mathbf{z}(\mathbf{x})$ and $\mathbf{H}(\mathbf{x})$ denote the drift vector and diffusion matrix, respectively.

The normalization condition for the probability density function is given by

$$\int p_X(\mathbf{x}, t)\, d\mathbf{x} = 1, \tag{5.2}$$

and the initial conditions are given by $p_X(\mathbf{x}_0, 0)$. Examples for initial conditions are, e.g., deterministic, given by the Dirac delta function

$$p_X(\mathbf{x}_0, 0) = \prod_{i=1}^{n} \delta\big((x_i - x_{0i})\big) \tag{5.3}$$

and the n-dimensional Gaussian distribution

$$p_X(\mathbf{x}_0, 0) = \frac{1}{(2\pi)^{n/2}|\boldsymbol{\Sigma}|^{1/2}} \exp\left(-\frac{1}{2}(\mathbf{x} - \boldsymbol{\mu})^T \boldsymbol{\Sigma}^{-1}(\mathbf{x} - \boldsymbol{\mu})\right) \tag{5.4}$$

in the random case.

At infinity, a zero-flux condition is imposed

$$p(x_i, t) \to 0 \quad \text{as } x_i \to \pm\infty, i = 1, 2, \ldots, n \tag{5.5}$$

Without loss of generality, and for a better comparison, the various methods introduced will be examined for the 2d linear case,

$$\begin{Bmatrix} \dot{x}_1 \\ \dot{x}_2 \end{Bmatrix} = \begin{bmatrix} x_2 \\ -2\xi\omega x_2 - \omega^2 x_1 \end{bmatrix} + \begin{bmatrix} 0 \\ 1 \end{bmatrix} w(t) \tag{5.6}$$

The corresponding FPE is

$$\frac{\partial p}{\partial t} = -\frac{\partial(x_2 p)}{\partial x_1} + \frac{\partial[(2\xi\omega x_2 + x_1)p]}{\partial x_2} + D\frac{\partial^2 p}{\partial x_2^2} \tag{5.7}$$

which, after application of the chain rule, becomes

$$\frac{\partial p}{\partial t} = -x_2\frac{\partial p}{\partial x_1} + (2\xi\omega x_2 + x_1)\frac{\partial p}{\partial x_2} + 2\xi\omega p + D\frac{\partial^2 p}{\partial x_2^2} \tag{5.8}$$

3 Finite Difference and Finite Element Methods

Many references deal with the application of FE and FD methods to the numerical solution of the Fokker–Planck equation (see e.g. [9, 19]).

3.1 Central Finite Differences

In terms of central finite differences, Eq. (5.8) becomes

$$\begin{aligned}\frac{p_{i,j}^{m+1}-p_{i,j}^{m}}{\Delta t} &= -x_2\frac{p_{i+1,j}^{m}-p_{i-1,j}^{m}}{2\Delta x_1} \\ &\quad + 2\xi\omega p_{i,j}^{m} + (2\xi\omega x_2)\frac{p_{i,j+1}^{m}-p_{i,j-1}^{m}}{2\Delta x_2} \\ &\quad + D\frac{p_{i,j+1}^{m}-2p_{i,j}^{m}+p_{i,j-1}^{m}}{\Delta x_2^2} \end{aligned} \tag{5.9}$$

and an explicit formulation is obtained for the probability density function

$$\begin{aligned} p_{i,j}^{m+1} &= p_{i,j}^{m} + \Delta t\bigg(-x_2\frac{p_{i+1,j}^{m}-p_{i-1,j}^{m}}{2\Delta x_1} \\ &\quad + 2\xi\omega p_{i,j}^{m} + (2\xi\omega x_2)\frac{p_{i,j+1}^{m}-p_{i,j-1}^{m}}{2\Delta x_2} \\ &\quad + D\frac{p_{i,j+1}^{m}-2p_{i,j}^{m}+p_{i,j-1}^{m}}{\Delta x_2^2}\bigg) \end{aligned} \tag{5.10}$$

The boundary conditions are given by $p_{i,j}=0$ for $i,\ j=1,N$. The discretization using central finite differences leads to an explicit scheme, which means that the values $p_{i,j}^{m+1}$ can be calculated directly from values $p_{i,j}^{m}$. Thus, the linear system of equations can be solved directly, and no inversion of the matrix relating $p_{i,j}^{m}$ to $p_{i,j}^{m+1}$ is required.

Explicit finite differences represent the simplest approximation; however, due to stability issues, implicit FD formulations are generally required.

Implicit, higher order finite difference schemes to solve Fokker–Planck equations have been developed by [20]. Higher order FD lead to more accurate results, but they are not used for comparison herein.

3.2 Alternating Directions Implicit Method

The alternating directions implicit method (ADI) is a finite difference scheme, for which the finite difference steps in each direction are resolved separately and in each step implicitly for one dimension and explicitly for the others, leading to a stable formulation. The main advantages are that the resulting operational matrix is tridiagonal and, thus, its inverse can be computed efficiently. Moreover, the dimensionality of the problem is reduced by one, and the problem is reduced to the solution of a series of problems of dimension of one order lower.

In Eq. (5.8), finite differences are first applied implicitly to the x_1-direction

$$\frac{p_{i,j}^{m+1/2} - p_{i,j}^{m}}{\Delta(t/2)} = -x_2 \frac{p_{i+1,j}^{m+1/2} - p_{i-1,j}^{m+1/2}}{2\Delta x_1} + 2\xi\omega p_{i,j}^{m} + (2\xi\omega x_2)\frac{p_{i,j+1}^{m} - p_{i,j-1}^{m}}{2\Delta x_2} + D\frac{p_{i,j+1}^{m} - 2p_{i,j}^{m} + p_{i,j-1}^{m}}{\Delta x_2^2} \tag{5.11}$$

and then to the x_2-direction.

$$\frac{p_{i,j}^{m+1} - p_{i,j}^{m+1/2}}{\Delta(t/2)} = -x_2 \frac{p_{i+1,j}^{m+1/2} - p_{i-1,j}^{m+1/2}}{2\Delta x_1} + 2\xi\omega p_{i,j}^{m+1/2} + (2\xi\omega x_2)\frac{p_{i,j+1}^{m+1} - p_{i,j-1}^{m+1}}{2\Delta x_2} + D\frac{p_{i,j+1}^{m+1} - 2p_{i,j}^{m+1} + p_{i,j-1}^{m+1}}{\Delta x_2^2} \tag{5.12}$$

Both Eq. (5.11) and Eq. (5.12) give $M-2$ tridiagonal linear systems of equations in x_1 for the $j=2,\dots,M-1$ values of x_2, and in case of Eq. (5.11) $M-2$ tridiagonal linear systems of equations in x_2 for the $i=2,\dots,M-1$ values of x_1.

The computational cost is mainly due to the n times N matrix inversions which are encountered in the n-loops solution for a full time step; n denotes the dimension of the problem and N the number of nodes per dimension.

3.3 Finite Element Method

Reduction of Eq. (5.1) to the weak form and the introduction of shape functions of C^0 continuity lead to

$$\mathbf{C}\dot{p} + Kp = 0 \tag{5.13}$$

where

$$C_{rs}^{e} = \int_{\Omega^e} N_r(\mathbf{x}) N_s(\mathbf{x})\, d\mathbf{x} \tag{5.14}$$

and

$$K_{rs}^{e} = \int_{\Omega^e} \left(\sum_{i=1}^{n} z_i(\mathbf{x}) N_s \frac{\partial}{\partial x_i} N_r\, d\mathbf{x} - \sum_{i=1}^{n}\sum_{j=1}^{n} \frac{\partial}{\partial x_i} N_r \frac{\partial}{\partial x_i} [H_{ij} N_s]\, d\mathbf{x} \right) \times N_r(\mathbf{x}) N_s(\mathbf{x})\, d\mathbf{x} \tag{5.15}$$

The integration over time can be performed in a suitable way using the Crank–Nicholson scheme ($\theta = 0.5$).

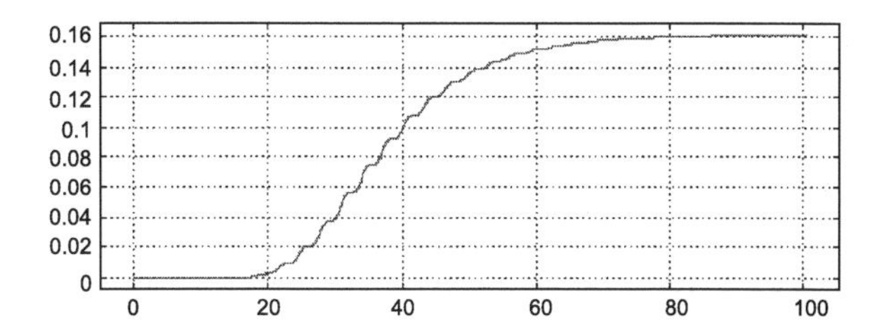

Fig. 5.1 Probability density function $p(0, 0, t)$ at central node over time

3.4 Multi-scale Finite Element Method

The multi-scale FEM used herein was introduced by [14] for the numerical treatment of advection-diffusion equations in fluid dynamics. Then, the methodology was extended by [13] to the special case of the Fokker–Planck equation. Finally, the method was applied to the numerical solution of the Fokker–Planck equation of a 3d linear system [12]; [5] provide an overview of stabilized finite element methods and recent developments of their application to the advection-diffusion equation.

For a description of the method the reader is referred to the aforementioned references. Basically, a multi-scale FEM means that an approximation of the error term from the traditional FE formulation is included at a fine scale into the formulation; the probability density function is then given by

$$p = \hat{p} + p' \tag{5.16}$$

where $\hat{p}$ represents the contribution of the coarse scale and p' the contribution of the fine scale.

3.5 Implementation Within COMSOL/FEAP

The FE code COMSOL provides the possibility to solve partial differential equations by finite differences. For an extensive discussion, refer to the COMSOL documentation [4]. Figure 5.1 shows the results obtained for the FPE for the 2d linear oscillator with parameters to be discussed later.

The multiscale finite element method was implemented by Masud and coworkers into the finite element code FEAP and is used herein for comparison of the 3d examples.

4 Numerical Examples

The numerical methods used in this comparison are:

1. central finite differences (FD)
2. alternating directions implicit method (ADI)

Table 5.1 Parameter for the linear oscillator

μ	σ	ξ	ω	D
[5, 5]	$\frac{1}{9}\mathbf{I}(2)$	0.05	1	0.1

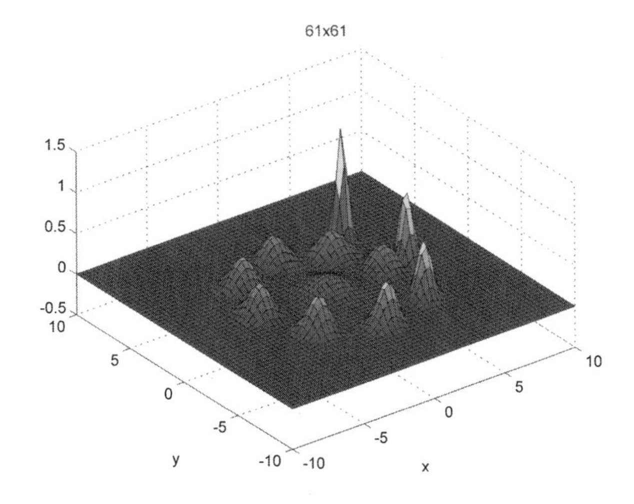

Fig. 5.2 FEM: 61 × 61—Probability density function $p(t)$ over time

3. Bubnov–Galerkin finite element method (FEM)
4. stabilized multiscale finite element method (MSFEM)

The methods 1–3 are coded in MATLAB and the analysis was carried out on a 64-bit Windows server (32 GB). The results for method 4 are obtained on 32 bit Linux or Windows machines with 2 GB memory using an implementation within FEAP [13].

4.1 2-d Linear Oscillator

The different methods are applied to solve the FPE for the linear oscillator (see Eq. (5.6)). The parameters of the oscillator are chosen according to [16] and are described in Table 5.1, were $\mathbf{I}(2)$ denotes the identity matrix in 2-d:

Finite element results obtained using a 61 × 61 mesh are shown in Fig. 5.2 and Fig. 5.3. All results are calculated with a time increment of $\Delta t = 0.001$ and a total time of $\tau = 20$ natural periods. The state space is discretized on the domain $[-10, 10] \times [-10, 10]$;

Figure 5.2 shows the evolution of the probability density function over time. In Fig. 5.3 the transient solution for the PDF at the origin is given. The exact stationary value at the origin is $p_{stat}(0, 0) = 1.5915e - 1$.

The accuracy of the numerical solutions are compared at stationarity (i.e., after $t = 20$ cycles) using two error measures. The first, the maximum norm

$$\|e\|_\infty = \left\| p^{ex} - p^{num} \right\|_\infty \tag{5.17}$$

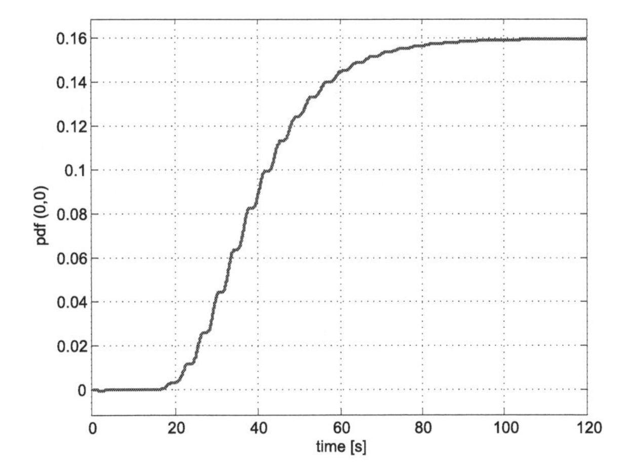

Fig. 5.3 FEM: 61 × 61—Probability density function at central node $p(0,0,t)$ over time

is a measure of the maximum error across the entire mesh. The second, the Euclidean norm

$$\|e\|_2 = \left\| p^{ex} - p^{num} \right\|_2 \tag{5.18}$$

can be used to describe the average nodal error $\overline{e} = \|e\|_2 / n_{nodes}$, where n_{nodes} is the total number of nodes.

Table 5.2 correctly visualizes the increasing accuracy for all methods when the mesh is refined. It can also be seen that the FD and ADI deliver similar results. The advantage of the ADI over FD is that the stability of the method allows one to use larger time steps. The FEM provides more accurate results for the same mesh refinement as the finite difference methods. The FEM is the preferable method to investigate the first passage problem in case small probabilities of failure are involved and a highly accurate method is required.

Alternatively, the accuracy of the solution at stationarity can also be represented by comparison of the exact and numerical covariance matrices K_{xx}, the latter computed from FE/FD results for the PDF.

The transient solution for the probability density at the center node can be obtained with all three methods as listed in Table 5.2. Figure 5.4 shows a comparison for the PDF at the origin using finite difference method and different mesh sizes.

Table 5.2 Comparison of the accuracy for the linear oscillator

Method/Mesh	61 × 61		81 × 81		101 × 101		121 × 121	
	$\|e\|_\infty$	$\|e\|_2$	$\|e\|_\infty$	$\|e\|_2$	$\|e\|_\infty$	$\|e\|_2$	$\|e\|_\infty$	$\|e\|_2$
FD	7.05e-3	2.99e-2	4.169e-3	2.405e-2	2.983e-3	2.201e-2	2.371e-3	2.199e-2
ADI	1.66e-2	4.61e-2	4.144e-3	2.375e-2	2.931e-3	2.149e-2	2.305e-3	2.124e-2
FEM	2.04e-3	1.00e-2	1.712e-3	1.152e-2	1.550e-3	1.370e-2	1.464e-3	1.613e-2

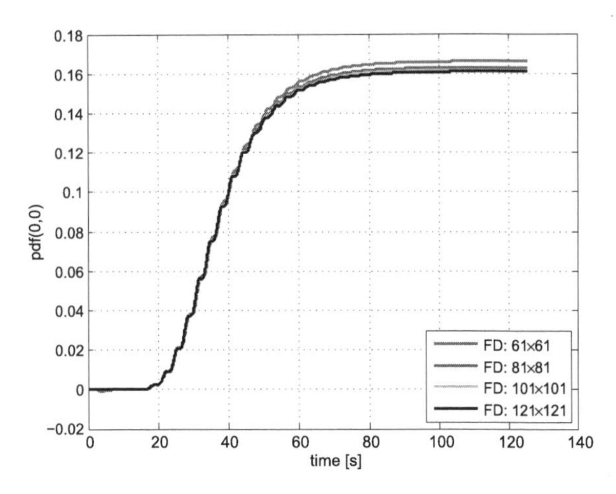

Fig. 5.4 FD—Comparison for probability density function at central node $p(0, 0, t)$ over time

Table 5.3 Parameters for the unimodal Duffing oscillator

μ	σ	ξ	ω	D	γ
[0, 10]	$\frac{1}{2}\mathbf{I}(2)$	0.2	1	0.4	0.1

4.2 2-d Duffing Oscillator

Both, the unimodal Duffing-oscillator and the bimodal Duffing-oscillator as well are investigated in the following.

4.2.1 Unimodal

The unimodal Duffing oscillator is considered next:

$$\begin{bmatrix} \dot{x}_1 \\ \dot{x}_2 \end{bmatrix} = \begin{bmatrix} x_2 \\ -2\xi\omega x_2 - \omega^2 x_1 - \omega^2\gamma x_1^3 \end{bmatrix} + \begin{bmatrix} 0 \\ 1 \end{bmatrix} w(t) \tag{5.19}$$

The parameters of the oscillator are chosen as in Table 5.3.

The state space is discretized on the domain $[-15, 15] \times [-15, 15]$.

It is known that central finite differences are not suitable in case of nonlinearities, but ADI can be utilized nonetheless. When the Duffing-oscillator is analyzed, it is found that the ADI can be used due to its implicit formulation with the largest time step Δt, thus providing a good compromise between accuracy and efficiency as can be seen from Table 5.4. The time steps used are $\Delta t = 1e - 2$ (ADI), $\Delta t = 1e - 3$ (FEM) and $\Delta t = 5e - 4$ (FEM: mesh 101).

Table 5.4 Comparison of the accuracy for the unimodal Duffing oscillator

Method/Mesh	61 × 61		81 × 81		101 × 101	
	$\|e\|_\infty$	$\|e\|_2$	$\|e\|_\infty$	$\|e\|_2$	$\|e\|_\infty$	$\|e\|_2$
ADI	1.7832e-2	4.9323e-2	9.6731e-3	3.5553e-2	6.1694e-3	2.8444e-2
FEM	2.6002e-3	1.1587e-2	1.3999e-3	8.5297e-3	9.3290e-4	6.8215e-3

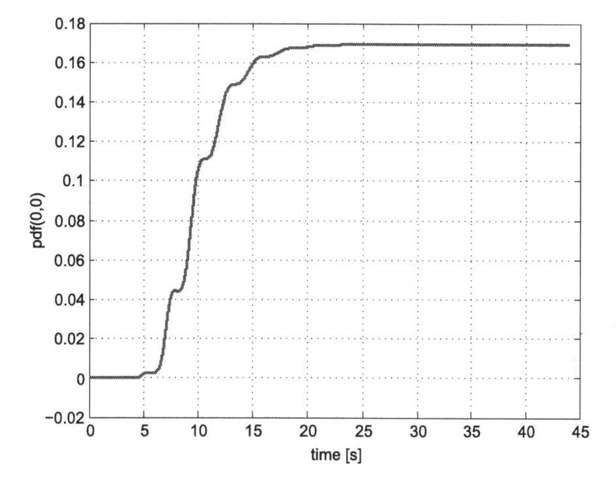

Fig. 5.5 FEM: 101 × 101—Probability density function at central node $p(0, 0, t)$ over time

The exact analytical expression for the stationary PDF of the unimodal Duffing oscillator of Eq. (5.19) is given by

$$\begin{aligned}\sigma_{x0}^2 &= \frac{\pi G_0}{4\xi\omega_0^3}\\ \sigma_{v0}^2 &= \omega_0^2\sigma_{x0}^2\\ p_X(\mathbf{x}) &= C\exp\left(-\frac{1}{2\sigma_{x0}^2}\left(x^2+\frac{\gamma}{2}x^4\right)-\frac{1}{2}\sigma_{v0}^2 v^2\right)\end{aligned} \tag{5.20}$$

The value of the stationary probability density function at the central node is $p_{stat}(0, 0) = 1.6851e - 1$. Figure 5.5 shows the evolution of the PDF at central node $p(0, 0, t)$ over time using the FEM and a mesh of 101 × 101.

4.2.2 Bimodal

The equations for the bimodal Duffing oscillator are characterized by the changed sign of the term $\omega^2 x_1$.

$$\begin{bmatrix}\dot{x_1}\\ \dot{x_2}\end{bmatrix} = \begin{bmatrix}x_2\\ -2\xi\omega x_2+\omega^2 x_1-\omega^2\gamma x_1^3\end{bmatrix}+\begin{bmatrix}0\\1\end{bmatrix}w(t) \tag{5.21}$$

Table 5.5 Parameter for the bimodal Duffing oscillator

μ	σ	ξ	ω	D	γ
[0, 10]	$\frac{1}{2}\mathbf{I}(2)$	0.2	1	0.4	0.1

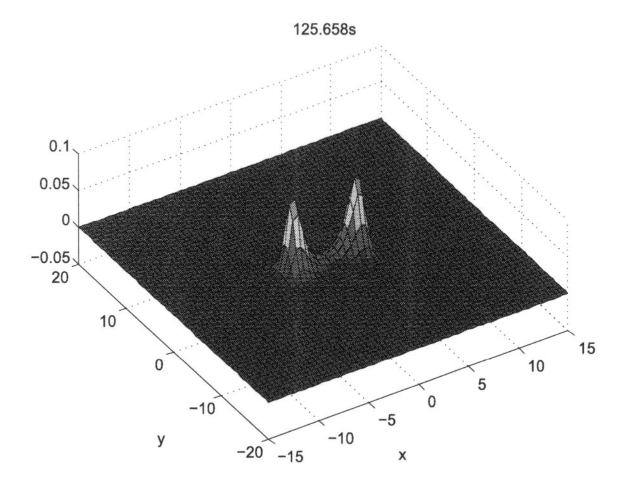

Fig. 5.6 61 × 61: Stationary probability density function p_{stat}

The parameters of the oscillator are chosen according to [16] and are given in Table 5.5.

The state space is discretized on the domain $[-15, 15] \times [-15, 15]$. Again, the ADI provides a tool for obtaining accurate results rather quickly.

In Fig. 5.6 the PDF is depicted for stationary conditions and a 61 × 61 mesh. A comparison of the evolution of the probability density function at the origin is shown in Fig. 5.7 for FEM and different meshes.

To compare the solution, the analytical expression according to [3] should be used. The exact analytical expression for the bimodal Duffing oscillator of Eq. (5.21) is given as

$$p_X(\mathbf{x}) = C \exp\left(-\frac{1}{2\sigma_{x_0}^2}\left(-x^2 + \frac{\gamma}{2}x^4\right) - \frac{1}{2}\sigma_{v_0}^2 v^2\right)$$

The value of the stationary PDF at the central node is $p_{stat}(0, 0) = 8.3161e - 3$. Table 5.6 shows a comparison of the accuracy for different mesh sizes for the bimodal Duffing oscillator. The maximum value of the stationary PDF of the bimodal oscillator at $x_{1,2} = \pm\sqrt{1/\gamma} = \pm 3.1623$ and $y_{1,2} = 0$ is $p_{stat}(x_{1,2}, 0) = 0.1013$. A comparison of the evolution of the probability density function at the mesh point $(x = 3,\ y = 0)$ which is closest to the maximum of the PDF is shown in Fig. 5.8 for FEM and different meshes; $p_{stat}(3, 0) = 0.0988$.

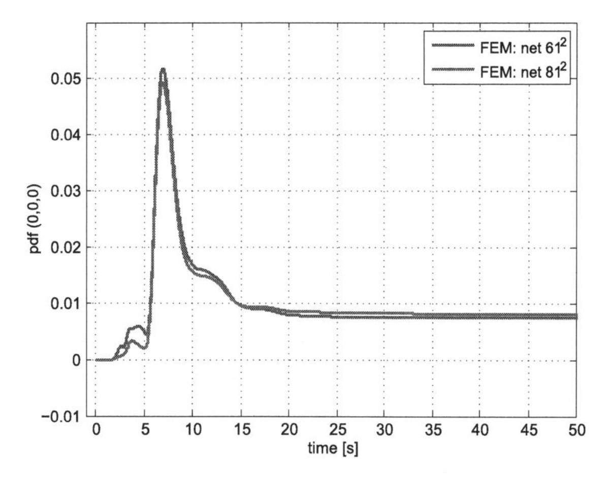

Fig. 5.7 FEM: Comparison of the probability density function at the central node $p(0, 0, t)$ over time

4.3 3-d Linear Oscillator

A 3-rd state variable is introduced in terms of a low pass filter for the white noise excitation which is applied to the linear 2d system.

$$\begin{bmatrix} \dot{x_1} \\ \dot{x_2} \\ \dot{x_3} \end{bmatrix} = \begin{bmatrix} x_2 \\ -2\xi\omega x_2 - \omega^2 x_1 + x_3 \\ -\alpha x_3 \end{bmatrix} + \begin{bmatrix} 0 \\ 0 \\ 1 \end{bmatrix} w(t) \tag{5.22}$$

The parameters of the 3d linear oscillator are given in Table 5.7.

Tables 5.8 and 5.9 show a comparison of the accuracy of the results for the linear oscillator. The time step is chosen to be $\Delta t = 0.01$, and only for FEM (net 81^3) $\Delta t = 0.001$ is required. The exact stationary value of the PDF at the origin is $p_{stat}(0, 0, 0) = 0.2409$. Figure 5.9 shows a comparison for the evolution of the PDF at the central node. The exact analytical solution is compared with the ADI and the FEM using various mesh sizes.

Table 5.6 Comparison of the accuracy for the bimodal Duffing oscillator

Method/Mesh	61 × 61		81 × 81		101 × 101	
	$\|e\|_\infty$	$\|e\|_2$	$\|e\|_\infty$	$\|e\|_2$	$\|e\|_\infty$	$\|e\|_2$
ADI	4.7186e-2	1.8229e-1	1.1899e-2	4.2491e-2	7.5077e-3	3.0640e-2
FEM	6.8085e-3	1.9074e-2	2.6024e-3	1.1265e-2	2.8419e-3	1.3709e-2

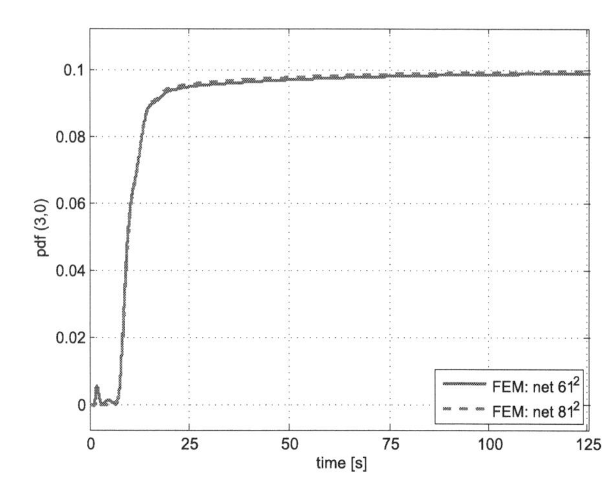

Fig. 5.8 FEM: Comparison of the probability density function at the node $p(3, 0, t)$ over time

Table 5.7 Parameter for the 3d linear oscillator

μ	σ	ξ	ω	D	α
[0, 0, 0]	$0.2\mathbf{I}(3)$	0.2	1	0.4	1

4.4 3-d Duffing Oscillator

$$\begin{bmatrix} \dot{x_1} \\ \dot{x_2} \\ \dot{x_3} \end{bmatrix} = \begin{bmatrix} x_2 \\ -2\xi\omega x_2 - \omega^2 x_1 - \omega^2 \gamma x_1^3 + x_3 \\ -\alpha x_3 \end{bmatrix} + \begin{bmatrix} 0 \\ 0 \\ 1 \end{bmatrix} w(t) \tag{5.23}$$

The parameters of the 3d Duffing oscillator are given in Table 5.10.

Table 5.8 Comparison of the accuracy for the 3-d linear oscillator

Method/Mesh	25^3		41^3	
	$\|e\|_\infty$	$\|e\|_2$	$\|e\|_\infty$	$\|e\|_2$
ADI	1.8009e-1	3.6649e+0	4.8336e-2	4.8748e-1
FEM	6.5802e-3	4.6480e-2	3.5574e-3	4.6357e-2
MSFEM	1.2533e-2	5.0262e-2	—	—

Table 5.9 Comparison of the accuracy for the 3-d linear oscillator

Method/Mesh	61^3		81^3		101^3	
	$\|e\|_\infty$	$\|e\|_2$	$\|e\|_\infty$	$\|e\|_2$	$\|e\|_\infty$	$\|e\|_2$
ADI	1.2542e-2	2.1170e-1	6.9165e-3	1.8067e-1	4.3899e-3	1.6063e-1
FEM	1.4081e-3	3.4554e-2	3.9016e-4	2.4920e-2	—	—

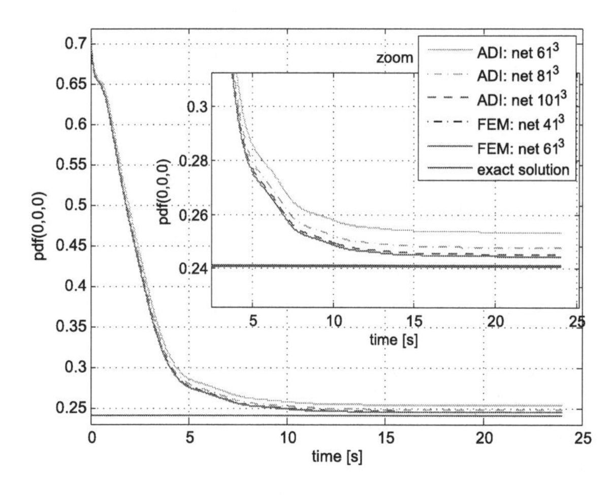

Fig. 5.9 Probability density function at the central node $p(0,0,0,t)$ over time

Table 5.10 Parameters for the 3d Duffing oscillator

μ	σ	ξ	ω	D	α	γ
[0, 0, 0]	$0.2\mathbf{I}(3)$	0.2	1	0.4	1	0.1

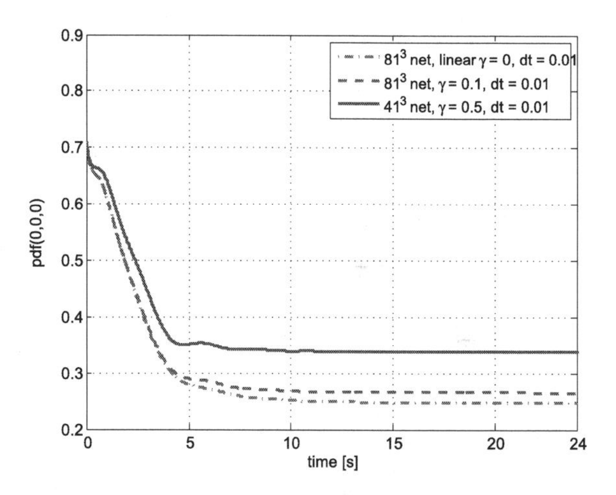

Fig. 5.10 Probability density function at the central node $p(0,0,0,t)$ over time

Figure 5.10 shows converged results for the evolution of the PDF at the origin over time using ADI for two different degrees of nonlinearity and for the corresponding linear system ($\gamma = 0$).

5 Discussion

Despite the greater numerical effort, the FEM is preferable over FD, because it yields more accurate results. However, at this time the FEM is only suitable for dimension ≤ 3. In the case of 3d and 4d problems, the stabilized multi-scale FEM provides a tool with a high order of accuracy, preserving numerical efficiency due to the fact that a coarser mesh size can be used.

The first effective numerical solution for 4d problems was reported by [21] in terms of high-order finite differences. The advantage of operator splitting methods including the ADI is the stability of the method, meaning that larger time steps (when compared to FEM) can be used, thus speeding up the analysis as the dimensionality of the problem is reduced by one.

The recently introduced PUFEM (see Kumar et al.) represents a possibility to obtain good results with coarse mesh sizes. The price paid is the computational overhead required in order to allow for the proposed coarse mesh size.

From the above discussion it is clear that future developments will be bounded by the so-called curse of dimensionality for some time.

Acknowledgements This research was partially supported by the Austrian Research Council FWF under Project No. J2989-N22 (LP, Schrödinger scholarship).

References

1. Bergman, L.A., Heinrich, J.C.: On the reliability of the linear oscillator and systems of coupled oscillators. Int. J. Numer. Methods Eng. **18**(9), 1271–1295 (1982)
2. Caughey, T.K.: Derivation and application of the Fokker–Planck equation to discrete nonlinear dynamic systems subjected to white random excitation. J. Acoust. Soc. Am. **35**(11), 1683–1692 (1963)
3. Caughey, T.K.: Nonlinear Theory of Random Vibrations. Elsevier, Amsterdam (1971)
4. COMSOL 5.5: Theory manual. COMSOL (2011)
5. Franca, L.P., Hauke, G., Masud, A.: Revisiting stabilized finite element methods for the advective-diffusive equation. Comput. Methods Appl. Mech. Eng. **195**(13–16), 1560–1572 (2006)
6. Kumar, M., Chakravorty, S., Junkins, J.L.: A homotopic approach to domain determination and solution refinement for the stationary Fokker–Planck equation. Probab. Eng. Mech. **24**(3), 265–277 (2009)
7. Kumar, M., Chakravorty, S., Junkins, J.L.: A semianalytic meshless approach to the transient Fokker–Planck equation. Probab. Eng. Mech. **25**(3), 323–331 (2010)
8. Kumar, M., Chakravorty, S., Singla, P., Junkins, J.L.: The partition of unity finite element approach with hp-refinement for the stationary Fokker–Planck equation. J. Sound Vib. **327**(1–2), 144–162 (2009)
9. Kumar, P., Narayana, S.: Solution of Fokker–Planck equation by finite element and finite difference methods for nonlinear systems. Sâdhana **31**(4), 445–461 (2006)
10. Langley, R.S.: A finite element method for the statistics of non-linear random vibration. J. Sound Vib. **101**(1), 41–54 (1985)
11. Lin, Y.K.: Probabilistic Theory of Structural Dynamics. Krieger, Melbourne (1976)
12. Masud, A., Bergman, L.A.: Solution of the four dimensional Fokker–Planck equation: still a challenge. In: Augusti, G., Schuëller, G.I., Ciampoli, M. (eds.) ICOSSAR 2005. Millpress, Rotterdam (2005)
13. Masud, A., Bergman, L.A.: Application of multi-scale finite element methods to the solution of the Fokker–Planck equation. Comput. Methods Appl. Mech. Eng. **194**(12–16), 1513–1526 (2005)

14. Masud, A., Khurram, R.A.: A multiscale/stabilized finite element method for the advection-diffusion equation. Comput. Methods Appl. Mech. Eng. **193**(21–22), 1997–2018 (2004)
15. Peaceman, D.W., Rachford, H.H.: The numerical solution of parabolic and elliptic differential equations. J. Soc. Ind. Appl. Math. **3**, 28–41 (1955)
16. Spencer, B.F., Bergman, L.A.: On the numerical solutions of the Fokker–Planck equations for nonlinear stochastic systems. Nonlinear Dyn. **4**, 357–372 (1993)
17. von Wagner, U., Wedig, W.V.: On the calculation of stationary solutions of multi-dimensional Fokker–Planck equations by orthogonal functions. Nonlinear Dyn. **21**(3), 289–306 (2000)
18. Wedig, W.V., von Wagner, U.: Extended Laguerre polynomials for nonlinear stochastic systems. In: Computational Stochastic Mechanics. Balkema, Rotterdam (1999)
19. Wijker, J.: Random Vibrations in Spacecraft Structures Design: Theory and Applications. Solid Mechanics and Its Applications, vol. 165. Springer, Berlin (2009)
20. Wojtkiewicz, S.F., Bergman, L.A., Spencer, B.F.: High fidelity numerical solutions of the Fokker–Planck equation. In: Seventh International Conference on Structural Safety and Reliability (ICOSSAR'97) (1998)
21. Wojtkiewicz, S.F., Bergman, L.A., Spencer, B.F., Johnson, E.A.: Numerical solution of the four-dimensional nonstationary Fokker–Planck equation. In: Narayanan, S., Iyengar, R.N. (eds.) IUTAM Symposium on Nonlinearity and Stochastic Structural Dynamics (1999)
22. Xie, W.-X., Cai, L., Xu, W.: Numerical simulation for a Duffing oscillator driven by colored noise using nonstandard difference scheme. Physica A **373**, 183–190 (2007)
23. Zorzano, M., Mais, H., Vázquez, L.: Numerical solution of two dimensional Fokker–Planck equations. Appl. Math. Comput. **98**, 109–117 (1999)

Chapter 6
A Comparative Study of Uncertainty Propagation Methods in Structural Problems

Manuele Corradi, Marco Gherlone, Massimiliano Mattone, and Marco Di Sciuva

Abstract Several uncertainty propagation algorithms are available in literature: (i) MonteCarlo simulations based on response surfaces, (ii) approximate uncertainty propagation algorithms, and (iii) non probabilistic algorithms. All of these approaches are based on some a priori assumptions about the nature of design variables uncertainty and on the models and systems behavior. Some of these assumptions could misrepresent the original problem and, consequently, could yield to erroneous design solutions, in particular where the prior information is poor or inexistent (complete ignorance). Therefore, when selecting a method to solve an uncertainty based design problem, several aspects should be considered: prior assumptions, non-linearity of the performance function, number of input random variables and required accuracy. It could be useful to develop some guidelines to choose an appropriate method for a specific situation.

In the present work some classical structural problems will be studied in order to investigate which probabilistic approach, in terms of accuracy and computational cost, better propagates the uncertainty from input to output data. The methods under analysis will be: Univariate Dimension Reduction methods, Polynomial Chaos Expansion, First-Order Second Moment method, and algorithms based on the Evidence Theory for epistemic uncertainty. The performances of these methods will be compared in terms of moment estimations and probability density function construction corresponding to several scenarios of reliability based design and robust design. The structural problems presented will be: (1) the static, dynamic and buck-

M. Corradi (✉) · M. Gherlone · M. Mattone · M. Di Sciuva
Department of Aeronautics and Space Engineering, Politecnico di Torino, Corso Duca degli Abruzzi 24, 10129 Torino, Italy
e-mail: manuele.corradi@polito.it

M. Gherlone
e-mail: marco.gherlone@polito.it

M. Mattone
e-mail: massimiliano.mattone@polito.it

M. Di Sciuva
e-mail: marco.disciuva@polito.it

M. Papadrakakis et al. (eds.), *Computational Methods in Stochastic Dynamics*, Computational Methods in Applied Sciences 26,
DOI 10.1007/978-94-007-5134-7_6, © Springer Science+Business Media Dordrecht 2013

ling behavior of a composite plate, (2) the reconstruction of the deformed shape of a structure from measured surface strains.

1 Introduction

The design for reliability, as well as robust design, is phased in over last decades in the structural design. Although these concepts are well-known in many engineering fields, the high computational cost of the mathematical approaches needed to perform these kinds of analysis, have set back their application in the aerospace structural design. Although in this last field, the problems that deal with the input variable uncertainties are known since the beginning of the aviation history, they are coped with deterministic methods based on the safety factor approach. The diffusion of components based on composite materials, in secondary and primary aerospace structures, and the dropping of aerospace and aviation companies' profit have reawaken the interest in design philosophies that deal with the uncertainty in a more effective way. For this reason, mathematicians and researchers have been urged on the study of new numerical approaches for an accurate Uncertainty Propagation (UP) from input to output data. Traditionally, both the reliability and the robustness of a design configuration have been studied using the MonteCarlo simulation; although it is the most accurate method, its computational cost could be prohibitive. For this reason several alternative approaches have been developed to face UP.

Most of the available UP algorithms have particular characteristics that make them appropriate for some specific problems but their capabilities are not fully exploited in all kinds of applications. First of all it is possible to distinguish between algorithms for the study of aleatory uncertainty and approaches that deal with the epistemic uncertainty. This classification can be based on the prior hypotheses needed to simulate the prior uncertainty. In order to model the epistemic uncertainty by means of probabilistic (aleatory) algorithms, some prior hypotheses should be adopted to transform the epistemic information into a probability distribution function (epistemic algorithms need not such assumptions). On the other hand, a probabilistic problem may be studied by means of an epistemic algorithm if the prior probability density functions are transformed into set-based information.

The UP algorithms based on the probability theory are usually classified into five categories [1]: (1) Simulation based methods: these techniques are based on the simulation of the problem in proper trial points, selected according to the stochastic characteristics of the input variables. MonteCarlo Simulation (MCS) is certainly the most known and used of these methods. (2) Local expansion based methods: these algorithms, also known as perturbation methods, are based on the local series expansion of output functions in terms of input random parameters. The methods based on Taylor expansion, such as the FOSM (First Order Second Moment) or the SOSM (Second Order Second Moment) methods, belong to this class. (3) Most Probable Points (MPP) based methods: this class includes the First and Second Order Reliability Methods (FORM and SORM, respectively). (4) Functional expansion based methods: they rely on a stochastic expansion of the performance function. The most known method of this class is the Polynomial Chaos Expansion (PCE). (5) Numerical integration based methods: these techniques are based on the numerical solving of integral equations for the statistical moments. These methods don't yield directly the performance joint probability function, but the corresponding statistical moments; by using the Pearson System and knowing the first four statistical moments, the probability distribution function can be obtained.

Several factors affect the choice of a suitable UP approach: (i) the identification and the classification of the input uncertainty, (ii) the definition of the required outputs (the first two statistical moments in robust design and the probability density function or the most probable points in a reliability based analysis), and (iii) the mathematical characteristics of the studied model (if the first order interactions cannot be neglected the Univariate Dimension Reduction method does not yield accurate prediction while if the performance function is non linear the methods based in Taylor local expansion are not accurate). This last information can be obtained using some numerical tools, such as the sensitivity analysis.

The main objective of this work is a comparative study of some of the most common and newest UP algorithms for both aleatory and epistemic uncertainties. As far as the first ones, the limits and merits of the Univariate Dimension Reduction method (UDR), of the Polynomial Chaos Expansion (PCE), and of the First Order Second Moments algorithm (FOSM) will be analyzed and discussed. These methods will be tested and compared on some numerical test functions and a classical structural problem: the probabilistic study of static, dynamic and buckling behavior of a composite plate. The sensitivity analysis has been performed in order to study the mathematical characteristics of the model. In the second part of the present work a probabilistic approach based on the UDR is compared with an epistemic approach based on the evidence theory. The structural application used as a test case for the comparison is an inverse problem: reconstruction of the deformed shape of a beam from measured surface strains using the inverse Finite elements Method (iFEM) [2, 3].

2 Uncertainty Propagation Algorithms

In this section, a review of some uncertainty propagation algorithms will be presented in order to set the framework for the assessment and comparison, through some structural applications, discussed in Sect. 3.

2.1 The Univariate Dimension Reduction Method (UDR)

This method involves an additive decomposition of a multidimensional integral function to multiple one-dimensional integral functions. The technique is suitable for calculating the stochastic moments of a system response function, as Rahman and Xu have shown [4–6].

The stochastic moments of a probability distribution may be calculated as follows

$$m_l = \zeta\left[Y^l(X)\right] = \int_{-\infty}^{+\infty} \cdots \int_{-\infty}^{+\infty} y^l(x) f_X(x)\, dx \quad l = 0, 1, \ldots, L \tag{6.1}$$

where m_l is the lth-order statistical moment (i.e., $l = 1$ is the mean value, $l = 2$ is the variance, etc.), $f_X(X)$ is the system response joint probability density function, $y(X)$ is the deterministic response when the input variables assume the values collected in the vector $X = \{x_1, \ldots, x_n\}^T$, and $Y(X) = y^l(X) f_X(X)$ is the performance function. The latter can be approximated as the sum of univariate functions, each one depending on only one random variable at a time and the other variables being fixed to nominal values

$$\begin{aligned} Y(x_1, \ldots, x_n) &\cong \tilde{Y}(X) \\ &\equiv \sum_{j=1}^{N} Y(\mu_1, \ldots, \mu_{j-1}, x_j, \mu_{j+1}, \ldots, \mu_N) - \cdots \\ &\quad + (N-1) \cdot y(\mu_1, \ldots, \mu_N) \end{aligned} \tag{6.2}$$

where $\mu_j$is the first moment of the stochastic variable x_j, $Y(\mu_1, \ldots, \mu_{j-1}, x_j, \mu_{j+1}, \ldots, \mu_N)$ is the stochastic response of the system only depending on the x_j random variable, and $y(\mu_1, \ldots, \mu_N)$ is the deterministic response of the system depending on the nominal value of the N input variables. Adopting the dimension-reduction procedure, the expression of statistical moments (6.1) can be rewritten as:

$$m_l = \int_{-\infty}^{+\infty} \cdots \int_{-\infty}^{+\infty} \sum_{j=1}^{N} \begin{pmatrix} Y(\mu_1, \ldots, \mu_{j-1}, x_j, \mu_{j+1}, \ldots, \mu_N) - \\ + (N-1) \cdot y(\mu_1, \ldots, \mu_N) \end{pmatrix}^l dx \tag{6.3}$$

To solve the univariate integration in the context of the UDR method, Xu and Rahman [4] suggest the use of the moment based quadrature rule. The evaluation of integration points x_j involves the solution of the following equation

$$x_j^n - r_{j,1} x_j^{n-1} + r_{j,2} x_j^{n-2} - \cdots + (-1)^n r_{j,n} = 0 \tag{6.4}$$

where the coefficients r_j are solution of the following linear system of equations

$$\begin{bmatrix} \mu_{j,n-1} & -\mu_{j,n-2} & \mu_{j,n-3} & \cdots & (-1)^{n-1}\mu_{j,0} \\ \mu_{j,n} & -\mu_{j,n-1} & \mu_{j,n-2} & \cdots & (-1)^{n-1}\mu_{j,1} \\ \mu_{j,n+1} & -\mu_{j,n} & \mu_{j,n-1} & \cdots & (-1)^{n-1}\mu_{j,2} \\ \cdots & \cdots & \cdots & \cdots & \cdots \\ \mu_{j,2n-2} & -\mu_{j,2n-3} & \mu_{j,2n-4} & \cdots & (-1)^{n-1}\mu_{j,n-1} \end{bmatrix} \begin{bmatrix} r_{j,1} \\ r_{j,2} \\ r_{j,3} \\ \cdots \\ r_{j,n} \end{bmatrix} = \begin{bmatrix} \mu_{j,n} \\ \mu_{j,n+1} \\ \mu_{j,n+2} \\ \cdots \\ \mu_{j,2n-1} \end{bmatrix} \tag{6.5}$$

$\mu_{j,i}$ $(i = 1, \dots, n)$ represents the ith stochastic moment of the jth input variable. Thus, the univariate integral can be numerically solved as

$$\int_{-\infty}^{\infty} y^l(\mu_1, \dots, x_j, \dots, \mu_N) f_{X_j}(x_j) \cdot dx_j$$

$$\cong \sum_{i=1}^{n} w_{j,i} y^l(\mu_1, \dots, x_j, \dots, \mu_N) \tag{6.6}$$

where f_{x_j} is the probability density function of input variable x_j. The weight $w_{j,i}$ appearing in Eq. (6.6) are evaluated using the following expression

$$w_{i,j} = \frac{\sum_{k=0}^{n-1}(-1)^k \mu_{j,(n-h-1)} \cdot q_{j,(ik)}}{\prod_{k=1,k\neq 1}^{n}(x_{j,i} - x_{j,k})} \tag{6.7}$$

$$q_{j,i_0} = 1; \qquad q_{j,ik} = r_{j,k} - x_{j,i} \cdot q_{j,i(k-1)}$$

2.2 The Polynomial Chaos Expansion (PCE)

The Polynomial Chaos Expansion was introduced by Wiener [7] and is based on the approximation of each random variable by means of a suitable polynomial expansion about centered normalized Gaussian variables.

Any set $X = \{x_1, \dots, x_n\}^T$ of independent variables (i.e. a set of Gaussian variables) can be expressed as function of a set $\xi = \{\xi_1, \dots, \xi_n\}$ of independent normal variables; this is also known as normalization process.

$$X = f(\xi) \tag{6.8}$$

Hence, a performance function $y = Y(X)$ could be transformed into a function expressed in terms of ξ and, afterwards, approximated by means of the Polynomial Chaos Expansion (PCE) on the vector space

$$Y(X) = a_0\Gamma_0 + \sum_{i_1=1}^{\infty} a_{i_1}\Gamma_1(\xi_1) + \sum_{i_1=1}^{\infty}\sum_{i_2=1}^{i_1} a_{i_1 i_2}\Gamma_2(\xi_{i_1}, \xi_{i_2}) +$$

$$+ \sum_{i_1=1}^{\infty}\sum_{i_2=1}^{i_1}\sum_{i_3=1}^{i_2} a_{i_1 i_2 i_3}\Gamma_3(\xi_{i_1}, \xi_{i_2}, \xi_{i_3}) + \cdots \tag{6.9}$$

where $a = [a_0, \dots, a_n]$ is the vector of the expansion unknown terms and $\Gamma_p(\xi_1, \dots, \xi_n)$ are the multidimensional Hermite polynomials (only if the input random variables are defined by a normal probability distribution) of order p.

Cameron and Martin have shown that this kind of series is convergent in the L_2-sense [8]. In order to simplify the notation a univocal relation between the functional

Γ and a new functional Ψ is defined. Hence, the PCE expansion, expressed by Eq. (6.9), can be rewritten as follows

$$Y(X) = \sum_{k=0}^{\infty} \beta_k \Psi_k\big(\xi(X)\big) \tag{6.10}$$

In the present work the classical convention is adopted:

- $\Psi_0 = 1$: is the 0th-order polynomial
- β_k are the constant coefficients of the expansion
- Ψ_k are multivariate Hermite polynomials, orthogonal in the L_2-space. These polynomials are the product of the proper set of univariate Hermite polynomials [9].

The expansion is normally truncated at a selected order P

$$Y(X) \approx \hat{Y}(X) = \sum_{k=0}^{P} \beta_k \Psi(\xi)_k \tag{6.11}$$

The number of unknown coefficients β_k (6.11) can be evaluated using the following expression

$$P + 1 = \frac{(p+n)!}{p!n!} \tag{6.12}$$

The procedure described above is general, but the Hermite polynomials can be used only in the case of input variables with Gaussian probability distribution function. Xiu and Karniadakis [9] have extended the PCE applicability to all kinds of input distribution function, adopting the Wiener–Askey scheme for non Gaussian input distribution. They have proposed to use the Askey scheme to combine the non Gaussian input distribution with orthogonal polynomial family; in this way the expansion convergence is guaranteed for all kind of input PDF. As well as the Hermite polynomials are orthogonal in the Hilbert space, in the same way all polynomials, adopted in the Wiener–Askey [9] scheme are orthogonal in the Hilbert space and form an Hilbert basis of the corresponding space.

The set $\beta = \{\beta_0, \ldots, \beta_n\}^T$ of the PCE unknown coefficients, can be approximated by a new vector $\hat{\beta}$, obtained solving the following least squares problem

$$\hat{\beta} = \arg\min_{\beta} \sum_{i=1}^{N} \left(Y(X_i) - \sum_{k=0}^{P} \beta_k \psi_k(\xi_i) \right)^2 \tag{6.13}$$

where N is the training points set size; generally, it is convenient that $N > p + 1$.

2.3 The First Order Second Moment Algorithm (FOSM)

In this approach a performance function $Y(X)$ is approximated by means of a first order Taylor-series expansion around the design point [10]

$$Y(X) \cong Y(\bar{X}) + \sum_{i=1}^{n} \left.\frac{\partial Y}{\partial x_i}\right|_{\bar{X}} (x_i - \bar{x}_i) \tag{6.14}$$

Substituting Eq. (6.14) in the expectation definition (mean)

$$E\big[Y(X)\big] = E\big[Y(\bar{X})\big] + E\left[\sum_{i=1}^{n} \left.\frac{\partial Y}{\partial x_i}\right|_{\bar{X}} (x_i - \bar{x}_i)\right] \tag{6.15}$$

and considering that:

$$E\left[\sum_{i=1}^{n} \left.\frac{\partial Y}{\partial x_i}\right|_{\bar{X}} (x_i - \bar{x}_i)\right] = \sum_{i=1}^{n} \left.\frac{\partial Y}{\partial x_i}\right|_{\bar{X}} E\big[(x_i - \bar{x}_i)\big] = 0 \tag{6.16}$$

$$E\big[(x_i - \bar{x}_i)\big] = E(x_i) - \bar{x}_i = \bar{x}_i - \bar{x}_i = 0 \tag{6.17}$$

the performance function mean value, estimated by means of FOSM, assumes the following expression

$$E\big[Y(X)\big] = Y(\bar{X}) \tag{6.18}$$

Now, given the variance definition

$$Var\big[Y(X)\big] = E\big[\big(Y(X) - E\big(Y(X)\big)^2\big)\big] \equiv \sigma^2\big[Y(X)\big] \tag{6.19}$$

and substituting in it Eq. (6.14), the variance assumes the following expression

$$\sigma\big[Y(X)\big] = \sum_{i=1}^{n}\sum_{j=1}^{m} \left.\frac{\partial Y}{\partial x_i}\right|_{\bar{x}_i} \left.\frac{\partial Y}{\partial x_j}\right|_{x_j} \cdot COV(x_i, x_j) \tag{6.20}$$

where $COV(x_i, x_j)$ is the covariance matrix.

2.4 The Evidence Theory

The Evidence Theory is a non probabilistic approach used to characterize the effect of epistemic uncertainty on a system.

Given a design variable x_1, the prior information, or evidence, consists of n intervals, obtained from s sources $[x^l_{1,i}, x^u_{1,i}]$ (with $i = 1, \ldots, s$) that enclose the supposed true value. Clearly, the traditional probability theory cannot handle this type of evidence, without making some assumptions that can misrepresent the nature of the information. Several combination rules have been formulated to handle this kind of prior information [11, 12]; in this work, the Dempster–Shafer combination rule is adopted.

When a source provides a set of information, this means that the variable can assume any value inside the interval. The probability that a variable x_1 assumes the value $\bar{x}_1$ is not defined by a probability distribution function but is included between a maximum probability (plausibility), and a minimum probability (belief). In order to define the plausibility and the belief, the basic probability assignment (m) must be introduced; m defines a mapping of the variable prior information. Formally the basic probability assignment function is defined by means of the following expressions

$$m : \mathrm{P}(x_1) \rightarrow [0, 1] \tag{6.21}$$

$$m(\varnothing) = 0 \tag{6.22}$$

$$m = 1 \quad \text{if } \bar{x}_1 \in S_i \text{ with } i = 1, \ldots, s \tag{6.23}$$

where $\mathrm{P}(x_1)$ represents the power set of x_1 (defined, according to the axiomatic set theory [12] as the set of all subset of S), while $\varnothing$ is the null set and S_i is the ith evidence set. According to the previous equations, the basic probability assignment (BPA) assumes any value included between 0 and 1; if $\bar{x}_1$ does not belong to any subset, the basic probability assignment assumes value 0 while if $\bar{x}_1$ belongs to every subset, it assumes the value 1. Once defined the basic probability assignment m, the plausibility and belief probability measures can be introduced. Given a set $C = [D(x_1)^-, \bar{x}_1]$, where $\bar{x}_1$ is a generic value of the variable x_1 on its domain $D(x_1)$, while $D(x_1)^-$ represents the lower domain boundary, the plausibility can be expressed by

$$Pl(C_1) = \sum_{C_1 \cap s^i_{x_1} \neq 0} m_{x_{1,i}}\left(S^i_{x_1}\right) \tag{6.24}$$

while the belief is defined as

$$Bel(C_1) = \sum_{S^i_{x_1} \subseteq C_1} m_{x_1}\left(S^i_{x_1}\right) \tag{6.25}$$

In other words, the plausibility is the sum of all BPA of the sets $S^i_{x_1}$ which intersect the set of interest C_1, hence it represents the maximum probability that a variable x_1 assumes a given value $\bar{x}_1$. On the other hand, the belief is defined as the sum of all BPA of the sets $S^i_{x_1}$ that $S^i_{x_1} \subseteq C_1$ hence it is a measure of the minimum probability that a variable x_1 assumes a given value $\bar{x}_1$.

The probability lies between the plausibility and the belief

$$Bel(C_1) \leq P(C_1) \leq Pl(C_1) \tag{6.26}$$

and, only when plausibility and belief are overlapped, it can be univocally defined.

In a problem with n input variables there is the need to transfer the BPA values m_{x_j}, evaluated for each variable, into an equivalent information in the n-dimensional

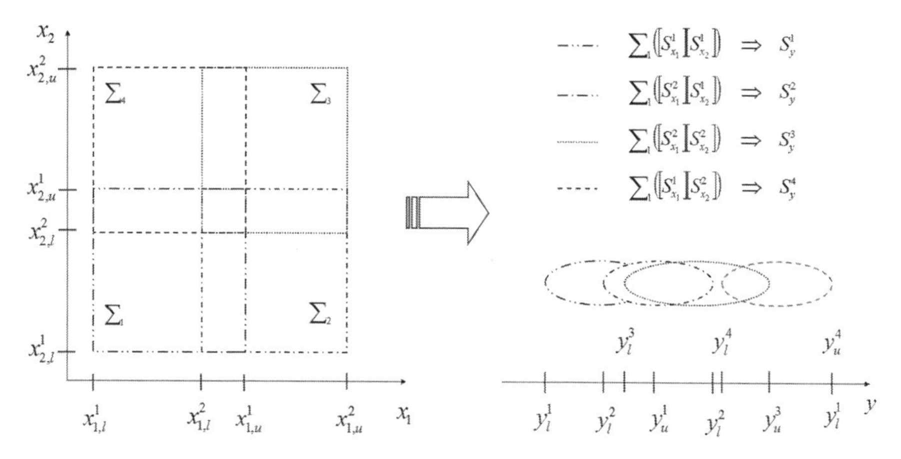

Fig. 6.1 Propagation of evidence from input to output space

design variables space. Assuming that all the variables are uncorrelated, the probability of each elementary set in the design variables space is defined by

$$m_{x_1,x_2,\ldots,x_n}\left(\left[S^j_{x_1}\right],\ldots,\left[S^j_{x_{n1}}\right]\right) = m_{x_1}\left(\left[S^j_{x_1}\right]\right)\cdot\cdots\cdot m_{x_1}\left(\left[S^j_{x_n}\right]\right) \tag{6.27}$$

Once defined the uncertainty acting on the design variables, its effects on the performance function can be evaluated. Given a generic function $y = Y(X)$, linking the output with the input variables $X = \{x_1,\ldots,x_n\}$, the evidence about y must be estimated from the joint body of evidence previously described in Eq. (6.27). By means of two optimization problems, for each evidence-set of the input variables space, the lower and upper boundary of the corresponding set into the output space are evaluated

$$\begin{aligned}
&\textit{find } \bar{X} = \{\bar{x}_1,\ldots,\bar{x}_n\} \in \Sigma_j\left(\left[S^j_{x_1}\right],\left[S^j_{x_2}\right],\ldots,\left[S^j_{x_n}\right]\right)\\
&t.c \ \max/\min y = f(X)\\
&\Downarrow\\
&S^j_y = \left[y^j_l, y^j_u\right]
\end{aligned} \tag{6.28}$$

The above optimization problems yield an evidence set on output-space S^j_y for each set Σ_j of the joint body of evidence (Fig. 6.1). Hence, in order to propagate the uncertainty from input to output, two optimizations for each set Σ_j have to be performed.

3 Numerical Examples

In this section some numerical examples will be presented in order to verify the accuracy of the UP methods described above. Some test functions and a structural

Table 6.1 Test functions used to test the UP methods

Function	PDF	PDF Parameters
$y = x_1^k x_2^k + 2x_3^4,\ k = 2, 3, 5$	Gaussian	$\mu = [1, 1, 1]$ $\sigma = 0.1, 0.2, 0.4, 0.8$
$y = \sin x_1 + a \sin^2 x_2 + b x_3^4 \sin x_1$	Gaussian	$\mu = [\pi/4, \pi/4, \pi/4]$ $\sigma = 0.05, 0.1, 0.2, 0.5$

problem (static, dynamic and buckling behavior of a composite plate) will be the test cases considered for assessing the methods for stochastic uncertainty. A comparison between stochastic and epistemic approaches when applied to a structural shape sensing problem will be then discussed.

3.1 Test Functions

Two test functions are used to compare the performance of the UP methods introduced in the previous paragraphs (Table 6.1).

The first example is a three-variate function, chosen to compare the performance of the UP algorithms against the first order effects of the input variables. In order to better understand this example and the capabilities of each method can be useful defined what are the main and high order effects. They deal with how the uncertainty on input parameters influences the output. In detail the main effects measure the influence of each single input parameter on the output, while the high order effects give information about the influence of each possible combination of input parameters (i.e. X_1X_2, X_1X_3) on the output. More details about the numerical methods used to evaluate these measures are beyond the purpose of this paper refer to [13] for further details.

The input variables follow a Gaussian distribution centered in $X = [1, 1, 1]$ and four values of standard deviation are tested ($\sigma = 0.1, 0.2, 0.4, 0.8$). In addition, the effect of the interactions among the variables is studied changing the value of k. The analysis of the accuracy of each method is performed comparing the predicted values of the statistical moments with those evaluated using a MonteCarlo Simulation, based on 10^6 observations. In this example the effect of an increasing input variability is combined with that of an increasing interaction effect.

In Table 6.2 the main effects and the interactions are listed for each value of k. These indices are evaluated by means of the Polynomial Chaos Expansion [14]. We can observe that changing the value of k the interaction x_1x_2 increases its effect on the output, becoming gradually the most important factor.

In Fig. 6.2 the errors in the estimation of the mean value are plotted as function of the input variables standard deviation for different values of k. As a general rule, when the input variability increases, all the UP methods here discussed become less accurate. This phenomenon is negligible if the first order interactions are marginal;

Table 6.2 Main and first order effects

	Main and First Order Effects		
	k = 2	k = 3	k = 5
X_1	0.011	0.0838	0.0575
X_2	0.011	0.0838	0.0572
X_3	0.9587	0.4734	0.0006
X_1X_2	0.01927	0.359	0.8808
X_1X_3	9.60E-25	8.78E-25	5.82E-04
X_2X_3	4.84E-26	8.34E-25	1.22E-03

on the contrary, in problems where the interaction effects are more important ($k = 3$ or $k = 5$) the results become more sensitive to the input variability.

As shown in Fig. 6.2(A), the UDR yields a good estimation of the mean values when the interaction between the variables is low ($k = 2$), also in the case of high

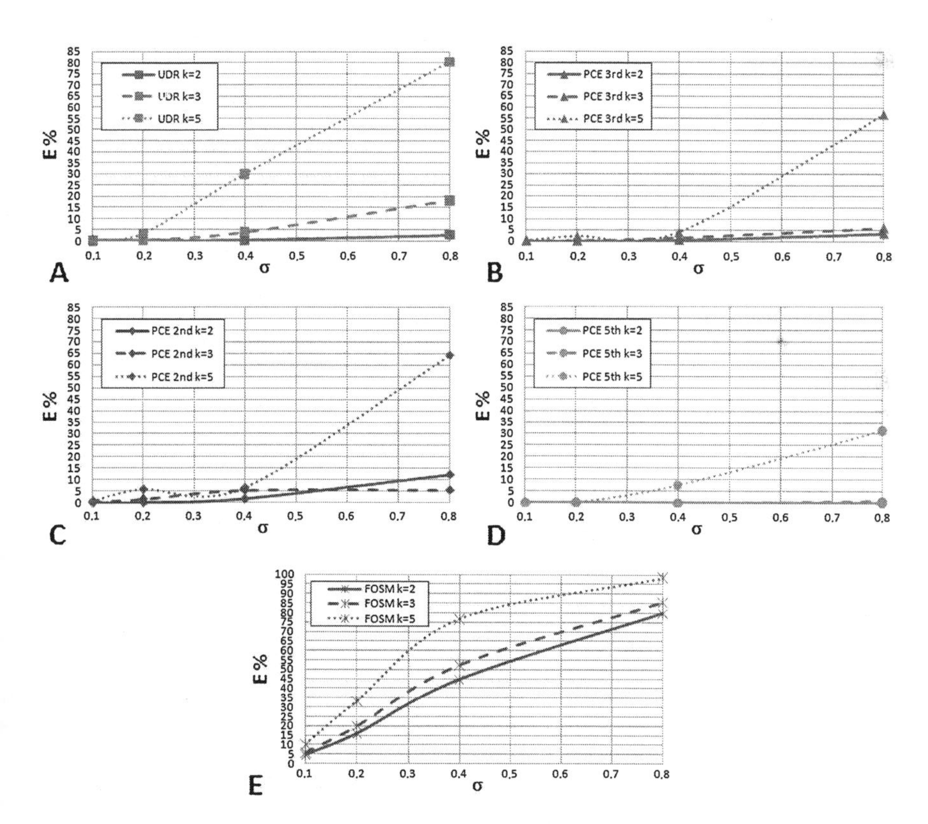

Fig. 6.2 Error in the estimation of the output mean value: (**A**) Error due to the UDR method (**B**)–(**D**) Error due to the PCE of 2nd, 3rd and 5th order, respectively. (**E**) Error due to the FOSM alghorithm

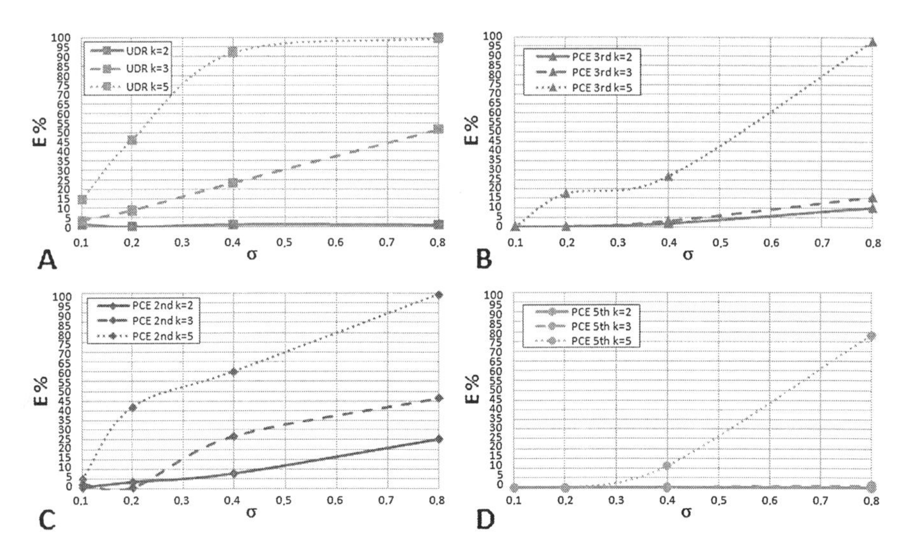

Fig. 6.3 Error in the estimation of the output variance: (**A**) Error due to the UDR method (**B**)–(**D**) Error due to the PCE of 2nd, 3rd and 5th order, respectively

input variability ($\sigma = 0.8$). Increasing the effect of interaction, the accuracy of this method greatly decays, in particular for higher values of input variability.

Similar behaviors are shown in Fig. 6.2(B)–(D); the output function is approximated with the Polynomial Chaos Expansion, truncated at different orders. Also in this case, for higher values of k and for a higher input uncertainty, the mean value is poorly approximated. When using the PCE, however, the reduced accuracy is not due to the interaction effects, but it is caused by the non-linearity of the output function: for example, if $k = 2$ we have a 4th order function, while if $k = 3$ we have a 6th order function. Hence, it is clear that a Polynomial Chaos Expansion truncated at the 5th order better describes the problem than an expansion truncated at the 2nd order, but, for $k = 5$, it does not guarantee adequate accuracy. Increasing the order of the expansion, the error in the prediction gradually vanishes.

Therefore, there is a substantial difference between the UDR and the PCE. In the UDR the lack of accuracy is inherent to its mathematical formulation and cannot be reduced. On the other hand, the accuracy of the PCE results can be improved increasing the order of the expansion.

In Fig. 6.2(E) the errors are shown on the output function mean value when computed using the FOSM algorithm. In this case the approximation is based on the hypothesis that the output function has a linear behavior in the studied domain; the errors are quite high also for small input variability levels.

The decay of UDR accuracy in the prediction of the statistical moments (due mainly to the first order effects) is more evident in the evaluation of the variance, Fig. 6.3(A) and of the higher order moments: Skewness Fig. 6.4(A) and Kurtosis Fig. 6.5(A). As already observed in the evaluation of mean value, for quite small

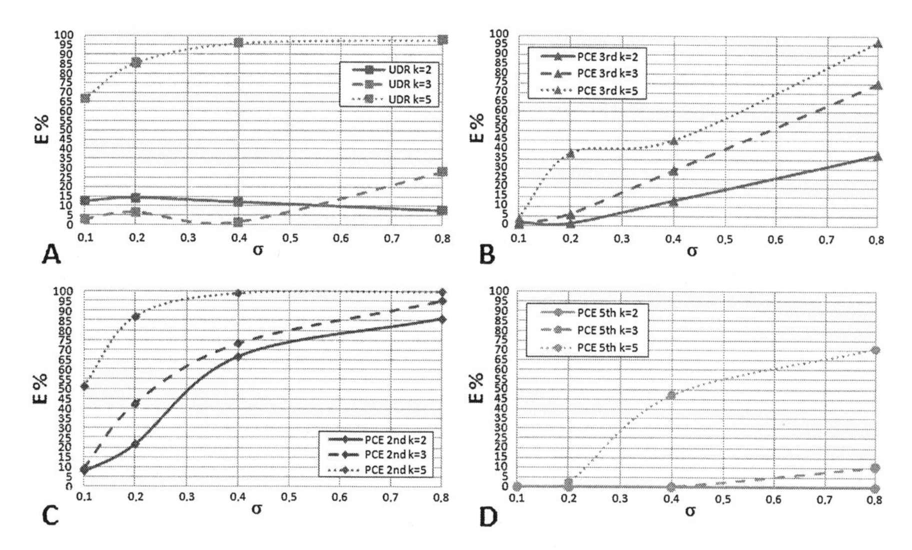

Fig. 6.4 Error in the estimation of the output skewness: (**A**) Error due to the UDR method (**B**)–(**D**) Error due to the PCE of 2nd, 3rd and 5th order, respectively

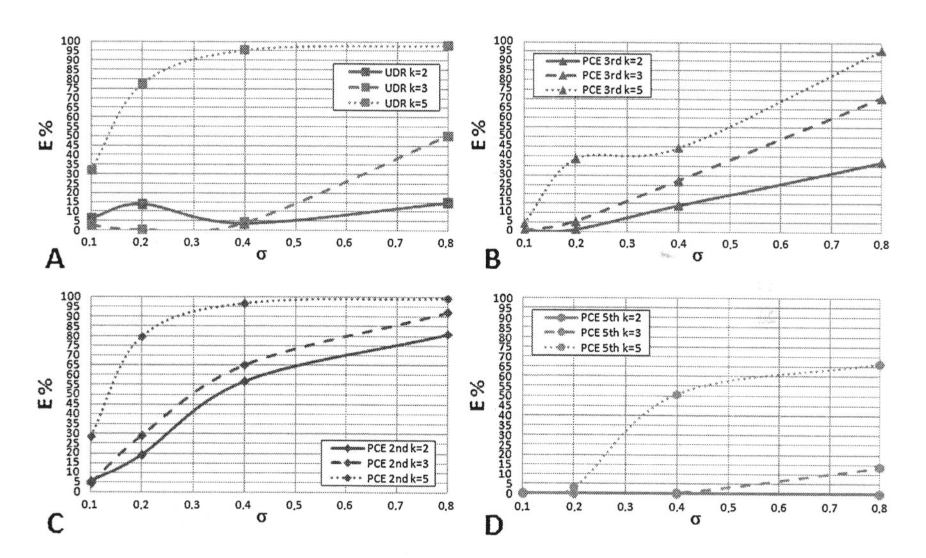

Fig. 6.5 Error in the estimation of the output kurtosis: (**A**) Error due to the UDR method (**B**)–(**D**) Error due to the PCE of 2nd, 3rd and 5th order, respectively

interaction effects ($k = 2$), the UDR approximation does not affect the accuracy of the results. This is not true for the higher order moments.

Results regarding higher order statistical moments (Figs. 6.3(B)–(D), 6.5(B)–(D), 6.6(B)–(D)) confirm that the interaction between the variables does not affect

Table 6.3 Number of observations for each method

FOSM	UDR	PCE 2nd	PCE 3rd	PCE 5th
49	16	31	61	168

the accuracy of the Polynomial Chaos Expansion; anyhow, a higher order expansion is required in order to have a good estimation of the variance, skewness and kurtosis.

In Table 6.3 the number of observation points needed to perform each analysis are listed. The UDR method needs only 16 observed data (it requires only $5n + 1$, where n are the stochastic input variables).

The UDR method is the cheapest one and, as seen in the present example, if there is a negligible interaction between the input variables, it yields a good estimation of the statistical moments. The computational cost of the PCE grows considerably increasing the order of expansion and the problem dimension. The number of unknown coefficients of PCE is $P + 1$, as given by Eq. (6.12); these coefficients are evaluated using a least-squares based method, Eq. (6.13). The number of training points must be at least equal to $P + 1$, but for and adequate accuracy of the solution, a higher number of points is usually adopted; this number (Table 6.3) is evaluated by a convergence study on the β values.

The second function here considered (see Table 6.1) is the Ishigami function, commonly used to test the uncertainty propagation algorithms and the sensitivity in order to understand their behavior with non-linear and non-monotonic functions. The three variables follow a Gaussian distribution, centered in $X = [\pi/4, \pi/4, \pi/4]$; the standard deviation ranges from 0.05 to 0.5. The accuracy of each method is assessed comparing the predicted values of the statistical moments with those evaluated using a MonteCarlo Simulation, based on 10^6 observations.

In Table 6.4 the percentage errors on the estimation of the mean value and variance are listed for different values of the input standard deviation. All methods yield a good estimation of the mean value (the error is always less than 1%). The differences between the methods are more evident when considering the variance evaluation. The FOSM method yields a very poor estimation in particular for high values of input variability: for example, the error with an input standard deviation of $\sigma = 0.2$ is around 12.3%, while with $\sigma = 0.5$ is around 56%. The UDR method leads to a good estimation of the variance (error around 2%); there is no evident correlation between the input variability and the estimation error. The PCE is very accurate, if the order of the expansion is sufficient to describe the problem; we can observe that a 5th order expansion is very accurate in the prediction of variance, and that the 7th order expansion yields the exact solution.

In a robust design framework it is important the accurate evaluation of the first two statistical moments (mean and variance), but for the evaluation of the reliability degree this information is not enough. Hence, the knowledge of the probability distribution function is needed. One of the main problems of the UDR approach is that it does not yield directly the probability distribution, but only the statistical moments. Anyway, it is possible to obtain the PDF, knowing the first four statistical moments, by means of the Pearson System. In Fig. 6.6 and Fig. 6.7 the probability

Table 6.4 Mean values and variance estimation errors

Standard Deviation Input	FOSM		UDR		PCE 2nd	
	Mean	Variance	Mean	Variance	Mean	Variance
0.05	3.336E-02	1.204E+00	2.515E-02	1.934E+00	2.363E-03	2.398E-01
0.1	7.351E-02	1.968E+00	4.734E-02	1.952E+00	2.837E-02	1.997E+00
0.2	1.609E-01	1.235E+01	7.092E-02	2.162E+00	1.754E-01	2.334E+00
0.5	5.364E-01	5.575E+01	3.560E-01	1.818E+00	5.596E-01	6.521E+00
Standard Deviation Input	**PCE 3rd**		**PCE 5th**		**PCE 7th**	
	Mean	Variance	Mean	Variance	Mean	Variance
0.05	0.000E+00	0.000E+00	0.000E+00	0.000E+00	0.000E+00	0.000E+00
0.1	0.000E+00	2.058E-02	0.000E+00	0.000E+00	0.000E+00	0.000E+00
0.2	4.731E-03	3.456E-02	0.000E+00	3.456E-02	0.000E+00	0.000E+00
0.5	1.189E-01	2.878E+00	4.749E-03	1.641E-02	0.000E+00	2.553E-02

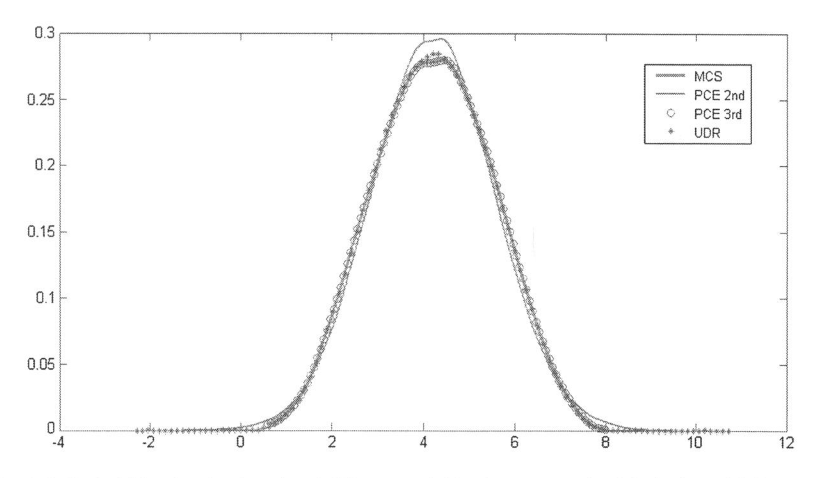

Fig. 6.6 Probability density function (all input variables have a standard deviation of 0.2)

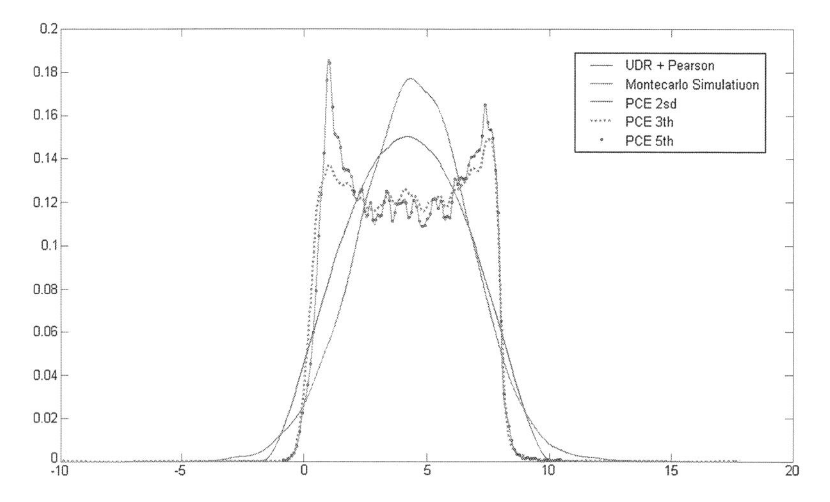

Fig. 6.7 Probability density function (all input variables have a standard deviation of 0.5)

distribution function, evaluated with the UDR and the Pearson System, is compared with the PDF obtained using the PCE with different expansion orders and the one obtained by means of MCS (10^6 training points). The curves plotted in Fig. 6.6 and in Fig. 6.7 are referred, respectively, to the case of an input standard deviation of $\sigma = 0.2$ and of $\sigma = 0.5$. In the first case (Fig. 6.5) a good agreement among all plotted curves can be observed. The UDR method coupled with the Pearson System yields a very good approximation of the probability density function. There is only a small discrepancy in the description of the tails: the tails of PDFs, obtained with the UDR method and the 2nd order PCE, end with an asymptotic behavior, while in the one obtained with MCS the tails are limited. On the contrary there is a perfect correspondence between the probability function obtained with a 3rd order PCE and the one obtained with the MonteCarlo Simulation.

In the second case ($\sigma = 0.5$) it is possible to appreciate a bigger discrepancy among the methods (Fig. 6.6(A)). Although the UDR method is able to predict with a good accuracy the mean and the variance of the output, it fails in the estimation of the PDF. This is mainly due to the fact that the higher order moments are predicted with low-accuracy and, as well known, the Pearson System is based on the relation between skewness and kurtosis. Also the 3rd order PCE gives not an accurate probability distribution. Only by means of a 5th order PCE a good PDF approximation can be obtained.

In this example we have seen that, although the UDR approach is adequately accurate to be used in a robust design problem, it cannot be used in a reliability based problem.

3.2 Composite Plate Mechanical Behavior

In this example the performances of UDR and PCE are tested on the static and dynamic response analysis of a symmetric composite plate ($0°/90°/90°/0°$) with all edges clamped. The material properties, the fiber orientation, and the plies thickness are considered affected by uncertainty and are described by means of Gaussian distributions. In Table 6.5 all plate properties are reported in terms of mean value and standard deviation. The stochastic moments of the maximum deflection ($\boldsymbol{w}$), the first natural frequency ($\boldsymbol{f}$), the maximum Von Misses stress ($\boldsymbol{\sigma}_{VM}$), the maximum transverse shear stresses, τ_{xz} and τ_{yz}, are evaluated by means of the UDR and of the PCE. The static and dynamic responses of the plate have been obtained using the Refined Zigzag Theory (RZT) for plates [15–19]; a Rayleigh–Ritz solution procedure has been adopted to find maximum deflection, first natural frequency and stresses distribution. The first four statistical moments, evaluated using the UDR and PCE, are compared with those obtained by a MonteCarlo Simulation based on 10^5 observations (Table 6.6). In Table 6.6 one can observe that approximately all approaches give the same results, the main difference between the UDR and the 2nd order PCE is in the computational cost needed to perform the analysis, indeed are needed 71 observations and 360 training points, respectively.

3.3 Structural Shape Sensing

The inverse Finite Element Method (iFEM), developed by Tessler for plate and shell structures [19] and specialized by Gherlone for beams and frames [2], is aimed at the reconstruction of the displacement field of a structure starting from in situ measurements of surface strains [2]; this represents an inverse problem [3]. A description of the iFEM approach is not within the scopes of the present paper, refer to [2, 3, 19] for further details.

Table 6.5 Plate properties: θ_1 is the fiber orientation of the first ply, t_1 is its thickness

		Mean	PDF	SD
Mechanical properties	E_{11} [Mpa]	1.58E+05	Gaussian	7.895
	E_{22} [Mpa]	9.58E+03	Gaussian	0.4792
	E_{33} [Mpa]	9.58E+03	Gaussian	0.4792
	G_{12} [Mpa]	5.93E+03	Gaussian	0.2965
	G_{13} [Mpa]	5.93E+03	Gaussian	0.2965
	G_{23} [Mpa]	3.23E+03	Gaussian	0.1613
	v_{12}	0.32	Deterministic	-
	v_{13}	0.32	Deterministic	-
	v_{23}	0.49	Deterministic	-
	ρ [T/mm^3]	1.90E-09	Deterministic	-
Fiber orientation	θ_1	0	Gaussian	3
	θ_2	90	Gaussian	3
	θ_3	90	Gaussian	3
	θ_4	0	Gaussian	3
Thickness	t_1 [mm]	1	Gaussian	0.05
	t_2 [mm]	1	Gaussian	0.05
	t_3 [mm]	1	Gaussian	0.05
	t_4 [mm]	1	Gaussian	0.05

In the present work a cantilevered aluminum beam (length L = 200 mm) with a circular thin-walled cross-section (radius r = 1 cm) and subjected to different load conditions has been studied. In lieu of the experimental measures of surface strains, high-fidelity direct FE analyses (MSC/NASTRAN) have been carried out for the example problem (Table 6.7). These results have also been used to verify the accuracy of the nodal displacements and rotations obtained by iFEM.

The position of a strain gauge, used to measure a surface strain, is defined by three coordinates: the first one, x, indicates the position along the longitudinal beam axis, the second one, θ, is an angle representing the circumferential position on the beam and the coordinate β indicates the strain gauge orientation (i.e., it represents the rotation of the strain gauge with respect to the beam axis (Fig. 6.8). For the current application, six strain gauges are used; their nominal positions are reported in Table 6.8 and their location is also represented in Fig. 6.9.

In this example three different load conditions are considered (Fig. 6.10(A)): (1) a shear force applied along y-axis, (2) a torque moment and (3) a bending moment around the z-axis. The free end displacements and rotations (Fig. 6.10(B)) are computed by means of the iFEM and are compared with the ones obtained using the direct MSC/NASTRAN FEM solution. Hence, the iFEM accuracy is evaluated by

Table 6.6 Stochastic moments: SD is the standard deviation, SKW is the skewness, KURT is the Kurtosis, w is the maximum deflection, f is the fundamental frequency, σ_{VM} is the maximum Von Mises stress, τ_{xz} and τ_{yz} are the maximum transverse shear stresses

		w	f	σ_{VM}	τ_{yz}	τ_{xz}
MCS	Mean	21.37	434.24	2194.65	28.76	14.01
	Variance	3.36	200.36	13570.68	0.70	0.45
	SD	1.84	14.15	116.49	0.84	0.67
	SKW	0.01	0.04	0.23	0.15	0.16
	KURT	3.05	3.03	3.06	3.01	3.03
UDR	Mean	21.36	434.24	2194.50	28.76	14.01
	Variance	3.35	201.67	13570.68	0.78	0.45
	SD	1.83	14.20	116.49	0.88	0.67
	SKW	0.00	0.00	0.23	0.15	0.16
	KURT	3.03	3.03	3.00	3.02	3.03
2nd PCE	Mean	21.37	434.24	2194.64	28.76	14.01
	Variance	3.34	205.67	13570.68	0.72	0.45
	SD	1.83	14.34	116.49	0.85	0.67
	SKW	0.01	0.00	0.23	0.15	0.16
	KURT	3.03	3.03	3.00	3.02	3.03

Table 6.7 High-fidelity FE discretization of the thin-walled beam problem

Element type (name)	N° of elements along the external circumference	N° of elements along the beam length	Total N° of elements	Total N° of nodes
Shell element (QUAD4)	114	360	41.040	41.156

means of the following error:

$$E \equiv \frac{Value(FEM) - Value(FEM)}{Value(FEM)} \tag{6.29}$$

The aim of the present application is to verify the robustness of the iFEM in evaluating the displacement field when the sensor positions are considered affected by uncertainty. For this purpose, a probabilistic approach is compared with a non-probabilistic method based on the evidence theory. The main issue is the definition of the uncertainty that affects the coordinate values describing the sensors position. In order to obtain this kind of information, three technicians have been interviewed They have given three different estimations of the error in the strain gauge location; all these experts are equally trusted. The second expert (see Table 6.9) defined the errors using disjoint sets.

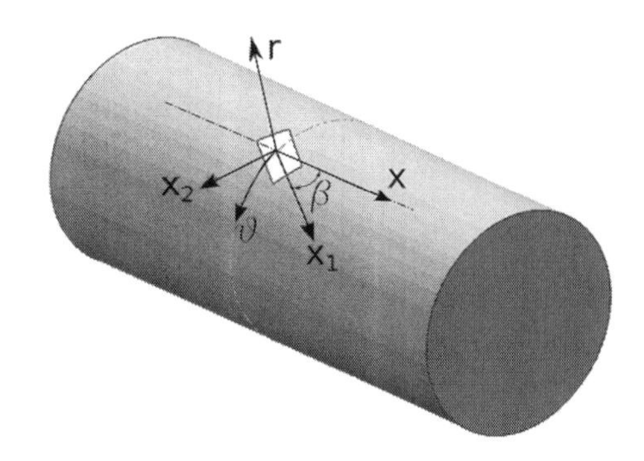

Fig. 6.8 Location of a strain gauge on the beam external surface [20]

Table 6.8 Strain gauge nominal positions

Strain gauges	x	θ	β
1	10	−120	0
2	10	−120	45
3	10	0	0
4	10	0	45
5	10	120	0
6	10	120	45

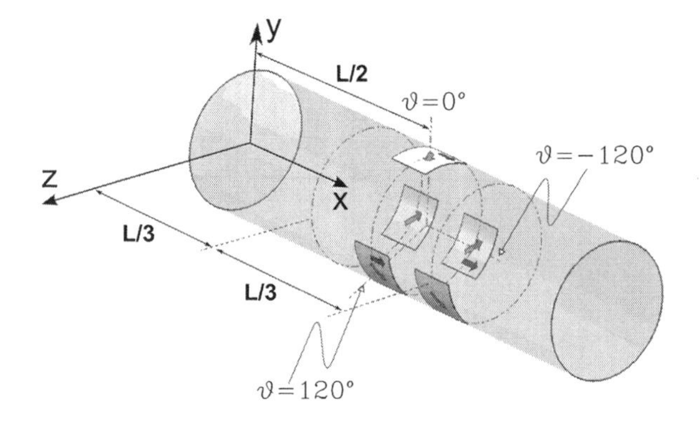

Fig. 6.9 Sensors position [20]

In order to use a probabilistic approach to propagate the uncertainty from input to output, there is the need to transform the input epistemic uncertainty into probabilistic information. Several hypotheses are then needed about the shape of the probability distribution and its standard deviation. In the present example we have assumed that the uncertainty in the sensor position is described by means of a Gaussian distribution, having the standard deviations listed in Table 6.10.

The information obtained by the sensitivity analysis [20] are used to select which input variables should be considered and which could be neglected during the uncertainty propagation process, performed both using the evidence theory and a probabilistic approach (the UDR method, having verified that there are not significant

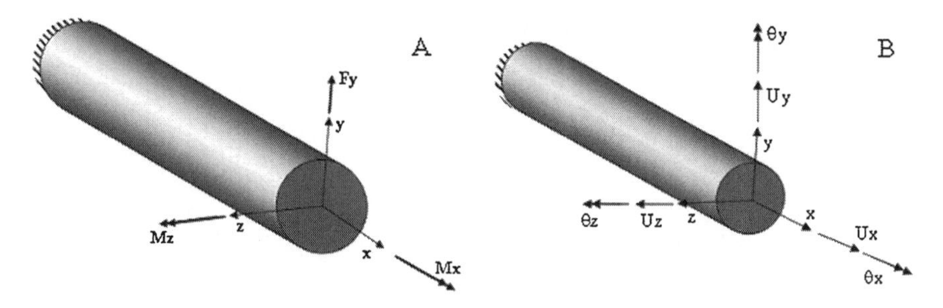

Fig. 6.10 (**A**) Applied load, (**B**) Studied degrees of freedom

Table 6.9 Sensor coordinates defined by means of interval sets

Expert	x [mm]		θ [°]		β [°]	
	LOWER	UPPER	LOWER	UPPER	LOWER	UPPER
1	−5	5	−5	5	−4	4
2	−5	−1	−5	5	−4	−1.5
	1	5	−5	5	1.5	4
3	−1	1	−5	5	−1.5	1.5

Table 6.10 Probabilistic assumptions about sensor position

Input variables	PDF	Standard Deviation
x	Gaussian	0.0233
θ	Gaussian	1.1666
β	Gaussian	1.3333

interactions between variables). Then, once the first four statistical moments are known, the corresponding probability distribution function is evaluated by means of the Pearson System.

The probability that the iFEM error w.r.t. the FEM reference displacements and rotations is greater than a given threshold value, is finally evaluated.

The evidence theory does not furnish a unique measure of the probability, but it gives two different probability curves: the plausibility, that describes the curve of the maximum reliability of the system, and the belief, that describes the minimum reliability of the system (Figs. 6.11, 6.12, and 6.13). According to what was said previously, the true reliability curve is included between the plausibility and the belief. Actually, the area included between the maximum reliability curve and the minimum one represents a region of uncertainty; this means that, without further information, no prediction about the actual behavior of the model can be made (we only know that the true error is included between the two probability curves). For this reason, the belief curve, that represents a conservative estimation of the model behavior, is used during the design phase. In this study we have compared the results

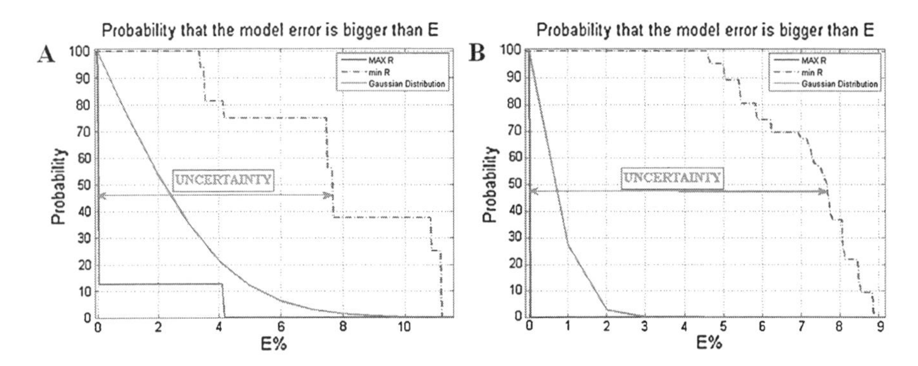

Fig. 6.11 Shear load (Fy): Probability that the error E about u_y (**A**) and θ_z (**B**) is bigger than a given threshold value. *Three curves* are plotted: the *first* one represents the maximum model reliability (labeled with *MAX R*), the *second* one represents the minimum reliability of the model (labeled with *min R*), the *red* one represents the curve obtained using the assumption of the Gaussian distribution

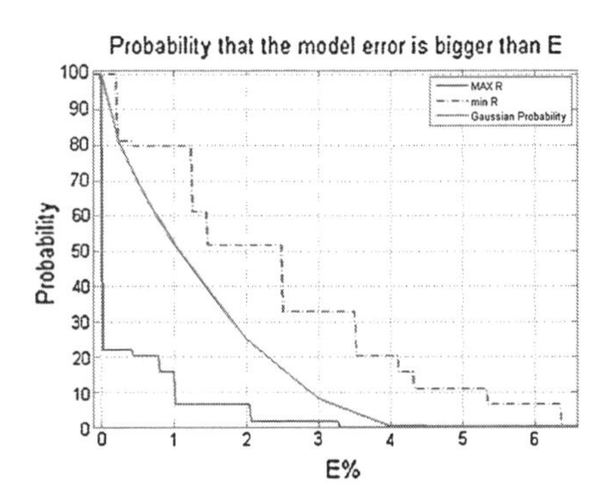

Fig. 6.12 Torque Moment: Probability that the error E about θ_x (**B**) is bigger than a given threshold value. *Three curves* are plotted: the *first* one represents the maximum model reliability (labeled with *MAX R*), the second one represents the minimum reliability of the model (labeled with *min R*), the *red* one represents the curve obtained using the assumption of the Gaussian distribution

obtained assuming that the position error is described by means of Gaussian distributions with those obtained assuming that each sensor is located inside an interval. In this last case no hypothesis has been made about the probability that a sensor is in a given point (inside the region). As shown in Figs. 6.11, 6.12, and 6.13, the maximum reliability curves give almost null prediction errors, whereas the minimum reliability curves indicate bigger probability to have large errors; in particular, the evaluation of the y-displacement and z-rotation is quite sensible to the sensor position uncertainty, Fig. 6.11 and Fig. 6.13. In most cases the reliability curves, based on the Gaussian distribution hypothesis, underestimate considerably the prediction errors. In particular the Gaussian hypothesis furnishes probability values close to those given by the maximum reliability curves.

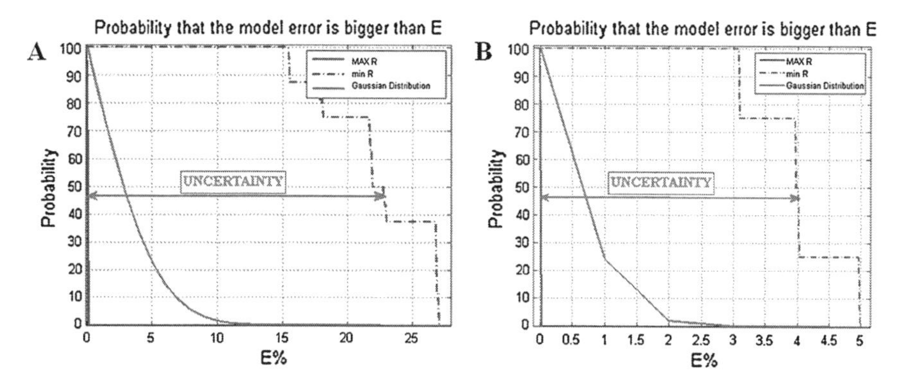

Fig. 6.13 Bending Moment around z: Probability that the error E about u_y (**A**) and θ_z (**B**) is bigger than a given threshold value. *Three curves* are plotted: the *first* one represents the maximum model reliability (labeled with *MAX R*), the second one represents the minimum reliability of the model (labeled with *min R*), the *red* one represents the curve obtained using the assumption of the Gaussian distribution

4 Conclusion

In the present work a comparative study of some uncertainty propagation algorithms is performed and discussed.

Methods for both aleatory and epistemic uncertainty are considered; in particular, a brief review of Univariate Dimension Reduction method (UDR), Polynomial Chaos Expansion (PCE), and First Order Second Moments algorithm (FOSM)—for aleatory uncertainty—and of Evidence Theory—for epistemic uncertainty—is presented.

Then, selected example problems are considered to assess and compare the available methods; some test functions are used as preliminary test cases, then structural applications are studied, ranging from the mechanical behavior of a composite plate to the shape sensing of a beam starting from measured surface strains. As for the latter application, an epistemic uncertainty propagation approach (Evidence Theory) has been compared with a probabilistic uncertainty propagation algorithm (UDR); the considered problem is a classical example of epistemic uncertainty, therefore probabilistic approaches may be applied after introducing some prior assumptions whose correctness may not be guaranteed.

Although this study is limited to some particular examples, interesting general conclusions can be drawn.

If there is no significant interaction between variables, the UDR is the most efficient method for statistical moments estimation. Its accuracy decreases when the interactions cannot be neglected; in particular, the evaluation of the 3rd and 4th statistical moments is more sensitive to the interaction effects and, therefore, also the evaluation of the corresponding Probability Distribution Function (PDF), by means of the Pearson System, can be compromised. The accuracy and the computational cost of the PCE depend on the truncation order of the expansion. However, the PCE is a useful approach when the knowledge of the PDF is desired. Moreover, the

UDR method leads to the best compromise between accuracy and computational cost when performing a probabilistic study of the mechanical behavior of a composite plate.

The transformation of the epistemic knowledge into a probabilistic knowledge could often cause a loss of information and consequently the underestimation of the uncertainty effects. The evidence theory, in the particular case of the shape sensing problem, seems to be a more robust and conservative approach. The use of a probabilistic approach is not wrong but it requires too strong prior assumptions. In other words, the correct use of the probabilistic approach would require the experimental probabilistic characterization of the sensors position.

References

1. Lee, S.H., Chen, W.: A Comparative study of uncertainty propagation methods for black-box-type problems. Struct. Multidiscip. Optim. **37**(3), 239–253 (2009)
2. Cerracchio, P., Gherlone, M., Mattone, M., Di Sciuva, M., Tessler, A.: Shape sensing of three-dimensional frame structures using the inverse finite element method. In: Proc. Fifth European Workshop on Structural Health Monitoring, Sorrento, Italy, pp. 615–620 (2010)
3. Cerracchio, P., Gherlone, M., Mattone, M., Di Sciuva, M., Tessler, A.: Inverse finite element method for three-dimensional frame structures. DIASP Report Politecnico di Torino 285 (2010)
4. Xu, H., Rahman, S.: A moment-based stochastic method for response moment and reliability analysis. In: Proc. Second MIT Conference on Computational Fluid and Solid Mechanics (2003)
5. Rahman, S., Xu, H.: A univariate dimension-reduction method for multi-dimensional integration in stochastic mechanics. Probab. Eng. Mech. **19**(1), 393–408 (2004)
6. Xu, H., Rahman, S.: A generalized dimension-reduction method for multidimensional integration in stochastic mechanics. Int. J. Numer. Methods Eng. **65**(13), 1992–2019 (2004)
7. Wiener, N.: The homogeneous chaos. Am. J. Math. **60**, 897–936 (1938)
8. Cameron, R.H., Martin, W.T.: The orthogonal development of nonlinear functionals in series of Fourier–Hermite functionals. Ann. Math. **48**(2), 385–392 (1947)
9. Xiu, D., Karniadakis, G.M.: The Wiener–Askey polynomial chaos for stochastic differential equations. SIAM J. Sci. Comput. **24**(2), 619–644 (2002)
10. Ticky, M.: Applied Methods of Structural Reliability. Springer, Berlin (1993)
11. Helton, J.C., Johnson, J.D., Oberkampf, W.L., Storli, C.B.: A sampling-based computational strategy for the representation of epistemic uncertainty in model predictions with evidence theory. Sandia report SAND2006-5557, Sandia National Laboratories (2006). http://prod.sandia.gov/techlib/access-control.cgi/2006/065557.pdf
12. Sentz, K., Ferson, S.: Combination of evidence in Dempster–Shafer theory. Sandia report SAND2002-083, Sandia National Laboratories (2002). http://www.sandia.gov/epistemic/Reports/SAND2002-0835.pdf
13. Saltelli, A., Chan, K., Scott, E.M.: Sensitivity Analysis. Wiley, New York (2008)
14. Crestaux, T., Le Maitre, O., Martinez, J.M.: Polynomial chaos expansion for sensitivity analysis. Reliab. Eng. Syst. Saf. **94**(7), 1161–1172 (2009)
15. Tessler, A., Di Sciuva, M., Gherlone, M.: Refined zigzag theory for laminated composite and sandwich plates. NASA/TP-2009-215561, Langley Research Center (2009)
16. Tessler, A., Di Sciuva, M., Gherlone, M.: A consistent refinement of first-order shear-deformation theory for laminated composite and sandwich plates using improved zigzag kinematics. J. Mech. Mater. Struct. **5**, 341–367 (2010). doi:10.2140/jomms.2010.5.341

17. Tessler, A., Di Sciuva, M., Gherlone, M.: A shear-deformation theory for composite and sandwich plates using improved zigzag kinematics. In: IX International Conference on Computational Structural Technology, 2–5 September, Greece, Proceedings on Cd, Paper 30 (2008)
18. Tessler, A., Di Sciuva, M., Gherlone, M.: Refined zigzag theory for homogeneous, laminate composite, and sandwich plates: a homogeneous limit methodology for zigzag function selection. Numer. Methods Partial Differ. Equ. **27**(1), 208–229 (2011)
19. Tessler, A., Spangler, J.L.: A least-squares variational method for full-field reconstruction of elastic deformations in shear-deformable plates and shells. Comput. Methods Appl. Mech. Eng. **94**, 327–339 (2005)
20. Corradi, M.: Uncertainty management techniques for aerospace structural design. Ph.D. thesis, Politecnico di Torino (2011)

Chapter 7
Fuzzy and Fuzzy Stochastic Methods for the Numerical Analysis of Reinforced Concrete Structures Under Dynamical Loading

Frank Steinigen, Jan-Uwe Sickert, Wolfgang Graf, and Michael Kaliske

Abstract This paper is mainly devoted to enhanced computational algorithms to simulate the load-bearing behavior of reinforced concrete structures under dynamical loading. In order to take into account uncertain data of reinforced concrete, fuzzy and fuzzy stochastic analyses are presented. The capability of the fuzzy dynamical analysis is demonstrated by an example in which a steel bracing system and viscous damping connectors are designed to enhance the structural resistance of a reinforced concrete structure under seismic loading.

1 Introduction

The numerical analysis of reinforced concrete (RC) structures under dynamical loads requires realistic nonlinear structural models and computational algorithms. Furthermore, the engineer/designer has to deal with uncertainty which results from variations in material parameters as well as incomplete knowledge about further excitations and the quality of the numerical model itself. The variations in material parameters may be assessed by the uncertainty measure probability. However, the stochastic model cannot be determined precisely because of rare information in most cases. Therefore, an imprecise probability approach is suggested in this contribution which is based on the uncertainty measure fuzzy probability resulting in a set of probability models assessed by membership values. Using this approach, input variables may be also modeled as fuzzy quantities and considered as a special case, if only subjective or linguistic assessments are available.

F. Steinigen (✉) · J.-U. Sickert · W. Graf · M. Kaliske
Institute for Structural Analysis, TU Dresden, Dresden 01062, Germany
e-mail: frank.steinigen@tu-dresden.de

J.-U. Sickert
e-mail: jan-uwe.sickert@tu-dresden.de

W. Graf
e-mail: wolfgang.graf@tu-dresden.de

M. Kaliske
e-mail: michael.kaliske@tu-dresden.de

M. Papadrakakis et al. (eds.), *Computational Methods in Stochastic Dynamics*, Computational Methods in Applied Sciences 26,
DOI 10.1007/978-94-007-5134-7_7, © Springer Science+Business Media Dordrecht 2013

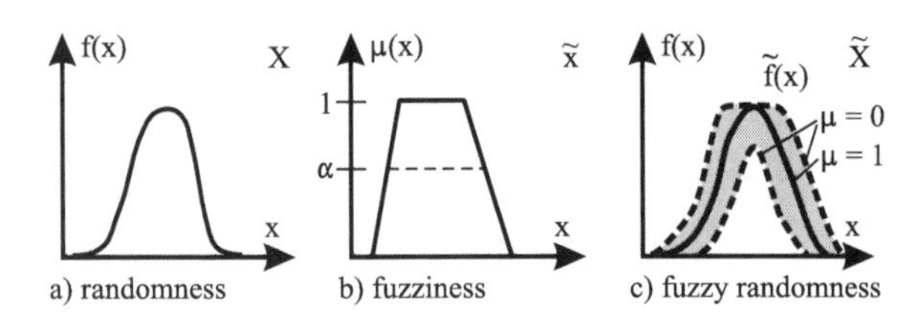

Fig. 7.1 Models of uncertainty

The incorporation of uncertain input variables within a dynamic structural analysis leads to uncertain structural responses representing the uncertain structural performance close to reality. The uncertain structural responses are determined using fuzzy stochastic dynamic analyses. Thereby, fuzzy stochastic processes result due to the specific dynamic actions. The fuzzy stochastic structural analysis of practical problems requires high performance computational tools in order to deal with a large number of uncertain input variables as well as complex nonlinear structural models. An efficient approach is introduced which is based on α-level optimization and Monte-Carlo simulation using meta-models which partly replace the dynamic structural analyses.

2 Uncertainty in Structural Dynamics

2.1 Data Models

The input variables—for geometry, material, load etc.—of the numerical simulations of structural behavior are generally uncertain. In order to describe this uncertainty, traditional stochastic and non-stochastic models are available [8]. In Fig. 7.1, the models randomness, fuzziness and fuzzy randomness are displayed. The choice of the model depends on the available data. If sufficient statistical data exist for a parameter and the reproduction conditions are constant, the parameter may be described stochastically. Thereby, the choice of type of probability distribution function affects the result considerably.

Overcoming the traditional probabilistic uncertainty model enables the suitable consideration of imprecision (epistemic uncertainty). Thereby, epistemic uncertainty is associated with human cognition, which is not limited to a binary measure. Advanced uncertainty concepts allow a gradual assessment of intervals. This extension can be realized with the uncertainty characteristic fuzziness. The combination of fuzziness and probabilistic leads to the generalized model fuzzy randomness.

2.1.1 Fuzzy Variables

Often, the uncertainty description for parameters is based on pure expert judgment or samples which are not validated statistically. Then, the description by the uncertainty model fuzziness is recommended. The model comprehends both objective and subjective information. The uncertain parameters are characterized with the aid of a membership function $\mu(x)$ (see Fig. 7.1b and Eq. (7.1)). The membership function $\mu_x(x)$ assesses the gradual membership of elements to a set. Fuzzy variables

$$\tilde{x} = \{(x; \mu_x(x)) \mid x \in X\}; \quad \mu_x(x) \geq 0 \ \forall x \in X \tag{7.1}$$

may be utilized to describe the imprecision of structural parameters directly as well as to specify the parameters of fuzzy random variables.

2.1.2 Fuzzy Random Variables

If, e.g. reproduction conditions vary during the period of observation or if expert knowledge completes the statistical description of data, an adequate uncertainty quantification succeeds with fuzzy random variables. The theory of fuzzy random variables is based on the uncertain data model fuzzy randomness representing a generalized model due to the combination of stochastic and non-stochastic characteristics. A fuzzy random variable $\tilde{X}$ is defined as the fuzzy set of their originals, whereby each original is a real-valued random variable X.

The representation of fuzzy random variables presented in this paper is based on [13]. The space of the random elementary events Ω is introduced. Here, e.g. the measurement of a structural parameter may be an elementary event ω. Each elementary event $\omega \in \Omega$ generates not only a crisp realization but a fuzzy realization $\tilde{x}(\omega) = \tilde{x}$, in which $\tilde{x}$ is an element of the set $\mathbf{F}(\mathbb{R})$ of all fuzzy variables on $\mathbb{R}$. Each fuzzy variable is defined as a convex, normalized fuzzy set, whose membership function $\mu_x(x)$ is at least segmentally continuous. Accordingly, a fuzzy random variable $\tilde{X}$ is a fuzzy result of the mapping given by

$$\tilde{X} : \Omega \mapsto \mathbf{F}(\mathbb{R}). \tag{7.2}$$

Based on this formal definition, a fuzzy random variable is described by its fuzzy cumulative distribution function (fuzzy cdf) $\tilde{F}(x)$. The function $\tilde{F}(x)$ is defined as the set of real-valued cumulative distribution functions F(x) which are gradually assessed by the membership $\mu_F(F(x))$. F(x) is the cdf of the original X and is referred to as trajectory of $\tilde{F}(x)$. As result, a fuzzy functional value $\tilde{F}(x_i)$ belongs to each value x_i (see Fig. 7.2). Thus, $\tilde{F}(x)$ represents a fuzzy function as defined in Sect. 2.2.1. A fuzzy probability density function

$$\tilde{f}(x) = \{(f(x); \mu_f(f(x))) \mid f \in \mathbf{f}\}; \quad \mu_f(f(x)) \geq 0 \ \forall f \in \mathbf{f} \tag{7.3}$$

is defined accordingly. In that, **f** represents the set of all probability density functions defined on X.

2.2 Uncertain Functions and Processes

2.2.1 Fuzzy Function

In case that fuzzy parameters depend on crisp or uncertain conditions, they are modeled as fuzzy functions $\tilde{x}(\underline{\tilde{t}}) = \tilde{x}(\underline{\tilde{\theta}}, \tilde{\tau}, \underline{\tilde{\varphi}})$ or in the special case of pure time depen-

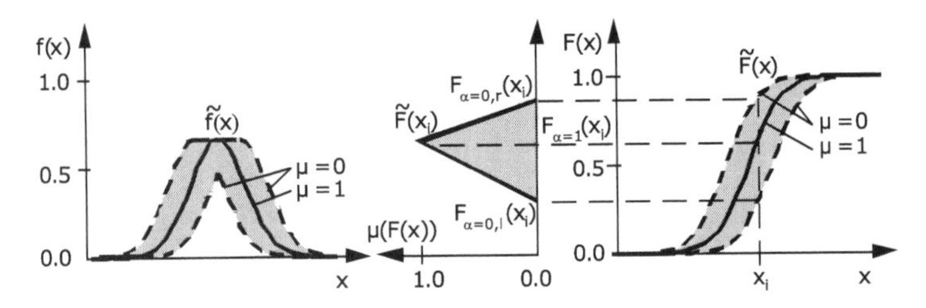

Fig. 7.2 Fuzzy probability density and cumulative distribution function

Fig. 7.3 Fuzzy process $\tilde{x}(\tau)$

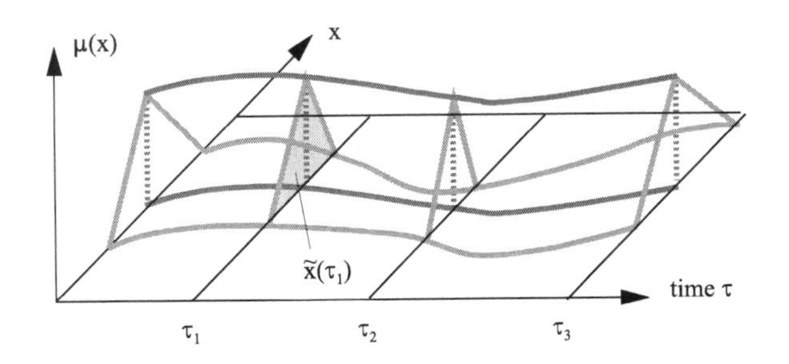

dency as fuzzy processes $\tilde{x}(\tilde{\tau})$. Arguments may be the time $\tilde{\tau}$, the spatial coordinates $\underline{\tilde{\theta}}$ and further parameters $\tilde{\varphi}$, e.g. temperature. A fuzzy function $\tilde{x}(\underline{\tilde{t}})$ enables the formal description of at least piecewise continuous uncertain structural parameters in $\mathbb{R}$. In the following, a definition of fuzzy functions is introduced. Given are

- the fundamental sets $\mathbf{T} \subseteq \mathbb{R}$ and $\mathbf{X} \subseteq \mathbb{R}$,
- the set $\mathbf{F}(\mathbf{T})$ of all fuzzy variables $\underline{\tilde{t}}$ on the fundamental set $\mathbf{T}$,
- the set $\mathbf{F}(\mathbf{X})$ of all fuzzy variables $\tilde{x}$ on the fundamental set $\mathbf{X}$.

Then, the uncertain mapping of $\mathbf{F}(\mathbf{T})$ into $\mathbf{F}(\mathbf{X})$ that assigns exactly one $\tilde{x} \in \mathbf{F}(\mathbf{X})$ to each $\underline{\tilde{t}} \in \mathbf{F}(\mathbf{T})$ is referred to as a fuzzy function denoted by

$$\tilde{x}(\underline{\tilde{t}}) : \mathbf{F}(\mathbf{T}) \mapsto \tilde{\to} \mathbf{F}(\mathbf{X}), \tag{7.4}$$

$$\tilde{x}(\underline{\tilde{t}}) = \left\{ \tilde{x}_t = \tilde{x}(\underline{\tilde{t}}) \; \forall \underline{\tilde{t}} \mid \underline{\tilde{t}} \in \mathbf{F}(\mathbf{T}) \right\}. \tag{7.5}$$

In Fig. 7.3, a fuzzy process $\tilde{x}(\tau)$ is presented, which assigns a fuzzy quantity $\tilde{x}(\tau_i)$ to each time τ_i. For the numerical simulation, a bunch parameter representation of a fuzzy function

$$x(\underline{\tilde{s}}, \underline{t}) = \tilde{x}(\underline{\tilde{t}}) \tag{7.6}$$

is applied. Therewith, the fuzziness of both $\tilde{x}$ and $\tilde{t}$ is concentrated in the bunch parameter vector $\underline{\tilde{s}}$.

For each crisp bunch parameter vector $\underline{s} \in \underline{\tilde{s}}$ with the assigned membership value $\mu(\underline{s})$, a crisp function $x(\underline{t}) = x(\underline{s}, \underline{t}) \in \tilde{x}(\underline{t})$ with $\mu(x(\underline{t})) = \mu(\underline{s})$ is obtained. The fuzzy function

$$\tilde{x}(\underline{t}) = \tilde{x}(\underline{\tilde{s}}, \underline{t}) = \left\{ \left(x(\underline{s}, \underline{t}), \mu\left(x(\underline{s}, \underline{t})\right) \right) \mid \mu\left(x(\underline{s}, \underline{t})\right) = \mu(\underline{s}) \; \forall \underline{s} \mid \underline{s} \in \underline{\tilde{s}} \right\} \tag{7.7}$$

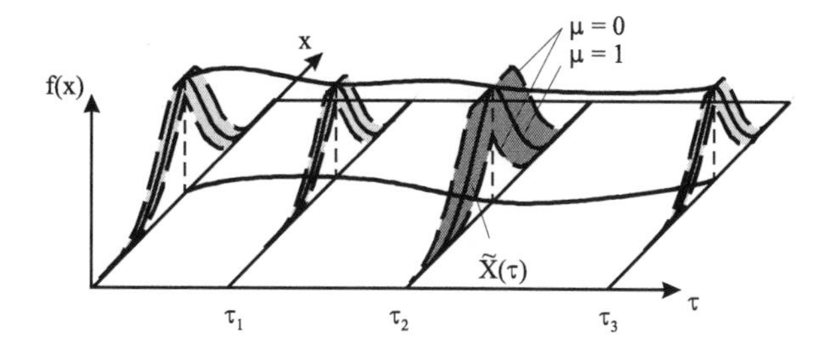

Fig. 7.4 Fuzzy random process $\tilde{X}(\underline{\theta}_j, \tau)$

may thus be represented by the fuzzy set of all real valued functions $x(\underline{s}, \underline{t})$ which may be generated from all possible real vectors $\underline{s} \in \underline{\tilde{s}}$. For every $\underline{t} \in \mathbf{T}$, each of the crisp functions $x(\underline{s}, \underline{t})$ yields values x_t which are contained in the associated fuzzy functional values $\tilde{x}_t$. The real functions $x(\underline{s}, \underline{t})$ of $\tilde{x}(\underline{t})$ are referred to as trajectories. Numerical processing of fuzzy functions $\tilde{x}(\underline{t}) = x(\underline{\tilde{s}}, \underline{t})$ demands the discretization of their arguments $\underline{t}$ in space and time.

2.2.2 Fuzzy Random Function

According to Eqs. (7.2) and (7.4), as well as Fig. 7.4, a fuzzy random function is the result of an uncertain mapping

$$\tilde{X}(\underline{t}) : \mathbf{F}(\mathbf{T}) \times \Omega \rightarrow \mathbf{F}(\mathbb{R}). \tag{7.8}$$

Thereby, $\mathbf{F}(\mathbf{X})$ and $\mathbf{F}(\mathbf{T})$ denote the sets of all fuzzy variables in $\mathbf{X}$ and $\mathbf{T}$ respectively [15, 16]. At a specific point $\underline{t}$, the mapping of Eq. (7.8) leads to the fuzzy random variable $\tilde{X}_t = \tilde{X}(\underline{t})$. Therefore, fuzzy random functions are defined as a family of fuzzy random variables

$$\tilde{X}(\underline{t}) = \left\{ \tilde{X}_t = \tilde{X}(\underline{t})\ \forall \underline{t} \mid \underline{t} \in \mathbf{T} \right\}. \tag{7.9}$$

For the numerical simulation, again the bunch parameter representation of a fuzzy random function is applied. For each crisp bunch parameter vector $\underline{s} \in \underline{\tilde{s}}$ with the assigned membership value $\mu(\underline{s})$, a real random function $X(\underline{t}) = X(\underline{s}, \underline{t}) \in \tilde{X}(\underline{t})$ with $\mu(X(\underline{t})) = \mu(\underline{s})$ is obtained. The fuzzy random function $\tilde{X}(\underline{t})$ may thus be represented by the fuzzy set of all real random functions $X(\underline{t}) \in \tilde{X}(\underline{t})$

$$X(\underline{\tilde{s}}, \underline{t}) = \left\{ \left(X(\underline{t}), \mu\left(X(\underline{t})\right)\right) \mid X(\underline{t}) = X(\underline{s}, \underline{t});\ \mu\left(X(\underline{t})\right) = \mu(\underline{s})\ \forall \underline{s} \mid \underline{s} \in \underline{\tilde{s}} \right\} \tag{7.10}$$

which may be generated from all possible real vectors $\underline{s} \in \underline{\tilde{s}}$. The real random function $X(\underline{t}) \in \tilde{X}(\underline{t})$ is defined for all $\underline{t} \in \mathbf{T}$ and referred to as original function. A numerical processing of a fuzzy random function $\tilde{X}(\underline{t}) = X(\underline{\tilde{s}}, \underline{t})$ requires the discretization of their arguments $\underline{t}$ in space and time.

3 Fuzzy Stochastic Analysis

Fuzzy stochastic analysis is an appropriate computational approach for processing uncertain data using the uncertainty model fuzzy randomness. Basic terms and def-

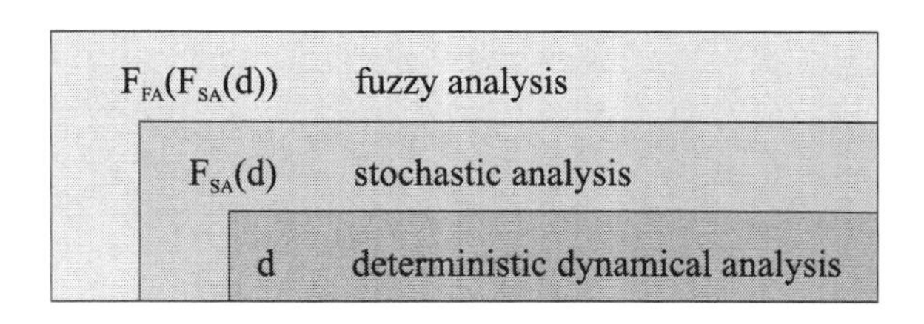

Fig. 7.5 Fuzzy stochastic analysis (FSA)

initions related to fuzzy randomness have been introduced, inter alia, by [13]. The formal description of fuzzy randomness chosen by [13] is however not suitable to formulating uncertainty encountered in engineering tasks. A suitable form of representation with the scope of numerical engineering tasks is given with the so-called α-discretization by [7] and [8].

The numerical simulation under consideration of fuzzy variables and fuzzy functions (fuzzy analysis) may formally be described by the mapping

$$F_{FA}(d) : \tilde{\underline{x}}(\underline{t}) \mapsto \tilde{\underline{z}}(\underline{t}). \tag{7.11}$$

According to Eq. (7.11), the fuzzy variables $\tilde{\underline{x}}$ and the fuzzy functions $\tilde{\underline{x}}(\underline{t})$ are mapped onto the fuzzy results $\tilde{\underline{z}}(\underline{t})$ with aid of the crisp analysis algorithm d. Every arbitrary deterministic fundamental solution may be used as algorithm d. On the basis of space and time discretization, fuzzy functional values $x(\tilde{\underline{s}}, \underline{\theta}_j, \tau_i, \underline{\varphi}_k)$ of the function $x(\tilde{\underline{s}}, \underline{\theta}, \tau, \underline{\varphi})$ are determined at points in space $\underline{\theta}_j$, time τ_i, and a realization of $\underline{\varphi}_k$.

The numerical simulation is carried out with the aid of the α-level optimization [7]. For fuzzy variables $\tilde{x}$ and fuzzy bunch parameter $\tilde{\underline{s}}$ of the fuzzy functions $x(\tilde{\underline{s}}, \underline{\theta}, \tau, \underline{\varphi})$, an input subspace $\underline{E}_\alpha$ is formed and assigned to the level α. By multiple application of the deterministic analysis, the extreme values $z_{\alpha,l}(\underline{\theta}_j, \tau_i, \underline{\varphi}_k)$ and $z_{\alpha,r}(\underline{\theta}_j, \tau_i, \underline{\varphi}_k)$ of the fuzzy result variable $\tilde{z}(\underline{\theta}_j, \tau_i, \underline{\varphi}_k)$ are computed. These points are interval bounds of the α-level sets and enable the numerical description of the convex membership function of the fuzzy result variable $\tilde{z}(\underline{\theta}_j, \tau_i, \underline{\varphi}_k)$. For the computation of $\tilde{z}(\underline{\theta}_j, \tau_{i+1}, \underline{\varphi}_k)$ at the time point τ_{i+1}, the procedure must be restarted at $\tau = 0$ due to the interaction within the mapping model.

Fuzzy stochastic analysis allows the mapping of fuzzy random input variables onto fuzzy random result variables. The fuzzy stochastic analysis can be applied for static and dynamic structural analysis and for assessment of structural safety, durability as well as robustness. Two different approaches for computation of the fuzzy random result variables have been developed. The first variant (Fig. 7.5) bases on the bunch parameter representation of fuzzy random variables by [16]. The second variant utilizes the $l_\alpha r_\alpha$-representation of fuzzy random variables. The variant to be preferred depends on the engineering task, the available uncertain data and the aspired results [10].

The fuzzy stochastic analysis is called fuzzy stochastic finite element method (FSFEM), if the deterministic dynamical analysis is based on a finite element (FE) model.

4 Deterministic Dynamical Analysis of RC Structures

4.1 1D-Beams

Plane and spatial beam structures are called 1D-structures. For the physical nonlinear analysis, the cross-sections of the beams are subdivided into layers (plane structures) or fibers (spatial structures). In contrast to the widespread finite element formulations, solutions based on the differential equations for the straight or imperfectly straight beam also exist. A respective approach for plane beam structures is presented here.

The geometrical and physical nonlinear analysis of plane reinforced concrete, prestressed concrete, and steel beam structures is chosen as fundamental model [14]. The beams are subdivided into integration sections, the cross-sections are subdivided into layers. On this basis, an incrementally formulated system of second order differential equations for the straight or imperfectly straight beam is solved

$$\left[\frac{d\Delta\underline{z}(\theta_1)}{d\theta_1}\right]_{(n)}^{[k]} = \underline{A}(\theta_1, \underline{z})_{(n-1)} \cdot \Delta\underline{z}(\theta_1)_{(n)}^{[k]} + \Delta\underline{b}(\theta_1)_{(n)}^{[k-1]}$$
$$+ \underline{d}(\theta_1)_{(n-1)} \cdot \Delta\dot{\underline{z}}(\theta_1)_{(n)}^{[k]} + \underline{m}(\theta_1)_{(n-1)} \cdot \Delta\ddot{\underline{z}}(\theta_1)_{(n)}^{[k]} \quad (7.12)$$

where $[k]$—counter of iteration steps; (n)—counter of increments; θ_1—bar coordinate; Δ—increment; $\underline{z} = \{\underline{z}_1, \underline{z}_2\} = \{uv\phi; NQM\}$—vector of structural responses; $\underline{A}$—matrix of coefficients (constant within the increment); $\underline{b}$—"right hand side" of the system of differential equations with loads and varying parts resulting from geometrical and physical nonlinearities as well as with forces from unbonded prestressing; $\underline{d}$—damping matrix; and $\underline{m}$—mass matrix.

The implicit nonlinear system of differential equations for the differential beam sections is linearized by increments. All geometrically and physically nonlinear components in the $\Delta\underline{b}$-vector are recalculated after every iteration step, and the $\underline{A}$-, $\underline{d}$-, and $\underline{m}$-matrix are recalculated after the completion of the iteration within the increment.

The solution of the system of differential equations by a Runge–Kutta integration results in the system of differential equations

$$\underline{K}_{T(n-1)} \cdot \Delta\underline{v}_{(n)}^{[k]} + \underline{D}_{T(n-1)} \cdot \Delta\dot{\underline{v}}_{(n)}^{[k]} + \underline{M}_{T(n-1)} \cdot \Delta\ddot{\underline{v}}_{(n)}^{[k]}$$
$$= \Delta\underline{P}_{(n)} - \Delta\overset{o}{\underline{F}}{}_{(n)}^{[k]} + \Delta\Delta\underline{F}_{(n-1)} \quad (7.13)$$

of the unknown incremental displacements $\Delta\underline{v}$, velocities $\Delta\dot{\underline{v}}$, and accelerations $\Delta\ddot{\underline{v}}$ of the nodes.

4.2 2D-Folded Plate RC Structures

Shells, folded plates, shear panels and plates are called 2D-structures. Here, we focus on folded plates which represent the general case for plane 2D structures. They

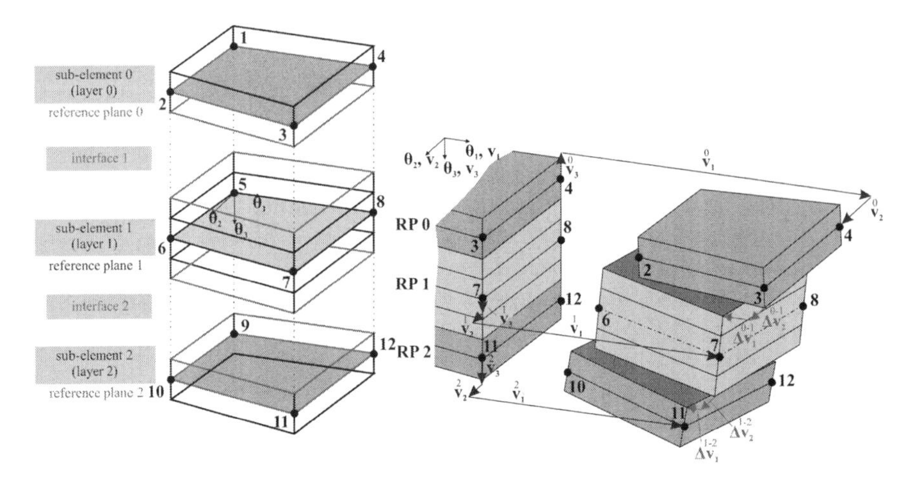

Fig. 7.6 MRM discretization and kinematics

can further be applied to approximate the shape of slightly curved structures. The internal forces are related to the reference plane, which is not stringently the midplane. The cross-section is subdivided into layers to describe the physical nonlinear behavior of reinforced concrete. Over the past years, a new strengthening technology for damaged RC structures has been developed. The thin strengthening layers consist of fine-grained concrete reinforced with textiles made of AR-glass or carbon, see e.g. [1] and [18]. The classical layered model with one reference plane for folded plate structures is enhanced to take into account the later applied strengthening layers.

An extended layer model with specific kinematics, the so-called multi-reference-plane model (MRM), is used to describe the load-bearing behavior of RC constructions with textile strengthening. The MRM consists of concrete layers and steel reinforcement layers of the old construction, the strengthening layers comprised of the inhomogeneous material textile concrete (TRC), and the interface layers (Fig. 7.6). This multilayer continuum has the following kinematic peculiarities. Due to the fact that the modification of the concrete layer thickness is very small and can be neglected, we have $\varepsilon_{33} = 0$. Furthermore, the transverse shear stresses in the concrete layers have no significant influence on the deformation, which means that $\varepsilon_{13} = 0$ and $\varepsilon_{23} = 0$ can be set to zero. The deformation state of the concrete layers may be described by Kirchhoff kinematics. The independent degrees of freedom are assigned to a reference plane which can be selected arbitrarily.

The very thin strengthening layers are subject to the same kinematic assumptions. Kirchhoff kinematic with a reference plane is also assigned to each strengthening layer. The independent degrees of freedom of the strengthening layer lie in the reference plane. The bond between the layers of reinforced concrete and an arbitrary strengthening layer is modeled by an interface. The interface is an immaterial layer of zero thickness. The bonding state is assessed with the help of the relative displacements Δv_1, Δv_2, Δv_3 between the contact surfaces. In conjunction with a bonding matrix, the relative displacements enable assumptions regarding delamination and shear failure.

The FE discretization of the MRM is based on the functional of the complementary energy extended by the static transition conditions $\Delta\underline{\tilde{p}} - \overset{+}{\underline{\tilde{p}}} = 0$ to O_p^r and the equilibrium conditions $\underline{\tilde{G}} \cdot \underline{\sigma}_{el}^e + \overset{+}{\underline{p}}{}^e - \rho^e \cdot \underline{\ddot{v}}^e = 0$ in V^e

$$\Pi_{mh} = \int_{\tau_1}^{\tau_2} \sum_{e=1}^{n} \Bigg\{ \int_{V^e} \left[w_c(\underline{\sigma}_{el}^e) + (\underline{\tilde{G}} \cdot \underline{\sigma}_{el}^e + \overset{+}{\underline{p}}{}^e - \rho^e \cdot \underline{\ddot{v}}^e)^T \cdot \underline{v}^e \right] dV + \int_{V^e} (\underline{\sigma}_{el}^e)^T \cdot \underline{\varepsilon}_0^e \, dV - \int_{O_p^{r,e}} (\underline{p}^{r,e} - \overset{+}{\underline{\tilde{p}}}{}^{r,e})^T \cdot \underline{v}^{r,e} \, dO - \int_{O_v^{r,e}} (\underline{p}^{r,e})^T \cdot \overset{+}{\underline{\tilde{v}}}{}^{r,e} \, dO \Bigg\} d\tau \tag{7.14}$$

with $w_c(\underline{\sigma}_{el}^e)$—internal complementary energy; $\underline{\tilde{G}}$—matrix of differential operators; $\overset{+}{\underline{\tilde{p}}}$—external forces in V^e; ρ^e—density in; $\underline{\ddot{v}}^e$—internal acceleration in V^e; ρ^e; $\underline{\varepsilon}_0^e$—initial strain; $\underline{p}^{r,e}$—internal forces in the boundary surface $O_p^{r,e}$; $\overset{+}{\underline{\tilde{p}}}{}^{r,e}$—external forces along the boundary surface $O_p^{r,e}$; $\underline{v}^{r,e}$—displacements of the boundary surface $O_p^{r,e}$; $\overset{+}{\underline{\tilde{v}}}{}^{r,e}$—prescribed displacements of the boundary surface $O_v^{r,e}$; τ—time.

After some transformations, the quasi-static part of the equilibrium conditions $(\underline{\tilde{G}} \cdot \underline{\sigma}_{el}^e + \overset{+}{\underline{p}}{}^e)$ and the kinetic energy become visible in the mixed hybrid functional

$$\Pi_{mh} = \int_{\tau_1}^{\tau_2} \sum_{e=1}^{n} \Bigg\{ \int_{V^e} \left[w_c(\underline{\sigma}_{el}^e) + (\underline{\tilde{G}} \cdot \underline{\sigma}_{el}^e + \overset{+}{\underline{p}}{}^e)^T \cdot \underline{v}^e + \frac{1}{2}\rho^e \cdot (\underline{\dot{v}}^e)^T \cdot \underline{\dot{v}}^e \right] dV + \int_{V^e} (\underline{\sigma}_{el}^e)^T \cdot \underline{\varepsilon}_0^e \, dV - \int_{O_p^{r,e}} (\underline{p}^{r,e} - \overset{+}{\underline{\tilde{p}}}{}^{r,e})^T \cdot \underline{v}^{r,e} \, dO - \int_{O_v^{r,e}} (\underline{p}^{r,e})^T \cdot \overset{+}{\underline{\tilde{v}}}{}^{r,e} \, dO \Bigg\} d\tau. \tag{7.15}$$

In extension to the static case [9, 18], this functional may be applied to a layered continuum with dynamic loads. Following the procedure described in [9, 18], the steady-state condition of the mixed hybrid functional

$$\delta\Pi_{mh,NC} = \frac{1}{2}\delta\left(d^2\Pi_{mh}\right) = \sum_{i=0}^{k} \delta\left(d^2\left({}^{(R_i)}\Pi_{mh,NC}\right)\right) + \sum_{j=1}^{k} \delta\left(d^2\left({}^{(I_j)}\Pi_{mh,NC}\right)\right) = 0 \tag{7.16}$$

with

$$
\begin{aligned}
{}^{(R_i)}\Pi_{mh,NC} = \int_{\tau_1}^{\tau_2} \Bigg\{ \sum_{e_i=1}^{n} \Bigg(\frac{1}{2} \sum_{m=0}^{s_i-1} \Bigg[\int_{V^{e_i,m}} \left(d\underline{\sigma}_{el}^{e_i,m}\right)^T \cdot d\underline{\varepsilon}_{el}^{e_i,m}\, dV \\
+ 2 \int_{V^{e_i,m}} \left(\underline{\tilde{G}} \cdot d\underline{\sigma}_{el}^{e_i,m} + d\,\overset{+}{\underline{\tilde{p}}}{}^{e_i,m}\right)^T \cdot d\underline{v}^{e_i,m}\, dV \\
+ \int_{V^{e_i,m}} \rho^{e_i,m} \cdot \left(d\underline{\dot{v}}^{e_i,m}\right)^T \cdot d\underline{\dot{v}}^{e_i,m}\, dV \\
+ 2 \int_{V^{e_i,m}} \left(d\underline{\sigma}_{el}^{e_i,m}\right)^T \cdot d\underline{\varepsilon}_{0}^{e_i,m}\, dV \Bigg] \\
- \int_{{}^{(R_i)}O_p^{r,e_i}} \left(d\underline{p}^{r,e_i} - d\,\overset{+}{\underline{\tilde{p}}}{}^{r,e_i}\right)^T \cdot d\underline{v}^{r,e_i}\, dO \\
- \int_{{}^{(R_i)}O_v^{r,e_i}} \left(d\underline{p}^{r,e_i}\right)^T \cdot d\,\overset{+}{\underline{\tilde{v}}}{}^{r,e_i}\, dO \Bigg) \Bigg\} d\tau
\end{aligned}
\tag{7.17}
$$

$$
\begin{aligned}
{}^{(I_j)}\Pi_{mh,NC} = \int_{\tau_1}^{\tau_2} \Bigg\{ \sum_{e_j=1}^{n} \int_{{}^{(I_j)}O_p^{e_j}} \left(d^{I}\underline{\sigma}^{e_j}\right)^T \\
\cdot \left({}^{(j|j)}d\underline{v}^{r,e_j} - {}^{(j-1|j)}d\underline{v}^{r,e_{j-1}}\right) dO \Bigg\} d\tau
\end{aligned}
\tag{7.18}
$$

for a layered continuum with k layers is obtained from Eq. (7.15). Equation (7.17) describes the functional for the sub-element R_i whereas Eq. (7.18) depicts the functional for the interface I_j. Compared to [9, 18] Eqs. (7.16), (7.17) and (7.18) are extended by inertial forces. In order to account for physical nonlinearities of the layered continuum, the layer i with the reference plane R_i is subdivided into s_i sub-layers in Eq. (7.17).

On the basis of Eq. (7.16), the differential equation of motion can be derived. Thereby, the same stress shape functions, the same boundary displacement shape functions and the same element displacement shape function are chosen within all layers of the continuum. The stress shape functions are chosen in such a way, that they fulfill strongly the quasi-static part of the equilibrium conditions

$$
\underline{\tilde{G}} \cdot d\underline{\sigma}_{el}^{e_i,m} + d\,\overset{+}{\underline{\tilde{p}}}{}^{e_i,m} = 0. \tag{7.19}
$$

The evaluation of the steady-state condition, Eq. (7.16) yields the MRM element and leads to the differential equation of motion

$$
\underline{\tilde{K}}_T \cdot d\underline{\tilde{q}} + \underline{\tilde{M}} \cdot d\underline{\ddot{\tilde{q}}} - d\underline{\tilde{R}} - d\underline{R}_K = 0 \tag{7.20}
$$

with $\underline{\tilde{K}}_T$—tangential system stiffness matrix, $\underline{\tilde{M}}$—system mass matrix, $d\underline{\tilde{R}}$, $d\underline{R}_K$—differential load contributions, and $\underline{\tilde{q}}$—nodal displacement degrees of freedom. The matrix $\underline{\tilde{K}}_T$ and the vectors $d\underline{\tilde{R}}$, and $d\underline{R}_K$ are identical to the corresponding values of the hybrid procedure in [9]. The system mass matrix $\underline{\tilde{M}}$ is specified in [19].

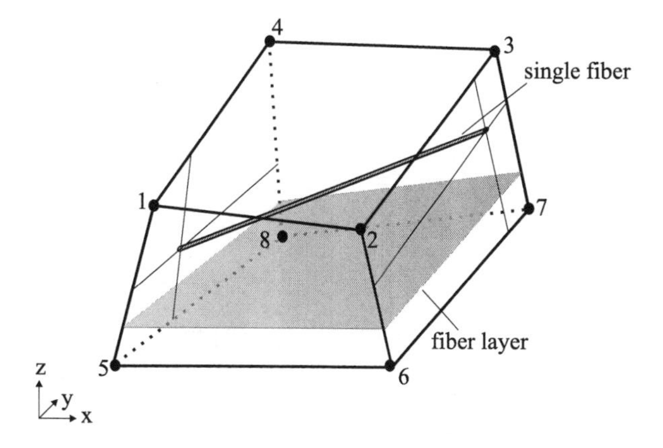

Fig. 7.7 Eight-node solid element with embedded reinforcements

4.3 3D-Compact RC Structures

Hybrid eight-node hexagonal solid elements for the physical linear static analysis are described in [12]. For the physically nonlinear analysis of reinforced concrete and textile reinforced concrete (TRC), respectively, two kinds of reinforcement are introduced—single fibers and fiber layers (see Fig. 7.7). The formulation of the hybrid eight-node hexagonal solid element with embedded (textile) reinforcement is outlined in the following.

Starting point is the functional of Hellinger–Reissner

$$\Pi_{HR} = \int_V \left(\underline{\sigma}^T \cdot (\underline{G} \cdot \underline{v}) - \frac{1}{2}\underline{\sigma}^T \cdot \underline{\varepsilon} - \overset{+}{\underline{p}}{}_V^T \cdot \underline{v} \right) dV - \int_{O_p} \overset{+}{\underline{p}}{}^T \cdot \underline{v}\, dO \tag{7.21}$$

with $\underline{\sigma}$, $\underline{\varepsilon}$, $\underline{v}$—stresses, strains and displacements in the volume V, $\overset{+}{\underline{p}}{}_V^T$—external forces in V and $\overset{+}{\underline{p}}{}^T$—external forces along the boundary surface O_p, and the matrix of differential operators $\underline{G}$.

Based on it, the Hamilton functional is build

$$H = \delta \int_{\tau_1}^{\tau_2} (K - \Pi_{HR})\, d\tau = \delta \int_{\tau_1}^{\tau_2} \left(\frac{1}{2} \int_V \rho \cdot (\dot{v})^T \cdot \dot{v}\, dV - \Pi_{HR} \right) d\tau \tag{7.22}$$

with the kinetic energy K.

The physical nonlinear analysis of reinforced concrete is a non-conservative problem arising e.g. from crack formation, nonlinear material behavior, bonding and damage. In order to solve this non-conservative problem, a differential load change is considered. Under such load change, the existence of a potential is assumed. The differential load change leads to a transition of the structure from the basic condition to a differentially adjacent neighboring condition (NC). The steady-state condition of the neighboring condition is therefore

$$\delta H_{NC} = \frac{1}{2}\delta\left(d^2 H\right) = 0 \tag{7.23}$$

with

$$H_{NC} = \int_{\tau_1}^{\tau_2} \left(\frac{1}{2} \int_V \rho \cdot (d\dot{v})^T \cdot d\dot{v}\, dV \right.$$
$$- \int_V \left(d\underline{\sigma}^T \cdot (\underline{G} \cdot d\underline{v}) - \frac{1}{2} d\underline{\sigma}^T \cdot d\underline{\varepsilon} - d\overset{+}{\underline{p}}{}_V^T \cdot d\underline{v} \right) dV$$
$$\left. + \int_{O_p} d\overset{+}{\underline{p}}{}^T \cdot d\underline{v}\, dO \right) d\tau. \tag{7.24}$$

The continuum is subdivided into n finite 3D elements. The volume V^e of one finite 3D element e consists of the matrix volume V_m^e and the reinforcement volume V_b^e. Single fibers (sf) and fiber layers (fl) are taken into account. The volume of the reinforcement V_b^e consists then of n_{sf} single fibers and n_{fl} fiber layers. For a function F (e.g. stresses, strains, displacements) holds

$$\int_{V^e} F\, dV = \int_{V_m^e} F_m\, dV + \int_{V_b^e} F_b\, dV = \int_{V^e} F_m\, dV + \int_{V_b^e} F_b\, dV - \int_{V_b^e} F_m\, dV$$
$$= \int_{V^e} F_m\, dV + \sum_{i=1}^{n_{sf}} \int_{V_{sf}^{e,i}} F_{sf}^{e,i}\, dV + \sum_{j=1}^{n_{fl}} \int_{V_{fl}^{e,j}} F_{fl}^{e,j}\, dV$$
$$- \sum_{i=1}^{n_{sf}} \int_{V_{sf}^{e,i}} F_m^{e,i}\, dV - \sum_{j=1}^{n_{fl}} \int_{V_{fl}^{e,j}} F_m^{e,j}\, dV. \tag{7.25}$$

With Eq. (7.25), the reinforcement is taken into account in Eq. (7.24).

5 Model Reduction

The computational cost of a fuzzy stochastic structural analysis of RC structures under dynamic loads is almost completely caused by the nonlinear FE analysis. Thus, the most effective measure to increase the numerical efficiency is to replace the costly deterministic computational model (innermost loop in Fig. 7.5) by a fast approximation solution based on a reasonable amount of initial deterministic computational results. The fuzzy stochastic analysis can then be performed with that surrogate model, which enables the utilization of an appropriate sample size for the simulation. The surrogate model is designed to describe a functional dependency between the structural parameters $\underline{x}$ and the structural responses $\underline{z}$ in the form of a response surface approximation

$$\underline{z} = f_{RS}(\underline{x}). \tag{7.26}$$

For response surface approximation, a variety of options exist (see [11, 17]). The suitability of the particular developments primarily depends on the properties of the computational model. Due to the very general properties of the FE analysis in structural analysis of textile strengthened RC structures, which can hardly be limited to convenient cases, a high degree of generality and flexibility of the approximation is demanded. In this context, an artificial neural network provides a

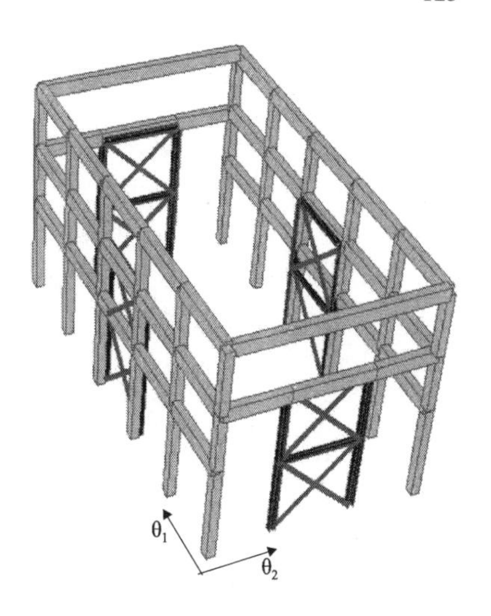

Fig. 7.8 3D pictorial view of the upgraded structure with scheme of the bracing system configuration

powerful basis for response surface approximation. This approach can extract information from initial deterministic computational results and can subsequently reproduce the structural response based on the extracted information only. According to the universal function approximation theorem, artificial neural networks are capable of uniformly approximating any kind of nonlinear functions over a compact domain of definition to any degree of accuracy. There is virtually no restriction for a response surface approximation with the aid of artificial neural networks.

In the case, that the global structural behavior is dominated from few eigen modes, the number of degrees of freedom can be reduced. In the following example, a simplified 2-DOF model is used as equivalent system for the whole structure.

6 Example

6.1 Investigated Structure

The investigated building (Fig. 7.8) has a rectangular plan whose dimensions are 10.80×20.40 m^2. The elevation of the first floor is 7.40 m, whereas the second one is at 11.10 m. It is characterized by a RC structure framed in the longitudinal direction only and is designed against vertical loads without account for seismic action. Columns and beams have rectangular 40×50 cm^2 and 40×70 cm^2 cross-sections, respectively. The T-shaped hollow tile RC floors have a 6 cm thick concrete slab, so that the total depth of the first floor is 36 cm, whereas the second, at the roof level, is 30 cm.

In [5], the results of the vulnerability evaluation have been published. Thereby, a three-dimensional FEM model with beam elements of the structure has been created considering floors like rigid diaphragms in the horizontal plane. Two nonlinear

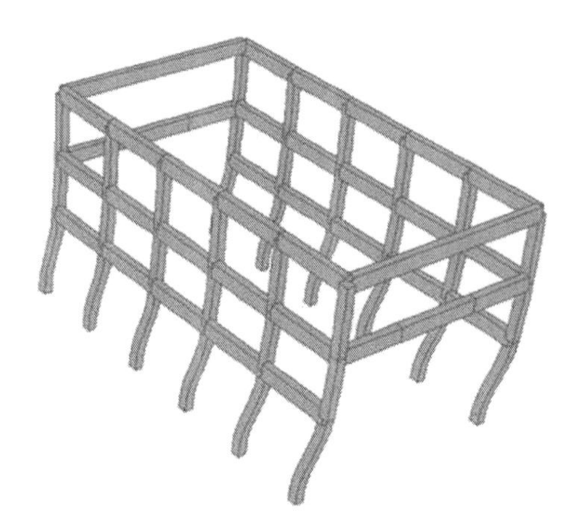

Fig. 7.9 Deformed shapes of the fundamental vibration modes for one principal direction

static analyses and a set of linear and nonlinear time history analyses have allowed to evaluate the vulnerability of the structure in the as-built condition and the effectiveness of the upgrading interventions. First of all, a calculation of the natural frequencies of the system has been carried out. Relevant values are 2.075 s^{-1} in Θ_1 direction (longitudinal, see Fig. 7.9) and 0.796 s^{-1} in Θ_2 direction (transverse). The mass participation factors are higher than 95% for such modes, so that the structure can be assumed as a matter of fact as made of two mutually independent SDOF systems in both Θ_1 and Θ_2 direction. This consideration assumes relevance in the determination of the optimal value of the damping devices. In fact, a design procedure for viscous devices based on simplified 2-DOF system can be used when the structural dynamic behavior can be interpreted through two SDOF systems [2, 6].

A peak ground acceleration (PGA) of 0.25 g has been assumed in the analysis, considering the combination of site effect and the importance of the structure with regard to collapse.

The time history analysis has shown an excessive deformability of the original structure, not compatible with the structural safety and immediate occupancy requirement after seismic events [3, 4]. The assumed upgrading interventions are aimed at reducing the lateral floor displacements of the structure by means of steel braces fitted with additional energy dissipation devices. Such devices connect the original structure at the first floor level with rigid steel braces and act due to the relative displacements occurring between the original structure and the steel braces. The study, presented in this paper, has been carried out considering the connection with purely viscous devices. As shown in [5], the reduction of horizontal floor displacements obtained thanks to the addition of this kind of devices is greater than the one obtained with a rigid connection of the original structure to the steel braces.

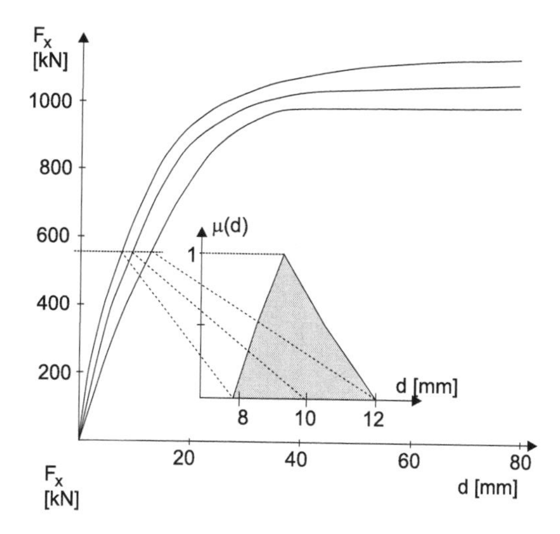

Fig. 7.10 Fuzzy load-displacement dependency of the existing RC frame structure

6.2 Uncertain Input Parameters

No technical documentation regarding the history of the structure is available, apart from the period of erection, which can be dated at the end of the 60's of XX century, on the basis of oral testimony.

Because of the lack of technical data and in order to find information about, the quality of structural materials, some characterization tests have been carried out on concrete core bored specimens and steel bars taken out of the structure. In result of the tests, the mechanical resistance of concrete is evaluated by means of fuzzy quantities. The concrete compressive and tensile strength are modeled as fuzzy triangular numbers $\tilde{f}_{ck} = \langle 14, 16.5, 20 \rangle$ N/mm^2 and $\tilde{f}_t = \langle 1.5, 2.0, 2.5 \rangle$ N/mm^2, respectively. A magneto-metric survey has been also carried out in order to locate the position and the diameter of steel bars in beams and columns. For the numerical study, twelve steel bars with a fuzzy cross-sectional area $\tilde{A} = \langle 2.69, 3.14, 3.21 \rangle$ cm^2 are considered. In order to assess the seismic vulnerability of the existing structure, nonlinear static analyses have been carried out under consideration of fuzzy resistance variables. The response of the as-built structure along both principal directions has then been evaluated in terms of fuzzy capacity curves F-d (Fig. 7.10). These curves have been represented in an approximate way by means of equivalent SDOF nonlinear relationships. Thereby, the kernel curve with $\mu(d(F)) = 1$ is scaled according to

$$\tilde{d}(F) = {}_{1.0}d(F) + \tilde{a} \cdot F \tag{7.27}$$

with $\tilde{a} = \langle -3.3, 0.0, 6.0 \rangle 10^{-3}$. The steel braces are also modeled as SDOF system with fuzzy stiffness $\tilde{K}$ and fuzzy mass $\tilde{M}$. Two variants are investigated especially: Variant 1 $\tilde{K}_1 = \langle 39, 40.8, 43 \rangle$ MN/m with $\tilde{M}_1 = \langle 1.1, 1.3, 1.5 \rangle$ t and Variant 2 $\tilde{K}_1 = \langle 50, 52.5, 55 \rangle$ MN/m with $\tilde{M}_1 = \langle 1.2, 1.55, 1.8 \rangle$ t. The uncertainty of the viscosity c_x of the connecting devices is with a fuzzy scaling factor according to $\tilde{c} = \tilde{b} \cdot c_x$ with $\tilde{b} = \langle 0.9, 1.0, 1.1 \rangle$.

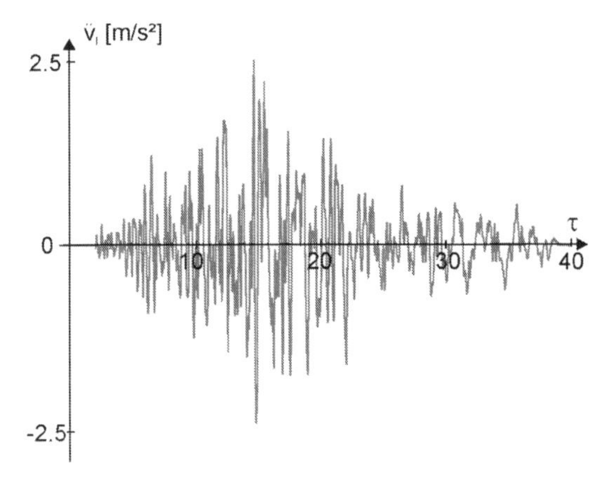

Fig. 7.11 Acceleration of the Taiwan earthquake scaled to PGA value of 0.25 g

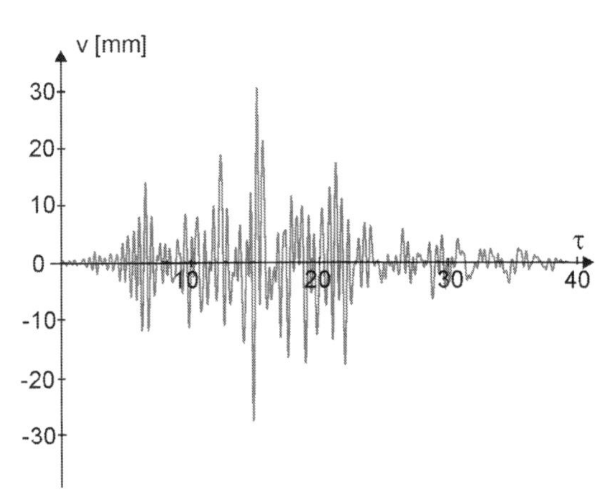

Fig. 7.12 Realization of the fuzzy displacement-time dependency due to the Taiwan earthquake

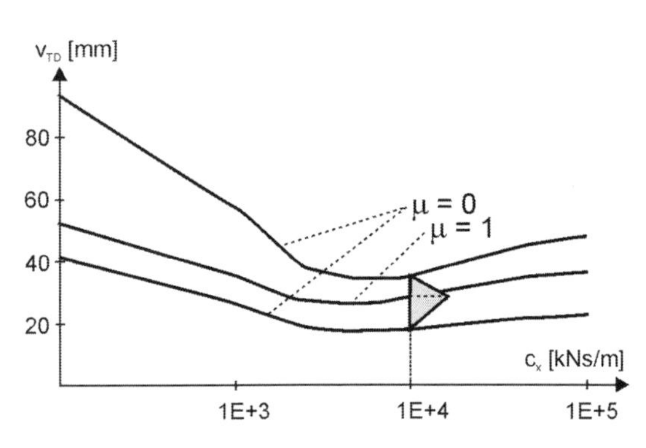

Fig. 7.13 Fuzzy top displacement in dependency of the viscosity

6.3 Fuzzy Structural Analysis

Nonlinear time-history analyses of the simplified 2-DOF system have then been performed considering the seismic input of Taiwan (1999) earthquake, scaled to PGA value of 0.25 g. Figure 7.11 displays the time-history of the ground acceleration of the Taiwan earthquake. The fuzzy maximum displacement $\tilde{v}_{TD}$ at the top of the structure has been calculated on the basis of the fuzzy displacement-time dependency, as shown for one realization in Fig. 7.12. The parameter study with variation of the viscosity of damping devices yields a fuzzy function $\tilde{v}_{TD}(c_x)$ as presented in Fig. 7.13 for the Taiwan earthquake.

Acknowledgements Authors gratefully acknowledge the support of the German Research Foundation (DFG) within the framework of the Collaborative Research Center (SFB) 528 and the contribution of Alberto Mandara (Second University of Naples).

References

1. Kaliske, M., Graf, W., Sickert, J.-U., Steinigen, F.: Numerische Prognose des Tragverhaltens textilverstärkter Stahlbetontragwerke. Bauingenieur **86**, 371–380 (2011)
2. Mandara, A., Mazzolani, F.M.: On the design of retro-fitting by means of energy dissipation devices. In: 7th International Seminar on Seismic Isolation, Passive Energy Dissipation and Active Control of Vibrations of Structures, Assisi (2001)
3. Mandara, A., Ramundo, F., Spina, G.: Seismic up-grading of an existing r.c. building by steel braces and energy dissipation devices. In: XXI National Congress of CTA, Catania (2007)
4. Mandara, A., Ramundo, F., Spina, G.: Steel bracing for the optimal seismic control of existing r.c. structures. In: 5th European Conference on Steel and Composite Structures, Graz (2008)
5. Mandara, A., Ramundo, F., Spina, G.: Seismic up-grading of r.c. structures with innovative bracing systems. In: Proceedings of PROHITEC, London (2009)
6. Mazzolani, F.M., Mandara, A.: Seismic up-grading of an old industrial building by dissipative steel roofing. In: International Seminar on Structural Analysis of Historical Constructions, Padova (2004)
7. Möller, B., Graf, W., Beer, M.: Fuzzy structural analysis using α-level optimization. Comput. Mech. **26**, 547–565 (2000)
8. Möller, B., Beer, M.: Fuzzy Randomness—Uncertainty in Civil Engineering and Computational Mechanics. Springer, Berlin (2004)
9. Möller, B., Graf, W., Hoffmann, A., Steinigen, F.: Numerical simulation of RC structures with textile reinforcement. Comput. Struct. **83**, 1659–1688 (2005)
10. Möller, B., Graf, W., Sickert, J.-U., Reuter, U.: Numerical simulation based on fuzzy stochastic analysis. Math. Comput. Model. Dyn. Syst. **13**, 349–364 (2007)
11. Myers, R.H., Montgomery, D.C.: Response Surface Methodology: Process and Product Optimization Using Designed Experiments. Wiley, New York (1995)
12. Pian, H.H.T., Wu, C.-C.: Hybrid and Incompatible Finite Element Methods. Chapman & Hall, Boca Raton (2006)
13. Puri, M.L., Ralescu, D.: Fuzzy random variables. J. Math. Anal. Appl. **114**, 409–422 (1986)
14. Schneider, R.: Stochastische Analyse und Simulation des nichtlinearen Verhaltens ebener Stabtragwerke mittels M-N-Q-Interaktionsmodell. Dissertation, Technische Universität Dresden, Veröffentlichungen des Lehrstuhls für Statik, Heft 2 (2001)
15. Sickert, J.-U.: Fuzzy-Zufallsfunktionen und ihre Anwendung bei der Tragwerksanalyse und Sicherheitsbeurteilung. Dissertation, Technische Universität Dresden, Veröffentlichungen Institut für Statik und Dynamik der Tragwerke, Heft 9 (2005)
16. Sickert, J.-U., Beer, M., Graf, W., Möller, B.: Fuzzy probabilistic structural analysis considering fuzzy random functions. In: 9th International Conference on Applications of Statistics and Probability in Civil Engineering. Millpress, Rotterdam (2003)
17. Simpson, T., Poplinski, J., Koch, P.N., Allen, J.: Metamodels for computer-based engineering design: survey and recommendations. Eng. Comput. **17**, 129–150 (2001)

18. Steinigen, F.: Numerische Simulation des Tragverhaltens textilverstärkter Bauwerke. Dissertation, Technische Universität Dresden, Veröffentlichungen Institut für Statik und Dynamik der Tragwerke, Heft 11 (2006)
19. Steinigen, F., Möller, B., Graf, W., Hoffmann, A.: Numerical simulation of textile reinforced concrete considering dynamic loading process. In: Hegger, J., Brameshuber, W., Will, N. (eds.) Textile Reinforced Concrete—Proceedings of the 1st International RILEM Conference, Aachen (2006)

Chapter 8
Application of Interval Fields for Uncertainty Modeling in a Geohydrological Case

Wim Verhaeghe, Wim Desmet, Dirk Vandepitte, Ingeborg Joris, Piet Seuntjens, and David Moens

Abstract In situ soil remediation requires a good knowledge about the processes that occur in the subsurface. Groundwater transport models are needed to predict the flow of contaminants. Such a model must contain information on the material layers. This information is obtained from in situ point measurements which are costly and thus limited in number. The overall model is thus characterised by uncertainty. This uncertainty has a spatial character, i.e. the value of an uncertain parameter can vary based on the location in the model itself. In other words the uncertain parameter is non-uniform throughout the model. On the other hand the uncertain parameter

W. Verhaeghe (✉) · W. Desmet · D. Vandepitte · D. Moens
Department of Mechanical Engineering, K.U. Leuven, Celestijnenlaan 300B, 3001 Heverlee, Belgium
e-mail: wim.verhaeghe@mech.kuleuven.be

W. Desmet
e-mail: wim.desmet@mech.kuleuven.be

D. Vandepitte
e-mail: dirk.vandepitte@mech.kuleuven.be

D. Moens
e-mail: david.moens@mech.kuleuven.be

I. Joris · P. Seuntjens
VITO, Boeretang 200, 2400 Mol, Belgium

P. Seuntjens
e-mail: piet.seuntjens@vito.be

P. Seuntjens
Department of Soil Management, Ghent University, Coupure Links 653, 9000 Gent, Belgium

P. Seuntjens
Department of Bioscience Engineering, University of Antwerp, Groenenborgerlaan 171, 2020 Antwerpen, Belgium

D. Moens
Dept. of Applied Engineering, Lessius Hogeschool—Campus De Nayer, K.U. Leuven Association, J. De Nayerlaan 5, 2860 Sint-Katelijne-Waver, Belgium

M. Papadrakakis et al. (eds.), *Computational Methods in Stochastic Dynamics*, Computational Methods in Applied Sciences 26,
DOI 10.1007/978-94-007-5134-7_8, © Springer Science+Business Media Dordrecht 2013

does have some spatial dependency, i.e. the particular value of the uncertainty in one location is not totally independent of its value in a location adjacent to it. To deal with such uncertainties the authors have developed the concept of interval fields. The main advantage of the interval field is its ability to represent a field uncertainty in two separate entities: one to represent the uncertainty and one to represent the spatial dependency. The main focus of the paper is on the application of interval fields to a geohydrological problem. The uncertainty taken into account is the material layers' hydraulic conductivity. The results presented are the uncertainties on the contaminant's concentration near a river. The second objective of the paper is to define an input uncertainty elasticity of the output. In other words, identify the locations in the model, whose uncertainties influence the uncertainty on the output the most. Such a quantity will indicate where to perform additional in situ point measurements to reduce the uncertainty on the output the most.

1 Introduction

In recent years, the study of uncertainties in numerical modeling has gained a lot of attention. Probabilistic and non-probabilistic methods were developed for dealing with scalar parameter uncertainties. However, scalar parameter uncertainties are not the only kind of uncertainties influencing numerical models. Often scalar parameter uncertainties represent uncertainties that have uncertainty on a smaller scale spatial dimension too. The spatial influence of such uncertainties is often neglected, as it is assumed captured by assumptions of uniformity and homogeneity. This neglect is not without reasons, for a thorough discretisation of an uncertain property over the spatial domain would result in an explosion of independent uncertainties and thus a drastic increase in the computation time for the uncertainty analysis. However, a go-between approach is possible when certain patterns describing the spatial behaviour of an uncertainty are available. Taking into account the patterns reduces the explosion of uncertainties in going from one spatially uniform uncertainty to a thorough discretisation of the spatial domain. The authors have developed an interval field approach [6] to formalize these notions.

The paper first presents the general problem of interval finite element analysis and the interval field approach to it. Secondly, a section details the choice of certain spatial patterns in the interval field approach, based on random field analogies. Next, the concept of input uncertainty elasticity of the output is introduced in the context op spatial uncertainties. In the next section the geohydrological problem is introduced and the obtained results are presented. The paper concludes with some suggestions for further research.

2 Interval Finite Element and Interval Field Analysis

This section first describes the general concept of Interval Finite Element (IFE) analysis and the method used to deal with it. Next the interval field concept is introduced to deal with dependent uncertain quantities.

2.1 Interval Finite Element Analysis

Generally an IFE problem can be represented by [5]:

$$\mathbf{y}^s = \left\{\mathbf{y} \mid \left(\mathbf{x} \in \mathbf{x}^I\right)\left(\mathbf{y} = f(\mathbf{x})\right)\right\} \tag{8.1}$$

with $\mathbf{x}^I$ the interval vector representing the bounds on the input uncertainties and $f(\mathbf{x})$ the function representing the input-output relationship. The solution is expressed as a set $\mathbf{y}^s$, rather than an interval vector $\mathbf{y}^I$ to stress that certain value combinations of components within a hypercubic approximation of the uncertain vector result $\mathbf{y}$ are not necessarily physically coherent. However in most cases the individual ranges of only some components of $\mathbf{y}$ are really of interest. Several implementation strategies for interval numerical analysis have been proposed. Because global optimisation based strategies yield physically correct results, they are more and more acknowledged as the standard approach for non-intrusive IFE analysis. The core of this analysis (the $f(\mathbf{x})$) is a black-box FE calculation which can roughly be any analysis (for example a static or dynamic structural analysis, but also a heat-conductivity problem, hydrogeological problem or vibro-accoustic problem), limited only by the capabilities of the FE solver. The global optimisation based solution strategies actively search in the non-deterministic input interval space for the combination that results in the minimum or maximum value of an output quantity. In theory, the global optimisation approach results in the exact interval vector.

However, despite the smooth behaviour of typical objective functions, the computational cost of the global optimisation based approach remains high. Hence, most research on this method focuses on fast approximate optimisation techniques. The approximating technique used in this paper starts by building a Kriging response surface based on a number of initial sample points. From this preliminary response surface the optimal additional samples are determined by focusing on the location of the possible extremes of the approximated output quantity in the uncertainty space [2]. The response surface is thus improved by additionally sampling the core FE-model till a pre-specified maximum number of samples are taken. Subsequently, global optimisation and anti-optimisation is performed on this response surface model to yield the bounds on the considered output quantity. For a thorough discussion of this adaptive response surface optimisation method, the interested reader is referred to [1].

For completeness the extension of an interval number to a fuzzy set is presented. A fuzzy set [12] is a set in which every member has a degree of membership, represented by the membership function $\mu_x(x)$, associated with it. If $\mu_x(x) = 1$, x is definitely a member of the fuzzy set. If $\mu_x(x) = 0$, x is definitely not a member of the fuzzy set. Analysis using fuzzy sets is very often done by using so-called α-cuts. An α-cut contains all the x for which $\mu_x(x) > \alpha$ is true. These α-cuts are essentially classical intervals, which means that the interval analysis is the basis of a fuzzy analysis.

2.2 Interval Fields

The interval field framework as developed in [6] has an explicit and an implicit implementation. For the application presented here the explicit implementation is needed.

For a spatially dependent uncertainty, the interval vector $\mathbf{x}^I$ containing an independent interval component for every spatial location is not a realistic description. Furthermore, it would result in an unfeasibly high dimensional optimisation problem. To describe spatially dependent variation, numerical modelling approaches often use some type of shape functions (e.g. the modes used to represent the dynamic behaviour of a structure using the modal superposition technique). The actual solution is a linear combination of these shape functions.

Accordingly, the explicit interval field $\mathbf{x}^F$ is defined as a superposition of n_b base vectors $\boldsymbol{\psi}_i$ using interval factors α_i^I:

$$\mathbf{x}^F = \sum_{i=1}^{n_b} \alpha_{x,i}^I \boldsymbol{\psi}_{x,i} \tag{8.2}$$

The base vectors represent a limited set of reference patterns over the spatial domain, each of which is scaled by an interval factor. The components of the interval fields themselves (the local value of the uncertainty) are coupled through the reference patterns. Once the reference patterns are chosen, the definition of the interval field requires the specification of the interval factors that define the field on x, which can be assembled in a classical (hypercubic) interval vector $\boldsymbol{\alpha}_x^I$. In matrix notation, the interval field is denoted as:

$$\mathbf{x}^F = [\boldsymbol{\psi}_x]\boldsymbol{\alpha}_x^I \tag{8.3}$$

The application of an explicit interval field on the input side of an analysis is rather straightforward. Since expert knowledge about the modelled system dominates the definition of the uncertainties, the freedom in choosing the base vectors is ideal to reflect this knowledge (for example: the sinusoidal (= base vector) deviation of the thickness of a rolled plate with uncertain amplitude (= interval factor)). The main limitation of the explicit interval field is that its definition only allows a linear relation between the base vectors and the interval factors.

The application of an explicit interval field on the output side of an analysis is less straightforward. The base vectors and interval factors are determined by the analysis itself. Furthermore, in order to obtain an explicit interval field that introduces no conservatism in its derived response variables (i.e. derivatives of the primary response variables), the output interval factors should be completely independent. An analysis of the application of the interval field approach to the output of static FE analysis is presented in [10].

Once the spatially dependent uncertainty on the input side of an analysis is defined by means of an explicit interval field, the dimensions of the uncertainty space are drastically reduced. This allows for the use of the adaptive response surface technique as described in the above subsection.

3 The Choice of Base Vectors

The use of the explicit interval field on the input side of an analysis requires the selection of appropriate base vectors and interval factors. This section first presents the factors influencing the selection of these base vectors and interval factors. The choice for base vectors and interval factors based on random field expansions is explained in the next subsection.

3.1 Factors Influencing the Choice of Base Vectors

- The bounds on the uncertainty on a model parameter $\mathbf{x}$ are specified by two functions of the spatial coordinate $\mathbf{r}$, one function for the upper bound $\overline{x}(\mathbf{r})$ and one for the lower bound $\underline{x}(\mathbf{r})$ of the uncertainty. The linear combination of the base vectors with the interval factors that makes up an interval field must remain within these bounds for any value of the interval factors.
- The base vectors must represent the expert's knowledge of the spatial dependency of the model parameter. Most often knowledge about this dependency is limited and the set of base vectors preferably allows for a range of small and large scale dependency.
- The number of base vectors and corresponding interval factors to represent the input uncertainty will influence the calculation time to get the output uncertainty.

3.2 Base Vectors Derived from Random Field Expansion

In an attempt to construct a base vector set that takes into account the above described factors, the expansion of a random field is studied.

The objective of a random field is to represent a spatial variation of a specific model property by a stochastic variable defined over the region on which the variation occurs [9]. A random field can thus be denoted as $H(\mathbf{r}, \theta)$ with $\mathbf{r}$ the spatial coordinate and θ the outcome of a random phenomenon. A random field is a random variable for a given $\mathbf{r_0}$ and is a realization of the field for a given θ_0. The specification of a random field generally comes down to the specification of the spatial evolution of the first two statistical moments of the field variable and a corresponding covariance function, expressing the spatial dependency of the field variable. In most cases the random field is considered to be weakly stationary, resulting in a constant for the first few statistical moments throughout the spatial domain (i.e. zero mean and unit variance). Furthermore the covariance function for weakly stationary random fields depends only on the distance between observation points, not on their actual location.

The application of the concept of random fields in a numerical modelling framework requires some sort of discretisation of the spatially varying stochastic field over the defined geometry. A good overview of methods can be found in the report by Sudret and Der Kiureghian [8]. The technique studied here is the Karhunen-Loève expansion [3] that has gained particular attention in literature. This approach is based on the spectral decomposition of the autocovariance function $C_{HH}(\mathbf{r_1}, \mathbf{r_2})$. The set of deterministic functions over which any realization of the field $H(\mathbf{r}, \theta_0)$ is expanded is defined by the eigenvalue problem:

$$\int_{\omega} C_{HH}(\mathbf{r_1}, \mathbf{r_2})\varphi_i(\mathbf{r_1})\, d\omega_{\mathbf{r_2}} = \lambda_i \varphi_i(\mathbf{r}) \tag{8.4}$$

with ω the spatial domain and $i = 1, \ldots$. Once the eigenfunctions are found, the zero-mean random field can be expressed as:

$$H(\mathbf{r}, \theta) = \sum_{i=1}^{\infty} \sqrt{\lambda_i} \xi_i(\theta)\varphi_i(\mathbf{r}) \tag{8.5}$$

with $\{\xi_i\}$ a set of orthonormal random variables. In stochastic analysis, this expansion is truncated after N terms to reduce the computational costs.

Several features of the random field expansion can be used in the interval field implementation after some adaptations. To begin, an off-set function $f_{mid}(\mathbf{r})$ to describe the mid value of the model parameter throughout the spatial domain is calculated

$$x_{mid}(\mathbf{r}) = \frac{\overline{x}(\mathbf{r}) + \underline{x}(\mathbf{r})}{2} \tag{8.6}$$

The eigenfunctions $\varphi_i(\mathbf{r})$ of the covariance function are then used as base vectors $\boldsymbol{\psi}_i(\mathbf{r})$ for the interval field with

$$\boldsymbol{\psi}_i(\mathbf{r}) = \lambda_i \varphi_i(\mathbf{r}) \big|\varphi_i(\mathbf{r})\big| \tag{8.7}$$

and replacing the orthonormal random variables $\{\xi_i\}$ by interval factors $\alpha_i^I \in \lfloor -1\ 1 \rceil$. These adaptations make sure that for $N \to \infty$ the interval field will assign a value from the interval $\lfloor -1\ 1 \rceil$ to the model parameter throughout the spatial domain. This unit interval is then scaled by the difference function

$$x_{dif}(\mathbf{r}) = \overline{x}(\mathbf{r}) - \underline{x}(\mathbf{r}) \tag{8.8}$$

describing the actual range of uncertainty on the model parameter for every location in the model. The description of the model parameter by the interval field is thus

$$\mathbf{x}^F = x_{mid}(\mathbf{r}) + \sum_{i=1}^{N} \big(\lambda_i \varphi_i(\mathbf{r}) \big|\varphi_i(\mathbf{r})\big| \alpha_i^I\big)\big(x_{dif}(\mathbf{r})\big) \tag{8.9}$$

With this equation the considerations from the first and last item in the list of influencing factors is accounted for. Next is the issue of uncertainty about the spatial dependency.

The base vectors taken from the expansion of a random field with a given autocovariance function only take into account the given correlation length L. In [11] a method is described to take into account interval correlation lengths with interval fields. Essentially the method relies on building an interval field description for the base vectors themselves in the correlation length space using a limited number of autocovariance functions. In this way the base vectors are depending on the correlation length and can be calculated by a simple matrix vector product. The resulting interval field for the model parameter can thus be summarised by

$$\mathbf{x}^F = x_{mid}(\mathbf{r}) + \sum_{i=1}^{N} \left(\lambda_i(L)\varphi_i(\mathbf{r}, L)\left|\varphi_i(\mathbf{r}, L)\right|\alpha_i^I\right)\left(x_{dif}(\mathbf{r})\right) \quad L \in \lfloor L_{min}\ L_{max} \rceil \tag{8.10}$$

This approach only introduces one additional interval to represent the uncertainty about the amount of spatial dependency. The solution strategy to find the uncertainty on the output remains the same, for example a response surface based optimisation and anti-optimisation, with only one additional dimension in the uncertainty space.

4 Input Uncertainty Elasticity of the Output

To assess the relative importance of an input uncertainty on an output uncertainty, the concept of input uncertainty elasticity of the output is introduced in general terms and then applied to the case of spatial uncertainty.

4.1 General Concept

As in economics, an elasticity R is defined as the ratio of the relative change (more precisely, the derivative with respect to some quantity) in one parameter Y to the relative change in an other parameter X

$$R_X^Y = \frac{\Delta Y}{\Delta X}\frac{X}{Y} \tag{8.11}$$

Let Y be the range of the uncertain output and X be the range of the uncertain input. The reduction (i.e. the Δ) on the range of the interval for the input X, will affect the range of the interval for the output Y to a greater or lesser extent. The relative magnitude of this influence is described by the input uncertainty elasticity of the output R_X^Y.

4.2 Spatial Uncertainty Context

In the context of spatial uncertainty, the influence of an input uncertainty on an output uncertainty has a spatial component. The influence of an uncertain input parameter will depend on the spatial distribution of its uncertainty. Figure 8.1 shows

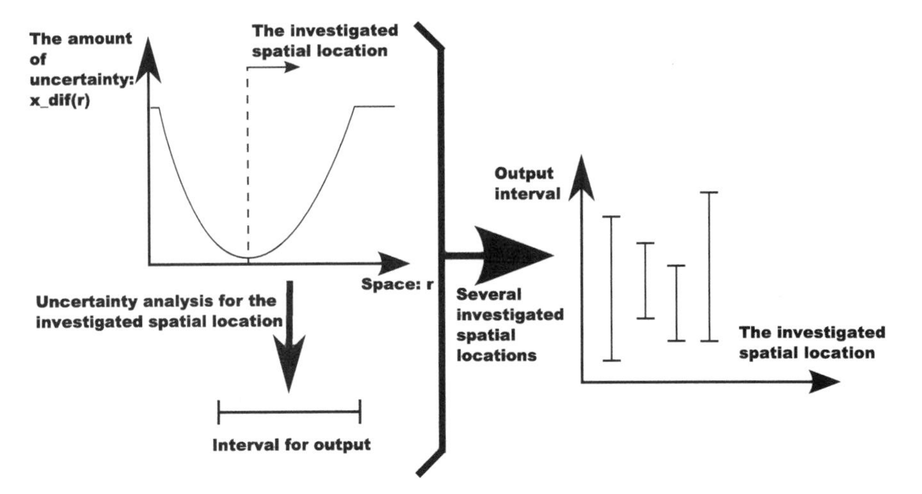

Fig. 8.1 Concept to determine R_X^Y in a spatial context

in a generic way the influence of the spatial uncertainty distribution. For an investigated spatial location the amount of uncertainty $x_{dif}(\mathbf{r})$ is reduced and some sort of coherent distribution of the uncertainty is assumed over the spatial domain (as illustrated at the top left in the figure). The uncertainty analysis is carried out for this spatial uncertainty distribution and a resulting uncertainty (an interval) for the output is found (bottom left). By repeating this for other investigated spatial locations, one finds the combined result which is shown at the right of the figure. It presents the different output uncertainties for several investigated spatial locations. This data is then used to calculate the input uncertainty elasticity of the output over the spatial domain. In this context the R_X^Y is in particular useful to identify the spatial location where an input uncertainty influences the output uncertainty the most. In allocating resources to reduce the uncertainty, the spatial location with the highest R_X^Y should get priority.

An appropriate selection of the $\overline{x}(\mathbf{r})$ and $\underline{x}(\mathbf{r})$ is needed to make a study over the spatial domain to give a scalar field of input uncertainty elasticities of the output. Important choices to be made in the selection of $\overline{x}(\mathbf{r})$ and $\underline{x}(\mathbf{r})$ to investigate a particular spatial location's uncertainty influence are listed below. Figure 8.2 shows $\overline{x}(\mathbf{r})$ and $\underline{x}(\mathbf{r})$ for three cases. The first case, at the left on the figure, is the reference case. The two other cases illustrate particular choices for $\overline{x}(\mathbf{r})$ and $\underline{x}(\mathbf{r})$ explained in the list below.

Important choices to be made in the selection of $\overline{x}(\mathbf{r})$ and $\underline{x}(\mathbf{r})$:

- the magnitude of the reduction of $x_{dif}(\mathbf{r})$ for the investigated spatial location. The second case in Fig. 8.2 shows a reduction of 50% for $x_{dif}(0.2)$, the third case shows a reduction of 90% for $x_{dif}(0.7)$.
- the magnitude of the reduction of $x_{dif}(\mathbf{r})$ in the local influence zone of the investigated spatial location. By reducing the amount of uncertainty for the investigated spatial location, the amount of uncertainty for the region around the investigated

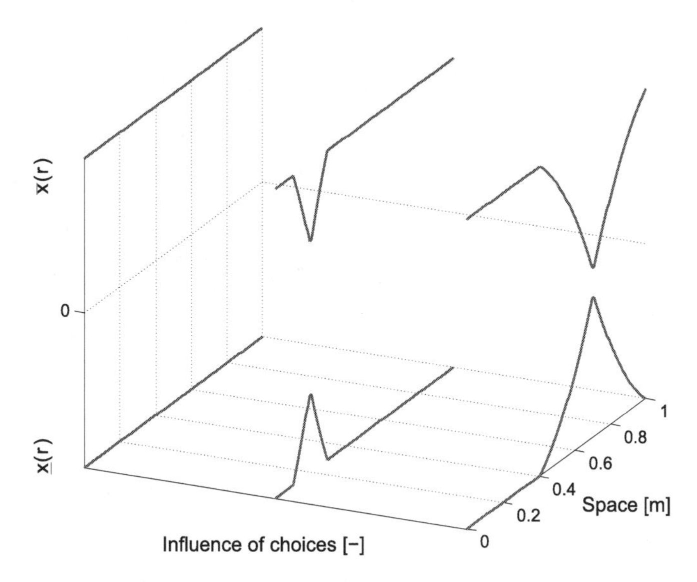

Fig. 8.2 Choices in the selection of $\overline{x}(\mathbf{r})$ and $\underline{x}(\mathbf{r})$

spatial location is also affected. In this so called local influence zone, a transition from the reduced amount of uncertainty to the reference amount of uncertainty is needed. In this paper a quadratic transition is suggested.

- the magnitude of the local influence zone of the investigated spatial location. The second case in Fig. 8.2 shows a zone of influence from -0.1 to $+0.1$ around the investigated spatial location 0.2. The third case shows a zone of influence from -0.3 to $+0.3$ around the investigated spatial location 0.7.
- the change in $x_{mid}(\mathbf{r})$. If $\overline{x}(\mathbf{r})$ and $\underline{x}(\mathbf{r})$ are not changed symmetrically with respect to $x_{mid}(\mathbf{r})$ in the reference case, then $x_{mid}(\mathbf{r})$ is affected as well. For simplicity this influence is not presented here.

These notions are explained further in the case study.

5 Geohydrological Case Study

A geohydrological case study was chosen to apply the above presented techniques. The case study deals with a groundwater pollution problem where benzene was spilled and is now being transported in groundwater to a river. To characterize the flow and transport of the benzene spill, a groundwater flow and transport model was built in HYDRUS3D. First, the problem together with its uncertainty is described and the results of a fuzzy analysis without taking into account the spatial dependency are presented. Next, the spatial dependency is introduced and an investigation of the input uncertainty elasticity of the output is performed.

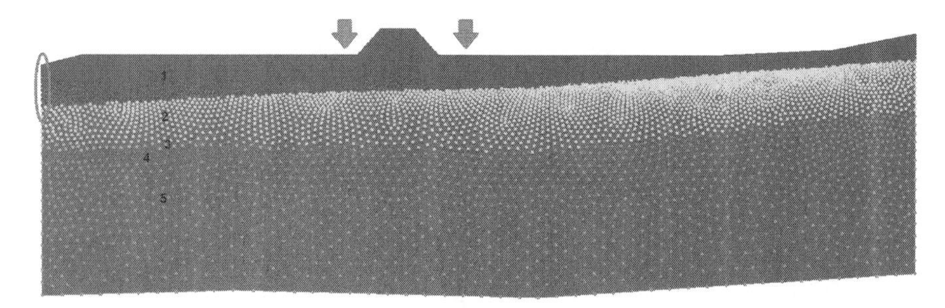

Fig. 8.3 FE-model for solute and ground water flow showing the five different material layers

Table 8.1 Intervals for the hydraulic conductivity K [m/day], ordered from top to bottom

Material Layer	Minimum K	Maximum K
1	1.4	2.1
2	8	12
3	3.6	5.4
4	2.6	3.9
5	4	6

5.1 Problem Description

The governing equation for solute transport in groundwater is a convection-diffusion equation based on conservation of mass. Convection is determined by groundwater flow which is based on the constitutive equation for variably saturated flow in porous media, called the Darcy Buckingham equation. For the solute (the contaminant: Benzene) and ground water flow problem at hand the following input was given:

- FE-model (14661 nodes) for the HYDRUS3D [7] solver (see Fig. 8.3). The dimensions of the problem are 1100 m in the length direction and between 32 and 36.5 m in the depth direction. In the time domain a period of 11000 days (approximately 30 years) is calculated. A deterministic run of this model takes 10 minutes.
- Intervals for the material properties, i.e. the saturated hydraulic conductivity K (see Table 8.1) of the five different material layers.

A river is situated at the left side of the domain (see the red ellipse on Fig. 8.3) and the two sources of the contaminant are in the middle of the domain (see the red arrows on Fig. 8.3). The requested output is the concentration of the contaminant over time at the river given the uncertainties on the material properties.

5.2 Fuzzy Analysis Without Spatial Uncertainty

In the first fuzzy analysis, the uncertainty on the material properties is represented by fuzzy numbers. The hydraulic conductivity of each layer is considered independent

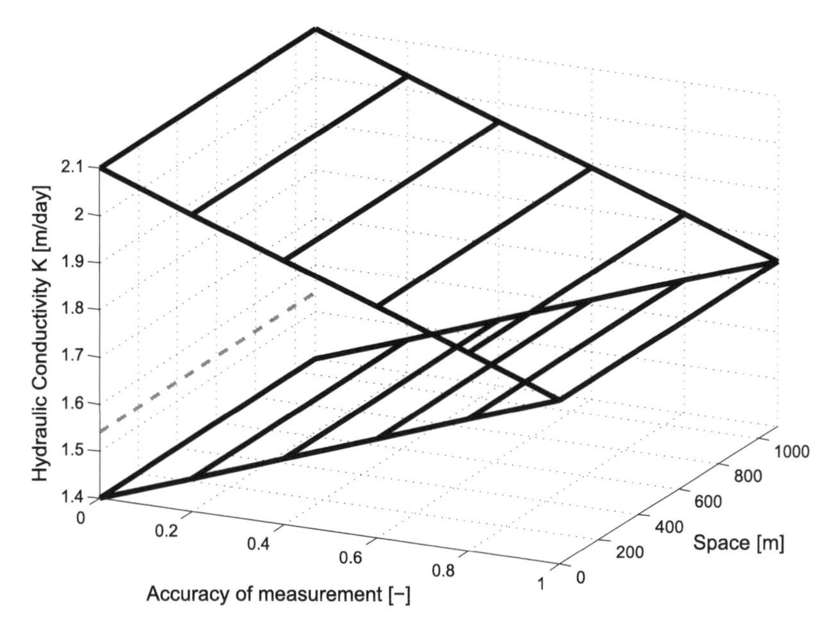

Fig. 8.4 The spatial fuzzy number in *black solid* line and a sample of it in *dashed grey* for the hydraulic conductivity of material layer 1, assuming homogeneity

and modelled as a triangular fuzzy number with the given intervals (see Table 8.1) as base and the mid value as the top of the triangle. In each material layer the hydraulic conductivity is considered homogeneous through space. Figure 8.4 shows for example the spatial fuzzy number in black and a possible sample of the fuzzy number in dashed grey for the hydraulic conductivity of material layer 1. Two types of fuzzy analyses were performed:

- Reduced Transformation Method (TM) [4] with 5 alpha-cuts, resulting in 161 samples.
- An optimisation on a Kriging response surface (ARSM) [1] that was built using 32 initial latinhypercube samples and 32 additional samples.

Additionally, a reference Monte Carlo Simulation using 200 samples was performed at each alpha-level, based on a uniform distribution within the interval at each alpha-level Fig. 8.5 shows the fuzzy concentration through time for location 11 (at the river, 3 m below the surface). In black solid line is the result of the reduced transformation method ($5 * 32 + 1$ samples); in dotted grey line is the result of the optimisation on a Kriging response surface (32 initial + 32 additional samples); in dashed grey line is the result of the Monte Carlo Simulation ($5 * 200$ samples). From these results it is clear that the TM and ARSM results are close to each other. The MCS result, despite being the computationally most expensive, does not yield good results for the maxima: the value given by the TM is an actual solution of the problem (i.e. a genuine sample) and results in higher maxima. The ARSM has problems identifying the proper minima since it tends to give negative (non-physical) results.

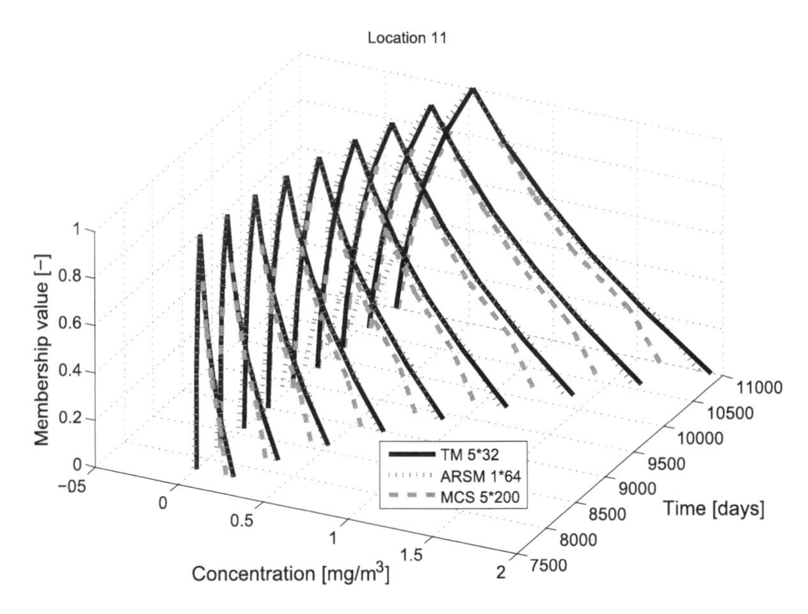

Fig. 8.5 The fuzzy concentration at the river, 3 m below the surface

5.3 Fuzzy Analysis with Spatial Uncertainty

In this fuzzy analysis, only the uncertainty on the hydraulic conductivity in material layer 1 is taken into account. The other hydraulic conductivities are set at their minimal value. To model the spatial uncertainty for the hydraulic conductivity of material layer 1, the following assumptions are made:

- The upper bound $\overline{x}(\mathbf{r})$ and lower bound $\underline{x}(\mathbf{r})$ are given by a constant, namely the maximum and minimum of the interval given in Table 8.1.
- The base vectors are derived from an exponential autocovariance function

$$C_{HH}(x_1, x_2) = e^{-|x_1 - x_2|/L} \tag{8.12}$$

  as described in Sect. 3.2. The first four eigenfunctions are used. Since a limited number of base vectors is used, the upper and lower bound on the uncertainty are not exactly satisfied throughout the domain. A scaling factor to adjust the maximal possible value of the interval field in the spatial domain to the requested bounds is applied. For a correlation length $L = 500$ m, the resulting base vectors are shown in Fig. 8.6.

To check the influence of taking into account the spatial uncertainty, two analyses (TM and ARSM) with the non-spatial uncertainty (i.e. uncertain, but homogeneous hydraulic conductivity of material layer 1) are performed as well. In total, the following types of analyses were performed:

- non-spatial uncertainty, Reduced Transformation Method (TM) with 5 alpha-cuts, resulting in 11 samples.

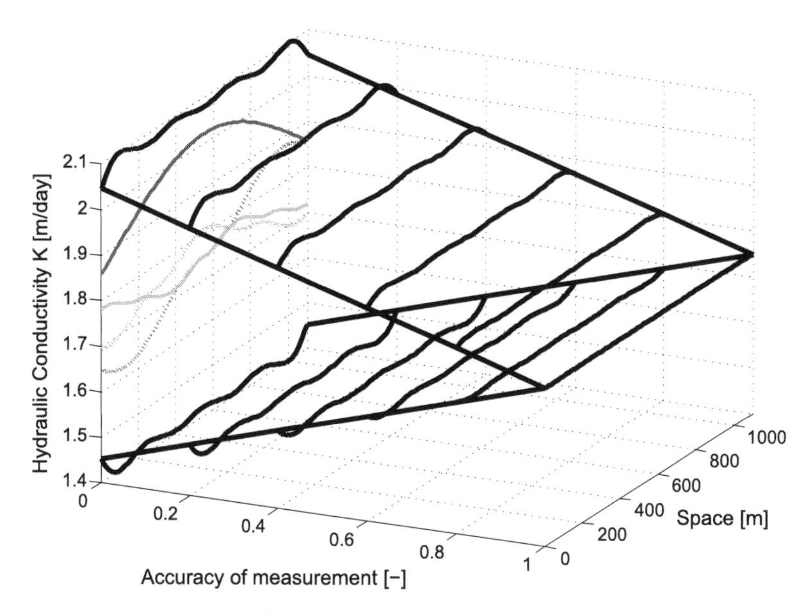

Fig. 8.6 The spatial fuzzy number in *black* and the four base vectors for membership level 0 in *grey* for the hydraulic conductivity of material layer 1, including spatial uncertainty

- non-spatial uncertainty, optimisation on a Kriging response surface (ARSM) that was built using 6 initial latinhypercube samples and 12 additional samples.
- spatial uncertainty, with correlation length between 500 and 2000 m, optimisation on a Kriging response surface (ARSM) that was built using 20 initial latinhypercube samples and 30 additional samples.

Figure 8.7 shows the fuzzy concentration through time for location 11 (at the river, 3 m below the surface). The influence of taking into account the spatial uncertainty results in slightly narrower fuzzy numbers. This suggests that assuming homogeneity for the hydraulic conductivity of material layer 1 gives conservative bounds on the contaminant's concentration for the studied case.

5.4 Input Uncertainty Elasticity of the Output

By performing an additional point measurement to determine the hydraulic conductivity in one location, the uncertainty on the contaminant's concentration will be reduced. To determine the optimal measurement location an input uncertainty elasticity of the output is calculated. The following assumptions, referring to Sect. 4.2 and Fig. 8.8, are made:

- The magnitude of the reduction of the uncertainty for the considered measurement location is a design parameter. By selecting a more accurate measurement

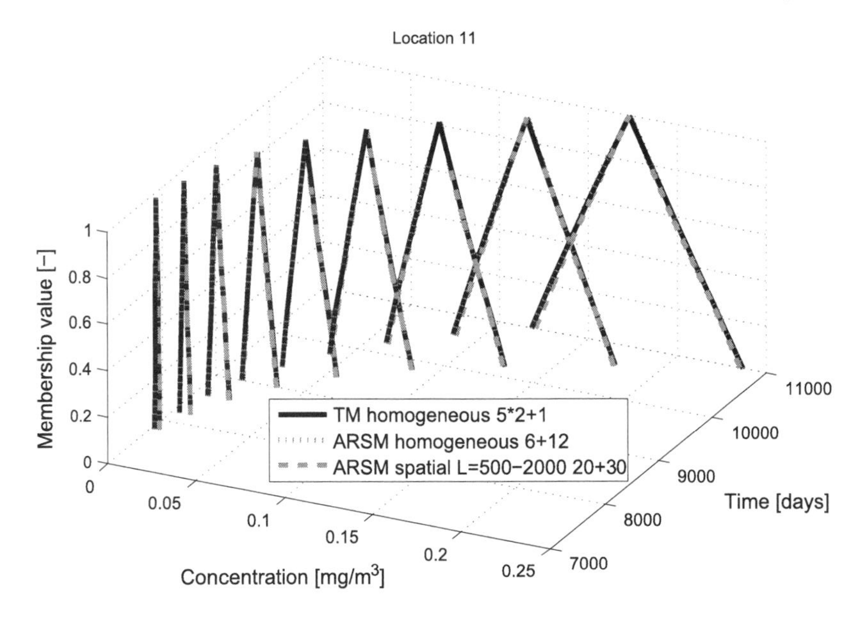

Fig. 8.7 The fuzzy concentration at the river, 3 m below the surface

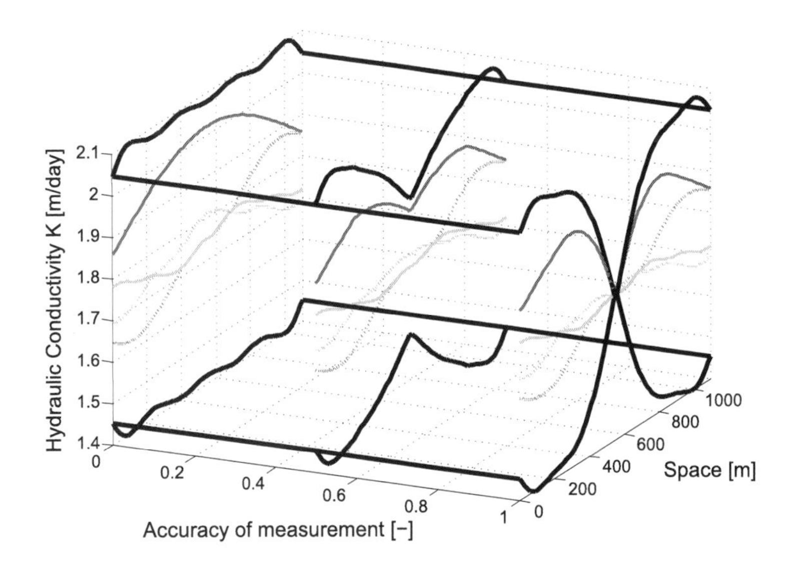

Fig. 8.8 The influence of a measurement in the middle of the domain on the bounds of the uncertainty (in *black*) and the base vectors (in *grey*), with a zone of influence of 330 m to both sides of the measurement location

device, the uncertainty remaining after measurement is a choice of the expert. In Fig. 8.8 the influence of an increasing measurement accuracy on the bounds and the base vectors is shown.

Table 8.2 Parameters in the input uncertainty elasticity of the output analysis

Design Parameter	Sampled values
Measurement location	110, 330 and 550 m from river
Measurement accuracy	50% reduction and 100% reduction of uncertainty
Uncertain Parameter	**Range**
Extent of influence	value chosen is 330 m
Correlation length	$\lfloor 500\ 2000 \rceil$ m

- The magnitude of the reduction of the uncertainty in the local influence zone is an uncertainty. $f_{dif}(\mathbf{r})$ increases from the value at the measurement location to the reference value at the end of the local influence zone. In the presented analysis a quadratic function of the distance to the measurement location is chosen.
- The magnitude of the local zone of influence is an uncertainty. What is the extent of the influence of a measurement in one location on the rest of the spatial domain? Since a comparison between the input uncertainty elasticities of the output for different locations is of interest, the magnitude of this local zone of influence is chosen to be a fixed value. In Fig. 8.8 the bounds and base vectors for an influence up to 330 m to both sides is shown, as it is used in the analysis.
- The actual outcome of the measurement gives a value for $f_{mid}(\mathbf{r})$ in the measurement location. Until the measurement is done, this is also an uncertainty that influences the actual bounds on the output uncertainty. In the presented analysis the value of $f_{mid}(\mathbf{r})$ is considered a constant and unchanged by a measurement.

To summarize: the measurement location and the accuracy of the measurement are design parameters, whereas the influence of the measurement and the spatial correlation length are uncertainties. For the influence of the measurement a fixed magnitude is assumed and the correlation length is modelled as an interval. The values used in the analysis are presented in Table 8.2. For a choice of the design parameters, the uncertainty analysis was carried out using the ARSM method with 20 initial samples and 30 additional samples. The results are presented in Fig. 8.9. The bounds on the contaminant's concentration are presented for location 11 (at the river, 3 m below the surface) at the end time of the simulation (approx. 30 years) as a function of the measurement location and the accuracy of the measurement. Based on this information the input uncertainty elasticity of the output is calculated using Eq. (8.11) with X and Y respectively the range on the hydraulic conductivity and the range on the contaminant's concentration. The results are presented in Table 8.3, the reference is the range on the uncertainty before measurement. From this Table 8.3 it becomes clear that performing an input uncertainty reduction (i.e. a measurement of the hydraulic conductivity in material layer 1) at 110 m from the river provides the greatest reduction in uncertainty on the output (i.e. the concentration of the contaminant at the considered location and time). Furthermore, for this measurement location increasing the uncertainty reduction from 50% to 100% will not decrease

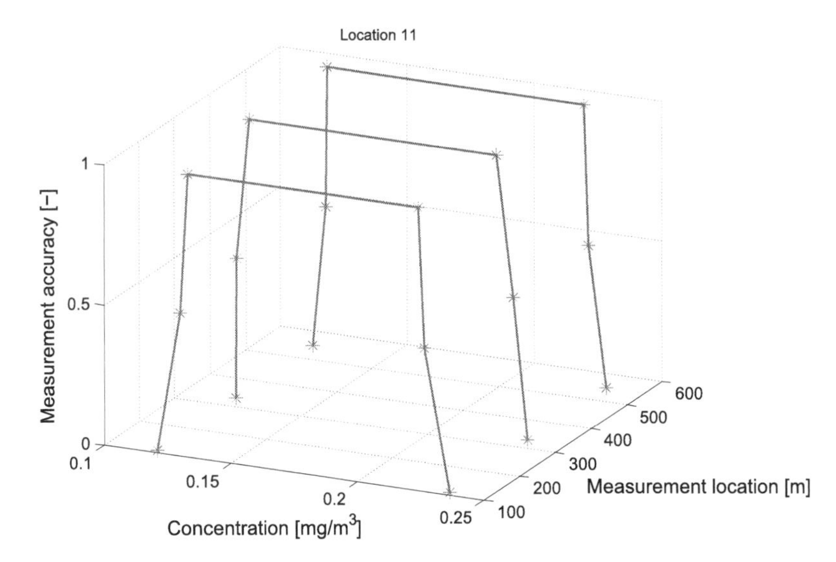

Fig. 8.9 The results of the uncertainty analysis to determine the input uncertainty elasticity of the output

Table 8.3 The input uncertainty elasticity of the output

Measurement accuracy [-]	Measurement location [m]		
	110	330	550
0.5	0.33	0.10	0.20
1.0	0.21	0.15	0.13

the uncertainty on the output with the same amount. In other words, a measurement with an uncertainty reduction of 50% will have a $\frac{0.33}{0.21} \simeq 1.5$ times higher relative uncertainty reduction on the output than a measurement with an uncertainty reduction of 100%. For a measurement at 330 m from the river the inverse is true: the extra effort of reducing the input uncertainty from 50% to 100% gives a 1.5 times higher relative uncertainty reduction on the output. Based on this information and knowledge of the actual costs of a measurement campaign an informed decision can be made concerning where and how accurate to measure.

6 Conclusion

From a methodological point of view, the paper introduces interval fields as an easy conceptual tool to deal with spatial uncertainty. The implementation of the interval field based on correlation length is made possible by deriving certain base vectors from the random field expansion technique. This allows for taking into account uncertainty on the correlation length. Furthermore, the concept of input uncertainty elasticity of the output is introduced in a spatial uncertainty context.

From an applied point of view, the paper shows the applicability of the interval field to a geohydrological problem of realistic complexity. The adaptive response surface technique proves to be very useful in practice. Certainty on the value of the correlation length often is a problem. The feasibility of dealing with the correlation length as an interval is shown. The concept of input uncertainty elasticity of the output in a spatial uncertainty context is proven to be useful to determine the optimal location of a measurement (i.e. an uncertainty reduction) to reduce uncertainty on the output.

Acknowledgements The support of the Flemish Government through IWT-SBO project no. 060043: Fuzzy Finite Element Method is gratefully acknowledged.

References

1. De Munck, M.: Efficient optimization approaches for interval and fuzzy finite element analysis. Ph.D. thesis, Katholieke Universiteit Leuven (2009)
2. De Munck, M., Moens, D., Desmet, W., Vandepitte, D.: An efficient response surface based optimisation method for non-deterministic harmonic and transient dynamic analysis. Comput. Model. Eng. Sci. **47**(2), 119–166 (2009)
3. Ghanem, R., Spanos, P.: Stochastic Finite Elements: A Spectral Approach. Springer, New York (1991)
4. Hanss, M.: Applied Fuzzy Arithmetic—An Introduction with Engineering Applications. Springer, Berlin (2005)
5. Moens, D., De Munck, M., Farkas, L., De Gersem, H., Desmet, W., Vandepitte, D.: Recent advances in interval finite element analysis in applied mechanics. In: Proceedings of the First Leuven Symposium on Applied Mechanics in Engineering LSAME.08, Leuven, pp. 553–568 (2008)
6. Moens, D., De Munck, M., Desmet, W., Vandepitte, D.: Numerical dynamic analysis of uncertain mechanical structures based on interval fields. In: IUTAM Symposium on Vibration Analysis of Structures with Uncertainties, Saint Petersburg (2009)
7. Sejna, M., Simunek, J., Van Genuchten, R.: Software package for simulating water, heat and solute movement in two- and three-dimensional variably saturated media. PC-Progress s.r.o, Prague. www.hydrus3D.com
8. Sudret, B., Der Kiureghian, A.: Stochastic finite element methods and reliability: a state-of-the-art report. Technical report, Department of Civil & Environmental Engineering, University of California, Berkley, Institute of Structural Engineering, Mechanics and Materials (2010)
9. Vanmarcke, E.: Random Fields: Analysis and Synthesis. MIT Press, Cambridge (1993)
10. Verhaeghe, W., Rousounelos, A., Desmet, W., Vandepitte, D., Moens, D.: Interval fields to represent uncertainty on input and output side of a FE analysis. In: Proceedings of the 3rd International Conference on Uncertainty in Structural Dynamics, pp. 5067–5078 (2010)
11. Verhaeghe, W., Desmet, W., Vandepitte, D., Moens, D.: Uncertainty assessment in random field representations: an interval approach. In: Proceedings of NAFIPS 2011, El Paso, TX (2011)
12. Zadeh, L.: Fuzzy sets. Inf. Control **8**, 338–353 (1965)

Chapter 9
Enhanced Monte Carlo for Reliability-Based Design and Calibration

Arvid Naess, Marc Maes, and Markus R. Dann

Abstract This paper extends the recently developed enhanced Monte Carlo approach to the problem of reliability-based design. The objective is to optimize a design parameter(s) so that the system, represented by a set of failure modes or limit states, achieves a target reliability. In a large majority of design and/or calibration contexts, the design parameter α itself can be used to parameterize the system safety margin $M(\alpha)$. The lower tail of this random variable behaves in a regular way and is therefore amenable to straightforward parametric analysis. In contrast to the original Naess et al. method (Naess et al. in Struct. Saf. 31:349–355, 2009), the intention is to estimate the value α_T that corresponds to a (very) small target system failure probability p_{fT}. Monte Carlo sampling occurs at a range of values for α that result in larger failure probabilities, and so the design problem essentially amounts to a statistical estimation of a high quantile. Bounds for α_T can easily be constructed. Several examples of the approach are given in the paper.

1 Introduction

A new Monte Carlo (MC) based method for estimating system reliability was recently developed in [1]. The aim of this method is to reduce computational cost while maintaining the advantages of crude MC simulation, specifically, its ease in dealing with complex systems. The key idea is to exploit the regularity of tail probabilities to enable an approximate prediction of far tail failure probabilities based on

A. Naess (✉)
Centre for Ships and Ocean Structures & Dept. of Mathematical Sciences, NTNU, Trondheim 7491, Norway
e-mail: arvidn@math.ntnu.no

M. Maes · M.R. Dann
University of Calgary, Calgary T2N 1N4, Canada

M. Maes
e-mail: mamaes@ucalgary.ca

M.R. Dann
e-mail: markusdann@gmx.de

M. Papadrakakis et al. (eds.), *Computational Methods in Stochastic Dynamics*, Computational Methods in Applied Sciences 26,
DOI 10.1007/978-94-007-5134-7_9,

small Monte-Carlo sample results obtained for much more moderate levels of reliability. The motivation behind this approach is that systems with multiple and complex failure modes or limit states are often exceedingly difficult to analyze using traditional methods of structural reliability. While direct MC does not suffer from this problem, it is computationally burdensome for small probabilities. Hence originates the idea of sampling in a different less reliable range and performing a statistical extrapolation unto the tail. A similar but somewhat different idea is presented in [2].

The fundamentals of the method proposed in [1] are as follows. A safety margin $M = G(X_1, \dots, X_n)$ expressed in terms of n basic variables, is extended to a parameterized class of safety margins using a scaling parameter λ $(0 \leq \lambda \leq 1)$:

$$M(\lambda) = M - (1 - \lambda)\mu_M. \tag{9.1}$$

The failure probability is then assumed to behave as follows:

$$p_f(\lambda) = \text{Prob}\big(M(\lambda) \leq 0\big) \underset{\lambda \to 1}{\approx} q(\lambda)\exp\big\{-a(\lambda - b)^c\big\}, \tag{9.2}$$

where the function $q(\lambda)$ is slowly varying compared with the exponential function $\exp\{-a(\lambda - b)^c\}$. It may be pointed out that the assumed behaviour of the failure probability applies to any safety margin for which FORM or SORM approximations can be used, but actually its range of applicability is much wider than that.

Clearly, the target failure probability $p_f = p_f(1)$ can be obtained from values of $p_f(\lambda)$ for $\lambda < 1$. It is now far easier to estimate the (larger) failure probabilities $p_f(\lambda)$ for $\lambda < 1$ accurately than the target value itself, since they are larger and hence require less simulations. Fitting the parametric function given by Eq. (9.2) for $p_f(\lambda)$ to the estimated values would then allow us to provide an estimate of the target value by extrapolation. The viability of this approach is demonstrated by both analytical and numerical examples in [1] and [3].

In the next sections, the Naess et al. [1] approach is extended to reliability-based design and calibration.

2 Using Enhanced Monte Carlo to Optimize a Design Parameter

First consider a typical component design, the reliability of which is governed by the safety margin:

$$M(\alpha) = G(X_1, \dots, X_n; \alpha) \tag{9.3}$$

with

$$p_f(\alpha) = \text{Prob}\big(M(\alpha) \leq 0\big), \tag{9.4}$$

where α acts as a design factor which "controls" the reliability of the component. The objective is now to determine the (assumed to be unique) value of $\alpha = \alpha_T$ that corresponds to a specified (target) component failure probability p_{fT}, i.e.:

$$\alpha_T : \text{Prob}\big(M(\alpha_T) \leq 0\big) = p_{fT}, \tag{9.5}$$

This assumes that the function $p_f(\alpha)$ is a monotonic function, that is, that the safety of the system either strictly increases or strictly decreases as the design factor

α increases and approaches α_T. In practical design situations, α may represent a safety factor, a partial load or resistance factor, or some exceedance level and, the condition of monotonicity is generally speaking satisfied, unless the problem relates to a poor or an unfeasible design.

A more general situation, typical in the context of calibration of design specifications, consists of having the safety margin controlled by a design check function $c(\alpha) = c(x_{1c}, \ldots, x_{nc}; \alpha)$ involving characteristic values x_{ic} of each basic variable X_i. Admissible designs are such that $c(\alpha) \leq 0$. Minimal acceptable designs are marked by $c(\alpha) = 0$, an assumption which is made throughout this paper. Often the design check function $c(\alpha)$ is selected to be the same mathematical function as G but this is not required—all that matters is that the resulting safety margin $M(\alpha) = G(X_1, \ldots, X_n | c(x_{1c}, \ldots, x_{nc}; \alpha) = 0)$ is monotonic with respect to α in its approach to the target α_T. Hence the objective is to determine α_T as follows:

$$\alpha_T : \text{Prob}\big(G\big(X_1, \ldots, X_n | c(x_{1c}, \ldots, x_{nc}; \alpha_T) = 0\big) \leq 0\big) = p_{fT}. \qquad (9.6)$$

Typically, p_{fT} is a very small target probability and hence the behavior of p_f as a function of α is similar to a deep tail estimation problem so that it is reasonable to assume that:

$$p_f(\alpha) \underset{\alpha \to 1}{\approx} q(\alpha) \exp\big\{-a(\alpha - b)^c\big\}, \qquad (9.7)$$

where $q(\alpha)$ is slowly varying compared to the exponential expression.

To illustrate this premise, consider a basic load and resistance safety margin $M(\alpha) = R(\alpha) - S$ controlled by a design check function $c(\alpha) = (r_c(\alpha)/\alpha) - s_c$, where r_c and s_c are characteristic values of a resistance R and a load S, and α acts as a partial resistance factor ($\alpha > 1$). Assume the load S is Weibull distributed with exponent d and scale parameter s_0, then the characteristic load s_c at its $(1-\theta)$ quantile, is equal to $s_c = (-\ln \theta)^{1/d} s_0 = k s_0$ where k is a known positive constant > 1. First consider the limiting case where the variance of R is zero, $\sigma_R^2 = 0$, hence $r_c(\alpha) = \alpha s_c$ such that $p_f(\alpha) = \text{Prob}(M(\alpha) \leq 0) = \exp(-(\alpha k)^d)$ which is fully consistent with Eq. (9.7) above. If the variance of $R(\alpha)$ now increases, then the mean resistance will shift even further down the tail since $r_c(\alpha)$ is a small quantile of R. But, the function $p_f(\alpha)$ will only be slightly "contaminated" by a much slower varying function of α; however, and this is certainly valid in the tail area as $\alpha \to \alpha_T$, the general form in Eq. (9.7) will persist and it is amenable to be fitted to data pairs $(p_f(\alpha), \alpha)$ obtained for (much) higher failure probabilities.

Once a satisfactory fit is achieved, the target value α_T corresponding to p_{fT} needs to be estimated, a problem which is similar to a high quantile estimation.

3 Extension to System Reliability

Using Monte Carlo methods for system reliability analysis has several attractive features, the most important being that the failure criterion is relatively easy to check almost irrespective of the complexity of the system. In order to limit the amount of computational effort that may be involved, it is useful to extend the above approach to systems.

Let $M_j(\alpha) = G_j(X_1, \ldots, X_n, \alpha)$, $j = 1, \ldots, m$ be a set of m given safety margins expressed in terms of n basic variables and a single design parameter α. The series system reliability expressed in terms of the failure probability can then be written as,

$$p_f(\alpha) = \text{Prob}\left(\bigcup_{j=1}^{m} \{M_j(\alpha) \leq 0\}\right), \tag{9.8}$$

while for the parallel system,

$$p_f(\alpha) = \text{Prob}\left(\bigcap_{j=1}^{m} \{M_j(\alpha) \leq 0\}\right). \tag{9.9}$$

In general, any system can be written as a series system of parallel subsystems. The failure probability would then be given as,

$$p_f(\alpha) = \text{Prob}\left(\bigcup_{j=1}^{l} \bigcap_{i \in C_j} \{M_i(\alpha) \leq 0\}\right), \tag{9.10}$$

Here each C_j is a subset of $1, \ldots, m$, for $j = 1, \ldots, l$. The C_js denote the index sets defining the parallel subsystems.

We then make the assumption that $p_f(\alpha)$ can also be represented as in Eq. (9.7) for the system reliability problems. Again, the objective is to determine the value α_T that achieves a stated overall system reliability.

4 Implementation

The method to be described in this section is based on the assumption expressed by Eq. (9.7). For practical applications it is implemented in the following form:

$$p_f(\alpha) \approx q(\alpha) \exp\{-a(\alpha - b)^c\}, \quad \text{for } \alpha_0 \leq \alpha \leq \alpha_T, \tag{9.11}$$

for a suitable value of α_0. An important part of the method is therefore to identify a suitable range for α so that the right hand side of Eq. (9.7) represents a good approximation of $p_f(\alpha)$ for $\alpha \in [\alpha_0, \alpha_T]$.

For a sample of size N of the vector of basic random variables $\mathbf{X} = (X_1, \ldots, X_n)$, let $N_f(\alpha)$ denote the number of outcomes of the random vector in the failure domain of $M(\alpha)$. The estimate of the failure probability is then

$$\hat{p}_f(\alpha) = \frac{N_f(\alpha)}{N}. \tag{9.12}$$

The coefficient of variation C_v of this estimator is

$$C_v\big(\hat{p}_f(\alpha)\big) = \sqrt{\frac{1 - \hat{p}_f(\alpha)}{\hat{p}_f(\alpha) N}}. \tag{9.13}$$

A fair approximation of the 95% confidence interval for the value $\hat{p}_f(\alpha)$ can be obtained as $\mathrm{CI}_{0.95} = (C^-(\alpha), C^+(\alpha))$, where

$$C^{\pm}(\alpha) = \hat{p}_f(\alpha)\big[1 \pm 1.96 C_v\big(\hat{p}_f(\alpha)\big)\big]. \tag{9.14}$$

Assuming now that we have obtained empirical Monte Carlo estimates of the failure probability, the problem then becomes one of optimal use of the information available. By plotting $\log|\log \hat{p}_f(\alpha)/q(\alpha)|$ versus $\log(\alpha - b)$, it is expected that an almost perfectly linear tail behavior will be obtained according to Eq. (9.11). Recalling that the function $q(\alpha)$ was assumed to be slowly varying compared with the exponential function $\exp\{-a(\alpha - b)^c\}$ for values of α close to α_T, it is now tentatively proposed to replace $q(\alpha)$ by a suitable constant value, q say, for tail values of α, say $\alpha > \alpha_1$ ($\geq \alpha_0$). Hence, we will investigate the viability of the following simpler version of Eq. (9.11):

$$p_f(\alpha) \approx q \exp\big\{-a(\alpha - b)^c\big\}, \quad \text{for } \alpha_1 \leq \alpha \leq \alpha_T, \tag{9.15}$$

for a suitable choice of α_1.

The problem of finding the optimal values of the parameters a, b, c, q is carried out by optimizing the fit on the log level by minimizing the following mean square error function with respect to all four arguments [4],

$$F(a, b, c, q) = \sum_{j=1}^{M} w_j \big(\log \hat{p}_f(\alpha_j) - \log q + a(\alpha_j - b)^c\big)^2, \tag{9.16}$$

where $\alpha_1 < \cdots < \alpha_M$ denotes the set of α values where the failure probability is empirically estimated. The w_j denote weight factors that put more emphasis on the more reliable data points, alleviating the heteroscedasticity of the estimation problem at hand. The choice of weight factor is to some extent arbitrary. In this paper, we use $w_j = (\log C^+(\alpha_j) - \log C^-(\alpha_j))^{-\theta}$ with the values $\theta = 1$ and 2, combined with a Levenberg–Marquardt least squares optimization method [5]. Note that the form of w_j puts some restriction on the use of the data. Usually, there is a level α_j beyond which w_j is no longer defined. Hence, the summation in Eq. (9.16) has to stop before that happens. Also, the data should be preconditioned by establishing the tail marker α_1 in a sensible way.

Although the Levenberg–Marquardt method as described above generally works well, it may be simplified by exploiting the structure of F. It is realized by scrutinizing Eq. (9.16) that if b and c are fixed, the optimization problem reduces to a standard weighted linear regression problem. That is, with both b and c fixed, the optimal values of a and $\log q$ are found using closed form weighted linear regression formulas in terms of w_j, $y_j = \log \hat{p}_f(\alpha_j)$ and $x_j = (\alpha_j - b)^c$.

It is obtained that the optimal values of a and q are given by the relations,

$$a^*(b, c) = -\frac{\sum_{j=1}^{M} w_j (x_j - \overline{x})(y_j - \overline{y})}{\sum_{j=1}^{M} w_j (x_j - \overline{x})^2}, \tag{9.17}$$

and

$$\log q^*(b, c) = \overline{y} + a^*(b, c)\overline{x}, \tag{9.18}$$

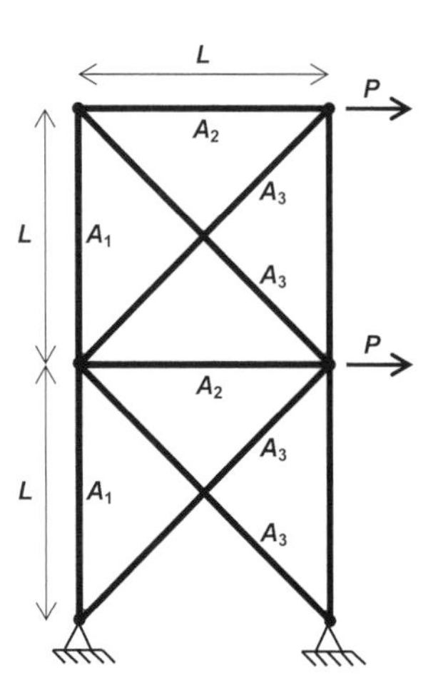

Fig. 9.1 Ten-bar truss structure

where $\overline{x} = \sum_{j=1}^{M} w_j x_j / \sum_{j=1}^{M} w_j$, with a similar definition of $\overline{y}$.

The Levenberg–Marquardt method may now be used on the function $\tilde{F}(b,c) = F(a^*(b,c), b, c, q^*(b,c))$ to find the optimal values b^* and c^*, and then the corresponding a^* and q^* can be calculated from Eqs. (9.17) and (9.18).

For estimation of the confidence interval for the predicted target quantile α_T provided by the optimal curve, the empirical confidence band is reanchored to the optimal curve. The range of fitted curves that stay within the reanchored confidence band will determine an optimized confidence interval of the predicted value.

5 Numerical Examples

The examples in the following two sections all have simple explicit limit state functions in terms of the basic random variables. The computational issue is therefore minor and no effort has been made to investigate the possibility of implementing more effective sampling strategies. If the proposed method were to be used in combination with computationally demanding procedures involving e.g. a FE method for calculating the sample, it would be necessary in general to use more effective sampling strategies than the brute force procedure used here.

5.1 Component Load Factor Calibration

In this first example, the 10-bar truss structure shown in Fig. 9.1 is studied. An enhanced Monte-Carlo reliability analysis of this truss is given in [1]. Here a load factor for a transversal load P is calibrated in order to achieve a target reliability of (10^{-6}) with respect to the horizontal sway of the truss. The ten truss members are cut from three different aluminum rods with cross-sectional areas A_1, A_2 and

Table 9.1 Basic variables

	Mean value	Coef. of var.	Prob. distr.	Char. value in (20)
A_1	10^{-2} m^2	0.05	Normal	1% quantile
A_2	$1.5 \cdot 10^{-3}$ m^2	0.05	Normal	1% quantile
A_3	$6.0 \cdot 10^{-3}$ m^2	0.05	Normal	1% quantile
B	1.0	0.10	Normal	mean
E	$6.9 \cdot 10^4$ MPa	0.05	Lognormal	1% quantile
P	based on Eq. (9.21)	0.10	Gumbel	95% quantile
d_0	0.1 m	-	-	-
L	9.0 m	-	-	-

A_3, as shown in Fig. 9.1. The structure is subjected to external loads P as shown in Fig. 9.1. The horizontal displacement D at the upper right hand corner of the truss structure can be written as [6]:

$$D = \frac{BPL}{A_1 A_3 E}\left\{\frac{4\sqrt{2}A_1^3(24A_2^2 + A_3^2) + A_3^3(7A_1^2 + 26A_2^2)}{D_T} + 4A_1A_2A_3\frac{20A_1^2 + 76A_1A_2 + 10A_3^2}{D_T} + 4\sqrt{2}A_1A_2A_3^2\frac{25A_1 + 29A_2}{D_T}\right\} \tag{9.19}$$

where $D_T = 4A_2^2(8A_1^2 + A_3^2) + 4\sqrt{2}A_1A_2A_3(3A_1 + 4A_2) + A_1A_3^2(A_1 + 6A_2)$ and E is Young's modulus. The random variable B accounts for model uncertainties. It is assumed that A_1, A_2, A_3, B, P, E are independent basic random variables. Their properties are summarized in Table 9.1. Also shown are the characteristic values used in the design check Eq. (9.21).

The safety margin

$$M(\alpha) = d_0 - D\big(A_1, A_2, A_3, B, E, P(\alpha)\big), \tag{9.20}$$

and the design check constraint is

$$c(\alpha) = d_0 - D\big(A_{1c}, A_{2c}, A_{3c}, B_c, E_c, \alpha P_c(\alpha)\big), \tag{9.21}$$

where α represents the transversal load factor.

Figures 9.2 and 9.3 show the optimized fitted parametric curve to the empirical data in a log plot for sample size 10^5 and for weighted regression coefficients $\theta = 2$ and $\theta = 1$, respectively. The difference between the two tail extrapolations is minimal. Applying the proposed procedure with a sample of size 10^5 gives the estimated value for α_T with the 95% confidence interval shown in Table 9.2 for both $\theta = 2$ and 1. Note that a crude Monte Carlo simulation verification of ($\alpha_T = 1.46$, $p_{fT} = 10^{-6}$) using $3 \cdot 10^9$ samples to within 2.5% at 95% confidence requires a computation time of about 24 h on a laptop computer. The CPU time for the results shown in Table 9.2 was only about 40 seconds on a standard laptop.

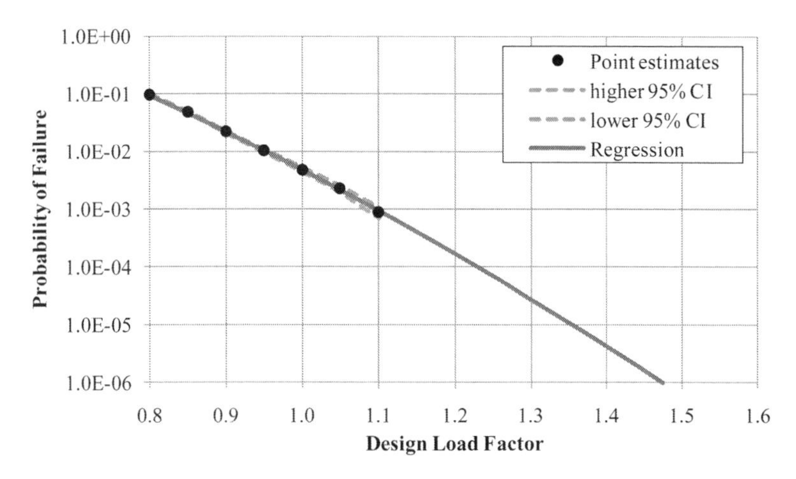

Fig. 9.2 Ten-bar truss structure. Sample size 10^5—weighted regression $\theta = 2$

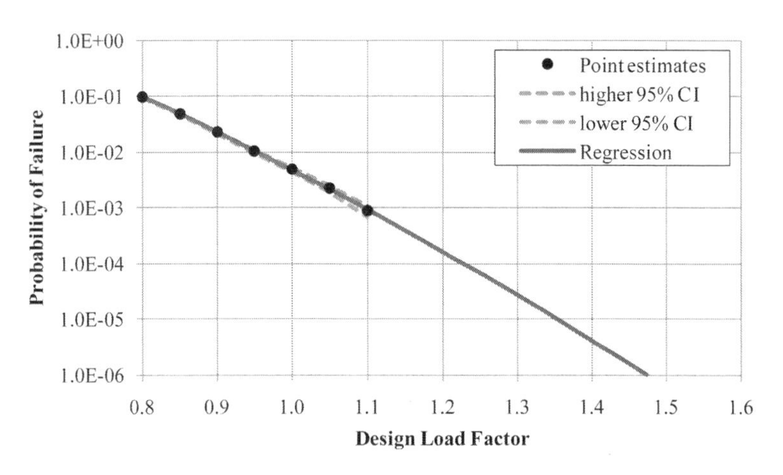

Fig. 9.3 Ten-bar truss structure. Sample size 10^5—weighted regression $\theta = 1$

Table 9.2 Optimal load factor α_T corresponding to $p_{fT} = 10^{-6}$ using sample size 10^5

	$\theta = 2$	$\theta = 1$
higher 95% CI	1.48	1.48
α_T	1.47	1.47
lower 95% CI	1.46	1.46

5.2 Design Resistance Safety Factor in a Series System

This example concerns the maximum internal forces in the members of a statically determinate 13-member truss structure subjected to external loading. The structure

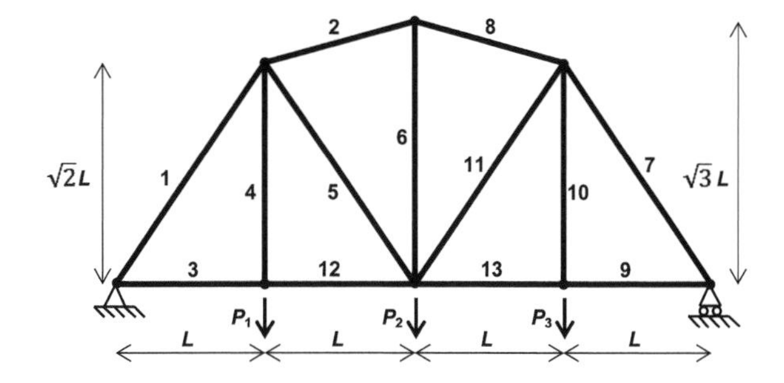

Fig. 9.4 Truss bridge example

is shown in Fig. 9.4, which also displays the numbering of the truss elements from 1 to 13.

The external loads P_1, P_2, P_3 which are acting on the structure as shown in Fig. 9.4, are modelled as independent Gaussian variables. The capacity for axial stress of truss element number j is expressed as $R_j = \sigma_{yj} A_j$ where σ_{yj} = the yield stress (MPa) and A_j = the cross-sectional area of this element (cm^2), and α is a resistance safety factor >1 used as a division factor in the design check equation below. It is assumed that $A_1 = A_7 = 18.7$, $A_2 = A_8 = 13.1$, $A_3 = A_9 = A_{12} = A_{13} = 11.7$, $A_4 = A_{10} = 11.3$, $A_5 = A_{11} = 3.3$, $A_6 = 8.0$. The 13 yield stresses are assumed to be independent Gaussian variables. The 16 basic random variables in this problem are listed in Table 9.3.

$$
\begin{aligned}
M_1 &= R_1 - 0.9186 P_1 - 0.6124 P_2 - 0.3062 P_3 \\
M_2 &= R_2 - 0.3029 P_1 - 0.6058 P_2 - 0.3029 P_3 \\
M_3 &= R_3 - 0.5303 P_1 - 0.3535 P_2 - 0.1768 P_3 \\
M_4 &= R_4 - P_1 \\
M_5 &= R_5 + 0.4186 P_1 - 0.3876 P_2 - 0.1938 P_3 \\
M_6 &= R_6 - 0.1835 P_1 - 0.3670 P_2 - 0.1835 P_3 \\
M_7 &= R_7 - 0.3062 P_1 - 0.6124 P_2 - 0.9186 P_3 \\
M_8 &= R_8 - 0.3029 P_1 - 0.6058 P_2 - 0.3029 P_3 \\
M_9 &= R_9 - 0.1768 P_1 - 0.3535 P_2 - 0.5303 P_3 \\
M_{10} &= R_{10} - P_1 \\
M_{11} &= R_{11} - 0.1938 P_1 - 0.3876 P_2 + 0.4186 P_3
\end{aligned}
\qquad (9.22)
$$

Table 9.3 The 16 basic variables

	Mean Value	Coef. of Var.	Prob. distr.	Char. value in Eq. (9.22)
P_j, $j = 1, 2, 3$	89 kN	0.15	Normal	99% quantile
σ_{yj}, $j = 1, \ldots, 13$	based on Eq. (9.23)	0.15	Normal	5% quantile
L	2.54 m	-	-	-

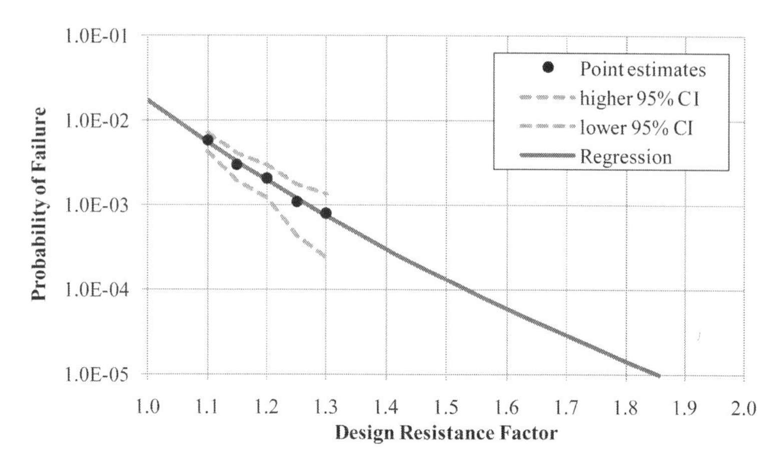

Fig. 9.5 Truss bridge. Sample size 10^4—weighted regression with $\theta = 1$

$$M_{12} = R_{12} - 0.5303P_1 - 0.3536P_2 - 0.1768P_3$$
$$M_{13} = R_{13} - 0.1768P_1 - 0.3536P_2 - 0.5303P_3$$

The 13 design check equations have the same mathematical set of 13 equations except that the deterministic characteristic values of Table 9.3 are used and a resistance safety factor is involved. The most severe constraint is the compressive stress in members 1 and 7 which therefore governs the design of the system as a whole:

$$\frac{\sigma_{yc}(\alpha)A_1}{\alpha} - 1.8372P_c = 0, \tag{9.23}$$

The objective is to find the value α_T such that the series system failure probability given by Eq. (9.8) is equal to a target $p_{fT} = 10^{-5}$. The log plot of $p_f(\alpha)$ versus α is shown in Figs. 9.5 and 9.6 for $\theta = 1$ and for samples of size 10^4 and 10^5, respectively. The estimated α_T corresponding to $p_f = 10^{-5}$ together with their CIs are shown in Table 9.4. A Winbugs script runs the entire analysis in under 1 min for 10^5 samples. As a contrast, crude Monte Carlo simulation with $5 \cdot 10^9$ samples confirms ($\alpha_T = 1.89$, $p_f = 10^{-5}$) for the series system accurate to within about 0.5% with 95% confidence, but requires a computation time of about 24 h on a laptop computer.

6 Conclusions

In this paper, we have described a Monte Carlo based method for a reliability-based calibration of design parameters such as load/resistance factors, safety factors or specification levels of structural systems. It has been shown that the method may provide good estimates of design factors for structural systems with a moderate computational effort. It has been pointed out that the use of Monte Carlo methods

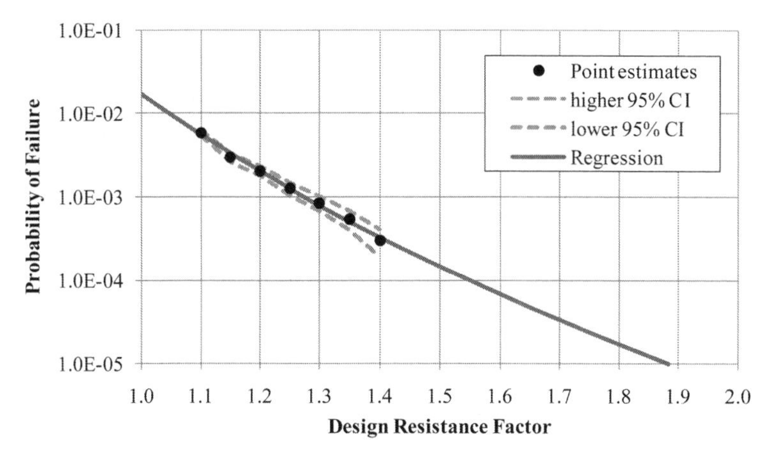

Fig. 9.6 Truss bridge. Sample size 10^5—weighted regression with $\theta = 1$

Table 9.4 Optimal resistance safety factor α_T corresponding to a system $p_{fT} = 10^{-5}$ using sample size 10^4 and 10^5 with $\theta = 1$

	$N = 10^4$	$N = 10^5$
higher 95% CI	1.95	1.92
α_T	1.85	1.88
lower 95% CI	1.69	1.80

for system reliability analysis has several very attractive features, the most important being that the failure criterion is usually relatively easy to check almost irrespective of the complexity of the system and the number of basic random variables.

Acknowledgements The first author is grateful for the financial support from the Research Council of Norway (NFR) through the Centre for Ships and Ocean Structures (CeSOS) at the Norwegian University of Science and Technology. The second and third authors gratefully acknowledge financial support from NSERC Canada.

References

1. Naess, A., Leira, B.J., Batsevych, O.: System reliability analysis by enhanced Monte Carlo simulation. Struct. Saf. **31**, 349–355 (2009)
2. Bucher, C.: Asymptotic sampling for high-dimensional reliability analysis. Probab. Eng. Mech. **24**, 504–510 (2009)
3. Naess, A., Leira, B.J., Batsevych, O.: Efficient reliability analysis of structural systems with a high number of limit states. In: Proceedings 29th International Conference on Offshore Mechanics and Arctic Engineering, pp. 809–814, OMAE–2010–21179. ASME, New York (2010)
4. Naess, A., Gaidai, O.: Estimation of extreme values from sampled time series. Struct. Saf. **31**, 325–334 (2009)
5. Gill, P., Murray, W., Wright, M.H.: Practical Optimization. Academic Press, London (1981)
6. Choi, S.-K., Grandhi, R.V., Canfield, R.A.: Reliability-Based Structural Design. Springer, London (2007)

Chapter 10
Optimal Design of Base-Isolated Systems Under Stochastic Earthquake Excitation

Hector A. Jensen, Marcos A. Valdebenito, and Juan G. Sepulveda

Abstract The development of a general framework for reliability-based design of base-isolated structural systems under uncertain conditions is presented. The uncertainties about the structural parameters as well as the variability of future excitations are characterized in a probabilistic manner. Nonlinear elements composed by hysteretic devices are used for the isolation system. The optimal design problem is formulated as a constrained minimization problem which is solved by a sequential approximate optimization scheme. First excursion probabilities that account for the uncertainties in the system parameters as well as in the excitation are used to characterize the system reliability. The approach explicitly takes into account all non-linear characteristics of the combined structural system (superstructure-isolation system) during the design process. Numerical results highlight the beneficial effects of isolation systems in reducing the superstructure response.

1 Introduction

There has been a growing interest during the last years in the application of base isolation techniques in order to improve the earthquake resistant performance of civil structures such as buildings, bridges, nuclear reactors, etc. [8, 10, 23, 30, 33]. In fact, the potential advantages of seismic isolation and the recent advancements in isolation-system products have lead to the design and construction of an increasing number of seismically isolated structural systems. Also, seismic isolation is extensively used for seismic retrofitting of existing structures [11, 26]. One of the difficulties in the design of base-isolated structural systems is the explicit consideration of the nonlinear behavior of the isolators during the design process. Similarly, the consideration of uncertainty about the structural model and the potential variability of future ground motions is a major challenge in the analysis and design of these

H.A. Jensen (✉) · M.A. Valdebenito · J.G. Sepulveda
Santa Maria University, Av. España 1680, Valparaiso, Chile
e-mail: hector.jensen@usm.cl

M.A. Valdebenito
e-mail: marcos.valdebenito@usm.cl

M. Papadrakakis et al. (eds.), *Computational Methods in Stochastic Dynamics*,
Computational Methods in Applied Sciences 26,
DOI 10.1007/978-94-007-5134-7_10,

systems. The goal of this work is the development of a general framework for reliability based design of base-isolated systems under uncertain conditions. In particular, base-isolated building structures subject to earthquake excitation are considered in this study. A probabilistic approach is adopted for addressing the uncertainties about the structural model as well as the variability of future excitations. The uncertain earthquake excitation is modeled as a non-stationary stochastic process with uncertain model parameters. Specifically, a point-source model characterized by the moment magnitude and epicentral distance is adopted in this formulation [6]. Isolation elements composed by hysteretic devices are used for the isolation system. The hysteretic behavior of the devices is characterized by a Bouc–Wen type model [5]. The model provides general parametric hysteresis rules that gives a smooth transition of the change of stiffness as the deformation of the nonlinear elements changes. The reliability-based optimization problem is formulated as the minimization of an objective function subject to multiple design requirements including reliability constraints. First excursion probabilities are used as measures of system reliability. Such probabilities are estimated by an adaptive Markov Chain Monte Carlo procedure [4]. A sequential optimization approach based on global conservative, convex and separable approximations is implemented for solving the optimization problem [14, 18, 21]. The approach explicitly takes into account all non-linear characteristics of the structural response and it allows for a complex characterization of structural systems and excitation models. The solution of the equation of motion of the combined system (superstructure-isolation system) required during the simulation process is computed by a modified Runge–Kutta scheme of fourth-order. A numerical example is presented in order to illustrate the applicability and effectiveness of the proposed framework for reliability-based design of base-isolated buildings.

2 Reliability-Based Design Problem

The optimal design problem is defined as the identification of a vector $\{\phi\}$ of design variables that minimizes an objective function, that is

$$\text{Minimize } f\big(\{\phi\}\big) \tag{10.1}$$

subject to design constraints

$$h_j\big(\{\phi\}\big) \leq 0, \quad j = 1, \ldots, n_c \tag{10.2}$$

and side constraints

$$\phi_i^l \leq \phi_i \leq \phi_i^u, \quad i = 1, \ldots, n_d \tag{10.3}$$

The objective function is defined in terms of quantities such as initial, construction, repair, or downtime costs. On the other hand, the design constraints are given in terms of reliability constraints and/or constrains related to deterministic design requirements. In a stochastic setting the reliability constraints are usually defined in terms of failure probabilities. These probabilities provide a measure of the plausibility of the occurrence of unacceptable behavior (failure) of the system, based on the available information. The probability of failure $P_{F_j}(\{\phi\})$ corresponding to a failure event F_j evaluated at the design $\{\phi\}$ can be expressed in terms of the multidimensional probability integral [13, 15]

$$P_{F_j}(\{\phi\}) = \int_{\Theta} I_{F_j}(\{\phi\}, \{\theta\}) q(\{\theta\}) \, d\{\theta\} \tag{10.4}$$

where $I_{F_j}(\{\phi\}, \{\theta\})$ is the indicator function for failure, which is equal to one if the system fails and zero otherwise, and $\{\theta\}$, θ_i, $i = 1, \ldots, n_u$ is the vector that represents the uncertain system parameters involved in the problem (structural parameters and excitation). The uncertain system parameters $\{\theta\}$ are modeled using a prescribed probability density function $q(\{\theta\})$ which incorporates available knowledge about the system. Note, that the failure probability function $P_{F_j}(\{\phi\})$ accounts for the uncertainty in the system parameters as well as the uncertainties in the excitation. A model prediction error, that is, the error between the response of the actual system and the response of the model, can also be considered in the formulation [12, 31]. In this case the prediction error may be modeled probabilistically by augmenting the vector $\{\theta\}$ to form an uncertain parameter vector composed of both the structural and excitation model parameters as well as the model prediction-error. The failure domain $\Omega_{F_j}(\{\phi\})$ corresponding to the failure event F_j evaluated at the design $\{\phi\}$ is typically described in terms of a performance function g_j as

$$\Omega_{F_j}(\{\phi\}) = \big\{\{\theta\} \mid g_j(\{\phi\}, \{\theta\}) \geq 0\big\} \tag{10.5}$$

Then, the probability of failure can also be expressed as the integral of the probability density function $q(\{\theta\})$ over the failure domain in the form

$$P_{F_j}(\{\phi\}) = \int_{\Omega_{F_j}(\{\phi\})} q(\{\theta\}) \, d\{\theta\} \tag{10.6}$$

With the previous notation, a reliability constraint can be written as $h_j(\{\phi\}) = P_{F_j}(\{\phi\}) - P^*_{F_j} \leq 0$, where $P^*_{F_j}$ is the target failure probability. The last inequality expresses the requirement that the probability of system failure must be smaller than an appropriate tolerance. It is noted that in the context of stochastic design a system that corresponds to a feasible design can not be certified with complete certainty, but with a tolerance $P^*_{F_j}$. In other words, the system will operate safely within the pre-specified probability of failure tolerance.

3 Structural Model

In general, base-isolated buildings are designed such that the superstructure remains elastic. Hence, the structure is modeled as a linear elastic system in the present formulation. The base and the floors are assumed to be infinitely rigid in plane. The superstructure and the base are modeled using three degrees of freedom per floor at the center of mass. Each nonlinear isolation element is modeled explicitly using the Bouc–Wen model. Let $\{x_s(t)\}$ be the n-th dimensional vector of displacements

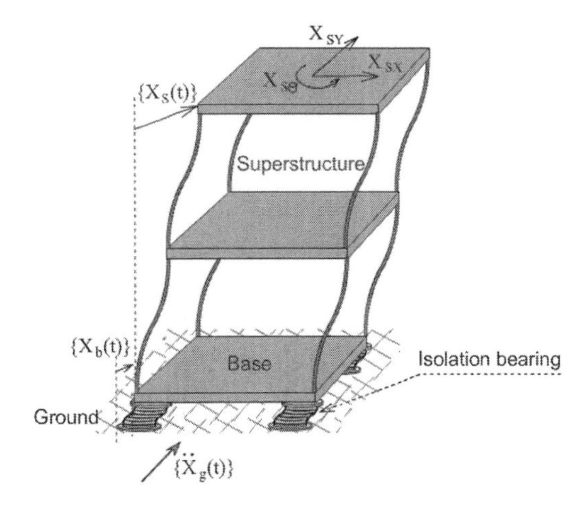

Fig. 10.1 Schematic representation of the base-isolated structural model

for the superstructure with respect to the base, and $[M_s]$, $[C_s]$, and $[K_s]$ be the corresponding mass, damping and stiffness matrices. Also, let $\{x_b(t)\}$ be the vector of base displacements with respect to the ground and $[G_s]$ be the matrix of earthquake influence coefficients of dimension $n \times 3$, that is, the matrix that couples the excitation components of the vector $\{\ddot{x}_g(t)\}$ to the degrees of freedom of the superstructure. The schematic representation of the base-isolated structural system as well as the displacement coordinates are shown in Fig. 10.1. The equation of motion of the elastic superstructure is then expressed in the form

$$\begin{aligned} &[M_s]\big\{\ddot{x}_s(t)\big\} + [C_s]\big\{\dot{x}_s(t)\big\} + [K_s]\big\{x_s(t)\big\} \\ &\quad = -[M_s][G_s]\big(\big\{\ddot{x}_b(t)\big\} + \big\{\ddot{x}_g(t)\big\}\big) \end{aligned} \tag{10.7}$$

where $\{\ddot{x}_b(t)\}$ is the vector of base accelerations relative to the ground. On the other hand, the equation of motion of the base can be written as

$$\begin{aligned} &\big([G_s]^T[M_s][G_s] + [M_b]\big)\big(\big\{\ddot{x}_b(t)\big\} + \big\{\ddot{x}_g(t)\big\}\big) \\ &\quad + [G_s]^T[M_s]\big\{\ddot{x}_s(t)\big\} + \{f_{is}\} = \{0\} \end{aligned} \tag{10.8}$$

where $[M_b]$ is the diagonal mass matrix of the rigid base, and $\{f_{is}\}$ is the vector containing the linear and nonlinear isolation elements forces (three components). The characterization of such forces is treated in a subsequent Section. Rewriting the previous equations, the combined equation of motion of the base-isolated structure system can be formulated in the form

$$\begin{aligned} &\begin{bmatrix} [M_s] & [M_s][G_s] \\ [G_s]^T[M_s] & [M_b] + [G_s]^T[M_s][G_s] \end{bmatrix} \begin{Bmatrix} \{\ddot{x}_s(t)\} \\ \{\ddot{x}_b(t)\} \end{Bmatrix} + \begin{bmatrix} [C_s] & [0] \\ [0]^T & [0] \end{bmatrix} \begin{Bmatrix} \{\dot{x}_s(t)\} \\ \{\dot{x}_b(t)\} \end{Bmatrix} \\ &\quad + \begin{bmatrix} [K_s] & [0] \\ [0]^T & [0] \end{bmatrix} \begin{Bmatrix} \{x_s(t)\} \\ \{x_b(t)\} \end{Bmatrix} \\ &= -\begin{Bmatrix} [M_s][G_s] \\ [M_b] + [G_s]^T[M_s][G_s] \end{Bmatrix} \{\ddot{x}_g(t)\} - \begin{Bmatrix} \{0\} \\ \{f_{is}(t)\} \end{Bmatrix} \end{aligned} \tag{10.9}$$

It is noted that elastic and viscous isolation elements can also be incorporated in the isolation model. Also, the above formulation can be directly extended to more complex cases, for example, to nonlinear models for the superstructure.

4 Earthquake Excitation Model

The ground acceleration is modeled as a non-stationary stochastic process. In particular, a point-source model characterized by the moment magnitude M and epicentral distance r is considered here [3, 6]. The model is a simple, yet powerful means for simulating ground motions and it has been successfully applied in the context of earthquake engineering. The time-history of the ground acceleration for a given magnitude M and epicentral distance r is obtained by modulating a white noise sequence by an envelope function and subsequently by a ground motion spectrum through the following steps: (1) generate a discrete-time Gaussian white noise sequence $\omega(t_j) = \sqrt{I/\Delta t}\theta_j,\ j = 1,\ldots,n_T$, where $\theta_j,\ j = 1,\ldots,n_T$, are independent, identically distributed standard Gaussian random variables, I is the white noise intensity, Δt is the sampling interval, and n_T is the number of time instants equal to the duration of the excitation T divided by the sampling interval; (2) the white noise sequence is modulated by an envelope function $h(t,M,r)$ at the discrete time instants; (3) the modulated white noise sequence is transformed to the frequency domain; (4) the resulting spectrum is normalized by the square root of the average square amplitude spectrum; (5) the normalized spectrum is multiplied by a ground motion spectrum (or radiation spectrum) $S(f,M,r)$ at discrete frequencies $f_l = l/T, l = 1,\ldots,n_T/2$; (6) the modified spectrum is transformed back to the time domain to yield the desired ground acceleration time history. Details of the characterization of the envelope function $h(t,M,r)$ and the ground acceleration spectrum $S(f,M,r)$ are provided in the subsequent sections. The probabilistic model for the seismic hazard at the emplacement is complemented by considering that the moment magnitude M and epicentral distance r are also uncertain. The uncertainty in moment magnitude is modeled by the Gutenberg–Richter relationship truncated on the interval [6.0, 8.0], which leads to the probability density function [24]

$$p(M) = \frac{be^{-bM}}{e^{-6.0b} - e^{-8.0b}}, \quad 6.0 \leq M \leq 8.0 \tag{10.10}$$

where b is a regional seismicity factor. For the uncertainty in the epicentral distance r, a lognormal distribution with mean value $\bar{r}$ (km) and standard deviation σ_r (km) is used. The point source stochastic model previously described is well suited for generating the high-frequency components of the ground motion (greater than 0.1 Hz). Low-frequency components can also be introduced in the analysis by combining the above methodology with near-fault ground motion models [25].

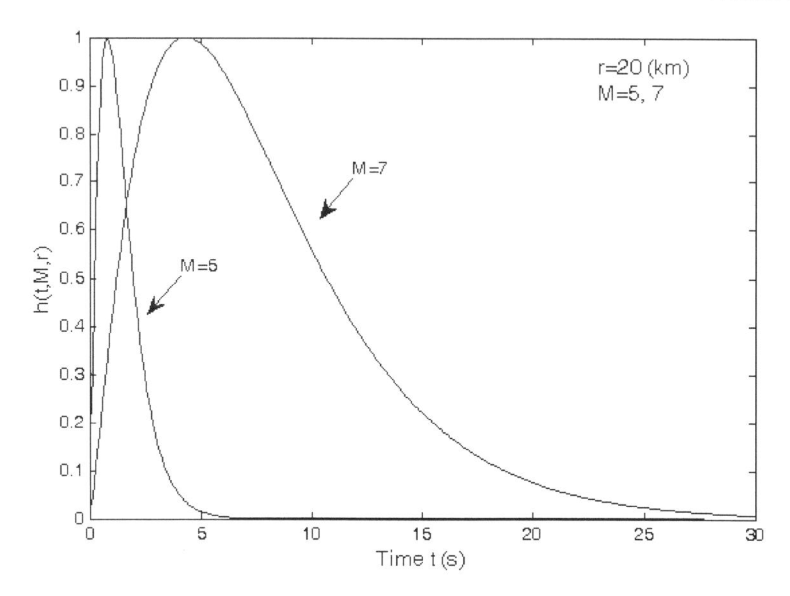

Fig. 10.2 Envelope function for epicentral distance $r = 20$ km and moment magnitudes $M = 5$ and $M = 7$

4.1 Envelope Function

The envelope function for the ground acceleration is represented by [6, 28]

$$h(t, M, r) = a_1 \left(\frac{t}{t_n}\right)^{a_2} e^{-a_3(t/t_n)} \tag{10.11}$$

where

$$a_2 = \frac{-0.2\ln(0.05)}{1+0.2(\ln(0.2)-1)}, \qquad a_3 = \frac{a_2}{0.2}, \qquad a_1 = \left(\frac{e^1}{0.2}\right)^{a_2} \tag{10.12}$$

The envelope function has a peak equal to unity when $t = 0.2t_n$, and $h(t, M, r) = 0.05$ when $t = t_n$, with $t_n = 2.0T_{gm}$, where T_{gm} is the duration of ground motion, expressed as a sum of a path dependent and source dependent component $T_{gm} = 0.05\sqrt{r^2 + h^2} + 0.5/f_a$, where r is the epicentral distance, and the parameters h and f_a (corner frequency) are moment dependent given by $\log(h) = 0.15M - 0.05$ and $\log(f_a) = 2.181 - 0.496M$ [3]. As an example Fig. 10.2 shows the envelope function for $r = 20$ km, and $M = 5$ and $M = 7$. Note that increasing the moment magnitude increases the duration of the envelope function, as expected.

4.2 Ground Motion Spectrum

The total spectrum of the motion at a site $S(f,M,r)$ is expressed as the product of the contribution from the earthquake source $E(f,M)$, path $P(f,r)$, site $G(f)$ and type of motion $I(f)$, i.e.

$$S(f,M,r) = E(f,M)P(f,r)G(f)I(f) \tag{10.13}$$

The source component is given by

$$E(f,M) = CM_0(M)S_a(f,M) \tag{10.14}$$

where C is a constant, $M_0(M) = 10^{1.5M+10.7}$ is the seismic moment, and the factor S_a is the displacement source spectrum given by [3]

$$S_a(f,M) = \frac{1-\varepsilon}{1+(f/f_a)^2} + \frac{\varepsilon}{1+(f/f_b)^2} \tag{10.15}$$

where the corner frequencies f_a and f_b, and the weighting parameter ε are defined, respectively, as $\log(f_a) = 2.181 - 0.496M$, $\log(f_b) = 2.41 - 0.408M$, and $\log(\varepsilon) = 0.605 - 0.255M$. The constant C is given by $C = UR_\Phi VF/4\pi\rho_s\beta_s^3 R_0$, where U is a unit dependent factor, R_Φ is the radiation pattern, V represents the partition of total shear-wave energy into horizontal components, F is the effect of the free surface amplification, ρ_s and β_s are the density and shear-wave velocity in the vicinity of the source, and R_0 is a reference distance.

Next, the path effect $P(f,r)$ which is another component of the process that affects the spectrum of motion at a particular site it is represented by functions that account for geometrical spreading and attenuation

$$P(f,r) = Z\big(R(r)\big)e^{-\pi f R(r)/Q(f)\beta_s} \tag{10.16}$$

where $R(r)$ is the radial distance from the hypocenter to the site given by $R(r) = \sqrt{r^2+h^2}$. The attenuation quantity $Q(f)$ is taken as $Q(f) = 180f^{0.45}$ and the geometrical spreading function is selected as $Z(R(r)) = 1/R(r)$ if $R(r) < 70.0$ km and $Z(R(r)) = 1/70.0$ otherwise [3]. The modification of seismic waves by local conditions, site effect $G(f)$, is expressed by the multiplication of a diminution function $D(f)$ and an amplification function $A(f)$. The diminution function accounts for the path-independent loss of high frequency in the ground motions and can be accounted for a simple filter of the form $D(f) = e^{-0.03\pi f}$ [2]. The amplification function $A(f)$ is based on empirical curves given in [7] for generic rock sites. An average constant value equal to 2.0 is considered. Finally, the filter that controls the type of ground motion $I(f)$ is chosen as $I(f) = (2\pi f)^2$ for ground acceleration. The particular values of the different parameters of the stochastic ground acceleration model are given in Table 10.1 (see Application Problem Section). For illustration purposes Fig. 10.3 shows the ground acceleration spectrum for a nominal distance $r = 20$ km, moment magnitudes $M = 5$ and $M = 7$, and model parameters

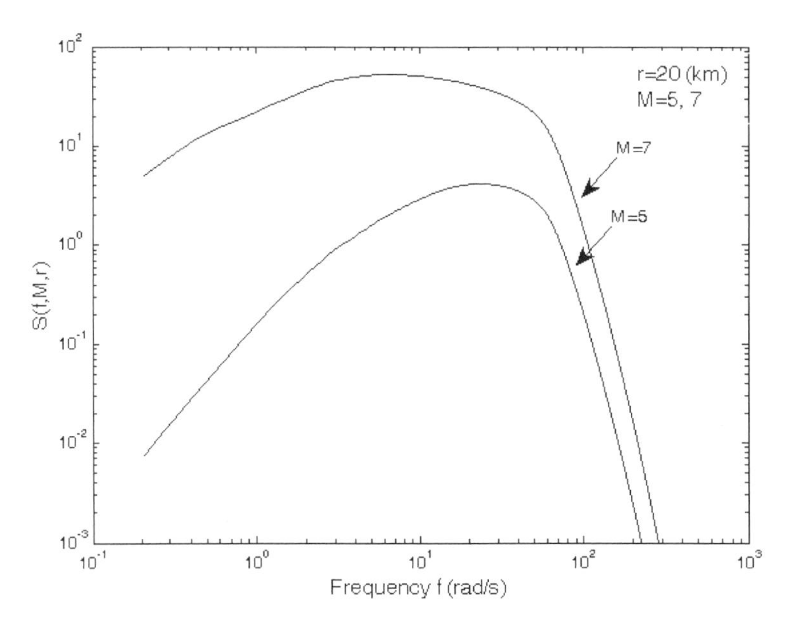

Fig. 10.3 Ground acceleration spectrum for epicentral distance $r = 20$ km and moment magnitudes $M = 5$ and $M = 7$

given in Table 10.1. As the moment magnitude increases, the spectral amplitude increases at all frequencies, with a shift of dominant frequency content towards the lower frequency regime, as anticipated.

5 Isolation Model

Several isolation elements can be used to model isolation systems. They include elastic, viscous, nonlinear fluid dampers, hysteretic (uniaxial or biaxial) elements for bilinear elastomeric bearings, hysteretic (uniaxial or biaxial) elements for sliding bearings, etc. Uniaxial elastomeric bearings with hysteretic behavior, such as lead rubber bearings, are considered in the present implementation. They are modeled using the Bouc–Wen model as [5]

$$U^y \dot{z}(t) = \begin{cases} \dot{x}_b(t)[\alpha - z^n(t)(\gamma \operatorname{sgn}(\dot{x}_b(t)z(t)) + \beta)] & \text{if n is even} \\ \dot{x}_b(t)[\alpha - z^n(t)(\gamma \operatorname{sgn}(\dot{x}_b(t)) + \beta \operatorname{sgn}(z(t)))] & \text{if n is odd} \end{cases} \tag{10.17}$$

where $z(t)$ is a dimensionless hysteretic variable, α, β, and γ are dimensionless quantities, U^y is the yield displacement, $x_b(t)$ and $\dot{x}_b(t)$ represent the base displacement and velocity, respectively, and $\operatorname{sgn}(\cdot)$ is the sign function. The forces activated in the elastomeric isolation bearing are modeled by an elastic-viscoplastic model with strain hardening

$$f_{is}(t) = k_p x_b(t) + c_v \dot{x}_b(t) + (k_e - k_p) U^y z(t) \tag{10.18}$$

where k_e is the pre-yield stiffness, k_p is the post-yield stiffness, c_v is the viscous damping coefficient of the elastomeric bearing, and U^y is the yield displacement. If the post-yield stiffness is written as $k_p = \alpha_L k_e$, where α_L is a factor which defines the extent to which the force is linear, the isolator forces can be expressed as

$$f_{is}(t) = \alpha_L k_e x_b(t) + c_v \dot{x}_b(t) + (1 - \alpha_L) k_e U^y z(t) \tag{10.19}$$

6 Sequential Approximate Optimization

The solution of the reliability-based optimization problem given by Eqs. (10.1)–(10.3) is obtained by transforming it into a sequence of sub-optimization problems having a simple explicit algebraic structure. Thus, the strategy is to construct successive approximate analytical sub-problems. To this end, the objective and the constraint functions are represented by using approximate functions dependent on the design variables. In particular, a hybrid form of linear, reciprocal and quadratic approximations is considered in the present formulation [14, 20, 27]. The approximate discrete sub-optimization problems take the form ($k = 1, 2, \ldots$)

$$\text{Minimize } \tilde{f}_k(\{\phi\}) \tag{10.20}$$

subject to

$$\tilde{h}_{jk}(\{\phi\}) \leq 0, \quad j = 1, \ldots, n_c \tag{10.21}$$

with side constraints

$$\phi_i^l \leq \phi_i \leq \phi_i^u, \quad i = 1, \ldots, n_d \tag{10.22}$$

where $\tilde{f}_k$ and $\tilde{h}_{jk}$, $j = 1, \ldots, n_c$ represent the approximate objective and constraint functions at the current point $\{\phi^k\}$ in the design space, respectively. The approximate objective function is obtained as

$$\tilde{f}_k(\{\phi\}) = f_{1k}(\{\phi\}) + f_{2k}(\{\phi\}) + f_{3k}(\{\phi\}) \tag{10.23}$$

where $f_{1k}(\{\phi\})$ is a linear function in terms of the design variables, $f_{2k}(\{\phi\})$ is a linear function with respect to the reciprocal of the design variables, and $f_{3k}(\{\phi\})$ is a quadratic function of the design variables. They are given by

$$f_{1k}(\{\phi\}) = \sum_{(i^+)} \frac{\partial f(\{\phi^k\})}{\partial \phi_i} \phi_i, \qquad f_{2k}(\{\phi\}) = -\sum_{(i^-)} \frac{\partial f(\{\phi^k\})}{\partial \phi_i} \frac{(\phi_i^k)^2}{\phi_i} \tag{10.24}$$

$$f_{3k}(\{\phi\}) = -2\chi^f \sum_{(i^-)} \frac{\partial f(\{\phi^k\})}{\partial \phi_i} \phi_i \left(\frac{\phi_i}{\phi_i^k} - 2 \right) \tag{10.25}$$

where (i^+) is the group that contains the variables for which the partial derivative of the objective function is positive at the expansion point $\{\phi^k\}$, (i^-) is the group that includes the remaining variables, and χ^f is a user-defined positive scalar that control the conservatism of the approximation [17, 18]. On the other hand, the constraint functions involving reliability measures (reliability constraints) are first transformed as $h^t_j(\{\phi\}) = \ln[P_{F_j}(\{\phi\})]$. Then the transformed constraint functions are approximated in the form

$$\tilde{h}^t_{jk}(\{\phi\}) = h^t_{j1k}(\{\phi\}) + h^t_{j2k}(\{\phi\}) + h^t_{j3k}(\{\phi\}) + \bar{h}^t_{jk}(\{\phi^k\}) \tag{10.26}$$

where

$$h^t_{j1k}(\{\phi\}) = \sum_{(i_j^+)} \frac{\partial h^t_j(\{\phi^k\})}{\partial \phi_i}\phi_i, \qquad h^t_{j2k}(\{\phi\}) = -\sum_{(i_j^-)} \frac{\partial h^t_j(\{\phi^k\})}{\partial \phi_i}\frac{(\phi_i^k)^2}{\phi_i} \tag{10.27}$$

$$h^t_{j3k}(\{\phi\}) = -2\chi^{h^t_j} \sum_{(i_j^-)} \frac{\partial h^t_j(\{\phi^k\})}{\partial \phi_i}\phi_i\left(\frac{\phi_i}{\phi_i^k} - 2\right) \tag{10.28}$$

$$\bar{h}^t_{jk}(\{\phi^k\}) = h^t_j(\{\phi^k\}) - \sum_{(i_j^+)} \frac{\partial h^t_j(\{\phi^k\})}{\partial \phi_i}\phi_i^k - \left(2\chi^{h^t_j} - 1\right)\sum_{(i_j^-)} \frac{\partial h^t_j(\{\phi^k\})}{\partial \phi_i}\phi_i^k \tag{10.29}$$

where $\sum_{(i_j^+)}$ and $\sum_{(i_j^-)}$ mean summation over the variables belonging to group (i_j^+) and (i_j^-), respectively, and $\chi^{h^t_j}$ is as before a user-defined positive scalar that control the conservatism of the approximations. Group (i_j^+) contains the variables for which $\partial h^t_j(\{\phi^k\})/\partial\phi_i$ is positive, and group (i_j^-) includes the remaining variables. The same type of approximations can be applied to the deterministic constraint functions. The explicit discrete sub-optimization problems (10.20)–(10.22) are solved by standard methods that treat the problem directly in the primal design variable space such as evolution-based optimization techniques [16]. The level of effectiveness of the above sequential optimization scheme depends on the degree of convexity of the functions involved in the optimization problem. For example, if the curvatures are not too large and relatively uniform throughout the design space the proposed algorithm converges within few iterations [9, 21, 29]. For more general cases methods based on trust regions and line search methodologies may be more appropriate [1, 19, 22].

7 Reliability and Sensitivity Assessment

The characterization of the sub-optimization problems (10.20)–(10.22) requires the estimation of first excursion probabilities and their sensitivities. In order to estimate the excursion probabilities at a given design high-dimensional integrals need to be evaluated. This difficulty favors the application of Monte Carlo Simulation as fundamental approach to cope with the probability integrals. However, in most engineering applications the probability that a particular system fails is expected to be small, e.g. between 10^{-4}–10^{-6}. Direct Monte Carlo is robust to the type and dimension of the problem, but it is not suitable for finding small probabilities. Therefore, advanced Monte Carlo strategies are needed to reduce the computational efforts. In particular a generally applicable method, called subset simulation, is implemented in this work [4]. On the other hand, the sensitivity of the failure probability functions with respect to the design variables is estimated by an approach recently introduced in [32]. The approach is based on the approximate local representation of two different quantities. The first approximation involves the performance functions that define the failure domains while the second includes the probability of failure in terms of the maximum response levels for safe system operation. For a detailed discussion of the approach the reader is referred to [22, 32].

8 Application Problem

8.1 Description

A four-story building with a base-isolation system under earthquake motion is considered as an application problem. The plan view, as well as the dimensions for each floor are shown in Fig. 10.4. The elevation of one resistant element (A-axis) is illustrated in Fig. 10.5. Each of the four floors is supported by 80 columns of square cross section. The first floor has a height equal to 3.5 m while the other floors have a constant height equal to 3.0 m, leading to a total height of 12.5 m.

As previously pointed out (see Structural Model Section) each floor is represented by three degrees of freedom, i.e. two translational displacements in the direction of the x axis and y axis, and a rotational displacement. The associated active masses in the x and y direction are taken constant for the first three floors and equal to 2.50×10^6 kg and 1.50×10^6 kg for the last floor. The corresponding mass moments of inertia are taken as 2.10×10^9 kg·m^2 and 1.20×10^9 kg·m^2, respectively. On the other hand, the mass of the base is equal to 6.0×10^6 kg, and its mass moment of inertia 5.00×10^9 kg·m^2. The Young's modulus and the modal damping ratios are treated as uncertain system parameters. The Young's modulus is modeled by a truncated normal random variable with most probable value $\bar{E} = 2.50 \times 10^{10}$ N/m^2 and coefficient of variation of 20%. Moreover, the damping ratios are modeled by independent Log-normal random variables with mean value $\bar{\zeta} = 0.03$ and coefficient of variation of 40%. The base isolation system is composed of 80 uniaxial lead rubber

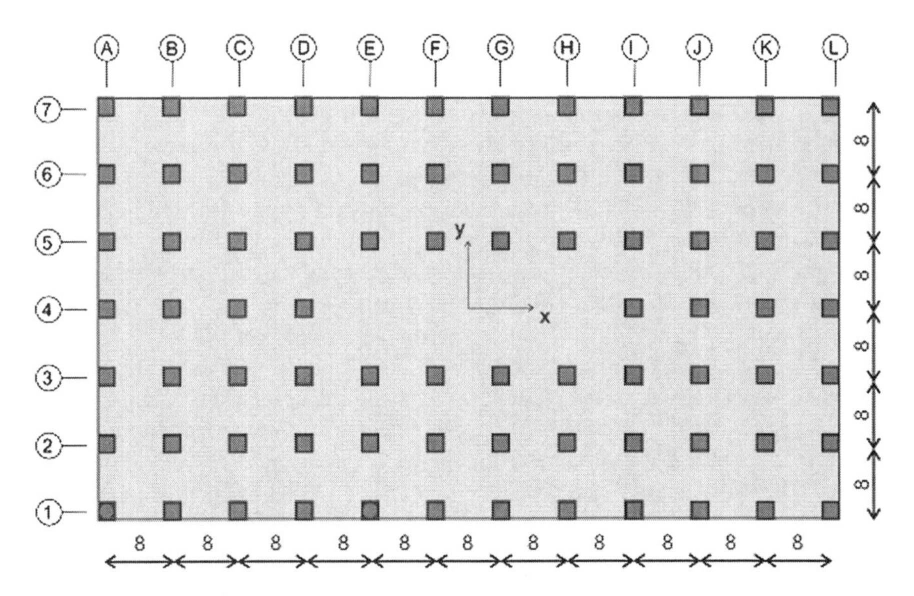

Fig. 10.4 Plan view of the structural model

Fig. 10.5 Elevation view of axis A

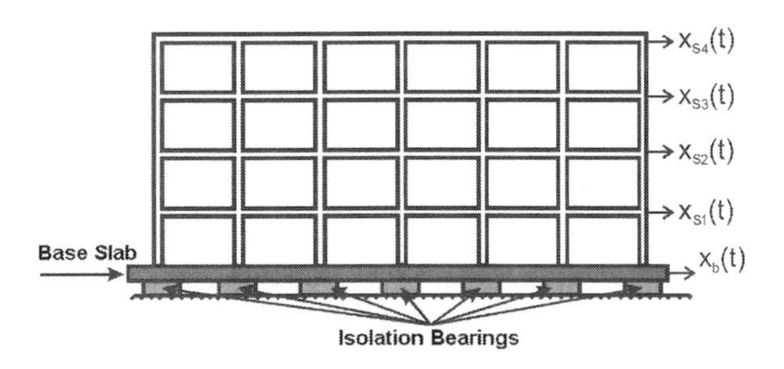

bearings with hysteretic behavior. The nonlinear behavior of these devices is modeled using the equations described in Sect. 5 with model parameters $n = 1$, $\alpha = 1.0$, $\beta = -0.65$, $\gamma = 0.5$, $U^y = 0.5$ cm, $\alpha_L = 0.1$, $k_e = 3 \times 10^6$ N/m, and $c_v = 0.0$. Figures 10.6 and 10.7 show a schematic representation of a lead rubber bearing and a typical displacement-restoring force curve of the isolation element, respectively. The structural system is excited horizontally by a ground acceleration applied in the y direction. The induced ground acceleration is characterized as in Sect. 4, with model parameters listed in Table 10.1.

8.2 Optimal Design Problem

The objective function f is defined as the volume of the column elements of the structural system. The design variables $\{\phi\}$ are chosen as the dimensions of the

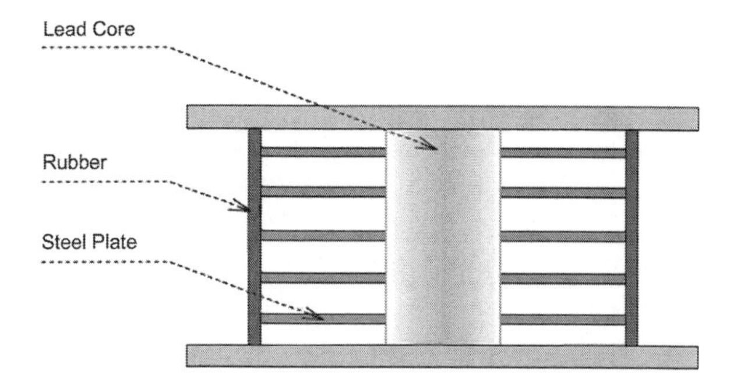

Fig. 10.6 Lead rubber bearing

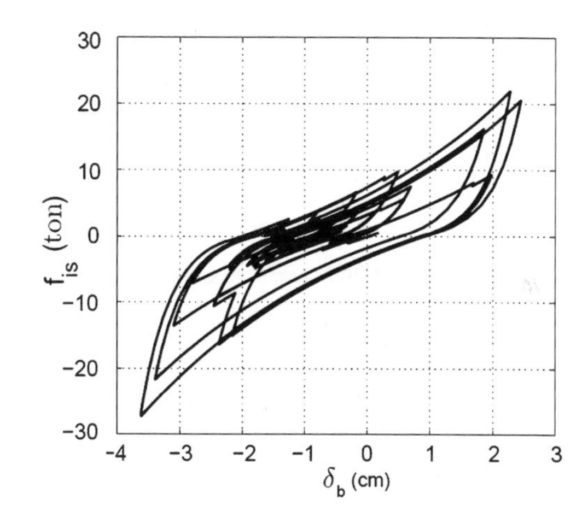

Fig. 10.7 Typical displacement-restoring force curve of the isolation element (lead rubber bearing)

columns throughout the height, grouped in four design variables, i.e. the dimensions of the columns of each floor constitute each of the design groups. The failure event is formulated as a first passage problem during the duration of the ground acceleration. The structural responses to be controlled are the 4 interstorey drift displacements. The threshold value is chosen equal to 0.2% of the floor height for the interstorey drift displacements. Thus, the failure domains evaluated at the design $\{\phi\}$ are given

Table 10.1 Parameters for the stochastic ground acceleration model

Parameter	Numerical Value	Parameter	Numerical Value
$\bar{r}$ (km)	20.0	σ_r (km)	9.0
b	1.8	U	10^{-20}
ρ_s (gm/cc)	2.8	β_s (km/s)	3.5
V	$1/\sqrt{2}$	R_{Φ}	0.55
F	2.0	R_0 (km)	1.0
T (s)	20.0	Δt (s)	0.01

Table 10.2 Initial and final designs

Design variable	Initial design	Final design	
		Problem 1	Problem 2
ϕ_1 (m)	0.90	0.68	0.85
ϕ_2 (m)	0.80	0.59	0.75
ϕ_3 (m)	0.75	0.57	0.72
ϕ_4 (m)	0.70	0.51	0.64
Normalized objective function	1.00	0.56	0.88

by

$$\Omega_{F_j}\big(\{\phi\}\big) = \Big\{\{\theta\} \mid \max_{t_k, k=1,\ldots,2001} \big|\delta_j\big(t_k, \{\phi\}, \{\theta\}\big)\big| - \delta^* \geq 0\Big\}, \quad j = 1, \ldots, 4 \tag{10.30}$$

where $\delta_j(t_k, \{\phi\}, \{\theta\})$ is the relative displacement between the $(j-1, j)$-th floor evaluated at the design $\{\phi\}$, t_k are the discrete time instants, δ^* is the critical threshold level, and $\{\theta\}$ is the vector that represents the uncertain system parameters (structural parameters and excitation). Note that more than two thousand random variables are involved in the characterization of the uncertain model parameters. The reliability-based optimization problem is defined as

$$\text{Min } f\big(\{\phi\}\big)$$

subject to

$$\begin{aligned} P_{F_j}\big(\{\phi\}\big) &\leq P_F^*, \quad j = 1, 2, 3, 4 \\ 0.30 \leq \phi_i &\leq 1.10, \quad i = 1, \ldots, 4 \end{aligned} \tag{10.31}$$

Two target failure probabilities are considered: $P_F^* = 10^{-2}$ and $P_F^* = 10^{-4}$. The first case can be interpreted as a design problem with a moderate level of reliability while the second case corresponds to a high level of reliability. In what follows the first case will be referred as Problem 1 while the second case as Problem 2.

8.3 Results

The initial and final designs of Problems 1 and 2 are given in Table 10.2. The results of the optimization process are presented in Figs. 10.8, 10.9 and 10.10 in terms of the evolution of the objective function and failure probabilities, respectively.

The objective function is normalized by its value at the initial design. It is observed that only a few optimization cycles are required for obtaining convergence. Moreover, most of the improvement of the objective function takes place in the first 3 iterations. It is also seen that the method generates a series of steadily improved

Fig. 10.8 Iteration history in terms of the objective function. Problem 1: moderate level of reliability. Problem 2: high level of reliability

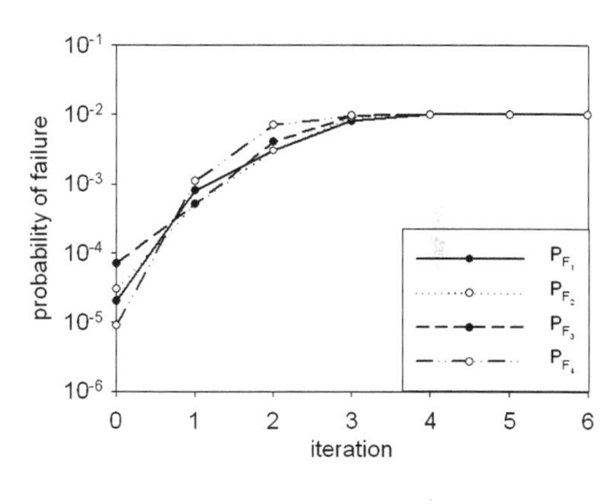

Fig. 10.9 Iteration history in terms of the reliability constraints. Problem 1

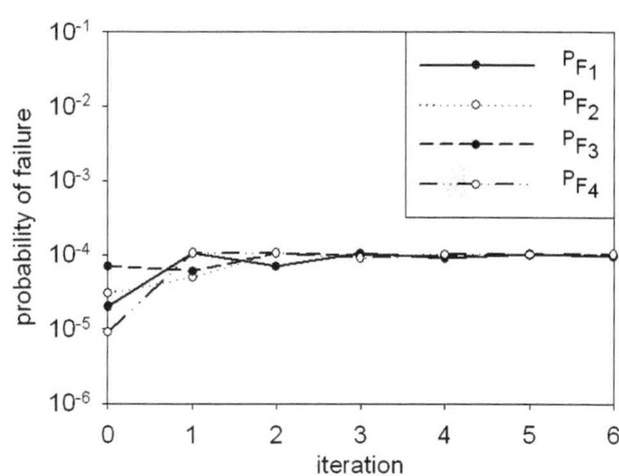

Fig. 10.10 Iteration history in terms of the reliability constraints. Problem 2

feasible designs that move toward the optimum. The results indicate that the value of the objective function at the final design of Problem 2 is greater than the corresponding value of Problem 1. This is turn implies that the structural components (columns) at the final design of Problem 2 are bigger than the corresponding components of Problem 1, as expected. The beneficial effects of the base isolation system are shown in Table 10.3. This table shows the value of the objective function at the

Table 10.3 Objective function value of models with and without the base isolation system

Model	Normalized objective function at the final design	
	Problem 1	Problem 2
With base isolation system	0.56	0.88
Without base isolation system	0.78	1.21
Difference	39%	38%

Table 10.4 Constraint violations

	Problem 2
P_{F_1}/P_F^*	103
P_{F_2}/P_F^*	55
P_{F_3}/P_F^*	20
P_{F_4}/P_F^*	2
$P_F^* = 10^{-4}$	

final designs of Problems 1 and 2 for models with and without the isolation system. The effect of the isolation system is clear from these results. The difference between the values of the objective functions is almost 40% in both Problems.

Finally, the effect of the base isolation system can also be observed from a constraint violation viewpoint. Table 10.4 shows the probability of occurrence of the failure events associated with the final design of Problem 2 (see Table 10.2) for the case where no base isolation is considered. The probability is normalized by the target failure probability $P_F^* = 10^{-4}$. It is seen for example that the probability of occurrence of failure event F_1 is more than 100 times greater than the target failure probability. Once again, the effect of the isolation system is evident from these results.

9 Conclusions

A general framework for reliability-based design of base-isolated buildings under uncertain conditions has been presented. The reliability-based design problem is formulated as an optimization problem with a single objective function subject to multiple reliability constraints. First excursion probabilities that account for the uncertainties in the system parameters as well as in the excitation are used to characterize the system reliability. The high computational cost associated with the solution of the optimization problem is addressed by the use of approximate reliability analyses during portions of the optimization process. The proposed approach takes into account all nonlinear characteristics of the structural response in the design process and it allows for a complex characterization of structural systems and excitation models. At the same time, uncertainties in structural and excitation model parameters are considered explicitly during the design process. The numerical results and additional validation calculations highlight the beneficial effects of base-isolation systems in reducing the superstructure response. This in turn implies more robust and safer designs.

Acknowledgements This research was partially supported by CONICYT (National Commission for Scientific and Technological Research) under grant 1110061. This support is gratefully acknowledged by the authors.

References

1. Alexandrov, N.M., Dennis, J.E. Jr., Lewis, R.M., Torczon, V.: A trust-region framework for managing the use of approximation models in optimization. Struct. Optim. **15**(1), 16–23 (1998)
2. Anderson, J.G., Hough, S.E.: A model for the shape of fhe Fourier amplitude spectrum of acceleration at high frequencies. Bull. Seismol. Soc. Am. **74**(5), 1969–1993 (1984)
3. Atkinson, G.M., Silva, W.: Stochastic modeling of California ground motions. Bull. Seismol. Soc. Am. **90**(2), 255–274 (2000)
4. Au, S.K., Beck, J.L.: Estimation of small failure probabilities in high dimensions by subset simulation. Probab. Eng. Mech. **16**(4), 263–277 (2001)
5. Baber, T.T., Wen, Y.: Random vibration hysteretic, degrading systems. J. Eng. Mech. Div. **107**(6), 1069–1087 (1981)
6. Boore, D.M.: Simulation of ground motion using the stochastic method. Pure Appl. Geophys. **160**(3–4), 635–676 (2003)
7. Boore, D.M., Joyner, W.B., Fumal, T.E.: Equations for estimating horizontal response spectra and peak acceleration from western North American earthquakes: a summary of recent work. Seismol. Res. Lett. **68**(1), 128–153 (1997)
8. Ceccoli, C., Mazzotti, C., Savoia, M.: Non-linear seismic analysis of base-isolated rc frame structures. Earthquake Eng. Struct. Dyn. **28**(6), 633–653 (1999)
9. Chickermane, H., Gea, H.C.: Structural optimization using a new local approximation method. Int. J. Numer. Methods Eng. **39**, 829–846 (1996)
10. Chopra, A.K.: Dynamics of Structures: Theory and Applications to Earthquake Engineering. Prentice Hall, New York (1995)
11. De Luca, A., Mele, E., Molina, J., Verzeletti, G., Pinto, A.V.: Base isolation for retrofitting historic buildings: evaluation of seismic performance through experimental investigation. Earthquake Eng. Struct. Dyn. **30**(8), 1125–1145 (2001)
12. Der Kiureghian, A.: Analysis of structural reliability under parameter uncertainties. Probab. Eng. Mech. **23**(4), 351–358 (2008)
13. Ditlevsen, O., Madsen, H.O.: Structural Reliability Methods. Wiley, New York (1996)
14. Fleury, C., Braibant, V.: Structural optimization: a new dual method using mixed variables. Int. J. Numer. Methods Eng. **23**(3), 409–428 (1986)
15. Freudenthal, A.M.: Safety and the probability of structural failure. Trans. Am. Soc. Civ. Eng. **121**, 1337–1397 (1956)
16. Goldberg, D.: Genetic Algorithms in Search, Optimization, and Machine Learning. Addison-Wesley, Reading (1989)
17. Groenwold, A.A., Etman, L.F.P., Snyman, J.A., Rooda, J.E.: Incomplete series expansion for function approximation. Struct. Multidiscip. Optim. **34**(1), 21–40 (2007)
18. Groenwold, A.A., Wood, D.W., Etman, L.F.P., Tosserams, S.: Globally convergent optimization algorithm using conservative convex separable diagonal quadratic approximations. AIAA J. **47**(11), 2649–2657 (2009)

19. Haftka, R.T., Gürdal, Z.: Elements of Structural Optimization, 3rd edn. Kluwer Academic, Norwell (1992)
20. Jensen, H.A.: Structural optimization of non-linear systems under stochastic excitation. Probab. Eng. Mech. **21**(4), 397–409 (2006)
21. Jensen, H.A., Sepulveda, J.G.: Structural optimization of uncertain dynamical systems considering mixed-design variables. Probab. Eng. Mech. **26**(2), 269–280 (2011)
22. Jensen, H.A., Valdebenito, M.A., Schuëller, G.I., Kusanovic, D.S.: Reliability-based optimization of stochastic systems using line search. Comput. Methods Appl. Mech. Eng. **198**(49–52), 3915–3924 (2009)
23. Kelly, J.M.: Aseismic base isolation: review and bibliography. Soil Dyn. Earthq. Eng. **5**(4), 202–216 (1986)
24. Kramer, S.L.: Geotechnical Earthquake Engineering. Prentince Hall, New York (2003)
25. Mavroeidis, G.P., Papageorgiou, A.S.: A mathematical representation of near-fault ground motions. Bull. Seismol. Soc. Am. **93**(3), 1099–1131 (2003)
26. Mokha, A.S., Amin, N., Constantinou, M.C., Zayas, V.: Seismic isolation retrofit of large historic building. J. Struct. Eng. **122**(3), 298–308 (1996)
27. Prasad, B.: Approximation, adaptation and automation concepts for large scale structural optimization. Eng. Optim. **6**(3), 129–140 (1983)
28. Saragoni, G.R., Hart, G.C.: Simulation of artificial earthquakes. Earthquake Eng. Struct. Dyn. **2**(3), 249–267 (1974)
29. Schittkowski, K., Zillober, C., Zotemantel, R.: Numerical comparison of nonlinear programming algorithms for structural optimization. Struct. Optim. **7**(1–2), 1–19 (1994)
30. Taflanidis, A.A.: Robust stochastic design of viscous dampers for base isolation applications. In: Computational Methods in Structural Dynamics and Earthquake Engineering (COMPDYN), 22–24 June, Rhodes, Greece (2009)
31. Taflanidis, A.A., Beck, J.L.: Stochastic subset optimization for optimal reliability problems. Probab. Eng. Mech. **23**(2–3), 324–338 (2008)
32. Valdebenito, M.A., Schuëller, G.I.: Efficient strategies for reliability-based optimization involving non linear, dynamical structures. Comput. Struct. **89**(19–20), 1797–1811 (2011)
33. Zou, X.-K., Wang, Q., Li, G., Chan, C.-M.: Integrated reliability-based seismic drift design optimization of base-isolated concrete buildings. J. Struct. Eng. **136**(10), 1282–1295 (2010)

Chapter 11
Systematic Formulation of Model Uncertainties and Robust Control in Smart Structures Using H_∞ and μ-Analysis

Amalia Moutsopoulou, Georgios E. Stavroulakis, and Anastasios Pouliezos

Abstract The influence of structural uncertainties on actively controlled smart beams is investigated in this paper. The dynamical problem of a model smart composite beam is based on a simplified modelling of the actuators and sensors, both being realized by means of piezoelectric layers. In particular, a practical robust controller design methodology is developed, which is based on recent theoretical results on H_∞ control theory and μ-analysis. Numerical examples demonstrate the vibration-suppression property of the proposed smart beams.

Keywords Uncertainty · Smart beam · Robust performance · Robust analysis · Robust synthesis

1 Introduction

The use of active control techniques in smart structures is a modern research area [1, 5, 6, 9, 11, 15, 16, 21]. Vibration control of beams may serve as a model problem, since the beam is a fundamental structural element [6, 11, 15, 21]. A number of different control schemes have been proposed, where the main class of controllers is based on linear feedback laws. In real life applications there are always differences between the physical plant that is controlled and the model on which the controller design is based (for instance, neglected higher frequency dynamics). Therefore, robustness must be an important goal for any applicable feedback controller design

A. Moutsopoulou (✉)
Department of Civil Engineering, Technological Educational Institute of Crete, Estavromenos, 71004, Heraklion, Greece
e-mail: amalia@staff.teicrete.gr

G.E. Stavroulakis · A. Pouliezos
Department of Production Engineering and Management, Technical University of Crete, Kounoupidiana, 73100, Chania, Greece

G.E. Stavroulakis
e-mail: gestavr@dpem.tuc.gr

A. Pouliezos
e-mail: tasos@dpem.tuc.gr

M. Papadrakakis et al. (eds.), *Computational Methods in Stochastic Dynamics*, Computational Methods in Applied Sciences 26,
DOI 10.1007/978-94-007-5134-7_11,

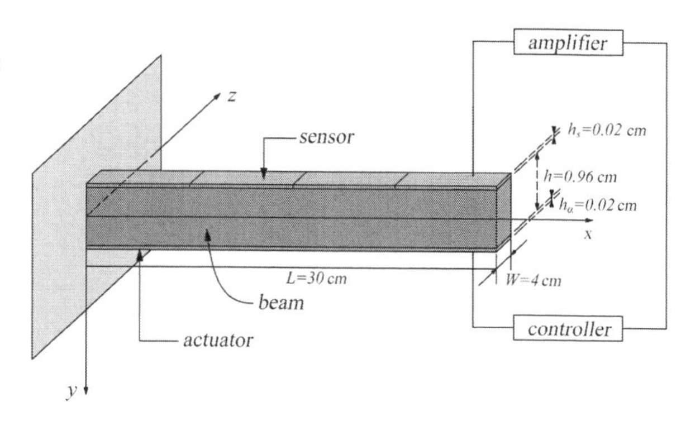

Fig. 11.1 Schematic picture with data used in the example

[5, 8, 10, 14, 16, 21]. The performance specifications, which the control system must fulfill and the class of uncertainties for which the control system must be robust against, determine the robust controller design methodology for any particular vibration control problem. In this study a vibration control problem in flexible structure (smart beam) is considered and the performance specification is stated in terms of a disturbance attenuation requirement for particular class of external disturbances acting on the structure. The paper illustrates H_∞ robust controller design techniques by considering the problem of active vibration control in a flexible cantilever beam using piezoelectric patches as sensors and actuators. This work demonstrates that the proposed robust control design schemes are suited to broadband vibration disturbances, which can be modelled as Gaussian white noise (e.g., in earthquake modelling), wind-like pressure, as well as structured uncertainties. Uncertainty denotes the difference between the model and the reality. By adopting the mechanical model described previously, we consider uncertainties in the parameters of the model. The H_∞ approach begins with an uncertain system model for the plant to be controlled [1, 5, 8, 9, 11]. In this section we will consider an uncertain system model whose primary purpose is to account for the uncertainty introduced by varying the nominal plant parameters.

First, for a simplified model, the governing equations of a beam with bonded piezoelectric sensors and actuators are formulated. After the finite element discretization, H_∞ robust control and m-analysis of the beam vibration is investigated. Numerical results obtained by using MATLAB routines demonstrate that these two robust control laws can effectively suppress the vibration of lower modes of the beam as well as avoid spillover from the higher frequency modes.

2 Mathematical Modelling

A cantilever slender beam with rectangular cross-section is considered. Four pairs of piezoelectric patches are embedded symmetrically at the top and the bottom surfaces of the beam, as shown in Fig. 11.1. The beam is made from graphite-epoxy T300–976 and the piezoelectric patches are $PZTG1195N$. The top patches act like

Table 11.1 Parameters of the composite beam

Parameters	Values
Beam length, L	0.3 m
Beam width, W	0.04 m
Beam thickness, h	0.0096 m
Beam density, ρ	1600 kg/m^3
Young's modulus of the beam, E	1.5×10^{11} N/m^2
Piezoelectric constant, d_{31}	254×10^{-12} m/V
Electric constant, ξ_{33}	11.5×10^{-3} Vm/N
Young's modulus of the piezoelectric element	1.5×10^{11} N/m^2
Width of the piezoelectric element	$b_S = b_a = 0.04$ m
Thickness of the piezoelectric element	$h_S = h_a = 0.0002$ m

sensors and the bottom like actuators. The resulting composite beam is modelled by means of the classical laminated technical theory of bending. Furthermore, we assume that the mechanical properties of both the piezoelectric material and the host beam are independent in time. The thermal effects are considered to be negligible as well [16].

The beam has length L, width b and thickness h. The sensors and the actuators have width b_S and b_A and thickness h_S and h_A, respectively. The electromechanical parameters of the beam used for the application of the method in this paper are given in Table 11.1.

2.1 Piezoelectric Equations

In order to derive the basic equations for piezoelectric sensors and actuators (S/As), we assume that:

- The piezoelectric S/A are bonded perfectly on the host beam;
- The piezoelectric layers are much thinner then the host beam;
- The piezoelectric material is homogeneous, transversely isotropic and linearly elastic;
- The piezoelectric S/A are transversely polarized (in the z-direction) [16].

Under these assumptions the three-dimensional linear constitutive equations are given by [5],

$$\begin{Bmatrix} \sigma_{xx} \\ \sigma_{xz} \end{Bmatrix} = \begin{bmatrix} Q_{11} & 0 \\ 0 & Q_{55} \end{bmatrix} \left(\begin{Bmatrix} \varepsilon_{xx} \\ \varepsilon_{xz} \end{Bmatrix} - \begin{bmatrix} d_{31} \\ 0 \end{bmatrix} E_z \right) \tag{11.1}$$

$$D_z = Q_{11} d_{31} \varepsilon_{xx} + \xi_{xx} E_z \tag{11.2}$$

where σ_{xx}, σ_{xz} denote the axial and shear stress components, D_z, denotes the transverse electrical displacement; ε_{xx} and ε_{xz} are axial and shear strain components; Q_{11}, and Q_{55}, denote elastic constants; d_{31}, and ξ_{33}, denote piezoelectric and dielectric constants, respectively. Equation (11.1) describes the inverse piezoelectric effect and equation (11.2) describes the direct piezoelectric effect. E_z, is the transverse component of the electric field that is assumed to be constant for the piezoelectric layers and its components in the xy-plain are supposed to vanish. If no electric field is applied in the sensor layer, the direct piezoelectric equation (11.2) is simplified to

$$D_z = Q_{11} d_{31} \varepsilon_{xx} \tag{11.3}$$

and it is used to calculate the output charge created by the strains in the beam [8].

2.2 Equations of Motion

We assume that:

- The beam centroidal and elastic axis coincides with the x-coordinate axis so that no bending-torsion coupling is considered;
- The axial vibration of the host beam is considered negligible;
- The displacement field $\{u\} = (u_1, u_2, u_3)$ is obtained based on the usual Timoshenko assumptions [11],

$$\begin{aligned} u_1(x,y,z) &\approx z\phi(x,t) \\ u_2(x,y,z) &\approx 0 \\ u_3(x,y,x) &\approx w(x,t) \end{aligned} \tag{11.4}$$

  where ϕ is the rotation of the beam's cross-section about the positive y-axis and w is the transverse displacement of a point of the centroidal axis ($y = z = 0$).

The strain displacement relations can be applied to equation (11.4) to give,

$$\varepsilon_{xx} = z\frac{\vartheta \phi}{\vartheta x} \qquad \varepsilon_{xz} = \phi + \frac{\vartheta w}{\vartheta x} \tag{11.5}$$

We suppose that the transverse shear deformation ε_{xx} is equal to zero [6].

In order to derive the equations of the motion of the beam we use Hamilton's principle [17],

$$\int_{t_2}^{t_1} (\delta T - \delta U + \delta W)\, dt = 0, \tag{11.6}$$

where T is the total kinetic energy of the system, U is the potential (strain) energy and W is the virtual work done by the external mechanical and electrical loads and moments. The first variation of the kinetic energy is given by,

$$\begin{aligned}\delta T &= \frac{1}{2}\int_V \rho \left\{\frac{\vartheta u}{\vartheta t}\right\}^r \left\{\frac{\vartheta u}{\vartheta t}\right\} dV \\ &= \frac{b}{2}\int_0^L \int_{-h/2-h_a}^{h/2+h_s} \rho\left(z\frac{\vartheta \phi}{\vartheta t}\delta\frac{\vartheta \phi}{\vartheta t} + \frac{\vartheta w}{\vartheta t}\delta\frac{\vartheta w}{\vartheta t}\right) dz\,dx \end{aligned} \tag{11.7}$$

The first variation of the kinetic energy is given by,

$$\begin{aligned}\delta U &= \frac{1}{2}\int_V \delta\{\varepsilon\}^T\{\sigma\} dV \\ &= \frac{b}{2}\int_0^L \int_{-h/2-h_a}^{h/2+h_s} \left[Q_{11}\left(z\frac{\vartheta w}{\vartheta x}\delta\right)\left(z\frac{\vartheta w}{\vartheta x}\right)\right] dz\,dx \end{aligned} \tag{11.8}$$

If the load consists only of moments induced by piezoelectric actuators and since the structure has no bending twisting couple then the first variation of the work has the form [17],

$$\delta W = b\int_0^L M^a \delta\left(\frac{\vartheta \phi}{\vartheta x}\right) dx \tag{11.9}$$

where M^a is the moment per unit length induced by the actuator layer and is given by,

$$\begin{aligned} M^a &= \int_{-h/2-h_a}^{-h/2} z\sigma_{xx}^a\,dz = \int_{-h/2-h_a}^{-h/2} zQ_{11}d_{31}E_z^a\,dz \\ &\left(E_z^a = \frac{V_a}{h_a}\right) \end{aligned} \tag{11.10}$$

2.3 Finite Element Formulation

We consider a beam element of length L_e, which has two mechanical degrees of freedom at each node: one translational ω_1 (respectively ω_2) in direction y and one rotational ψ_1 (respectively ψ_2), as it is shown in Fig. 11.2. The vector of nodal displacements and rotations q_e is defined as [5],

$$q_e = [\omega_1, \psi_1, \omega_2, \psi_2] \tag{11.11}$$

The transverse deflection $\omega(x,t)$ and rotation $\psi(x,t)$ along the beam are continuous and they are interpolated by Hermitian linear shape functions H_i^{ω} and H_i^{ψ} as

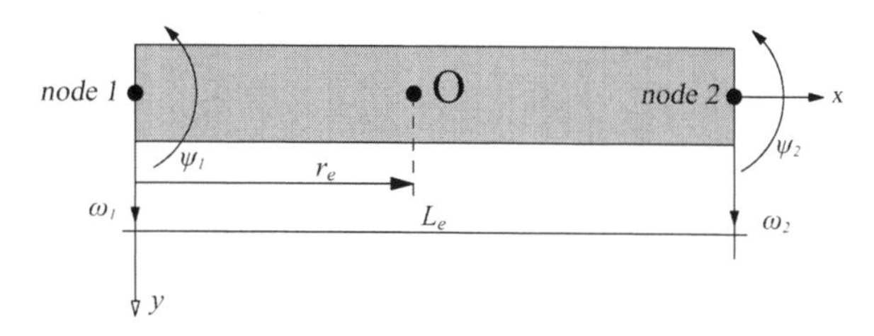

Fig. 11.2 Beam finite element

follows [10],

$$\omega(x,t) = \sum_{i=1}^{4} H_i^{\omega}(x) q_i(t)$$
$$\psi(x,t) = \sum_{i=1}^{4} H_i^{\psi}(x) q_i(t) \tag{11.12}$$

This classical finite element procedure leads to the approximate (discretized) problem. For a finite element the discrete differential equations are obtained by substituting the discretized expressions (11.12) into equations (11.7) and (11.8) to evaluate the kinetic and strain energies. Integrating over spatial domains and using the Hamilton's principle (11.6) the equation of motion for a beam element are expressed in terms of nodal variable q as follows [5, 6, 14],

$$M\ddot{q}(t) + D\dot{q}(t) + Kq(t) = f_m(t) + f_e(t) \tag{11.13}$$

where M is the mass matrix, D is the viscous damping matrix, K is the stiffness matrix, f_m is the external loading vector and f_e is the generalized control force vector produced by electromechanical coupling effects. The independent variable vector $q(t)$ is composed of transversal deflections ω_i and rotations ψ_i, i.e. [21],

$$q(t) = \begin{bmatrix} \omega_1 \\ \psi_1 \\ \vdots \\ \omega_n \\ \psi_n \end{bmatrix} \tag{11.14}$$

where n is the number of nodes used in the analysis. Vectors w and f_m are positive upwards. For the state-space control transformation,we are presented with,

$$\dot{x}(t) = \begin{bmatrix} q(t) \\ \dot{q}(t) \end{bmatrix} \tag{11.15}$$

Furthermore to express $f_e(t)$ in the form of $Bu(t)$ we write it as the product $f_e^* u$, where f_e^* is the piezoelectric force for a unit applied on the corresponding actuator, and u represents the voltages on the actuators. Finally, $d(t) = f_m(t)$ is the disturbance vector [15]. Then,

$$\dot{x}(t) = \begin{bmatrix} 0_{2n\times 2n} & I_{2n\times 2n} \\ -M^{-1}K & -M^{-1}D \end{bmatrix} x(t) + \begin{bmatrix} 0_{2n\times n} \\ M^{-1}f_e^* \end{bmatrix} u(t) + \begin{bmatrix} 0_{2n\times 2n} \\ M^{-1} \end{bmatrix} \tag{11.16}$$

$$= Ax(t) + Bu(t) + Gd(t) = Ax(t) + \begin{bmatrix} B & G \end{bmatrix} \begin{bmatrix} u(t) \\ d(t) \end{bmatrix}$$

$$= Ax(t) + \tilde{B}\tilde{u}(t) \tag{11.17}$$

The previous description of the dynamical system will be augmented with the output equation, under the assumption that only displacements are measured [10]

$$y(t) = \begin{bmatrix} x_1(t) & x_3(t) & \cdots & x_{n-1}(t) \end{bmatrix}^T = Cx(t) \tag{11.18}$$

In this formulation u is $n \times 1$ (at most, but can be smaller), while d is $2n \times 1$. The units used are compatible for instance m, rad, s and N [5, 14].

3 Design Objectives and System Specifications

The structured singular value of the transfer function is defined as,

$$\mu(M) = \begin{cases} \frac{1}{\min_{k_m}\{\det(I-k_m M\Delta)=0, \bar{\sigma}(\Delta)\leq 1\}} \\ 0, \quad \Delta \det(I - M\Delta) = 0 \end{cases} \tag{11.19}$$

This quantity defines the smallest structured $\mu(M)$ (measured in terms of $\bar{\sigma}(\Delta)$) which makes $\det(I - M\Delta) = 0$: then $\mu(M) = \frac{1}{\bar{\sigma}(\Delta)}$. It follows that values of μ smaller than 1 are desired [2, 4].

The design objectives fall into two categories:

1. Stability of closed loop system (plant+controller).
   a. Disturbance attenuation with satisfactory transient characteristics (overshoot, settling time).
   b. Small control effort.
2. Robust performance

Stability of closed loop system (plant+controller) should be satisfied in the face of modelling errors [3].

In order to obtain the required system specifications with respect to the above objectives we need to represent our system in the so-called—Δ structure. Let us start with the simple typical diagram of Fig. 11.3 [7, 12].

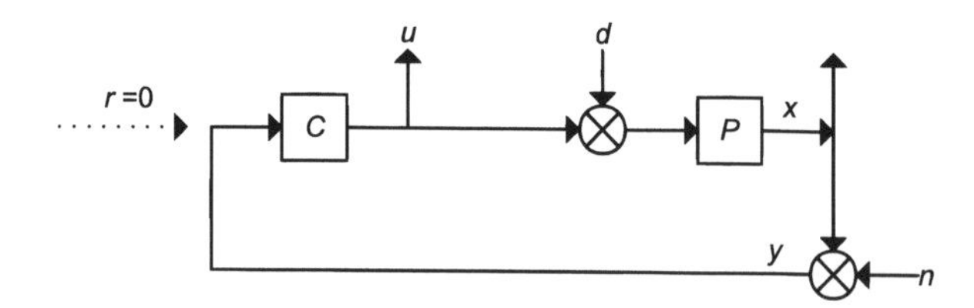

Fig. 11.3 Classical control block diagram (P: plant dynamical system, C: controller)

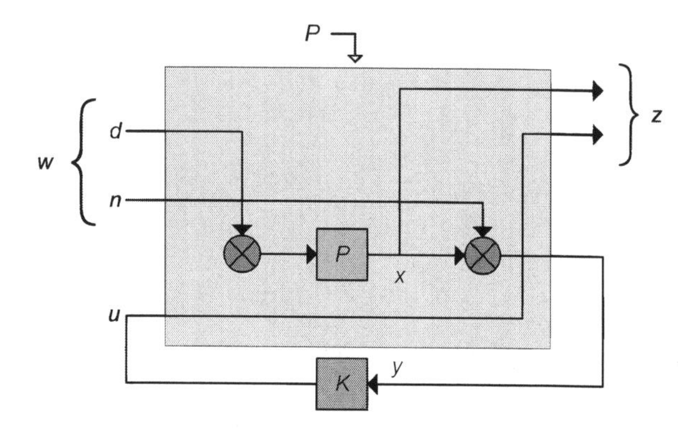

Fig. 11.4 Detailed two-port diagram (with a linear feedback control K)

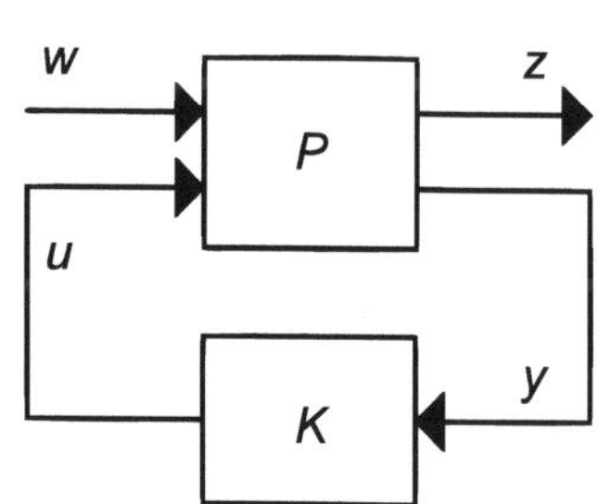

Fig. 11.5 Two-port diagram

In this diagram there are two inputs, d and n, and two outputs, u and x. In what follows it is assumed that,

$$\left\| \begin{matrix} d \\ n \end{matrix} \right\|_2 \leq 1, \qquad \left\| \begin{matrix} x \\ u \end{matrix} \right\|_2 \leq 1 \tag{11.20}$$

If this is not the case, appropriate frequency-dependent weights can transform original signals so that the transformed signals have this property. The details of the system are given in Fig. 11.4 or in less details Fig. 11.5:

In this description,

$$z = \begin{bmatrix} u \\ x \end{bmatrix}, \qquad w = \begin{bmatrix} d \\ n \end{bmatrix} \tag{11.21}$$

where z are the output variables to be controlled, and w the exogenous inputs.

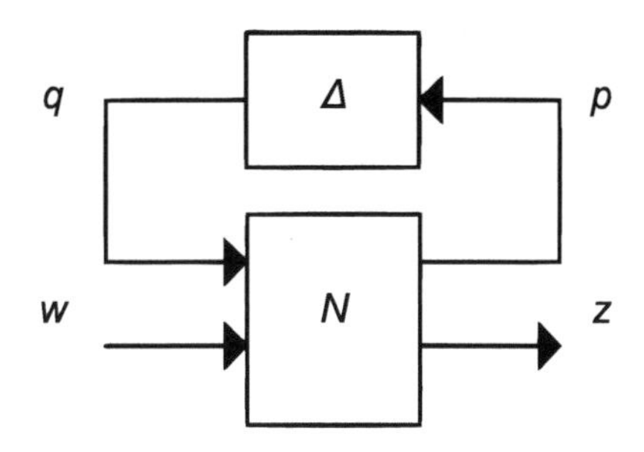

Fig. 11.6 Two port diagram with uncertainty

Given that P has two inputs and two outputs it is, as usual, naturally partitioned as,

$$\begin{bmatrix} z(s) \\ y(s) \end{bmatrix} = \begin{bmatrix} P_{zw}(s) & P_{zu}(s) \\ P_{yw}(s) & P_{yu}(s) \end{bmatrix} \begin{bmatrix} w(s) \\ u(s) \end{bmatrix} = P(s) \begin{bmatrix} w(s) \\ u(s) \end{bmatrix} \tag{11.22}$$

In addition the controller is written,

$$u(s) = K(s)y(s) \tag{11.23}$$

Substituting (11.22) in (11.23) gives the closed loop transfer function $N_{zw}(s)$,

$$N_{zw}(s) = P_{zw}(s) + P_{zu}(s)K(s)\big(I - P_{yu}(s)K(s)\big)^{-1} P_{yw}(s) \tag{11.24}$$

To deduce robustness specifications one more diagram is needed, namely that of Fig. 11.6: where N is defined by (11.24) and the uncertainty modelled in Δ satisfies $\|\Delta\|_\infty \leq 1$ (details are given later on in this paper). Here,

$$z = \mathcal{F}_u(N, \Delta)w = \big[N_{22} + N_{21}\Delta(I - N_{11}\Delta)^{-1}N_{12}\big]w = Fw \tag{11.25}$$

Given this structure we can state the following definitions:

$$\begin{aligned} &\text{Nominal stability } (NS) \quad \Leftrightarrow \quad N \text{ internally stable} \\ &\text{Nominal performance } (NP) \quad \Leftrightarrow \quad \|N_{22}(j\omega)\|_\infty \leq 1 \forall\omega \text{ and } NS \\ &\text{Robust stability } (RS) \quad \Leftrightarrow \quad F = \mathcal{F}_u(N, \Delta) \text{ stable } \forall\Delta, \|\Delta\|_\infty < 1 \text{ and } NS \\ &\text{Robust performance } (RP) \quad \Leftrightarrow \quad \|F\|_\infty < 1, \forall\Delta, \|\Delta\|_\infty < 1 \text{ and } S \end{aligned} \tag{11.26}$$

It has been proved that the following conditions hold in the case of block-diagonal real or complex perturbations Δ:

1. The system is nominally stable if M is internally stable.
2. The system exhibits nominal performance if $\bar{\sigma}(N_{22}(j\omega)) < 1$
3. The system (M, Δ) is robustly stable if and only if,

$$\sup_{\omega\in\mathbb{R}} \mu_\Delta\big(N_{11}(j\omega)\big) < 1 \tag{11.27}$$

where μ_Δ is the structured singular value of N given the structured uncertainty set Δ. This condition is known as the generalized small gain theorem.

4. The system (N, Δ) exhibits robust performance if and only if,

$$\sup_{\omega\in\mathbb{R}} \mu_{\Delta_a}\big(N(j\omega)\big) < 1 \tag{11.28}$$

where,

$$\Delta_a = \begin{bmatrix} \Delta_p & 0 \\ 0 & \Delta \end{bmatrix} \tag{11.29}$$

and Δ_p is full complex, has the same structure as Δ and dimensions corresponding to w, z [12].

Unfortunately, only bounds on μ can be estimated.

3.1 Controller Synthesis

All the above results support the analysis problem and provide tools to judge the performance of any controller or to compare different controllers. However it is possible to approximately synthesize a controller that achieves given performance in terms of the structured singular value μ.

In this procedure, which is called $(D, G - K)$ iteration [20] the problem of finding an μ-optimal controller K such that $\mu(\mathcal{F}_u(F(j\omega)), K(j\omega)) \leq \beta$, $\forall\omega$ is transformed into the problem of finding transfer function matrices $D(\omega) \in \mathcal{D}$ and $G(\omega) \in \mathcal{G}$, such that,

$$\sup_{\omega} \bar{\sigma}\left[\left(\frac{D(\omega)\mathcal{F}_u(F(j\omega), K(j\omega))D^{-1}(\omega)}{\gamma} - jG(\omega)\right) \times \left(I + G^2(\omega)\right)^{-1/2}\right] \leq 1, \quad \forall\omega \tag{11.30}$$

Unfortunately this method does not guarantee even finding local maxima. However for complex perturbations a method known as $D - K$ iteration is available (implemented in MATLAB) [20]. It combines H_∞ synthesis and μ-analysis and often yields good results. The starting point is an upper bound on μ in terms of the scaled singular value,

$$\mu(N) \leq \min_{D\in\mathcal{D}} \bar{\sigma}\left(DND^{-1}\right) \tag{11.31}$$

The idea is to find the controller that minimizes the peak over the frequency range namely,

$$\min_K \left(\min_{D\in\mathcal{D}} \|DN(K)D^{-1}\|_\infty\right) \tag{11.32}$$

by alternating between minimizing $\|DN(K)D^{-1}\|_\infty$ with respect to either K or D (while holding the other fixed).

1. *K-step*. Synthesize an $\mathcal{H}_\infty$ controller for the scaled problem $\min_K \|DN(K) \times D^{-1}\|_\infty$ with fixed $D(s)$.
2. *D-step*. Find $D(j\omega)$ to minimize at each frequency $\bar{\sigma}(DND^{-1}(j\omega))$ with fixed N.
3. Fit the magnitude of each element of $D(j\omega)$ to a stable and minimum phase transfer function $D(s)$ and got to Step 1 [20].

3.2 System Uncertainty

Let us assume uncertainty in the mass M and K matrices of the form,

$$\begin{aligned} K &= K_0(I + k_p I_{2n\times 2n}\delta_K) \\ M &= M_0(I + m_p I_{2n\times 2n}\delta_M) \end{aligned} \tag{11.33}$$

Alternatively, since in general the Rayleigh damping assumption is,

$$D = aK + \beta M \tag{11.34}$$

D could be expressed similarly to K, M, as,

$$D = D_0(I + d_p I_{2n\times 2n}\delta_D) \tag{11.35}$$

In this way we introduce uncertainty in the form of percentage variation in the relevant matrices. More detailed correlation of uncertainty with certain properties of the structures (e.g., material constants, flexibility of joints, cracks or delaminations) is possible and will be investigated in the future.

Here it will be assumed,

$$\|\Delta\|_\infty \overset{def}{=} \left\| \left[\begin{array}{c|c} I_{n\times n}\delta_K & 0_{n\times n} \\ \hline 0_{n\times n} & I_{n\times n}\delta_M \end{array} \right] \right\|_\infty < 1 \tag{11.36}$$

hence m_p, k_p are used to scale the percentage value and the zero subscript denotes nominal values (it is reminded here that the norm for a matrix $A_{n\times n}$ is calculated through $\|A\|_\infty = \max_{1\le j\le m} \sum_{j=1}^{n} |a_{ij}|$).

With these definitions Eq. (11.13) becomes,

$$\begin{aligned} M_0(I + m_p I_{2n\times 2n}\delta_M)\ddot{w}(t) &+ K_0(I + k_p I_{2n\times 2n}\delta_K)w(t) \\ &+ \big[D_0 + 0.0005[K_0 k_p I_{2n\times 2n}\delta_K + M_0 m_p I_{2n\times 2n}\delta_M]\big]\dot{w}(t) \\ &= f_m(t) + f_e(t) \\ \Rightarrow\quad & M_0\ddot{w}(t) + D_0\dot{q}(t) + K_0 w(t) \end{aligned}$$

$$= -\left[M_0 m_p I_{2n\times 2n}\delta_M \ddot{w}(t) + 0.0005[K_0 k_p I_{2n\times 2n}\delta_K + M_0 m_p I_{2n\times 2n}\delta_M]\dot{w}(t) + K_0 k_p I_{2n\times 2n}\delta_K w(t)\right]$$

$$= f_m(t) + f_e(t)$$

$$\Rightarrow \quad M_0\ddot{w}(t) + D_0\dot{w}(t) + K_0 w(t) = \tilde{D}q_u(t) + f_m(t) + f_e(t) \tag{11.37}$$

where,

$$q_u(t) = \begin{bmatrix} \ddot{w}(t) \\ \dot{w}(t) \\ w(t) \end{bmatrix} \tag{11.38}$$

$$\begin{aligned}\tilde{D} &= -\begin{bmatrix} M_0 m_p & K_0 k_p \end{bmatrix} \left[\begin{array}{c|c} I_{2n\times 2n}\delta_M & 0_{2n\times 2n} \\ \hline 0_{2n\times 2n} & I_{2n\times 2n}\delta_K \end{array}\right] \\ &\quad \times \begin{bmatrix} I_{2n\times 2n} & 0.0005 I_{2n\times 2n} & 0_{2n\times 2n} \\ 0_{2n\times 2n} & 0.0005 I_{2n\times 2n} & I_{2n\times 2n} \end{bmatrix} \\ &= G_1 \cdot \Delta \cdot G_2 \end{aligned} \tag{11.39}$$

Writing (11.37) in state space form, gives,

$$\begin{aligned}\dot{x}(t) &= \begin{bmatrix} 0_{2n\times 2n} & I_{2n\times 2n} \\ -M^{-1}K & -M^{-1}D \end{bmatrix} x(t) + \begin{bmatrix} 0_{2n\times 2n} \\ M^{-1} f_e^* \end{bmatrix} u(t) \\ &\quad + \begin{bmatrix} 0_{2n\times 2n} \\ M^{-1} \end{bmatrix} d(t) + \begin{bmatrix} 0_{2n\times 6n} \\ M^{-1}F_1 \cdot \Delta \cdot G_2 \end{bmatrix} q_u(t) \\ &= Ax(t) + Bu(t) + Gd(t) + G_u G_2 q_u(t) \end{aligned} \tag{11.40}$$

In this way we treat uncertainty in the original matrices as an extra uncertainty term. To express our system in the form of Fig. 11.6, consider Fig. 11.7.

The matrices E_1, E_2 are used to extract,

$$q_u(t) \stackrel{def}{=} \begin{bmatrix} \ddot{w}(t) \\ \dot{w}(t) \\ w(t) \end{bmatrix} \tag{11.41}$$

Since,

$$\gamma = \begin{bmatrix} \dot{w}(t) \\ \ddot{w}(t) \end{bmatrix} \qquad \beta = \int \begin{bmatrix} \dot{w}(t) \\ \ddot{w}(t) \end{bmatrix} dt = \begin{bmatrix} w(t) \\ \dot{w}(t) \end{bmatrix} \tag{11.42}$$

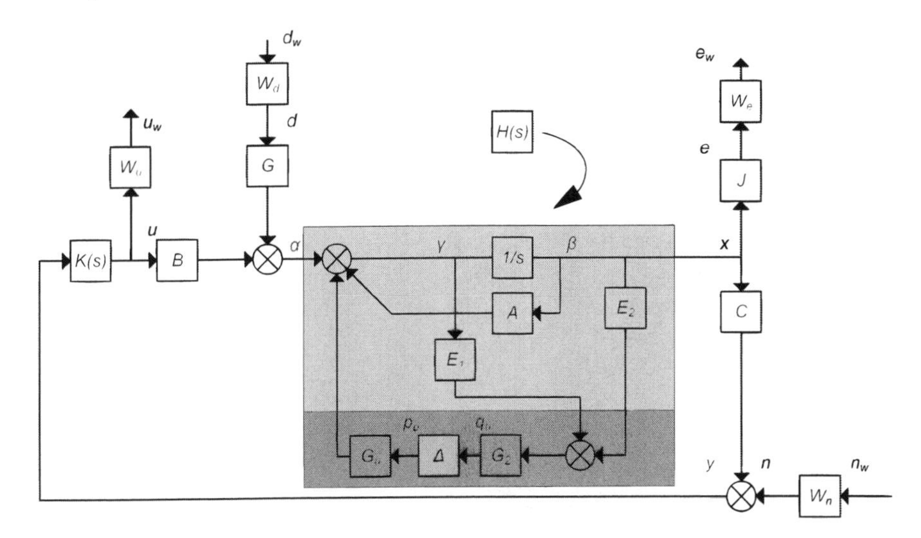

Fig. 11.7 Uncertainty block diagram

appropriate choices for E_1, E_2 are,

$$E_1 = \begin{bmatrix} 0_{2n\times 2n} & \vdots & I_{2n\times 2n} \\ \cdots & \vdots & \cdots \\ I_{2n\times 2n} & \vdots & 0_{2n\times 2n} \\ \cdots & \vdots & \cdots \\ 0_{2n\times 2n} & \vdots & 0_{2n\times 2n} \end{bmatrix}, \qquad E_2 = \begin{bmatrix} 0_{2n\times 2n} & \vdots & 0_{2n\times 2n} \\ \cdots & \vdots & \cdots \\ 0_{2n\times 2n} & \vdots & 0_{2n\times 2n} \\ \cdots & \vdots & \cdots \\ I_{2n\times 2n} & \vdots & 0_{2n\times 2n} \end{bmatrix} \tag{11.43}$$

The idea is to find an N such that,

$$\begin{bmatrix} q_u \\ \cdots \\ e_w \\ u_w \end{bmatrix} = N \begin{bmatrix} p_u \\ \cdots \\ d_w \\ n_w \end{bmatrix},$$

$$N = \begin{bmatrix} N_{p_u q_u} & \vdots & N_{d_w q_u} & N_{n_w q_u} \\ \cdots & \vdots & \cdots & \cdots \\ N_{p_u e_w} & \vdots & N_{d_w e_w} & N_{n_w e_w} \\ N_{p_u u_w} & \vdots & N_{d_w u_w} & N_{n_w u_w} \end{bmatrix} = \begin{bmatrix} N_{11} & N_{12} \\ N_{21} & N_{22} \end{bmatrix} \tag{11.44}$$

or in the notation of Fig. 11.6

$$\begin{bmatrix} q_u \\ w \end{bmatrix} = N \begin{bmatrix} p_u \\ z \end{bmatrix} \tag{11.45}$$

Now $N_{d_w e_w}$, $N_{n_w e_w}$, $N_{n_w u_w}$ are known. For the rest we will use a methodology known as "pulling out the Δ's". To this end, break the loop at points p_u, q_u (which will be used as additional inputs/outputs respectively) and use the auxiliary signals a, β, γ. To get the transfer function $N_{d_w q_u}$ (from d_w to q_u):

$$q_u = G_2(E_2\beta + E_1\gamma) = G_2\left(E_2\frac{1}{s} + E_1\right)\gamma \tag{11.46}$$

$$\gamma = GW_d d_w + Bu + A\frac{1}{s}\gamma = GW_d d_w + BKC\frac{1}{s}\gamma + A\frac{1}{s}\gamma \tag{11.47}$$

$$\Rightarrow \quad \gamma = \left(I - BKC\frac{1}{s} - A\frac{1}{s}\right)^{-1} GW_d d_w \tag{11.48}$$

Hence,

$$N_{d_w q_u} = G_2\left(E_2\frac{1}{s} + E_1\right)\left(I - BKC\frac{1}{s} - A\frac{1}{s}\right)^{-1} GW_d \tag{11.49}$$

Now, $N_{p_u q_u}$, $N_{p_u e_w}$, $N_{p_u u_w}$, are similar to $N_{d_w q_u}$, $N_{d_w e_w}$, $N_{d_w u_w}$, with GW_d replaced by G_u, i.e.,

$$\begin{aligned} N_{p_u q_u} &= G_2\left(E_2\frac{1}{s} + E_1\right)\left(I - BKC\frac{1}{s} - A\frac{1}{s}\right)^{-1} G_u \\ N_{p_u e_w} &= W_y JH\big[I + B\big[K(I - CHBK)^{-1}CH\big]\big]G_u \\ M_{p_u u_w} &= W_u K(I - CHBK)^{-1}CHG_u \end{aligned} \tag{11.50}$$

Finally to find $N_{n_w q_u}$,

$$q_u = G_2(E_2\beta + E_1\gamma) = G_2\left(E_2\frac{1}{s} + E_1\right)\gamma \tag{11.51}$$

$$\begin{aligned} \gamma &= Bu + A\frac{1}{s}\gamma = BK(W_n n_w + y) + A\frac{1}{s}\gamma \\ &= BKW_n n_w + BKC\frac{1}{s}\gamma + A\frac{1}{s}\gamma \end{aligned} \tag{11.52}$$

$$\Rightarrow \quad \gamma = \left(I - BKC\frac{1}{s} - A\frac{1}{s}\right)^{-1} BKW_n n_w \tag{11.53}$$

Hence,

$$N_{n_w q_u} = G_2\left(E_2\frac{1}{+}E_1\right)\left(I - BKC\frac{1}{s} - A\frac{1}{s}\right)^{-1} BKW_n \tag{11.54}$$

Collecting all the above yields N:

$$N = \begin{bmatrix} N_{11} & N_{12} & N_{13} \\ N_{21} & N_{22} & N_{23} \\ N_{31} & N_{32} & N_{33} \end{bmatrix} \tag{11.55}$$

where

$$N_{11} = G_2\left(E_2\frac{1}{s} + E_1\right)\left(I - BKC\frac{1}{s} - A\frac{1}{s}\right)^{-1} G_u,$$

$$N_{12} = G_2\left(E_2\frac{1}{s} + E_1\right)\left(I - BKC\frac{1}{s} - A\frac{1}{s}\right)^{-1} GW_d,$$

$$N_{13} = G_2\left(E_2\frac{1}{s} + E_1\right)\left(I - BKC\frac{1}{s} - A\frac{1}{s}\right)^{-1} BKW_u,$$

$$N_{21} = W_e JH\left[I + BK(I - CHBK)^{-1}CF\right]G_u,$$

$$N_{22} = W_e J(I - HBKC)^{-1}HGW_d,$$

$$N_{23} = W_e J(I - HBKC)^{-1}HBKW_u,$$

$$N_{31} = W_u K(I - CHBK)^{-1}CFG_u,$$

$$N_{32} = W_u(I - KCHB)^{-1}KCHGW_d,$$

$$N_{33} = W_u(I - KCHB)^{-1}KW$$

Having obtained N for the beam problem, all proposed controllers $K(s)$ can be compared using the structured singular value relations [1, 18, 19].

4 Robustness Issues

The superiority of H_∞ control lies in its ability to take explicitly into account the worst effect of unknown disturbances and noise in the system. Furthermore, at least in theory, it is possible to synthesize an H_∞ controller that is robust to a prescribed amount of modeling errors. Unfortunately, this last possibility is not implementable in some cases, as it will be subsequently illustrated [9, 13].

In what follows, the robustness to modeling errors of the designed H_∞ controller will be analyzed. Furthermore an attempt to synthesize a μ-controller will be presented, and comparisons between the two will be made.

In all simulations, routines from Matlab's Robust Control Toolbox will be used. In particular:

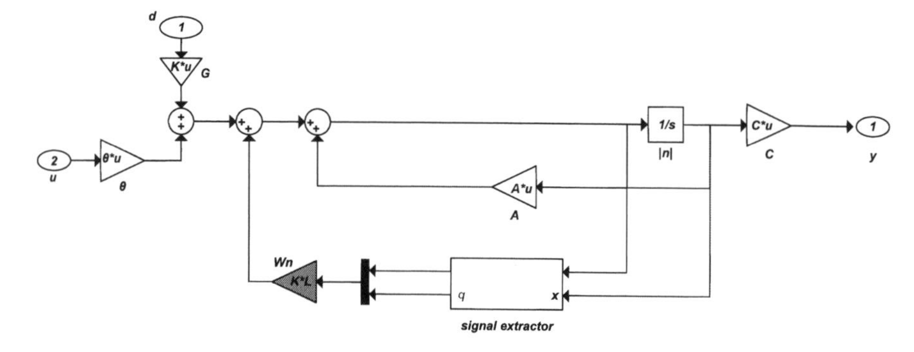

Fig. 11.8 Simulink diagram of uncertain plant

1. For uncertain elements,
2. To calculate bounds on the structured singular value,
3. To calculate a μ-controller,

Numerical models used in all simulations, are implemented in three ways:

1. Through Eq. (11.56)

$$\begin{aligned} K &= K_0(I + k_p I_{2n\times 2n}\delta_K) \\ M &= M_0(I + m_p I_{2n\times 2n}\delta_M) \\ D &= D_0 + 0,0005[K_0 k_p I_{2n\times 2n}\delta_K + M_0 m_p I_{2n\times 2n}\delta_M] \end{aligned} \tag{11.56}$$

   and subsequent evaluation of matrix N for specific values of k_p, m_p.
2. By use of Matlab's "uncertain element object". As explained, this form is needed in the D-K robust synthesis algorithm.
3. By Simulink implementation of Fig. 11.8.

4.1 Inputs-Loading

Loading corresponds to the wind excitation. The function $y(t)$ has been obtained from the wind velocity record, through the relation

$$f_m(t) = y(t), \quad \text{where } y(t) = 0.5pv^2(t)C_v E \tag{11.57}$$

Where $p = 1.125$ kg/m^3 is the air density, C_v is a coefficient that depends on beam cross section, for rectangular cross section $C_v = 1.5$, $v(t)$ is wind velocity and E represents the beam surface area that is exposed to wind (Fig. 11.9).

Moreover, in all simulations random noise has been introduced within a probability interval of $\pm 1\%$ to measurements at the system output locations. Due to small displacements of system nodal points noise amplitude is taken to be small of the order of $5 \cdot 10^{-5}$ (Fig. 11.10).

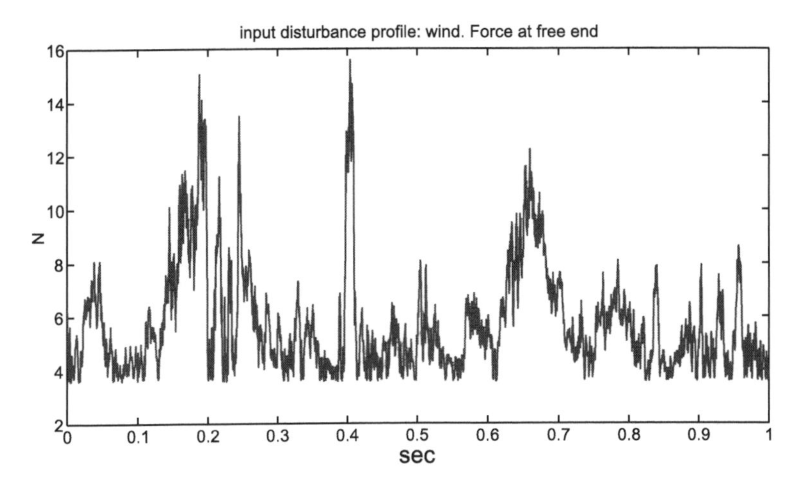

Fig. 11.9 Wind force

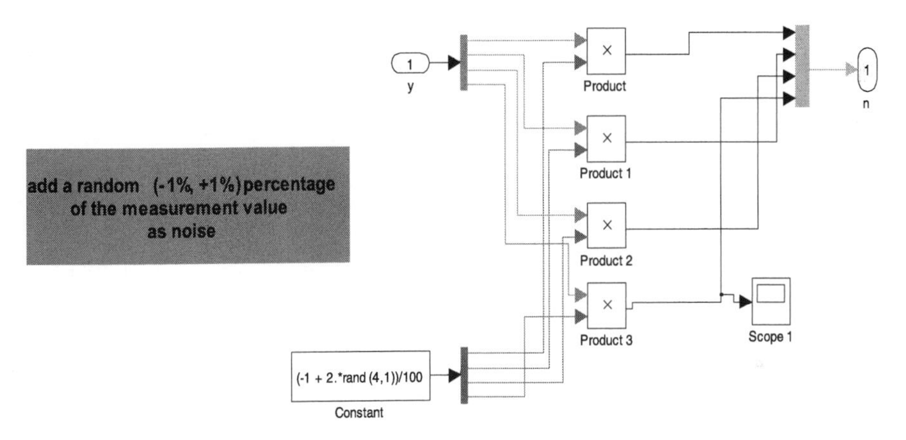

Fig. 11.10 Simulink diagram of noise

4.2 Robust Analysis—Results

Robust analysis is carried out through the relations:

$$\sup_{\omega \in \mathbb{R}} \mu_{\Delta}\big(N_{11}(j\omega)\big) < 1 \tag{11.58}$$

for robust stability, and,

$$\sup_{\omega \in \mathbb{R}} \mu_{\Delta_a}\big(N(j\omega)\big) < 1 \tag{11.59}$$

for robust performance.

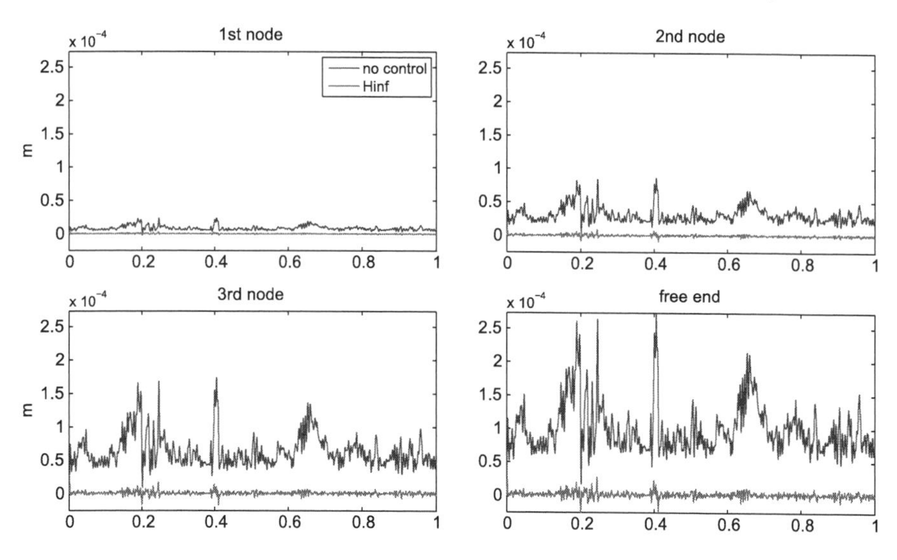

Fig. 11.11 Displacement response with and without $H\infty$ control for the four rodes of the beam

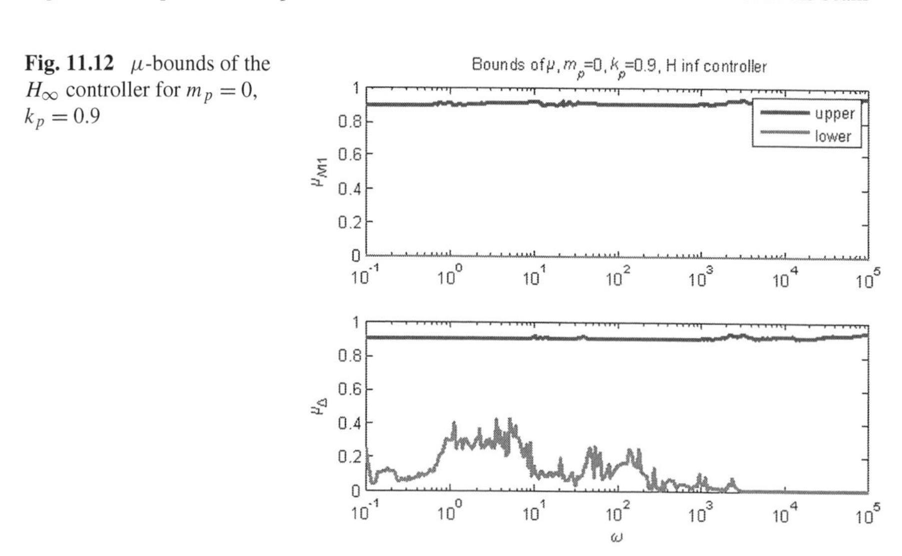

Fig. 11.12 μ-bounds of the H_∞ controller for $m_p = 0$, $k_p = 0.9$

For the H_∞ found, robust analysis was performed for the following values of m_p, k_p.

1. $m_p = 0$, $k_p = 0.9$. This corresponds to a $\pm 90\%$ variation from the nominal value of the stiffness matrix K.

   In Fig. 11.11 the displacement responses for this controller for the mechanical input are shown. In Fig. 11.12 are shown the bounds on the μ values. As seen the system remains stable and exhibits robust performance, since the upper bounds of both values remain below 1 for all frequencies of interest. This result is validated

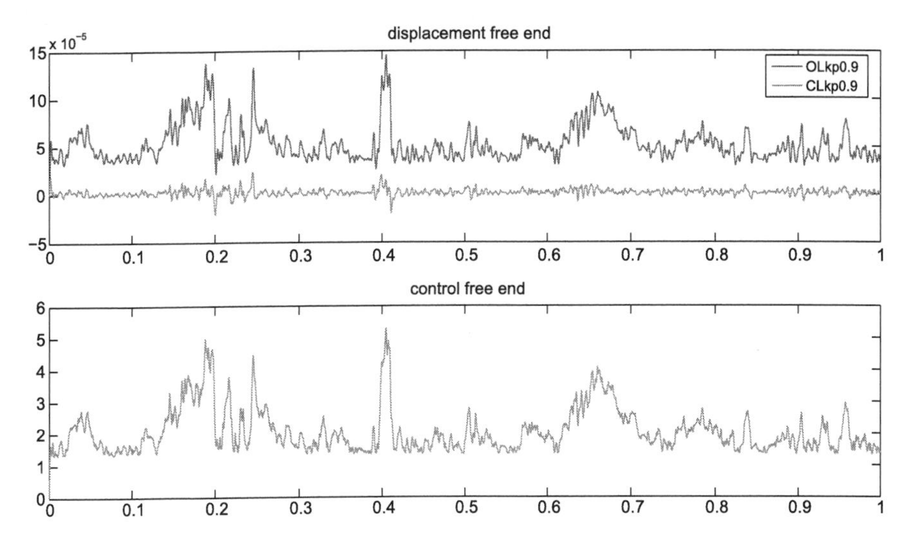

Fig. 11.13 Displacement and control at free end for the H_∞ controller with $m_p = 0$, $k_p = 0.9$ (extreme values)

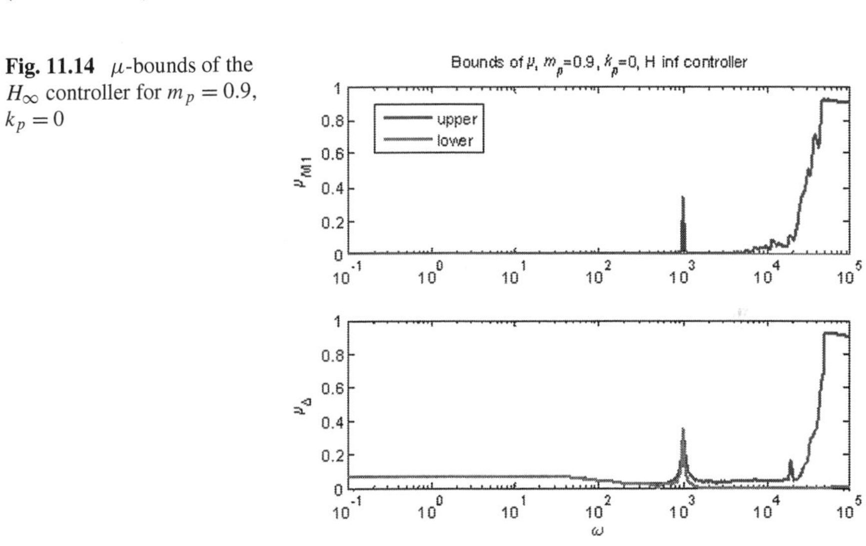

Fig. 11.14 μ-bounds of the H_∞ controller for $m_p = 0.9$, $k_p = 0$

in Fig. 11.13, where the displacement of the free end and the voltage applied are shown at the extreme uncertainty. Comparison with the open loop response for the same plant shows the good performance of the nominal controller.

2. $m_p = 0.9$, $k_p = 0$. This corresponds to a $\pm 90\%$ variation from the nominal value of the mass matrix M.

   In Fig. 11.14 are shown the bounds on the μ values. As seen the system remains stable and exhibits robust performance, since the upper bounds of both values remain below 1 for all frequencies of interest. This result is validated in

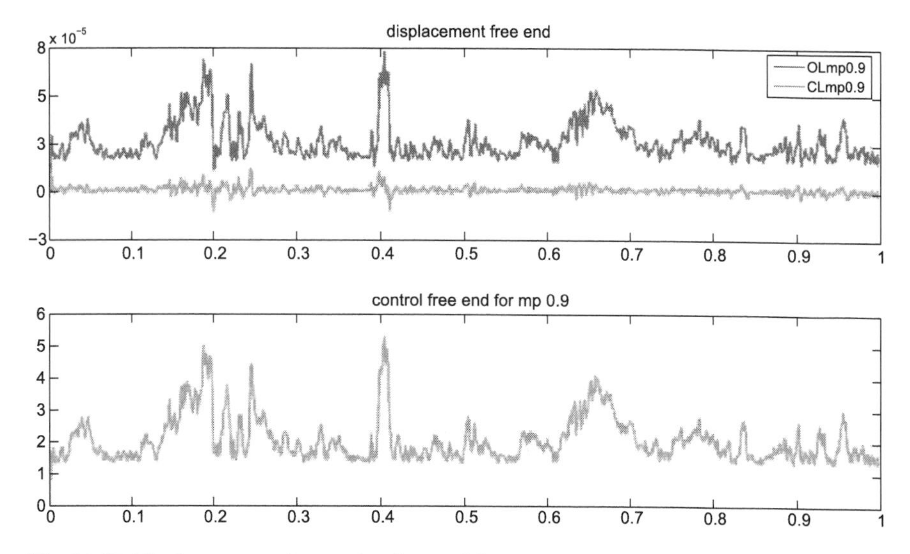

Fig. 11.15 Displacement and control at free end for the H_∞ controller with $m_p = 0.9$, $k_p = 0$ (extreme values)

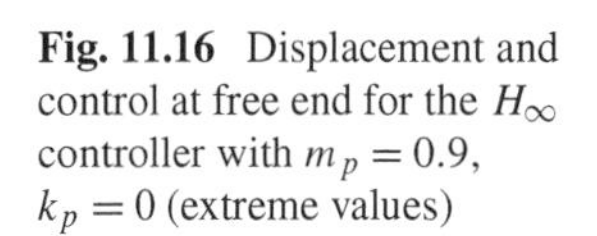

Fig. 11.16 Displacement and control at free end for the H_∞ controller with $m_p = 0.9$, $k_p = 0$ (extreme values)

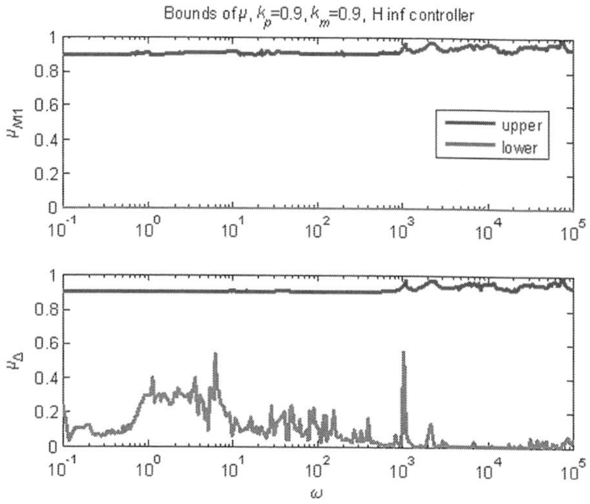

Fig. 11.15, where the displacement of the free end and the voltage applied are shown. Comparison with the open loop response for the same plant shows the good performance of the nominal controller.

3. $m_p = 0.9$, $k_p = 0.9$. This corresponds to a ±90% variation from the nominal values of both the mass and stiffness matrices M, K.

In Fig. 11.16 are shown the bounds on the μ values. As seen the system remains stable and exhibits robust performance, since the upper bounds of both values remain below 1 for all frequencies of interest. This result is validated in

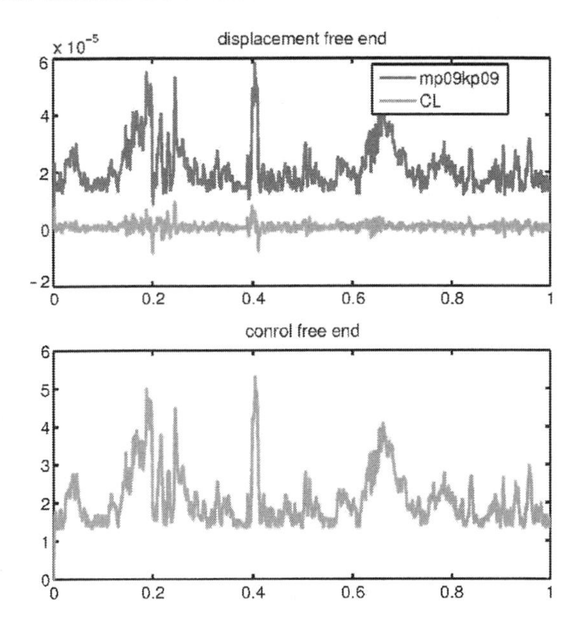

Fig. 11.17 Displacement and control at free end for the H_∞ controller with $m_p = 0.9$, $k_p = 0$ (extreme values)

Fig. 11.17, where the displacement of the free end and the voltage applied are shown. Comparison with the open loop response for the same plant shows the good performance of the nominal controller.

4.3 Robust Synthesis: μ-Controller—Results

A μ-controller can be synthesized via the procedure of D-K iteration As explained, this is an approximate procedure, providing bounds on the μ-value. To facilitate comparison with the H_∞ controller, similar bounds for the uncertainty will be used. In all the simulations that follow the disturbance is 10 N at the free end of the beam.

1. $m_p = 0$, $k_p = 0.9$. This corresponds to a $\pm 90\%$ variation from the nominal value of the stiffness matrix K. In Fig. 11.18 -values of the calculated controller are shown. As seen the controller is robust in most frequencies. In Fig. 11.19 performance of the μ and H_∞ controllers is compared at the free end (this is indicative of overall performance). As seen the H_∞ controller performs better at the expense of increased control effort. Figure 11.20 (the top two panels) verifies this result, where it is seen that the H_∞ controller fares better at the extreme value. This could be due to numerical difficulties in the calculation of the μ-controller arising from the bad condition number of the plant. It could also be due to the high order of the μ-controller. In any case, further investigation is needed.

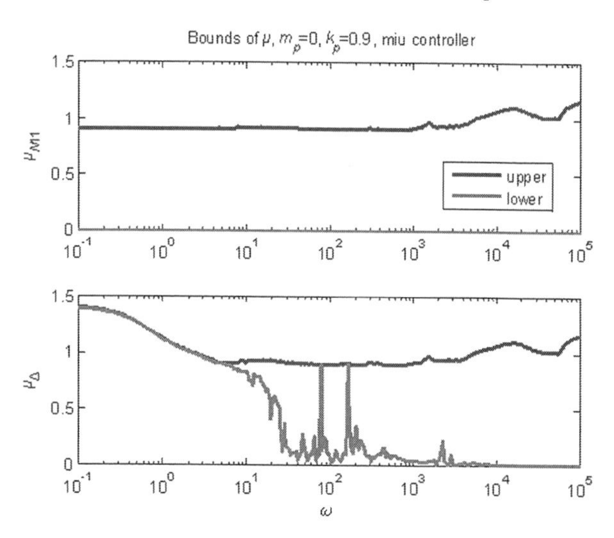

Fig. 11.18 μ-bounds of the μ-controller for $mp = 0$, $kp = 0.9$

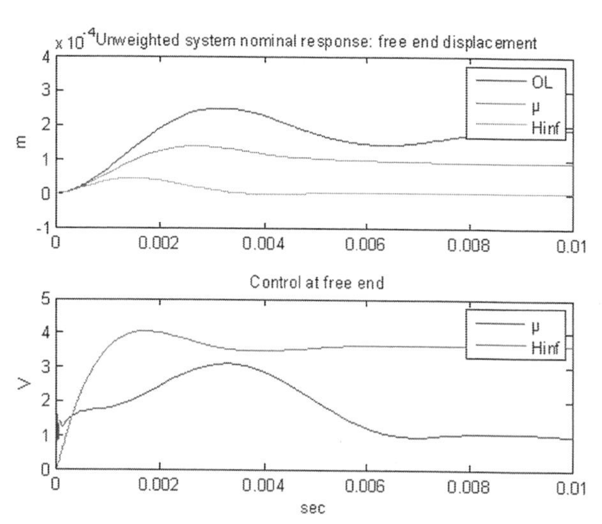

Fig. 11.19 Comparison of H_∞ and μ-controller

5 Conclusions

A finite element based model for a representative smart structure, namely a smart beam, including uncertainties has been presented. Based on this model H_∞ and μ-controller has been designed which effectively suppress the vibrations of the smart beam under stochastic load. The advantage of the H_∞ criterion is its ability to take into account the worst influence of uncertain disturbances or noise in the system. It is possible to synthesize a H_∞ controller which will be robust with respect to a prespecified number of errors in the model. The vibration suppression is achieved by the application of H_∞ controller. The system remains stable and exhibits robust performance, for all frequencies of interest. In addition a ro-

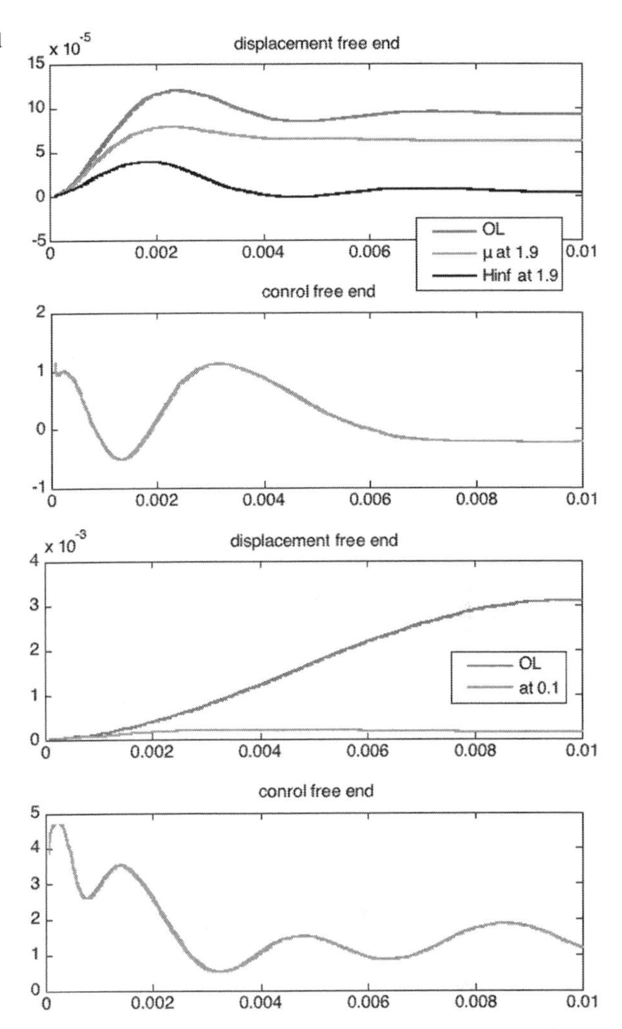

Fig. 11.20 Displacement and control at free end for μ-controller with $m_p = 0$, $k_p = 0.9$ (extreme values)

bust μ-controller was analyzed and synthesized, using the D-K Iterative method. The results are compared and commented. The results are very good: the oscillations were suppressed, with the voltages of the piezoelectric components' lying within their endurance limits. The above findings indicate that modern robust control techniques combined with classical finite element modelling provide us a powerful tool for applicable structural control design. Further work in this direction, including specialization for given damage-related uncertainties, investigation of different excitations as well as experimental verification, seems to be very promising.

References

1. Arvanitis, K.G., Zacharenakis, E.C., Soldatos, A.G., Stavroulakis, G.E.: New trends in optimal structural control. In: Selected Topics in Structronic and Mechatronic System, pp. 321–415. World Scientific, Singapore (2003)
2. Bosgra, O., Kwakernaak, H.: Design methods for control systems. Course notes, Dutch Institute for Systems and Control, p. 69 (2001)
3. Burke, J.V., Henrion, D., Lewis, A.S., Overton, M.L.: HIFOO—a MATLAB package for fixed-order controller design and H_∞ optimization. In: Proceedings of the IFAC Symposium on Robust Control Design, Toulouse, France (2006). www.cs.nyu.edu/overton/software/hifoo
4. Burke, J.V., Henrion, D., Lewis, A.S., Overton, M.L.: Stabilization via nonsmooth, nonconvex optimization. IEEE Trans. Autom. Control **5**(11), 1760–1769 (2006)
5. Foutsitzi, G., Marinova, D., Hadjigeorgiou, E., Stavroulakis, G.: Finite element modelling of optimally controlled smart beams. In: Venkov, G., Marinov, M. (eds.) 28th Summer School: Applications of Mathematics in Engineering and Economics, Sozopol, Bulgaria (2002)
6. Foutsitzi, G., Marinova, D., Hadjigeorgiou, E., Stavroulakis, G.: Robust H2 vibration control of beams with piezoelectric sensors and actuators. In: Proceedings of Physics and Control Conference (PhyCon03), St. Petersburg, Russia, 20–22 August, vol. I, pp. 158–163 (2003)
7. Hou, M., Muller, P.C.: Design of observers for linear systems with unknown inputs. IEEE Trans. Autom. Control **37**, 871–875 (1992)
8. Huang, W.S., Park, H.C.: Finite element modelling of piezoelectric sensors and actuators. AIAA J. **31**, 930–937 (1993)
9. Marinova, D., Stavroulakis, G.E., Foutsitzi, D., Hadjigeorgiou, E., Zacharenakis, E.C.: Robust design of smart structures taking into account structural defects. In: Indeitsev, D.A. (ed.) Summer School Conference Advanced Problems in Mechanics, Russian Academy of Sciences, pp. 288–292 (2004)
10. Miara, B., Stavroulakis, G., Valente, V. (eds.): Topics on Mathematics for Smart Systems: Proceedings of the European Conference, Rome, Italy, 26–28 October 2006. World Scientific, Singapore (2007)
11. Moutsopoulou, A., Stavroulakis, G., Pouliezos, A.: Model uncertainties in smart structures. In: Papadrakakis, M., Fragiadakis, M., Plevris, V. (eds.) 3rd Thematic Conference on Computational Methods in Structural Dynamics and Earthquake Engineering, Corfu, Greece (2011)
12. Packard, A., Doyle, J., Balas, G.: Linear, multivariable robust control with a perspective. J. Dyn. Syst. Meas. Control **115**(2b), 310–319 (1993)
13. Pouliezos, A.: MIMO control systems. Class notes (2008). http://pouliezos.dpem.tuc.gr
14. Shahian, B., Hassul, M.: Control System Design Using MATLAB. Prentice Hall, New York (1994)
15. Sisemore, C., Smaili, A., Houghton, R.: Passive damping of flexible mechanism system: experimental and finite element investigation. In: The 10th World Congress of the Theory of Machines and Mechanisms, Oulu, Finland, vol. 5, pp. 2140–2145 (1999)
16. Stavroulakis, G.E., Foutsitzi, G., Hadjigeorgiou, E., Marinova, D., Baniotopoulos, C.C.: Design and robust optimal control of smart beams with application on vibrations suppression. Adv. Eng. Softw. **36**(11–12), 806–813 (2005)
17. Tiersten, H.F.: Linear Piezoelectric Plate Vibrations. Plenum Press, New York (1969)
18. Tits, A.L., Yang, Y.: Globally convergent algorithms for robust pole assignment by state feedback. IEEE Trans. Autom. Control **41**, 1432–1452 (1996)
19. Ward, R.C.: Balancing the generalized eigenvalue problem. SIAM J. Sci. Stat. Comput. **2**, 141–152 (1981)
20. Young, P., Newlin, M., Doyle, J.: Practical computation of the mixed problem. In: Proceedings of the American Control Conference, pp. 2190–2194 (1992)
21. Zhang, N., Kirpitchenko, I.: Modelling dynamics of a continuous structure with a piezoelectric sensor/actuator for passive structural control. J. Sound Vib. **249**, 251–261 (2002)

Chapter 12
Robust Structural Health Monitoring Using a Polynomial Chaos Based Sequential Data Assimilation Technique

George A. Saad and Roger G. Ghanem

Abstract With the recent technological advances and the evolution of advanced smart systems for damage detection and signal processing, Structural Health Monitoring (SHM) emerged as a multidisciplinary field with wide applicability throughout the various branches of engineering, mathematics and physical sciences. However, significant challenges associated with modeling the physical complexity of systems comprising these structures remain. This is mainly due to the fact that numerous uncertainties associated with modeling, parametric and measurement errors could be introduced. In cases where these uncertainties are significant, standard identification and damage detection techniques are either unsuitable or inefficient. This study presents a robust data assimilation approach based on a stochastic variation of the Kalman Filter where polynomial functions of random variables are used to represent the inherent process uncertainties. The presented methodology is combined with a non-parametric modeling technique to tackle structural health monitoring of a four-story shear building. The structure is subject to a base motion specified by a time series consistent with the El-Centro earthquake and undergoes a preset damage in the first floor. The purpose of the problem is localizing the damage in both space and time, and tracking the state of the system throughout and subsequent to the damage time. The application of the introduced data assimilation technique to SHM enhances the latter's applicability to a wider range of structural problems with strongly nonlinear dynamical behavior and with uncertain and complex governing laws.

G.A. Saad (✉)
American University of Beirut, Beirut, Lebanon
e-mail: george.saad@aub.edu.lb

R.G. Ghanem
University of Southern California, Los Angeles, CA, USA
e-mail: ghanem@usc.edu

M. Papadrakakis et al. (eds.), *Computational Methods in Stochastic Dynamics*, Computational Methods in Applied Sciences 26,
DOI 10.1007/978-94-007-5134-7_12,

1 Introduction

With the recent technological advances and the evolution of advanced smart systems for damage detection and signal processing, Structural Health Monitoring (SHM) emerged as a multidisciplinary field with wide applicability throughout the various branches of engineering, mathematics and physical sciences. Typically, the SHM problem can be addressed as a statistical pattern recognition paradigm with three main components:

1. A numerical model that accurately represents the governing system dynamics
2. Real-time data acquisition and management system
3. A sequential data assimilation technique that relies on a set of observational data to calibrate and update the underlying dynamic principles governing the system under observation.

In such context, numerous uncertainties associated with modeling, parametric and measurement errors could be introduced. In cases where these uncertainties are significant, standard identification and damage detection techniques are either unsuitable or inefficient. Therefore, the need rises for robust system identification algorithms that can tackle the aforementioned challenges. This has been a very active research area over the past decade [3–5, 8, 9, 11, 12].

Sequential data assimilation has been widely used for structural health monitoring and system identification problems. Many extensions of the Kalman Filter were developed as adaptations to important classes of these problems. While the Extended Kalman Filter may fail in the presence of high non-linearities, Monte Carlo based Kalman Filters usually give satisfactory results. The Ensemble Kalman Filter (EnKF) [1, 2] was recently used for damage detection in strongly nonlinear systems [4], where it is combined with non-parametric modeling techniques to tackle structural health monitoring for non-linear systems. The EnKF uses a Monte Carlo Simulation scheme for characterizing the noise in the system, and therefore allows representing non-Gaussian perturbations. Although this combination gives good results, it requires a relatively accurate representation of the non-linear system dynamics. It also requires a large ensemble size to quantify the non-Gaussian uncertainties in such systems and consequently imposes a high computational cost.

This study presents a system identification approach based on coupling robust non-parametric non-linear models with the Polynomial Chaos methodology in the context of the Kalman Filtering techniques [10]. The proposed approach uses a Polynomial Chaos expansion [6, 7] of the nonparametric representation of the system's non-linearity to statistically characterize the system's behavior. A filtering technique that allows the propagation of a stochastic representation of the unknown variables using Polynomial Chaos is used to identify the chaos coefficients of the unknown parameters in the model. The introduced filter is a modification of the EnKF that uses the Polynomial Chaos methodology to represent uncertainties in the system. This allows the representation of non-Gaussian uncertainties in a simpler, less taxing way without the necessity of managing a large ensemble. It also allows obtaining the probability density function of the model state or parameters at any instant in time by simply simulating the Polynomial Chaos basis.

2 The Polynomial Chaos Kalman Filter (PCKF)

The Kalman Filter is an optimal sequential data assimilation method for linear dynamics and measurement processes with Gaussian error statistics. The PCKF builds on the mathematics of the original Kalman Filter to allow the propagation of a stochastic representation of the unknown variables using Polynomial Chaos. In the PCKF, the model state is given by,

$$x = \sum_{i=0}^{P} x_i \psi_i(\xi), \tag{12.1}$$

where, $P+1$ is the number of terms in the Polynomial expansion of the state vector, $\{\psi_{\mathrm{i}}\}$ is set of Hermite polynomials function of the Gaussian random variable, ξ. Consequently, the covariance matrix of the model state is defined around the mean, the zero order term, of the stochastic representation,

$$\begin{aligned} \mathbf{P} &\approx \left\langle \left(\sum_{i=0}^{P} x_i \psi_i - x_0\right)\left(\sum_{i=0}^{P} x_i \psi_i - x_0\right)^T \right\rangle \\ &\approx \left\langle \left(\sum_{i=1}^{P} x_i \psi_i\right)\left(\sum_{i=1}^{P} x_i \psi_i\right)^T \right\rangle \\ &\approx \sum_{i=1}^{P} x_i x_i^T \langle \psi_i^2 \rangle, \end{aligned} \tag{12.2}$$

where, $\mathbf{P}$ is the covariance matrix, and $\langle\rangle$ denotes the mathematical expectation. The Polynomial Chaos representation depicts all the information available through the complete probability density function, and therefore allows the propagation of all the statistical moments of the unknown parameters and variables.

The observations are also treated as random variables represented via a Polynomial Chaos expansion with a mean equal to the first-guess observations. Since the model and measurement errors are assumed to be independent, the latter is represented as a Markov process.

2.1 Analysis Scheme

For computational efficiency, the dimensionality and order of the Polynomial Chaos expansion are homogenized through out the solution. These parameters are initially defined based on the uncertainty within the problem at hand and are assumed to be constant thereafter. Since the model state and measurement vectors are assumed independent, the Polynomial Chaos representation of these variables has a sparse structure.

Let **A** be the matrix holding the chaos coefficients of the state vector x,

$$\mathbf{A} = (x_0, x_1, \ldots, x_P) \in R^{n\times(P+1)}, \tag{12.3}$$

where $P+1$ is the total number of terms in the Polynomial Chaos representation of x and n is the size of the model state vector. The mean of x is stored in the first column of **A** and is denoted by x_0. The state perturbations are given by the higher order terms stored in the remaining columns. Consequently, the state error covariance matrix **P** is defined as:

$$\mathbf{P} = \sum_{i=1}^{P} x_i x_i^T \langle \psi_i^2 \rangle \in R^{n\times n} \tag{12.4}$$

Given a vector of measurements $d \in R^m$, with m being the number of measurements at each occurrence, a Polynomial chaos representation of the measurements is defined as

$$d = \sum_{j=0}^{P} d_j \psi_j(\xi), \tag{12.5}$$

where the mean d_0 is given by the actual measurement vector, and the higher order terms represent the measurement uncertainties. The representation d can be stored in matrix form as:

$$\mathbf{B} = (d_0, d_1, \ldots, d_P) \in R^{m\times(P+1)}. \tag{12.6}$$

Based on Eq. (12.5), the measurement error covariance matrix, **R**, is defined as:

$$\mathbf{R} = \sum_{i=1}^{P} d_i d_i^T \langle \psi_i^2 \rangle \in R^{m\times m} \tag{12.7}$$

The Kalman Filter forecast step is carried out by employing a stochastic Galerkin scheme, and the assimilation step consists of the traditional Kalman Filter correction step applied on the Polynomial Chaos expansion of the model state vector,

$$\sum_{i=0}^{P} x_i^a \psi_i = \sum_{i=0}^{P} x_i^f \psi_i + \mathbf{PH}^T \left(\mathbf{HPH}^T + \mathbf{R}\right)^{-1} \left(\sum_{i=0}^{P} d_i \psi_i - \mathbf{H} \sum_{i=0}^{P} x_i^f \psi_i \right) \tag{12.8}$$

where, **H** is the observation matrix, and the superscripts f and a represent the forecast and analysis states respectively. Projecting the above equation on an approximating space spanned by the Polynomial Chaos $\{\psi_i\}_{i=0}^{P}$ yields,

$$x_i^a = x_i^f + \mathbf{PH}^T \left(\mathbf{HPH}^T + \mathbf{R}\right)^{-1} \left(d_i - \mathbf{H} x_i^f\right) \quad \forall i = 0, 1, \ldots, P. \tag{12.9}$$

In matrix form, the assimilation step is expressed as:

$$\mathbf{A}^a = \mathbf{A}^f + \mathbf{PH}^T \left(\mathbf{HPH}^T + \mathbf{R}\right)^{-1} \left(\mathbf{B} - \mathbf{HA}^f\right) \tag{12.10}$$

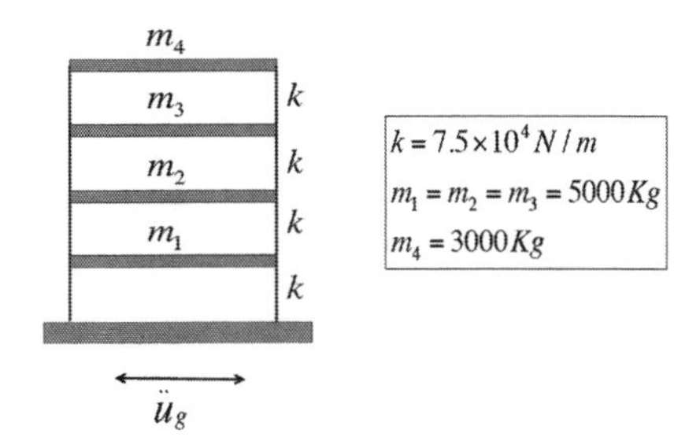

Fig. 12.1 Shear building under analysis

Table 12.1 Bouc–Wen model coefficients

Model Coefficient	Pre-Change	Post-Change
α	0.15	0.15
β	0.1	10
n	1	1
Υ	0.1	10
A	1	1

3 Numerical Example

The efficiency of the presented method is assessed by applying it to the structural health monitoring of the four-story shear building shown in Fig. 12.1. This model has a constant stiffness on each floor and a 5% damping ratio in all modes. All structural elements of this frame are assumed to involve hysteretic behavior, and it is supposed that a change in the hysteretic loop of the first floor element occurs at some point. It is of utmost importance to localize that point in time and track the state of the system throughout and subsequent to that point.

A synthetically generated dataset representing measurements of the displacements and velocities at each floor is obtained by representing the hysteretic restoring force by the Bouc–Wen model, which is therefore considered as the exact hysteretic behavior of the system. Thus, the equation of motion of the system is given by,

$$\mathbf{M}\ddot{u}(t) + \mathbf{C}\dot{u}(t) + \alpha\mathbf{K}_{\text{el}}u(t) + (1-\alpha)\mathbf{K}_{\text{in}}z(x,t) = -\mathbf{M}\tau\ddot{u}_g(t) \tag{12.11}$$

where, $\mathbf{M}$, $\mathbf{C}$, $\mathbf{K}_{\text{el}}$, and $\mathbf{K}_{\text{in}}$ are the mass, damping, elastic and inelastic stiffness matrices respectively; α is the ratio of the post yielding stiffness to the elastic stiffness, τ is the influence vector, u is the displacement vector, x is the inter-story drift vector, and z is an n-dimensional evolutionary hysteretic vector whose i^{th} component is give by the Bouc–Wen model as,

$$\dot{z}_i = A_i\dot{x}_i - \beta|\dot{x}_i||z_i|^{n_i-1} - \gamma_i\dot{x}_i|z_i|^{n_i}, \quad i = 1,\ldots,n \tag{12.12}$$

A, β, and Υ are the Bouc–Wen model parameters. The adopted values for these parameters are shown in Table 12.1.

The structure is subject to a base motion specified by a time series consistent with the 1940 El-Centro earthquake shown in Fig. 12.2, and a change of the first

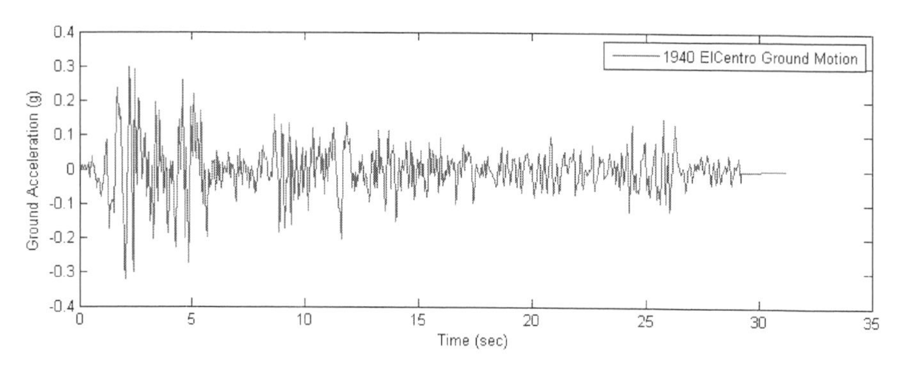

Fig. 12.2 The 1940 ElCentro excitation applied to the structure

floor hysteretic behavior is assumed to take place five seconds after the excitation. A monitoring scenario where it is assumed that measurements are available every 5 time steps is adopted. A nonparametric representation of the system nonlinearity is adopted, and the filtering technique is used to characterize the latter representation in order to capture any ambiguous behavior of the structure examined.

4 Non-parametric Representation of the Non-linearity

The proposed filtering methodology is combined with a non-parametric modeling technique to tackle structural health monitoring of non-linear systems but instead of adopting a deterministic nonparametric representation of the non-linearity, a stochastic representation via Polynomial Chaos is used. The basic idea behind the non-parametric identification technique used is to determine an approximating analytical function $\hat{F}$ that approximates the actual system non-linearities, with the form of $\hat{F}$ including suitable basis functions that are adapted to the problem at hand [8]. For general non-linear systems, a suitable choice of basis would be the list of terms in the power series expansion in the doubly indexed series,

$$S = \sum_{i=0}^{i_{max}} \sum_{j=0}^{j_{max}} u^i \dot{u}^j \tag{12.13}$$

where u and $\dot{u}$ are used to represent the system's displacement and velocity respectively. Therefore, if $i_{max} = 3$ and $j_{max} = 3$, the basis functions become:

$$\begin{aligned} basis = \{&1, \dot{u}, \dot{u}^2, \dot{u}^3, u, u\dot{u}, u\dot{u}^2, u\dot{u}^3, u^2, u^2\dot{u}, u^2\dot{u}^2, u^2\dot{u}^3, \\ &u^3, u^3\dot{u}, u^3\dot{u}^2, u^3\dot{u}^3\} \end{aligned} \tag{12.14}$$

In the proposed method the displacements and velocities are stochastic processes represented by their Polynomial Chaos expansion. Thus, the approximating function is also expressed as a stochastic process via a Polynomial Chaos representation. The model adopted within the Kalman Filter is hence given by

$$M\ddot{u}(t) + F(u, \dot{u}) = -M\tau\ddot{u}_g(t) \tag{12.15}$$

where, F is the non-parametric representation of the non-linearity whose i^{th} floor component is given by

$$\begin{aligned}
F^i &\approx \sum_j F^i_j(u, \dot{u})\psi_j \\
&\approx \sum_j a^i_j \psi_j \Big(\sum_k (u_k - u^{i-1}_k)\psi_k\Big) + \sum_j a^{i+1}_j \psi_j \Big(\sum_k (u^i_k - u^{i+1}_k)\psi_k\Big) \\
&\quad + \sum_j b^i_j \psi_j \Big(\sum_k (u^i_k - u^{i-1}_k)\psi_k\Big)^2 + \sum_j b^{i+1}_j \psi_j \Big(\sum_k (u^i_k - u^{i+1}_k)\psi_k\Big)^2 \\
&\quad + \sum_j c^i_j \psi_j \Big(\sum_k (\dot{u}^i_k - \dot{u}^{i-1}_k)\psi_k\Big) + \sum_j c^{i+1}_j \psi_j \Big(\sum_k (\dot{u}^i_k - \dot{u}^{i+1}_k)\psi_k\Big) \\
&\quad + \sum_j d^i_j \psi_j \Big(\sum_k (u^i_k - u^{i-1}_k)\psi_k\Big)\Big(\sum_l (\dot{u}^i_l - \dot{u}^{i-1}_l)\psi_l\Big) \\
&\quad + \sum_j d^{i+1}_j \psi_j \Big(\sum_k (u^i_k - u^{i+1}_k)\psi_k\Big)\Big(\sum_l (\dot{u}^i_l - \dot{u}^{i+1}_l)\psi_l\Big)
\end{aligned} \tag{12.16}$$

In the above equation, $\{a_j\}$, $\{b_j\}$, $\{c_j\}$, and $\{d_j\}$ represent the chaos coefficients of the unknown parameters to be identified. The fourth order Runge–Kutta method is used for the time stepping and a stochastic Galerkin approach is employed to solve the system at each time step.

5 Results

In the numerical example, it is assumed that observations of displacements and velocities from all floors are available. The noise signals perturbing both the model and measurements are modeled as first order, one dimensional, independent, Polynomial Chaos expansions having zero-mean and an RMS of 0.05 and 0.001 respectively. The parametric uncertainties on the other hand, are modeled as second order, one dimensional, Polynomial Chaos expansions whose coefficients are to be determined in accordance with the available observations. This is done to incorporate the possibility that the unknown parameters may deviate from Gaussianity. Furthermore, it is assumed that the first floor undergoes a change in its hysteretic behavior 5 seconds after the ground excitation. The purpose of the application is to detect this behavioral change.

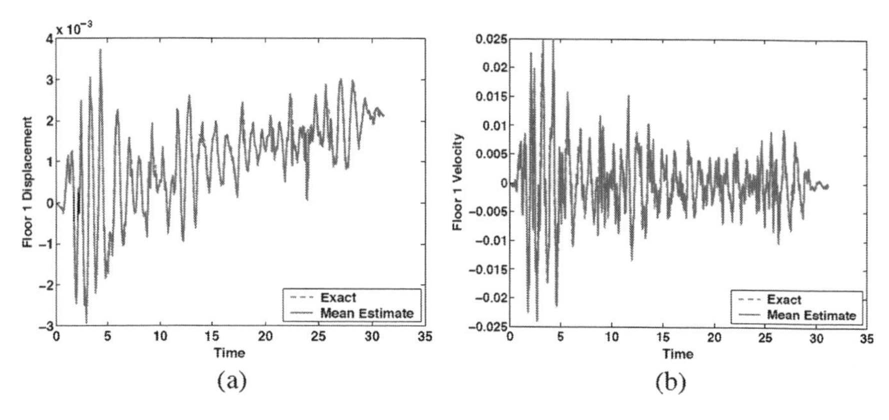

Fig. 12.3 Estimate of the first floor parameters, (**a**) displacement, (**b**) velocity

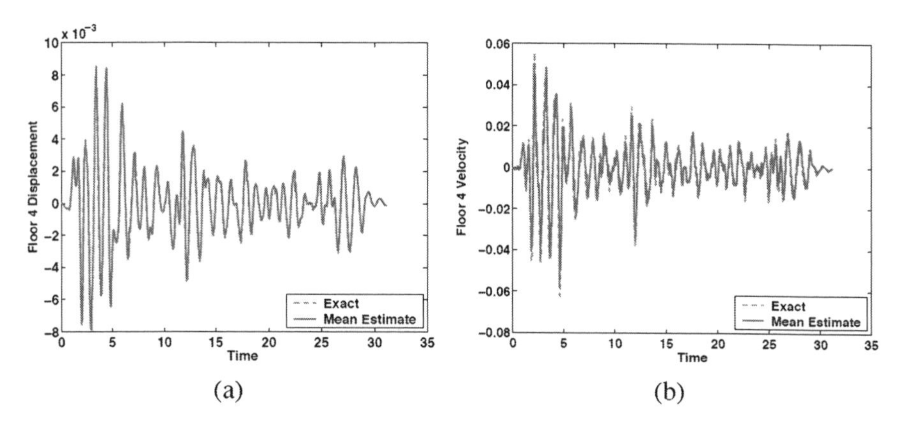

Fig. 12.4 Estimate of the fourth floor parameters, (**a**) displacement, (**b**) velocity

Fig. 12.3 and Fig. 12.4 describe the tracking of the displacement and velocity for the first and fourth floor respectively. Excellent match between the results estimated using the Polynomial Chaos based Kalman Filter and the true state is observed.

Fig. 12.5 presents the evolution of the mean of the unknown parameters identified by the proposed filtering technique. Error bars representing the scatter in the estimated parameters are also present in Fig. 12.5. The different jumps within the parameters are associated with the perks in the corresponding excitation.

Further investigation of the parameters indicates that the main changes take place in the first floor following the 5 s time interval. Note that the parameters a and c in floors 1 and 2 undergo the greatest jumps since they are associated with inter-story drift and velocity, respectively. One of the main advantages of using the Polynomial Chaos Kalman filter is that it provides a scatter around the estimated parameters. This is represented by the probability density functions corresponding to each of

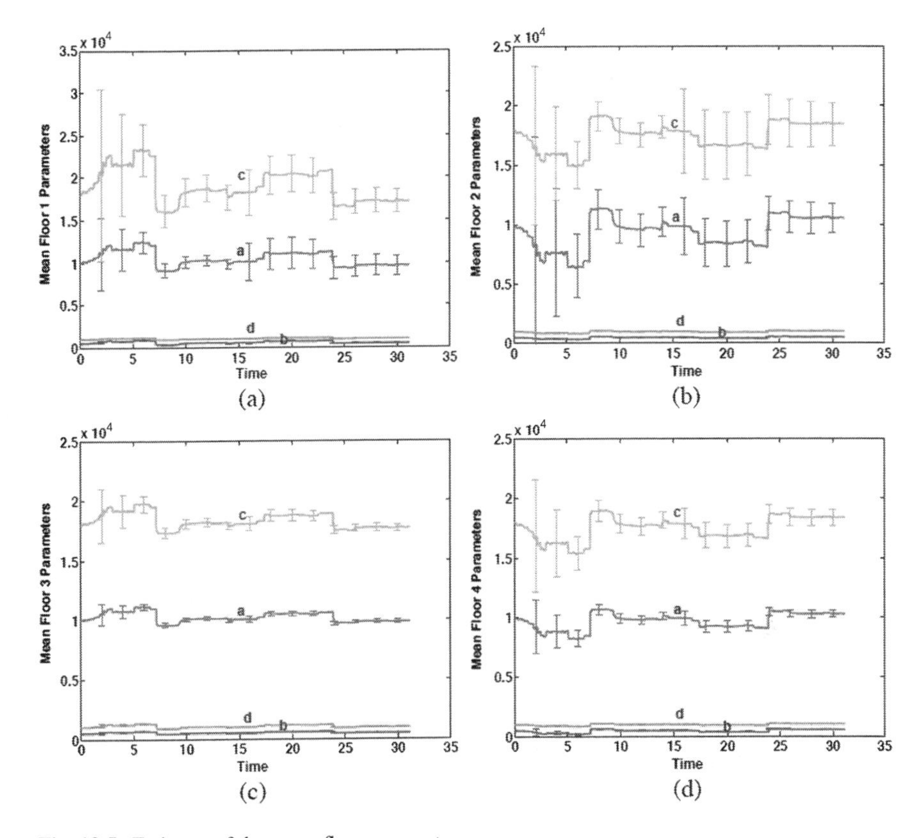

Fig. 12.5 Estimate of the mean floor parameters

the estimated parameters. Fig. 12.6 presents the probability density functions of the estimated floor 1 parameters.

6 Conclusions

The combination of Polynomial Chaos with the Ensemble Kalman Filter renders an efficient data assimilation methodology that competes with other Kalman Filtering techniques while maintaining a relatively low computational cost. Although the proposed method employs traditional Kalman Filter updating schemes, it preserves all the error statistics, and hence allows the computation of the probability density function of the uncertain parameters and variables at all time steps. This is achieved by simply simulating the Polynomial Chaos representation of these parameters. Together with the non-parametric representation of the nonlinearities, the approach constitutes an effective system identification technique that accurately detects any changes in the systems behavior. The Polynomial Chaos representation of the non-parametric model for the nonlinearities is a robust innovative approach that per-

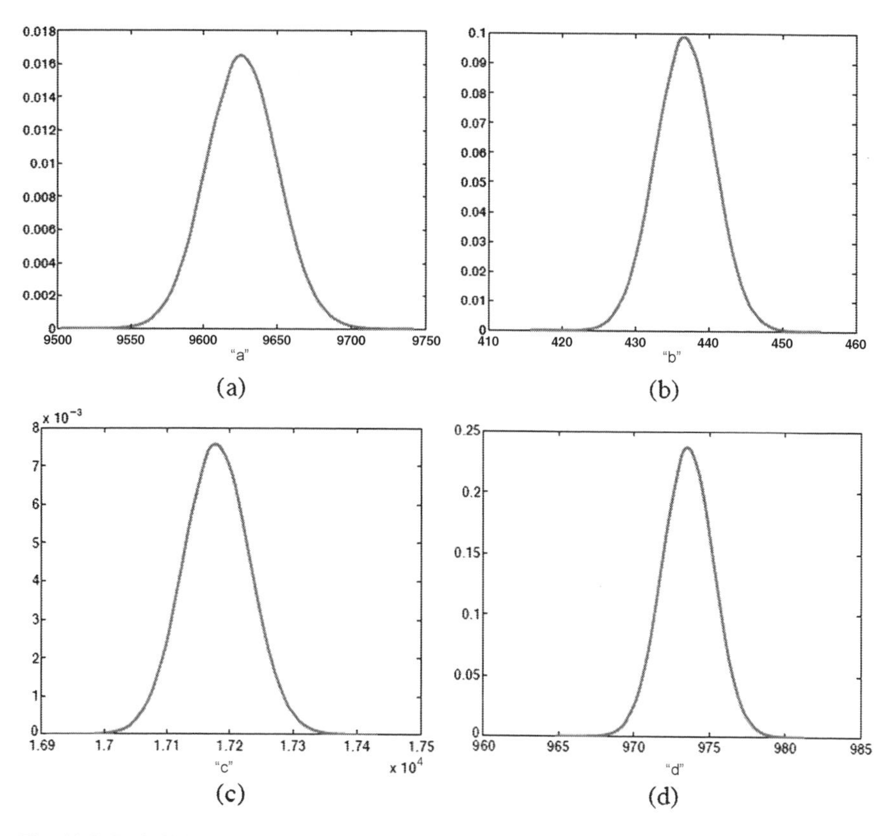

Fig. 12.6 Probability density functions of the estimated floor 1 parameters

mits damage identification and tracking the dynamical state beyond that point. Using Polynomial Chaos, the uncertainty associated with the assumed non-parametric model is inherently present and thus represents the actual nonlinearity in a more accurate way.

References

1. Evensen, G.: Sequential data assimilation with a nonlinear quasi-geostrophic model using Monte Carlo to forecast error statistics. J. Geophys. Res. **99**, 10143–10162 (1994)
2. Evensen, G.: The ensemble Kalman filter: theoretical formulation and practical implementation. Ocean Dyn. **53**, 343–367 (2003)
3. Franco, G., Betti, R., Lus, S.: Identification of structural systems using an evolutionary strategy. J. Eng. Mech. **130**, 1125–1139 (2005)
4. Ghanem, R., Ferro, G.: Health monitoring for strongly non-linear systems using the ensemble Kalman filter. Struct. Control Health Monit. **13**, 245–259 (2002)
5. Ghanem, R., Shinozuka, M.: Structural systems identification I, theory. J. Eng. Mech. **121**, 255–264 (1995)

6. Ghanem, R., Spanos, P.: Stochastic Finite Elements: A Spectral Approach, revised edn. Dover, New York (2003)
7. Ghanem, R., Saad, G., Doostan, A.: Efficient solution of stochastic systems: application to the embankment dam problem. Struct. Saf. **29**, 238–251 (2007)
8. Masri, S., Caffrey, J.P., Caughey, T.K., Smyth, A.W., Chassiakos, A.G.: Identification of the state equation in complex non-linear systems. Int. J. Non-Linear Mech. **39**, 1111–1127 (2004)
9. Masri, S., Ghanem, R., Arrate, F., Caffrey, J.P.: A data based procedure for analyzing the response of uncertain nonlinear systems. Struct. Control Health Monit. **16**, 724–750 (2009)
10. Saad, G., Ghanem, R.: Characterization of reservoir simulation models using a polynomial chaos based ensemble Kalman filter. Water Resour. Res. **45**, W04417 (2009)
11. Saad, G., Ghanem, R., Masri, S.: Robust system identification of strongly non-linear dynamics using a polynomial chaos-based sequential data assimilation technique. In: 48th AIAA/ASME/ASCE/AHS/ASC Structures, Structural Dynamics, and Materials Conference, Honolulu, Hawaii (2007)
12. Zhang, H., Foliente, G., Yang, Y., Ma, F.: Parametric identification of elastic structures under dynamic loads. Earthquake Eng. Struct. Dyn. **31**, 1113–1130 (2002)

Chapter 13
Efficient Model Updating of the GOCE Satellite Based on Experimental Modal Data

B. Goller, M. Broggi, A. Calvi, and G.I. Schuëller

Abstract The accurate prediction of the structural response of spacecraft systems during launch and ascent phase is a crucial aspect in design and verification stages which requires accurate numerical models. The enhancement of numerical models based on experimental data is denoted model updating and focuses on the improvement of the correlation between finite element (FE) model and test structure. In aerospace industry the examination of the agreement between model and real structure involves the comparison of the modal properties of the structure. Model updating techniques have to handle several difficulties, like incomplete experimental data, measurement errors, non-unique solutions and modeling uncertainties. To cope with the computational challenges associated with the large-scale FE-models involving up to over one million degrees of freedom (DOFs), enhanced strategies are required. A large-scale numerical example, namely a satellite model, will be used for demonstrating the applicability of the employed updating procedure to complex aerospace structures.

1 Introduction

The dynamic loads acting on a spacecraft during the launch and ascent phase are modeled by the spacecraft-launcher coupled dynamic analysis. The accuracy of the

---

G.I. Schuëller is deceased.

B. Goller (✉)
Institute of Engineering Mechanics, University of Innsbruck, Technikerstr. 13, 6020 Innsbruck, Austria
e-mail: barbara.goller@uibk.ac.at

M. Broggi
Institute for Risk and Uncertainty, University of Liverpool, Warrington, WA4 4AD, UK
e-mail: matteo.broggi@liverpool.ac.uk

A. Calvi
Structures Section TEC-MSS, European Space Agency/ESTEC, P.O. Box 299, 2200 AG Noordwijk, The Netherlands
e-mail: adriano.calvi@esa.int

M. Papadrakakis et al. (eds.), *Computational Methods in Stochastic Dynamics*, Computational Methods in Applied Sciences 26,
DOI 10.1007/978-94-007-5134-7_13,

structural response in this low-frequency mechanical environment depends on the quality of the underlying mechanical model of the spacecraft. Therefore, it is mandatory to ensure that the FE-model represents the real structure accurately enough. This level of accuracy to be reached for aerospace structures is defined in [1] and is based on the agreement of experimentally and computationally determined modal properties, respectively. Possible sources for discrepancies between test data and respective computed values are e.g. uncertainties in the modeling process arising from inadequate theory for some system behaviors, simplifying assumptions made in order to reduce the complexity of the model and uncertainties about model parameter values. Hence, the need for improving the mechanical model based on experimental data arises which is referred to as model updating and the consecutive corroboration of the model by means of modal properties is denoted by validation [2–4].

The use of deterministic updating procedures does not allow for a quantification of the involved uncertainties in the design and verification processes which will subsequently affect the accuracy of the predictive structural response. Probabilistic methods for model updating provide a means for tackling these problems and for avoiding a wrong conclusion about the fit of the experimental data and analytical results [5]. A significant obstacle in the consideration of uncertainties when performing model updating of complex structures is posed by the associated computational efforts. Therefore, the most frequently used approaches for model updating performed by industry are deterministic approaches (see e.g. [6–9]). While stochastic methods have been developing successfully in this research field, applications in industry are relatively limited (see e.g. [10]). Hence, in this work it is aimed at a computationally efficient application of a stochastic model updating procedure to complex aerospace models.

The thereby adopted updating process is the Bayesian approach which is based on updating the initial engineering knowledge about the ranges of the adjustable parameters using experimental data [11–13]. In this way, a revised information about the parameters is obtained, which is expressed by posterior probability density functions. Probability is therefore not interpreted in the usual frequentist sense, where it refers to the relative frequency of occurrence in case of many events, but it is based on the idea of reasonable expectation, i.e. probability is interpreted as a measure of plausibility of the hypothesis. This interpretation makes it possible to extend the application of probability theory to fields where the frequentist interpretation may not be directly intuitive, as it is the case for one-of-a-kind structures, where no ensemble exists, and also in the case of limited data, where classical statistics is of limited applicability. Therefore, Bayesian statistics makes it possible to deal with the usual situation in aerospace industry, where a large amount of experimental data is infeasible due to the enormous costs associated with test campaigns, and it provides a means for making decisions based on limited, incomplete information.

The computational tools for the Bayesian updating procedure are sampling-based algorithms, where a multi-level Markov chain Monte Carlo algorithm is adopted in this approach [14]. As a remedy for the large computational efforts associated with the Bayesian updating procedure, the application of a surrogate model (a so-called "meta-model") is proposed. This meta-model is formulated with respect to the repeated analysis tasks, which are the eigensolutions in case of model updating based on modal data. Hence, a simple relation between the input data and output quantity of interest has to be established in order to replace the computationally intensive evaluation of a full finite element analysis by a function evaluation at low computational costs. Several techniques, e.g. linear or polynomial regression, kriging and the radial basis functions have been developed in this context (see e.g. [15]), where in this work neural networks [16, 17] will be adopted in order to approximate the modal properties of the structure.

This manuscript will demonstrate the feasibility of the application of Bayesian model updating procedures on spacecraft structures using eigenfrequencies and mode shape vectors. Section 2 is devoted to the presentation of the basic steps of Bayesian model updating, and in Sect. 3 the algorithm for the generation of samples in the solution space is summarized. Computational aspects will be addressed in the following (Sect. 4), where the basic concepts of neural networks are discussed. In order to apply these outlined concepts to an FE-model of a spacecraft structure, the use of a surrogate model within the updating process is adopted for the FE-model of the GOCE satellite (Sect. 5).

2 Bayesian Model Updating

2.1 Introduction

The fundamental rule that governs the Bayesian updating procedure is Bayes' Theorem, which is formulated in general terms as [18]

$$P(H|\mathscr{D}, I) = \frac{P(\mathscr{D}|H, I)P(H|I)}{P(\mathscr{D}|I)}, \tag{13.1}$$

where H is any hypothesis to be tested, $\mathscr{D}$ denotes the data and I is the available background information. Bayes' Theorem provides a means to update the prior probability density function (PDF) of H, $P(H|I)$, by using the data in the likelihood function $P(\mathscr{D}|H, I)$ in order to obtain the posterior distribution of H, $P(H|\mathscr{D}, I)$. The denominator $P(\mathscr{D}|I)$ is a normalizing constant and does not affect the shape of the posterior PDF. All probabilities in Eq. (13.1) are conditional on I, which means that the outcome of the updating procedure depends on the available information.

2.2 Bayesian Updating Using Modal Data

If applying Bayes' Theorem for structural model updating [11, 12], the hypothesis H is interpreted as the vector of unknown (i.e. adjustable) parameters, which will be referred to as θ in the following, and $\mathscr{D}$ denotes the experimental data, which consist of the measured modal properties of the investigated structure. The available information I is interpreted as the experience and knowledge of the engineer which is reflected by the established model itself and is therefore denoted as $\mathscr{M}$. This leads to the following form of Eq. (13.1):

$$p(\theta|\mathscr{D},\mathscr{M}) = \frac{p(\mathscr{D}|\theta,\mathscr{M})p(\theta|\mathscr{M})}{p(\mathscr{D}|\mathscr{M})}. \tag{13.2}$$

The prior distribution $p(\theta|\mathscr{M})$ expresses the initial knowledge about the adjustable parameters. The choice of the distribution can be based on the principle of maximum entropy [19]. In this case, the PDF used for describing the initial uncertainty maximizes the uncertainty subject to the prescribed constraints, which can be given by e.g. imposing moment constraints. The likelihood function $p(\mathscr{D}|\theta,\mathscr{M})$ gives a measure of the agreement between the system data and the corresponding structural model output. This measure of the data fit of each model defined by the parameters vector θ, is given by the probability model established for the system output. The derivation of the likelihood function for modal data will be summarized in Sect. 2.3. The posterior distribution $p(\theta|\mathscr{D},\mathscr{M})$ expresses the revised knowledge about the parameters θ conditional on the initial knowledge and the experimental data.

2.3 Formulation of the Likelihood Function for Modal Data

In general terms, the connection between the model output $q(\theta)$ and the corresponding system output y is given by the prediction error e in the form of

$$y = q(\theta) + e. \tag{13.3}$$

The choice for the probability model of the prediction error e, which is the difference between the model output for a certain value of θ and the corresponding system output, is based on the maximum entropy principle [19] which yields a multidimensional Gaussian distribution with zero mean and covariance matrix Σ. The Gaussian PDF arises because it gives the largest amount of uncertainty among all probability distributions for a real variable whose first two moments are specified. Hence, the predictive PDF for the system output conditional on the parameter vector θ is given by

$$p(y|\theta,\mathscr{M}) = \frac{1}{(2\pi)^{N/2}|\Sigma|^{1/2}} \exp\left[-\frac{1}{2}\big(y-q(\theta)\big)^T \Sigma^{-1}\big(y-q(\theta)\big)\right], \tag{13.4}$$

where N denotes the length of the vector y, i.e. the number of observed points. If a set of measured output $\mathscr{D} = \{y_j : j = 1,\ldots,N_s\}$ is available, then the likelihood function can be constructed as $p(\mathscr{D}|\theta,\mathscr{M}) = \prod_{j=1}^{N_s} p(y_j|\theta,\mathscr{M})$ if the prediction errors are modeled as statistically independent.

The formulation of the likelihood function using modal data is derived in [20] and is summarized in the following. The experimental data $\mathscr{D}$ from the structure is assumed to consist of N_s sets of modal data, $\mathscr{D} = \{\hat{\omega}_{1,j}\cdots\hat{\omega}_{N_m,j}, \hat{\Psi}_{1,j}\cdots\hat{\Psi}_{N_m,j}\}_{j=1}^{N_s}$ comprised of N_m modal frequencies $\hat{\omega}_r$ and N_m incomplete mode shape vectors $\hat{\Psi}_r \in \mathbb{R}^{N_0}$, where N_0 is the number of observed DOFs. The model output $q(\theta)$ is the set of corresponding modal properties of the structural model, i.e. eigenfrequencies $\omega_r(\theta)$ and partial eigenvectors $\psi_r(\theta)$, $r = 1, \ldots, N_m$, defined by the parameter vector $\theta \in \Theta \in \mathbb{R}^{N_p}$.

The probability model conditional on the parameter vector θ is chosen to have statistical independence between the mode shape vectors and modal frequencies, between the different modes, and between one data set to another. Therefore, the likelihood function can be written as the product of the probability density functions for the modal frequencies and mode shape components:

$$p(\mathcal{D}|\theta, \mathcal{M}) = \prod_{j=1}^{N_s} \prod_{r=1}^{N_m} p\left(\hat{\omega}_{r,j}^2|\theta, \mathcal{M}\right) p(\hat{\psi}_{r,j}|\theta, \mathcal{M}), \tag{13.5}$$

where $p(\hat{\omega}_{r,j}^2|\theta, \mathcal{M})$ and $p(\hat{\psi}_{r,j}|\theta, \mathcal{M})$, $r = 1, \ldots, N_m$ and $j = 1, \ldots, N_s$, are, respectively, the PDFs for the squared modal frequency and the mode shape vector of the rth mode in the jth data set.

In the first step, the likelihood function for the mode shape vectors is formulated by rewriting Eq. (13.3) as

$$\hat{\psi}_{r,j} = a_r \psi_r(\theta) + e_{\psi_r}, \tag{13.6}$$

where a_r is an optimal scaling factor to relate the scaling of the model mode shape vector $\psi_r(\theta)$ to that of the experimental mode shape vector $\hat{\psi}_{r,j}$, which is assumed to be normalized so that its Euclidean norm $\|\hat{\psi}_{r,j}\| = 1$. Since the latter is usually constituted by an incomplete set of observed DOFs N_0, the corresponding model mode shape vector is given by $\psi_r = \Gamma \phi_r$, where the matrix Γ picks the observed degrees of freedom from the complete model eigenvector ϕ_r. Using a Gaussian distribution for the probabilistic characterization of the prediction error for the mode shape vector, the likelihood function for the mode shape vector, after some algebraic manipulation, may be written as:

$$p(\hat{\psi}_{r,j}|\theta, \mathcal{M}) = c_1 \exp\left(\frac{\psi_r^T (I - \hat{\psi}_{r,j} \hat{\psi}_{r,j}^T) \psi_r}{2\delta_r^2 \|\psi_r\|^2}\right) \tag{13.7}$$

$$= c_1 \exp\left(\frac{1}{2\delta_r^2}\left[1 - \frac{|\psi_r^T \hat{\psi}_{r,j}|^2}{(\psi_r^T \psi_r)^2}\right]\right) \tag{13.8}$$

where I is the identity matrix of dimension N_m and $\delta_r^2 I$ denotes the mode shape prediction error covariance matrix for the rth mode. The equality in Eq. (13.7) shows that the probability density function for $\hat{\psi}_{r,j}$ involves the MAC (Modal Assurance Criterion) between $\hat{\psi}_{r,j}$ and $\psi_r(\theta)$, the experimental and model partial mode shapes of the rth mode, respectively.

Secondly, Eq. (13.3) is formulated for the squared modal frequencies, which yields

$$\hat{\omega}_{r,j}^2 = \omega_r^2(\theta) + e_{\omega_r^2}. \tag{13.9}$$

Using again a Gaussian probability model for the prediction error of the modal frequencies, the likelihood function for the modal frequencies is given by

$$p\left(\hat{\omega}_{r,j}^2|\theta,\mathcal{M}\right) = c_2 \exp\left[-\frac{1}{2}\left(\frac{1-\hat{\omega}_{r,j}^2/\omega_r^2}{\varepsilon_r}\right)^2\right], \tag{13.10}$$

where ε_r^2 denotes the variance of the prediction error of the squared r-th eigenfrequency, i.e. of $e_{\omega_r^2}$. Using the probability distributions for the mode shape vectors and modal frequencies given in Eqs. (13.7) and (13.10), the likelihood function in Eq. (13.5) can be written as

$$p(\mathcal{D}|\theta,\mathcal{M}) = c_3 \exp\left(-\frac{1}{2}\sum_{r=1}^{N_m} J_r(\theta)\right), \tag{13.11}$$

where the modal measure of fit $J_r(\theta)$ is defined by

$$J_r(\theta) = \sum_{j=1}^{N_s}\left[\left(\frac{1-\hat{\omega}_{r,j}^2/\omega_r^2}{\varepsilon_r}\right)^2 + \left(1 - \frac{|\psi_r^T\hat{\psi}_{r,j}|^2}{(\psi_r^T\psi_r)^2}\right)/\delta_r^2\right] \tag{13.12}$$

3 Transitional Markov Chain Monte Carlo Algorithm

The evaluation of Eq. (13.2) requires the computation of high-dimensional integrals for the determination of the normalizing constant of the posterior PDF, which cannot be tackled analytically or numerically. Recently, efficient stochastic simulation algorithms have been proposed which generate samples of the posterior distribution and which hence identify the parameter regions with the highest posterior probability mass. In this work, the so-called Transitional Markov Chain Monte Carlo (TMCMC) algorithm [14] is applied whose basic steps are discussed in the following.

The main idea of this algorithm is to iteratively proceed from the prior to the posterior distribution. It starts with the generation of samples from the prior PDF in order to populate the space in which also the most probable regions of the posterior distribution lie. Then, some intermediate PDFs are defined, where the shape does not change remarkably from the intermediate PDF $p[j]$ to the next $p[j+1]$. The small change of the shape makes it possible to efficiently sample according to $p[j+1]$ if samples according to $p[j]$ have been generated. The intermediate distributions are defined by

$$p[j+1] \propto p(\mathcal{D}|\theta,\mathcal{M})^{\beta_j}\, p(\theta|\mathcal{M}), \tag{13.13}$$

with $j = 0,\dots,m$ as the step index and $0 = \beta_0 < \beta_1 < \cdots < \beta_m = 1$. Hence, the exponent β_j can be interpreted as the percentage of the total information provided by the experimental data which is incorporated in the jth iteration of the updating procedure: $\beta_0 = 0$ corresponds to the prior distribution and for $\beta_m = 1$ the samples are generated from the posterior distribution.

Samples of the subsequent intermediate distribution $p[j+1]$ are obtained by generating Markov chains where the lead samples are selected from the distribution $p[j]$ by computing their probability weights with respect to $p[j+1]$, which are given by

$$w\left(\theta_j^{(l)}\right) = \frac{p(\mathscr{D}|\theta,\mathscr{M})^{\beta_{j+1}}\, p(\theta|\mathscr{M})}{p(\mathscr{D}|\theta,\mathscr{M})^{\beta_j}\, p(\theta|\mathscr{M})} = p(\mathscr{D}|\theta,\mathscr{M})^{\beta_{j+1}-\beta_j}, \tag{13.14}$$

where the upper index $l = 1, \ldots, N_j$ denotes the sample number in the jth iteration step. Each sample of the current step is generated using the Metropolis–Hastings algorithm [21, 22]: the starting point of a Markov chain is a sample from the previous step that is selected according to the probability equal to its normalized weight

$$\overline{w}\left(\theta_j^{(l)}\right) = \frac{w(\theta_j^{(l)})}{\sum_{l=1}^{N_j} w(\theta_j^{(l)})} \tag{13.15}$$

and the proposal density for the Metropolis–Hastings algorithm is a Gaussian distribution centered at the preceding sample of the chain and with a covariance matrix Σ_0 which is equal to the scaled version of the estimated covariance matrix of the current intermediate PDF:

$$\Sigma_0 = c^2 \sum_{l=1}^{N_j} \overline{w}\left(\theta_j^{(l)}\right)\left(\theta_j^{(l)} - \overline{\theta}_j\right)^T \left(\theta_j^{(l)} - \overline{\theta}_j\right), \tag{13.16}$$

where

$$\overline{\theta}_j = \sum_{l=1}^{N_j} \overline{w}\left(\theta_j^{(l)}\right)\theta_j^{(l)}. \tag{13.17}$$

The parameter c is a scaling parameter that is used to control the rejection rate of the Metropolis–Hastings algorithm at each step. These steps are repeated until $\beta_j = 1$ is reached, i.e. until the samples are generated from the posterior distribution.

4 Computational Aspects

4.1 General Remarks

Due to the repeated execution of the normal mode analysis of the FE-model, the computational effort of the Bayesian updating method might become infeasible for large FE-models. Hence, in order to reduce the wall clock time, i.e. the time between submitting the updating analysis and its completion, a strategy based on the reduction of the computational efforts associated with the normal mode analysis of the full FE-model is applied in this manuscript. This strategy is based on the use of neural networks for the modal parameters, which will be addressed in the following.

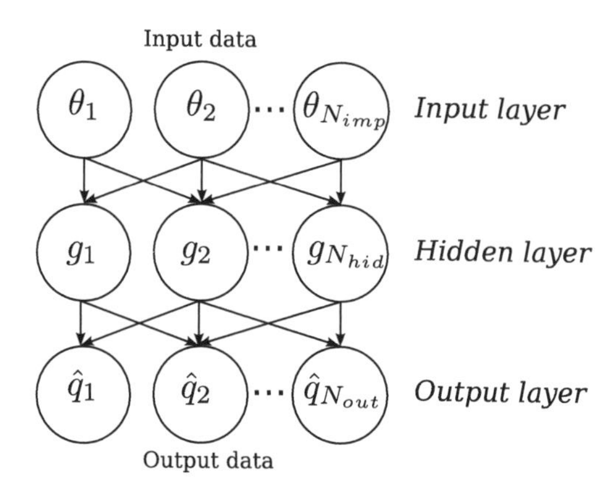

Fig. 13.1 Schematic representation of the concept of neural networks

4.2 Neural Networks

An Artificial Neural Network (ANN) is a machine-learning algorithm that tries to simulate the structure and functional aspects of biological networks of neurons in order to approximate a relation $f : \theta \rightarrow q$ by a simple mathematical model at low computational efforts. It consists of an interconnected group of computational units, called neurons or nodes, and processes information using consecutively connected layers of neurons. In the following, the most widely used neural network, namely the so-called feed-forward neural network, is discussed (see e.g. [23, 24]). It is composed of a multi-layered structure, with a first layer of nodes, called input layer, one or more intermediate layers, called hidden, and a final output layer. For simplicity, but without loss of generality, the scheme is discussed when using one single hidden layer.

Each layer is characterized by a different number of neurons, indicated by N_{inp}, N_{hid} and N_{out} for the input, hidden and output layers, respectively. Each node of the hidden layer receives as input a linear combination $\sum_{i=1}^{N_{inp}} w_{i,j}\theta_i$ of the input values θ_i of all the nodes of the input layer, scaled by a so-called connection weight $w_{i,j}$, where j denotes the number of the hidden node. Then, the node proceeds this function value through a non-linear function K, which is called activation function and which is of the form $g_j = K(\sum_{i=1}^{N_{inp}} w_{i,j}\theta_i)$. This collection of function values $(g_1, \ldots, g_{N_{hid}})$ is then sent to all the nodes of the subsequent, i.e. output, layer, where the approximated model output $\hat{q}_k(\theta) = K(\sum_{j=1}^{N_{hid}} w_{j,k}y_j)$, $k = 1, \ldots, N_{out}$ is evaluated. A schematic representation of this algorithm is shown in Fig. 13.1.

Hence, the connection weights act as parameters of the meta-model which have to be adapted through a calibration procedure of supervised learning, called also training, by means of e.g. the error back-propagation algorithm. In this calibration phase, the network, which is initialized with random weights, is fed with a set of input/output values which is called calibration set and which is obtained from the target physical model. The network processes the inputs and produces then an estimation of the outputs; such outputs are compared with the real outputs through a predefined error measurement (typically, a sum of the squared errors of each output). The training consists thus of an optimization problem which aims to minimize the error of the network in the output prediction. Such optimization is carried out by computing the gradient of the error with respect to the connection weights, and it is interrupted when a target error is reached or when a certain number of input/output pairs have been processed.

An important indicator of the goodness of the network after training is the coefficient of determination R^2, defined as

$$R^2 = 1 - \frac{\sum_{i=1}^{N_{data}} (q_i - \hat{q}_i)^2}{\sum_{i=1}^{N_{data}} (q_i - \overline{q}_i)^2} \tag{13.18}$$

where q_i are the real outputs of the physical model, $\overline{q}_i = \frac{1}{N_{data}} \sum_{i=1}^{N_{data}} q_i$ and $\hat{q}_i$ are the output values predicted by the meta-model. The accuracy of the output prediction of the neural network can be judged by the closeness of the value R^2 to the target value of 1.0, which expresses an exact match of the network prediction and the output of the full model. This quantity is computed both using the calibration set and a new set of input/output values, called validation or verification set. In the latter case, a qualitative indication of the generalization capabilities of the network is obtained.

The freely available Fast Artificial Neural Network (FANN) library [25], which is an implementation of the here discussed Neural Network and learning algorithm, has been used for the approximation of the modal properties in the following numerical example.

5 Numerical Example: GOCE Satellite

5.1 Problem Statement

The use of a meta-model within the Bayesian updating procedure is illustrated using the Gravity Field and Steady-State Ocean Circulation Explorer (GOCE) satellite. The mission of the GOCE satellite is to determine the geoid and to measure the gravitational field of Earth with a very high degree of accuracy in a low Earth orbit. The particularities of the GOCE are its arrow-shape with winglets and its ion propulsion engine, used to compensate the air-drag induced orbit-decay. The total length of the satellite is 5.3 m, and the mass amounts to approximately 1,000 kg including the fuel of the propulsion system.

Figure 13.2 shows the FE model of the satellite, provided by Thales Alenia Space (Italy) for use within the commercial FE code MSC.Nastran [26]. Approximately 360,000 DOFs and 74,000 elements compose the FE model, with half of the elements used in the main satellite platform and half in the gravitational gradiometer.

In the main GOCE platform, quadrilateral (QUAD4) and triangular (TRIA3) shell elements are used to model the body panels, the wings and the winglets, the

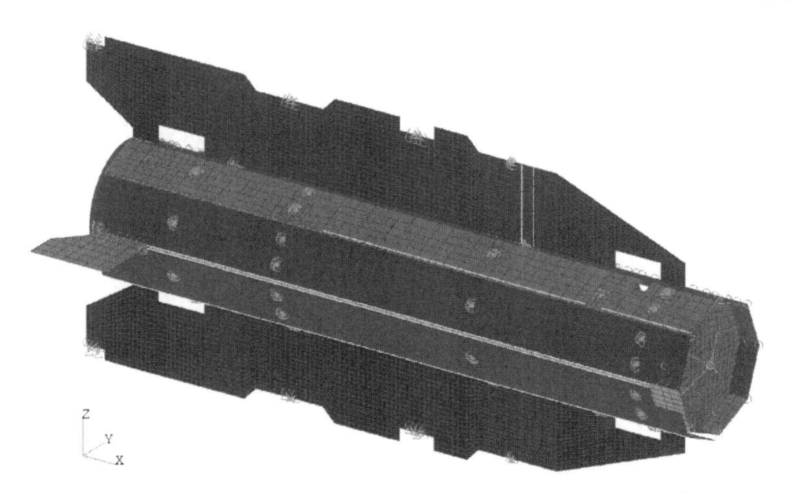

Fig. 13.2 FE model of the GOCE satellite (courtesy of Thales Alenia Space Italy)

internal floors and the solar panels. Beam elements (BAR, ROD and BEAM) constitute the connections of the wings to the main structure and of the instrumentation to the floors. Solid elements (HEXA and PENTA) are used in the Launch Vehicle Adapter (LVA) ring, and scalar spring elements (CELAS2) represent the connection between the solar panels and the structure, as well as the fixing of the wing to the main octagonal body.

A total number of 18 groups combining 3047 structural parameters are defined according to the type and location of the respective materials or geometric specifications (see Table 13.1). This grouping is carried out with the purpose of remarkably reducing the number of parameters to be used within the Bayesian updating procedure since an independent processing of all involved parameters might become infeasible. Hence, the updating procedure is carried out with the goal of identifying as to which changes have to be performed to the single parameter groups in order to obtain a better agreement of the numerical model with the real satellite structure.

5.2 Experimental Modal Data

The experimental data used for model updating consists of $N_s = 1$ set of 7 modal frequencies and partial mode shapes vectors (83 components) which have been determined from the vibration responses during the GOCE structural model qualification test. The dynamic qualification test was performed on the multi-axis vibration test facility of ESA/ESTEC in Noordwijk (The Netherlands). The correlation of the experimental and computed eigenvectors of the initial model is shown by means of the MAC matrix in Fig. 13.3 and the initial comparison of the eigenfrequencies can be found in Table 13.2. The large discrepancies, especially for the first two modes, arise the need for model updating which will be discussed in the following.

Table 13.1 Definition of the groups of parameters of the GOCE satellite

Group no.	Parameters
1	Young's modulus of isotropic materials
2	Poisson's ratio of isotropic materials
3	Young's modulus in the principal direction of orthotropic materials
4	Young's modulus in the secondary direction of orthotropic materials
5	Poisson's ratio of orthotropic materials
6	In-plane shear modulus of orthotropic materials
7	First out-of-plane shear modulus of orthotropic materials
8	Second out-of-plane shear modulus of orthotropic materials
9	Densities of the materials
10	Thicknesses of the shells
11	Linear elastic connections of panels to the main satellite structure
12	Linear elastic connections of panels to the satellite wings
13–18	Linear elastic connections of the wings to the main satellite structure

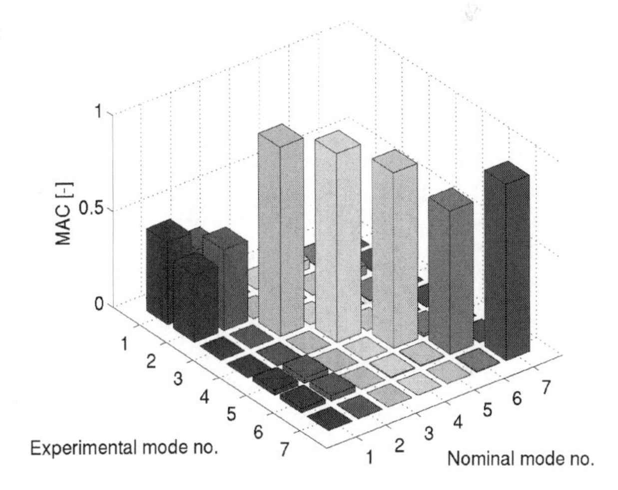

Fig. 13.3 Initial MAC values obtained with the nominal model and the experimental data $\hat{\psi}$

5.3 Accuracy Analysis of the Neural Network

A set of $N_{data} = 2{,}000$ finite element simulations have been performed using Gaussian distributions with the mean values equal to the nominal values and coefficients of variation of 10%. Out of these samples, 1,900 have been dedicated to calibrate the neural networks, and 100 have been kept to verify the generalization capabilities of the networks after training. Each of the 14 neural networks, constituted by $N_{inp} = 18$ inputs and $N_{out} = 1$ output, predicts either one of the eigenfrequencies or one of the diagonal terms of the MAC matrix.

Table 13.2 Comparison of the analytical eigenfrequencies f_a and the experimental data f_e

Mode no.	f_a [Hz]	f_e [Hz]	Δ [%]
1	18.43	15.98	13.29
2	18.37	16.40	10.69
3	28.84	29.59	−2.60
4	34.96	33.33	4.66
5	46.90	48.91	−4.28
6	49.13	51.61	−5.04
7	65.81	61.35	6.77

An automated training procedure has been implemented such that various network topologies are tested and the best networks, characterized by the highest R^2 value (see Eq. (13.18)) obtained with the verification data, are kept. As an indication of the accuracy of the network, Fig. 13.4 shows the regression plots for the neural network of the first eigenfrequency, using the calibration and verification data, respectively. Moreover, the values of R^2 of all the networks obtained with the verification data are listed in Table 13.3.

5.4 Bayesian Model Updating

The prior distributions assigned to the 18 groups of parameters to be updated are all Gaussian with the moments as specified for the calibration of the neural networks. For these ranges, the neural networks of the considered modal properties show high accuracy as discussed in the previous section and are therefore applied for substituting the full FE-analysis when evaluating the likelihood function within the updating process.

The results of the updating procedure are shown exemplary for three parameter groups, namely the thickness, the Young's moduli of the orthotropic materials and the group of stiffnesses of joints between the wings and the main structure (groups no. 10, 3 and 14 in Table 13.1). The prior and posterior histograms of these parameters, which are all transformed in standard normal space, are depicted in Figs. 13.5, 13.6 and 13.7. This representation in standard normal space is advantageous due to the fact that these figures do not show one single parameter each but a parameter group, where the members of each group may have different initial values. Hence, in order to obtain the posterior values in physical space, a back-transformation has to be performed for each parameter of a group, which is achieved by

$$\theta_i = \theta_i^* \sigma_i + \theta_{i,\mathrm{nom}}, \qquad (13.19)$$

where θ_i denotes the i-th parameter in the physical space, θ_i^* the value in the standard normal space, σ_i the prior standard deviation and $\theta_{i,\mathrm{nom}}$ the prior (nominal) mean of this parameter.

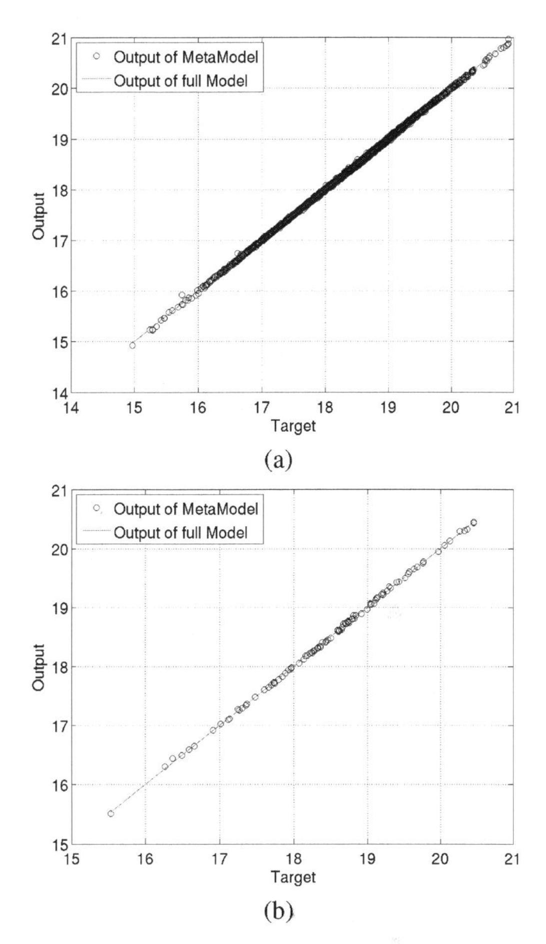

Fig. 13.4 Regression plot of the output of the neural network of the first eigenfrequency for (**a**) calibration data and (**b**) verification data

In Fig. 13.5, the prior and posterior histograms of the thickness of shell elements are depicted. This figure leads to the conclusion that the information contained in the experimental data suggests a decrease of these values. A reduction of approximately 10% of the mean value of the prior distribution leads to a better fit with the experimental modal properties. Also for the Young's modulus of the orthotropic material in the longitudinal direction (material card MAT8 in the MCS.Nastran input file), the updating process suggests a decrease of the mean values (see Fig. 13.6). As opposed to these two parameter groups, all other 16 out of 18 parameter groups used in the updating process show small changes if compared to the prior distribution. As an example, the stiffness values of the joints of the main structure to the wings (used for the specification of the CELAS2 elements in the MCS.Nastran input file) is shown in Fig. 13.7. Due to the information extracted from the experimental values, the prior uncertainty about these parameters could be reduced which is visible through the smaller variation of the posterior distribution.

Table 13.3 R^2 values of the 14 neural networks

Neural network output	R^2 of verification data
Frequency 1	0.99960
Frequency 2	0.99966
Frequency 3	0.99968
Frequency 4	0.99957
Frequency 5	0.88299
Frequency 6	0.97804
Frequency 7	0.95822
$MAC_{1,1}$	0.99886
$MAC_{2,2}$	0.99782
$MAC_{3,3}$	0.99577
$MAC_{4,4}$	0.99861
$MAC_{5,5}$	0.94947
$MAC_{6,6}$	0.97623
$MAC_{7,7}$	0.82779

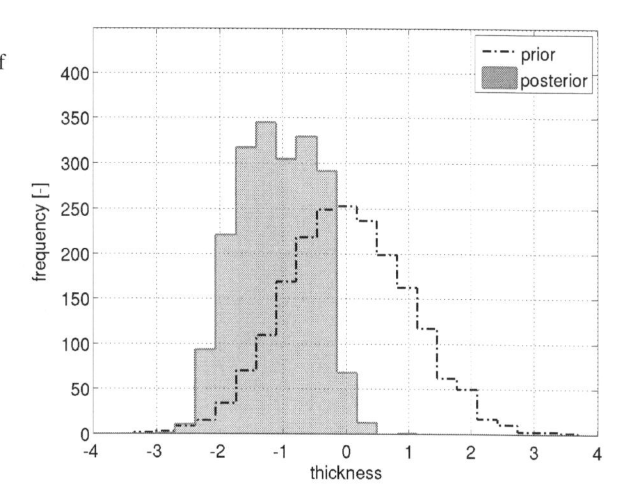

Fig. 13.5 Histograms of the prior and posterior samples of the group of thicknesses (group no. 10 in Table 13.1)

The effect of the choice of the prior on the posterior distribution has not been investigated in this example, however a few remarks will be added in this context: in general, the influence of the prior distribution on the results decreases with increasing amount of experimental data. This is due to the fact that the likelihood function becomes the dominant term in comparison to the prior distribution in Eq. (13.2). In this case, also values in the tails of the prior distribution can be identified and only values with zero probability (e.g. values out of the interval of uniform distributions) cannot be reached since they are excluded from the possible solution space due to the prior knowledge. In case of limited data the selection of the prior distribution clearly has an influence on the results. The prior distribution can therefore be seen

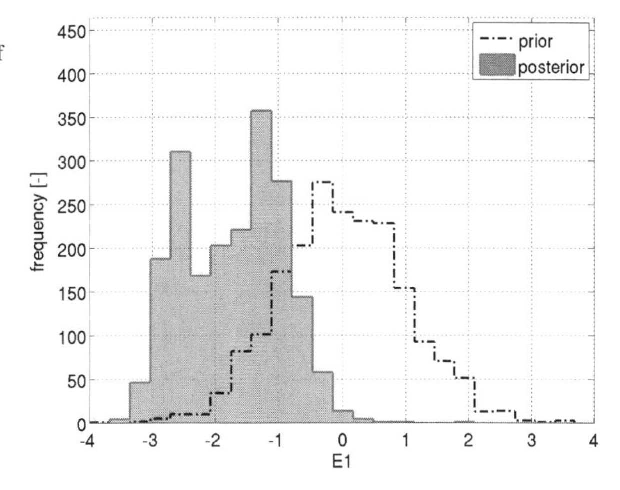

Fig. 13.6 Histograms of the prior and posterior samples of the group of Young's moduli of the orthotropic materials (group no. 3 in Table 13.1)

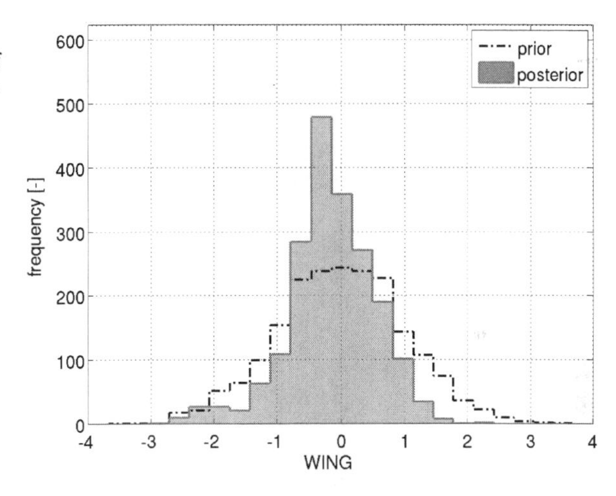

Fig. 13.7 Histograms of the prior and posterior samples of the group of stiffnesses of the joints between wings and main structure (group no. 14 in Table 13.1)

as a means to incorporate initial knowledge about parameter ranges into the identification process and it is subjective in the sense that people with different experience may use different priors leading to broader ranges of the solution in case smaller amount of prior information is available. The selection can therefore be seen as part of the modeling process since also the model itself is affected by a certain amount of subjectivity of the designer. However, the probability content of the prior PDF is updated by the data and if one felt uncomfortable with the choice of the prior distribution the effect of different prior PDFs on the posterior PDF can be studied.

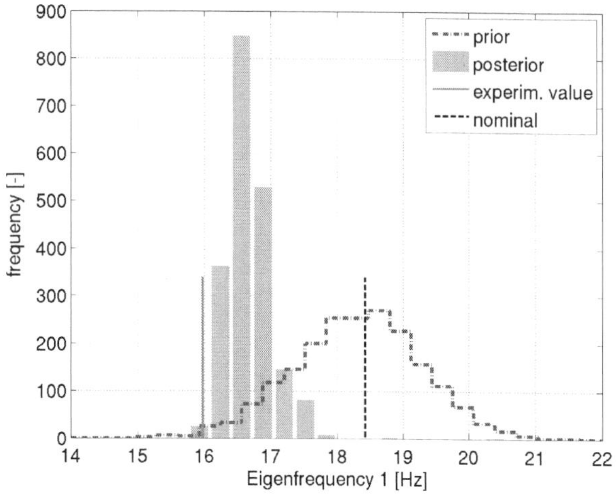

Fig. 13.8 Prior (*dashed-dotted line*) and posterior (*shaded bars*) histograms of the 1st eigenfrequency with experimental value (*solid line*) and nominal value (*dashed line*)

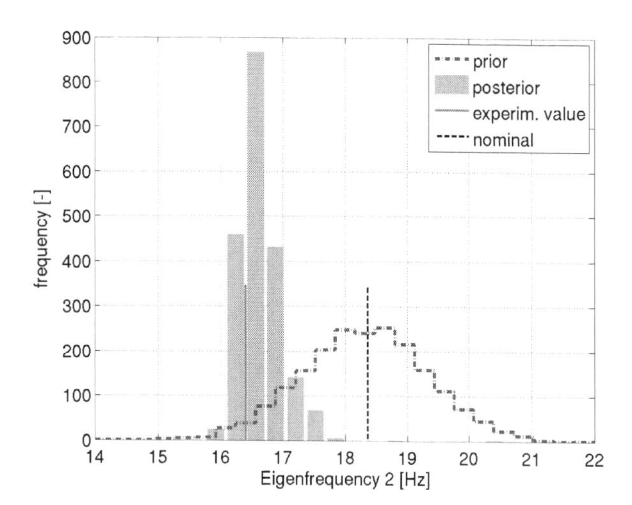

Fig. 13.9 Prior (*dashed-dotted line*) and posterior (*shaded bars*) histograms of the 2nd eigenfrequency with experimental value (*solid line*) and nominal value (*dashed line*)

5.5 Effect on the Correlation of Modal Data

The effect of the updating procedure on the correlation of the experimental and computed modal properties is shown exemplary for the two lowest and highest considered modes. Figures 13.8, 13.9, 13.10 and 13.11 are devoted to the eigenfrequencies and Fig. 13.12 to the corresponding diagonal MAC-values. The figures show that a successful shift of the PDFs towards the experimental values could be achieved, as it can be seen for modes 1, 2 and 7, where it shall be annotated that the correlation of the eigenvector no. 7 with respect to the corresponding experimental data reveals to be high already for the initial model (initial $\mathrm{MAC}_{7,7} = 0.92$).

However, for the 6th eigenfrequency and eigenvector no improvement could be achieved. The reasons might lie in the fact that there is no parameter combination possible which affects an improvement of the fit with respect to all 14 target values

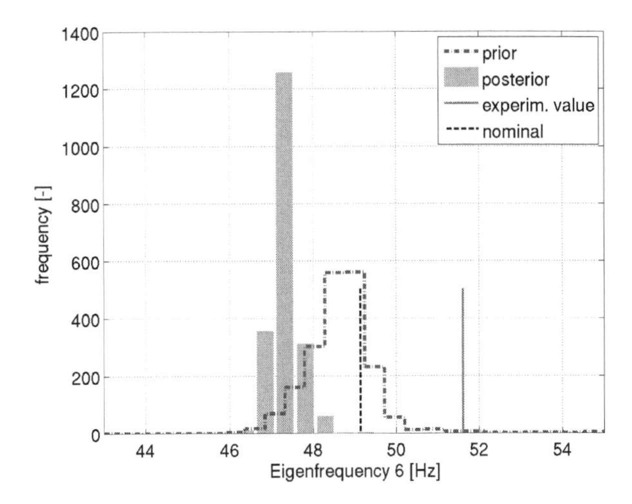

Fig. 13.10 Prior (*dashed-dotted line*) and posterior (*shaded bars*) histograms of the 6th eigenfrequency with experimental value (*solid line*) and nominal value (*dashed line*)

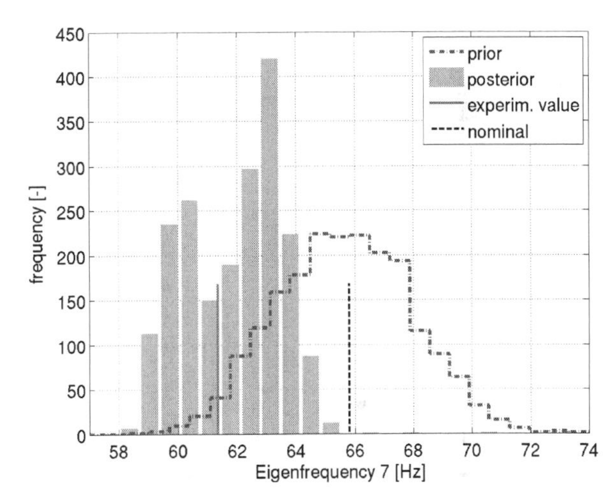

Fig. 13.11 Prior (*dashed-dotted line*) and posterior (*shaded bars*) histograms of the 7th eigenfrequency with experimental value (*solid line*) and nominal value (*dashed line*)

defined by the first 7 modal frequencies and mode shape vectors. Only a revision of the model itself might lead to the situation where the prior distributions span the full solution space, meaning that posterior samples provoke a high correlation with all targets. This example uses real experimental data in the Bayesian updating procedure, thus a fit with respect to all targets might not be possible without revising the FE-model itself.

5.6 Computational Aspects

In this example, the eigensolution of the full FE-model is replaced by an approximate relation at low computational costs which is given by a neural network for

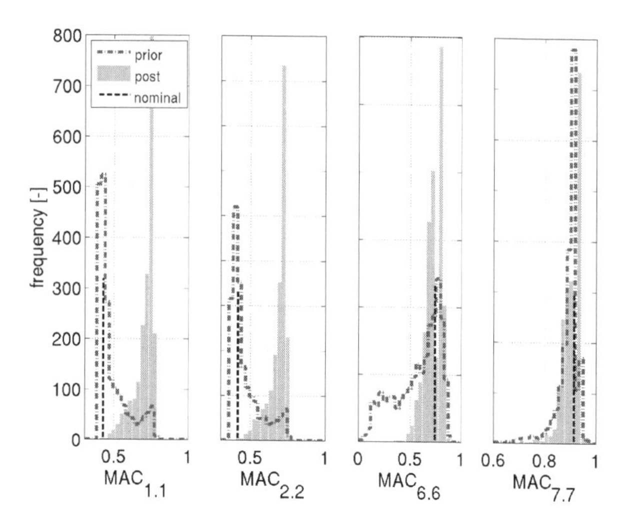

Fig. 13.12 Prior (*dashed-dotted line*) and posterior (*shaded bars*) histograms of the diagonal terms of the MAC matrix corresponding to modes no. 1, 2, 6 and 7 and nominal values (*dashed lines*)

each modal property. If considering that on the above described dual quad-core Xeon server (i) the replacement of the nominal parameter values θ_{nom} by the current value θ in the FE-input file, the normal mode analysis of the full model performed with MSC.Nastran and the import of the modal quantities into Matlab [27] requires 220 s (please see the Appendix for details on the interaction of Matlab with MSC.Nastran), (ii) the updating process of the present model involves approximately 32 iterations $\times$ 2,000 eigensolutions and (iii) the remaining part of the updating process lasts for about 96 min, then the total time amounts up to a theoretical value of

$$t_{\text{full}} = 64{,}000 \cdot 220 \text{ s} + 96 \cdot 60 \text{ s} \approx 160 \text{ days}.$$

The computational time of 220 s of a normal mode analysis of the full FE-model is replaced by the evaluation of the neural networks which takes 0.0014 s. Hence, in this way the analysis time can be remarkably reduced to

$$t_{\text{NN}} = 64{,}000 \cdot 0.0014 \text{ s} + 96 \cdot 60 \text{ s} \approx 100 \text{ min}.$$

However, it shall be noted that in addition 2,000 calibration samples have been generated (see Sect. 5.3), which require evaluations of the full FE-model lasting for a total time of approximately 5 days if performed sequentially (please refer to the Appendix for the interaction with 3rd party software).

6 Conclusions

In this manuscript, the basic steps of model updating within the Bayesian framework using modal data have been summarized and strategies for reducing the analysis time are proposed. The numerical example shows the applicability of the Bayesian updating procedure on complex aerospace structures. It demonstrates that

```
MAT1, 1, <cossan name="E" format="%8.2e" original="7.e+10"/>,,
<cossan name="nu" format="%8.2e" original="0.33"/>,
<cossan name="rho" format="%8.2e" original="2000."/>,
2.40E-5, 20.0000
```

Fig. 13.13 Excerpt of a master input file with identifiers

```
MAT1, 1, 6.13e+10,, 3.98e-01, 2.22e+03, 2.40E-5, 20.0000
```

Fig. 13.14 Excerpt of a stochastic analysis input file with sampled values

ill-conditioned inverse problems in the high-dimensional parameter space can be tackled and that the limited, incomplete data can be used for reducing the initial uncertainty about the adjustable parameters. As a remedy for the large computational efforts of model updating the establishment of a surrogate model has been proposed which approximates the modal properties at a low computational cost. In this way, model updating of a finite element model of a full satellite structure of a size of approximately 360,000 DOFs becomes feasible as shown in the numerical example.

Acknowledgements This research was partially supported by the European Space Agency (ESA) under Contract No. 20829/07/NL/EM, which is gratefully acknowledged by the authors. The authors thank Thales Alenia Space Italy for the FE-model of the GOCE satellite and the experimental modal data. The first author is a recipient of a DOC-fForte-fellowship of the Austrian Academy of Science at the Institute of Engineering Mechanics (University of Innsbruck).

Appendix: Interaction with 3rd Party Software

FE models are defined uniquely by one or more ASCII input files. These files contain the definition of the nodes and elements constituting the model, as well as the structural parameters and boundary and loading conditions in form of fixed numerical values. However, in a stochastic analysis some of these values change, since they are samples from a given probability distribution function. Thus, it is envisioned to automatically manipulate the input files such that in each simulation the respective sample values are inserted into the FE-input file. For this purpose, XML-like tags, called identifiers, are inserted into the master input files in order to define the parameters which have to be changed in each simulation, as shown in Fig. 13.13. An identifier defines the name of the random variable used within the stochastic analysis, the format in which the number is written into the file as well as the original value of the parameter.

The code used to drive the simulation is COSSAN-X, a software for computational stochastic structural analysis [28]. This code parses the master input files in order to identify the positions and the insertion formats of all variables. In each analysis, these identifiers are replaced by sampled numerical values, obtaining a valid input file which is then used for the finite element analysis (see Fig. 13.14). It shall be noted that this software is not restricted to a particular FE-code, but it is applicable to any FE-solver which uses ASCII input files.

References

1. European Cooperation for Space Standardization: Space engineering: modal survey assessment. ECSS-E-ST-32-11C (2008)
2. The American Institute of Aeronautics and Astronautics: Guide for the verification and validation of computational fluid dynamics simulations. AIAA Standards Series G-077 (1998)
3. The American Society of Mechanical Engineers: Guide for verification and validation in computational solid mechanics. ASME V&V (10) (2006)
4. Göge, D., Link, M.: Assessment of computational model updating procedures with regard to model validation. Aerosp. Sci. Technol. **7**, 47–61 (2003)
5. Calvi, A., Garcia de Paredes, S., Roy, N., Lefevre, Y.: On the development of a stochastic approach for the validation of spacecraft structural dynamic models. In: Proceedings of the European Conference on Spacecraft Structures, Materials and Mechanical Testing (CD-ROM), Toulouse, France (2002)
6. Friswell, M., Mottershead, J.: Finite Element Model Updating in Structural Dynamics. Kluwer Academic, Norwell (1995)
7. Girard, A., Roy, N.: Structural Dynamics in Industry. Wiley, New York (2008)
8. Buffe, F.: Application of updating methods on the finite element model of Picard. In: European Conference on Spacecraft Structures, Materials and Mechanical Testing (ECSSMMT 2009 CD-ROM), Toulouse, France (2009)
9. Göge, D., Link, M.: Results obtained by minimizing natural frequencies and mode shape errors of a beam model. Mech. Syst. Signal Process. **17**(1), 21–27 (2003)
10. Calvi, A.: Uncertainty-based loads analysis for spacecraft: finite element model validation and dynamic responses. Comput. Struct. **83**(14), 1103–1112 (2005)
11. Beck, J., Katafygiotis, L.: Updating models and their uncertainties. I: Bayesian statistical framework. J. Eng. Mech. **124**(4), 455 (1998)
12. Katafygiotis, L., Beck, J.: Updating models and their uncertainties. II: model identifiability. J. Eng. Mech. **124**(4), 463 (1998)
13. Yuen, K.-V.: Bayesian Methods for Structural Dynamics and Civil Engineering. Wiley, New York (2010)
14. Ching, J., Chen, Y.-C.: Transitional Markov chain Monte Carlo method for Bayesian updating, model class selection, and model averaging. J. Eng. Mech. **133**, 816–832 (2007)
15. Kleijnen, J., Sargent, R.: A methodology for fitting and validating metamodels in simulation. Eur. J. Oper. Res. **124**(1), 14–29 (2000)
16. Rumelhart, D., McClelland, J.: Parallel Distributed Processing, Exploration in the Microstructure of Cognition, vol. 1. MIT Press, Cambridge (1986)
17. Rumelhart, D., McClelland, J.: Parallel Distributed Processing, Exploration in the Microstructure of Cognition, vol. 2. MIT Press, Cambridge (1986)
18. Bayes, T.: An essay towards solving a problem in the doctrine of chances. Philos. Trans. R. Soc. Lond. **53**, 370–418 (1763)
19. Jaynes, E.: Probability Theory: The Logic of Science. Cambridge University Press, Cambridge (2003)

20. Vanik, M., Beck, J., Au, S.-K.: Bayesian probabilistic approach to structural health monitoring. J. Eng. Mech. **126**, 738–745 (2000)
21. Metropolis, N., Rosenbluth, A., Rosenbluth, M., Teller, A., Teller, E.: Equations of state calculations by fast computing machines. J. Chem. Phys. **21**(6), 1087–1092 (1953)
22. Hastings, W.: Monte Carlo sampling methods using Markov chains and their applications. Biometrika **57**(1), 97–109 (1970)
23. Anderson, J.: Introduction to Neural Network. MIT Press, Cambridge (1995)
24. Bishop, C.: Neural Networks for Pattern Recognition. Oxford University Press, London (1995)
25. Nissen, S.: Implementation of a fast artificial neural network library (fann). Technical report, Department of Computer Science University of Copenhagen (DIKU) (2003)
26. MSC.Software Corporation: MSC.NASTRAN, version 2007.1.0. Santa Ana, CA, USA (2007)
27. The MathWork: Matlab R2009b. Natick, MA, USA (2009)
28. COSSAN-X: COmputational Stochastic Structural ANalysis. Chair of Engineering Mechanics, University of Innsbruck, Innsbruck, Austria, EU (2010)

Chapter 14
Identification of Properties of Stochastic Elastoplastic Systems

Bojana V. Rosić and Hermann G. Matthies

Abstract This paper presents the parameter identification in a Bayesian setting for the elastoplastic problem, mathematically speaking the variational inequality of a second kind. The inverse problem is formulated in a probabilistic manner in which unknown quantities are embedded in a form of the probability distributions reflecting their uncertainty. With the help of the stochastic functional analysis the update procedure is introduced as a direct, purely algebraic way of computing the posterior, which is comparatively inexpensive to evaluate. Such formulation involves the process of solving the convex minimisation problem in a stochastic setting for which the extension of classical optimization algorithm in predictor-corrector form as the solution procedure is proposed. A validation study of identification procedure is done through a series of virtual experiments taking into account the influence of the measurement error and the order of approximation on the posterior estimate.

Keywords Linear Bayesian update · Stochastic Galerkin method · Stochastic elastoplasticity · Stochastic convex minimisation

1 Introduction

In recent years several mathematical models have been proposed to predict the yielding and elastoplastic behavior of heterogeneous materials. Even though these models carry some confidence as to own fidelity, they can not be taken as realistic as the most of quantities entering the model are only incompletely known, i.e. uncertain. In order to give the more reliable description we try to identify these quantities from the given experimental data (system response). However such identification is often regarded as ill-posed due to limited size of the measurement data. In order to resolve this problem various approaches have been proposed, from which the most often uti-

B.V. Rosić (✉) · H.G. Matthies
Institut für Wissenschaftliches Rechnen, TU Braunschweig, Braunschweig, Germany
e-mail: bojana.rosic@tu-bs.de

H.G. Matthies
e-mail: wire@tu-bs.de

M. Papadrakakis et al. (eds.), *Computational Methods in Stochastic Dynamics*, Computational Methods in Applied Sciences 26,
DOI 10.1007/978-94-007-5134-7_14,

lized is the Bayesian regularisation technique described in [23] together with its all possible variants.

In this work we use the so-called Bayesian type of regularisation, a probabilistic approach employing an additional—prior—information on the material property q next to the measurement data. The prior information is usually posed in a form of a distribution function obtained from the maximum entropy principle [26] under given constraints—known properties of q (e.g. positive definiteness, boundness etc.). In this manner the unknown parameter q is modelled as a random variable whose probabilistic description is further altered with the help of the measurement data to the so-called posterior model.

In order to extract the information from the posterior most estimates take the form of expectations (integrals) w.r.t. the posterior. Higdon et al. [6], Gamerman et al. [3] and Tarantola et al. [27] estimate these integrals with the help of the Markov chain Monte Carlo (MCMC) method. By letting the Markov chain to run sufficiently long time the posterior distribution is approached in an asymptotic manner. Regarding this the asymptotic approach is relatively simple and straightforward, though the obtained samples are not any more independent. As opposite to simplicity the MCMC computational efficiency is not so satisfactory due to slow convergence rates; and hence for an efficient run the method requires fewer simulations of the prior model. This can be achieved by a polynomial chaos (PC) [5, 28] or a Karhunen–Loève (KL) approximation [4, 7, 13] of the prior distribution and corresponding observations as presented in [10, 12].

The approaches mentioned above require a large number of samples in order to obtain satisfactory results. In contrast to this the main idea here is to do the Bayesian update directly on the polynomial chaos expansion (PCE) without any sampling [15, 17, 22, 23]. This idea has appeared independently in [1] in a simpler context, whereas in [24] it appears as a variant of the Kalman filter (e.g. [8]). A PCE for a push-forward of the posterior measure is constructed in [16].

The paper is organized as follows: in Sect. 2 we briefly describe the Bayesian formulation of the inverse problem, which is then reduced to a linear update formula in Sect. 3 with the help of the theory of conditional expectations and minimum variance estimation. Such update procedure is based on the polynomial chaos approximation of the system response obtained via stochastic Galerkin method as presented in Sect. 4. Finally in Sect. 5 we test the update procedure on two numerical examples in plain strain conditions.

2 Bayesian Updating

The elastoplastic system is modelled by an evolution equation for its state:

$$\frac{\partial}{\partial t}u(t) + A\big(p; u(t)\big) = f(p; t), \tag{14.1}$$

where $u(t) \in \mathcal{U}$ describes the state of the system at time $t \in [0, T]$ lying in a Hilbert space $\mathcal{U}$ (for the sake of simplicity), A is the nonlinear operator modelling the physics of the system, and $f \in \mathcal{U}^*$ is some external influence (action/excitation/loading). The model depends on a set of parameters $p \in \mathcal{P}$ with corresponding subset $q \subset p$ representing the material properties such as yield stress σ_y, bulk K and shear G modulus.

The process of identifying q by observing a function of the state $Y(u(q), q) \in \mathcal{Y}$ (e.g. the stress, strain etc.) is called the inverse problem. As one can only observe a finite number of quantities the space $\mathcal{Y}$ is finite dimensional and hence the mapping $q \mapsto Y(q)$ is usually not invertible. However, in practice that is not the case as the parameter set q is only incompletely known (uncertain). Regarding this we may model it as a mapping $q(\omega) : \Omega \to \mathcal{Q}$ (i.e. random variable) on a probability space $(\Omega, \mathfrak{A}, \mathbb{P})$ with Ω being the set of all events, $\mathfrak{A}$ a σ-algebra of subsets of Ω and $\mathbb{P}$ a probability measure. This a priori information originates from the maximum entropy principle [26] based on the available information we have on properties of q (i.e. K, G and σ_y). In addition to this, the measurement data y in real experiments are "polluted" by some kind of noise ε, often assumed to be of additive type, i.e. $y = z + \varepsilon$, where $z := Y(q)$.

Only with the previous assumptions in mind the inverse problem becomes well-posed and reduces to a comparison of the forecast obtained from the forward problem (the system response on prior q) with the actual information—so-called Bayesian inference. Its practical realisation generally classifies into two groups: the one performing by changing the probability measure $\mathbb{P}$ and leaving the mapping $q(\omega)$ as it is, whereas the other set of methods leaves the probability measure unchanged and updates the function $q(\omega)$. See [23] for synopsis.

3 Linear Bayesian Update

The probability of an event is the same as the expected value of the indicator variable for that event, which may help us to reformulate the full Bayesian update

$$\mathbb{P}(I_q | M_y) = \frac{\mathbb{P}(M_y | I_q)}{\mathbb{P}(M_y)} \mathbb{P}(I_q), \tag{14.2}$$

in terms of conditional expectations. According to [9] the conditional probability $\mathbb{P}(I_q | M_y)$ is equal to conditional expectation $\mathbb{E}(\chi_q | y)$ where M_y is the information provided by a measurement and χ_q is the characteristic function of some subset of possible q's. Defining the conditional expectation $\mathbb{E}(q | \sigma(Y))$ measurable w.r.t. $\sigma(Y)$ for the sub-σ-algebra $\mathfrak{S} = \sigma(Y)$ generated by Y, we may state that $\mathbb{E}(q | \sigma(Y)) = H(Y)$ for some measurable $H \in \mathcal{L}(\mathcal{Y}; \mathcal{Q}) \subset L_0(\mathcal{Y}; \mathcal{Q})$ according to *Doob–Dynkin* lemma [2]. Here we limit ourselves to the vector space $\mathcal{L}(\mathcal{Y}; \mathcal{Q})$ of linear measurable maps from $\mathcal{Y}$ to $\mathcal{Q}$. The more general case is considered in [23].

Following previous statements the linear approximation of the full Bayesian update derived in [17, 22, 23] can be written as the orthogonal projection $P_{\mathcal{Q}_l}$ of q onto the subspace $\mathcal{Q}_l = \overline{\text{span}}\{H(Y(q)) \in \mathcal{Q} | H \in \mathcal{L}(\mathcal{Y}; \mathcal{Q})\} \subset \mathcal{Q} := \mathcal{Q} \otimes L_2(\Omega)$, i.e.:

$$\begin{aligned} K &= \mathbb{E}\big(q|\sigma(Y)\big) = P_{\mathcal{Q}_l}(q) \\ &= \arg\min_{\tilde{q}\in\mathcal{Q}_l} \|q - \tilde{q}\|^2_{\mathcal{Q}} \\ &= \arg\min_{H\in\mathcal{L}(\mathcal{Y},\mathcal{Q})} \big\|q - H\big(Y(q)\big)\big\|^2_{\mathcal{Q}}. \end{aligned} \tag{14.3}$$

The optimal K is not hard to find by taking the derivative in Eq. (14.3) w.r.t. the linear map H (see e.g. [8, 11]) and requiring the derivative to vanish. This further leads to the formula (see e.g. [18, 22]):

$$q_a = q_f + K(y - z_f) \tag{14.4}$$

representing the so-called linear Bayesian update. Note that Eq. (14.4) in the mean reduces to the familiar Kalman filter formula [8, 18].

Finally, the update in Eq. (14.4) employs the measurement data y and assimilates it with both, the prior (forecast) information q_f and the measurement forecast z_f, to the posterior value q_a through the Kalman gain K. The gain[1] is computed as

$$K = C_{q,z}(C_z + C_\varepsilon)^\dagger \tag{14.5}$$

where the corresponding covariances are given as

$$\begin{aligned} C_{q,z} &= \mathbb{E}\big(\big(q - \mathbb{E}(q)\big) \otimes \big(Y(q,u) - \mathbb{E}\big(Y(q,u)\big)\big)\big) \\ C_z &= \mathbb{E}\big(\big(Y(q,u) - \mathbb{E}\big(Y(q,u)\big)\big) \otimes \big(Y(q,u) - \mathbb{E}\big(Y(q,u)\big)\big)\big) \end{aligned} \tag{14.6}$$

together with C_ε being the covariance of the measurement noise ε. The last one is often assumed to be of Gaussian type i.e. $C_\varepsilon = \sigma_\varepsilon^2 I$.

3.1 Sampling Free Update

In order to numerically compute the linear formula in Eq. (14.4) one has to discretise the space $\mathcal{Q} := \mathcal{Q} \otimes \mathcal{S}$, $\mathcal{S} := L_2(\Omega)$. This is performed by taking the finite element discretisation $\mathcal{Q}_M$ of $\mathcal{Q}$ and a finite subset $\mathcal{S}_J = \text{span}\{H_{\boldsymbol{\alpha}} : \boldsymbol{\alpha} \in \mathcal{J}\}$ of $\mathcal{S}$, where $\mathcal{J}$ is the finite set of multi-indices with cardinality $J = |\mathcal{J}|$ and $H_{\boldsymbol{\alpha}}$ the multivariate Hermite polynomial in Gaussian random variables $\boldsymbol{\theta}$. The orthogonal projection P_J onto $\mathcal{S}_J$ is then simply

$$P_J : \mathcal{Q}_M \otimes \mathcal{S} \ni \sum_{\boldsymbol{\alpha}\in\mathcal{N}} \boldsymbol{q}^{\boldsymbol{\alpha}} H_{\boldsymbol{\alpha}} \mapsto \sum_{\boldsymbol{\alpha}\in\mathcal{J}} \boldsymbol{q}^{\boldsymbol{\alpha}} H_{\boldsymbol{\alpha}} \in \mathcal{Q}_M \otimes \mathcal{S}_J, \tag{14.7}$$

[1] The Moore–Penrose pseudo-inverse † is used as a general inverse in case $C_z + C_\varepsilon$ is not invertible or close to singularity.

where $\sum_{\alpha\in\mathcal{N}} \boldsymbol{q}^{\alpha} H_{\alpha}$ denotes the polynomial chaos expansion (PCE) of a random variable q. Elements of the discretised space $\mathcal{Q}_{M,J} = \mathcal{Q}_M \otimes \mathcal{S}_J \subset \mathcal{Q}$ thus may be written as $\sum_{m=1}^{M} \sum_{\alpha\in\mathcal{J}} q^{\alpha,m} \rho_m H_{\alpha}$ and the tensor representation of parameter set as $\mathbf{q} := (q^{\alpha,m}) = \sum_{\alpha\in\mathcal{J}} \boldsymbol{q}^{\alpha} \otimes \boldsymbol{e}^{\alpha}$, where $\boldsymbol{e}^{\alpha}$ are the unit vectors in $\mathbb{R}^J$. With the previous notation the update Eq. (14.4) is simply computed in the PCE representation without any sampling as:

$$\mathbf{q}_a = \mathbf{q}_f + \mathbf{K}(\mathbf{y} - \mathbf{z}_f), \tag{14.8}$$

where $\mathbf{K} = \boldsymbol{K} \otimes \boldsymbol{I}$. The gain $\mathbf{K}$ follows from the formula given in Eq. (14.5), where the covariance $\boldsymbol{C}_{q,z}$ is evaluated via PCE in the following manner

$$\boldsymbol{C}_{q,z} = \sum_{\alpha\in\mathcal{N},\alpha\neq 0} (\alpha!)\boldsymbol{q}^{\alpha} \otimes \boldsymbol{z}^{\alpha} \approx \sum_{\alpha\in\mathcal{J},\alpha\neq 0} (\alpha!)\boldsymbol{q}^{\alpha} \otimes \boldsymbol{z}^{\alpha}, \tag{14.9}$$

and similarly C_z and C_{ε}.

4 Elastoplastic Problem

Let be given the state variable $w = (u, \varepsilon_p, \nu) \subset \mathcal{Z} := \mathcal{U} \times \mathcal{P} \times \mathcal{C}$ describing the infinitesimal quasi-static von Mises elasto-plastic behaviour with mixed hardening. Here u denotes the displacement vector, ε_p the plastic deformation and ν the appropriate internal hardening variable. Their spaces of definition are $\mathcal{U} := U \otimes \mathcal{S} = H_0^1(\mathcal{G}) \otimes L_2(\Omega)$, $\mathcal{P} \subset \mathcal{E} = L_2(\mathcal{G}) \otimes \mathcal{S}$ and $\mathcal{C} \subset \mathcal{E}$ respectively. In this notation the variational form of a quasi-static problem in Eq. (14.1) is described by a $\mathcal{Z}$-elliptic and bounded bilinear form

$$a(w, v) = \langle\!\langle \boldsymbol{A} : \big(\varepsilon(u) - \varepsilon_p\big), \varepsilon(u_1) - \varepsilon_p \rangle\!\rangle + \langle\!\langle \boldsymbol{H} : \eta, \mu \rangle\!\rangle, \tag{14.10}$$

where $\langle\!\langle \cdot, \cdot \rangle\!\rangle$ denotes the duality pairing

$$\langle\!\langle \boldsymbol{H} : \eta, \mu \rangle\!\rangle = \int_{\Omega} \int_{\mathcal{G}} \boldsymbol{H} : \eta : \mu \, \mathrm{d}x \mathbb{P}(\mathrm{d}\omega), \tag{14.11}$$

ε the total deformation, $\boldsymbol{A}$ the elastic and $\boldsymbol{H}$ the hardening positive-definite constitutive tensor. In a similar manner after multiplication of Eq. (14.1) with the test functions and integration its right hand side transforms to the following functional of a linear type:

$$\ell : \mathcal{Z} \to \mathbb{R} : \quad \ell(v) = \langle\!\langle f, v \rangle\!\rangle. \tag{14.12}$$

The definitions Eq. (14.11) and Eq. (14.12) correspond to the conservation law of the momentum balance. However, additionally to this law the system has to satisfy the second law of thermodynamics describing the energy of a system. In a Clausius–Duhelm form the energy law introduces the dissipation functional $j(v)$ also supposed to be the support functional of a closed, non-empty, convex set $\mathcal{K} \subset \mathcal{Z}$ [20].

Thus, following the definition of its sub-differential:

$$\partial j(\dot{w}) := \{w^* \in \mathcal{Z}^* : j(v) \geq j(w(t)) + \langle\!\langle w^*(t), v - w(t) \rangle\!\rangle, \forall v \in \mathcal{Z}\} \tag{14.13}$$

after few mathematical steps one may arrive to the mixed variational formulation of the elastoplastic problem (see [20]):

Proposition 1 *Find unique functions* $w \in H^1(\mathcal{T}, \mathcal{Z}^*)$ *and* $w^* \in H^1(\mathcal{T}, \mathcal{Z}^*)$ *with* $w(0) = 0$ *and* $w^*(0) = 0$ *such that the equilibrium equation*

$$a(w(t), v) + \langle\!\langle w^*(t), v \rangle\!\rangle = \langle\!\langle f(t), v \rangle\!\rangle \tag{14.14}$$

and the flow rule

$$\forall v^* \in \mathcal{K} : \quad \langle\!\langle \dot{w}(t), v^* - w^*(t) \rangle\!\rangle \leq 0 \tag{14.15}$$

are satisfied almost surely on Ω *and for all* $t \in \mathcal{T}$.

Proof The proof of existence, uniqueness and stability of the solution, as well as complete derivation can be found in Rosić et al. [20]. □

Here $w^* \in H^1(T, \mathcal{Z}^*)$ denotes the dual variable $(g, \sigma, \chi) \in \mathcal{U}^* \times \mathcal{R} \times \mathcal{C}$ with g being the force-like variable, σ the Cauchy stress and $\chi := (\varsigma, \zeta)$ the hardening stress with the back-stress ς (kinematic hardening) and the isotropic stress ζ as components.

4.1 Minimisation

The variational inequality Eq. (14.14) may be equivalently formulated as a standard minimisation problem of a convex objective function Φ. In particular we look at a continuous (or bounded $a(v_1, v_2) \leq c\|v_1\|\|v_2\|$), symmetric and $\mathcal{Z}$-elliptic ($a(v, v) \geq c\|v\|^2$) bilinear form $a : \mathcal{Z} \times \mathcal{Z} \to \mathbb{R}$ and an element $\varrho \in \mathcal{Z}^*$ such that the solution w^* in $\mathcal{K}$ is the closest point to ϱ in the a^* metric (*closest point projection*) [20]:

$$w^* = \arg\min_{v^* \in \mathcal{K}} \Phi = \arg\min_{v^* \in \mathcal{K}} \frac{1}{2} a^*(\varrho - v^*, \varrho - v^*) \tag{14.16}$$

and that there exists some $w \in \mathcal{Z}$ satisfying

$$\forall v \in \mathcal{Z} : \quad a(w, v) = \langle\!\langle \varrho - w^*, v \rangle\!\rangle. \tag{14.17}$$

As a and ϱ are continuous and Gâteaux-differentiable, and as a is $\mathcal{Z}$-elliptic, Φ has all desired properties. To handle the dissipation, we have to allow for a second convex functional j on $\mathcal{Z}$, which may not be Gâteaux differentiable everywhere.

4.2 Discretisation

As the elastoplastic problem is time dependent the implicit Euler procedure is used for its discretisation by taking the time step $h := t_n - t_{n-1}$ to be constant over time. Regarding to this the total deformation ε, the Cauchy stress σ, the hardening stress χ and the plastic deformation ε_p are assumed to be known at time t_{n-1}. The goal is to find those variables at time t_n by initially assuming that the increment of the total strain $\Delta\varepsilon_n$ is purely elastic. Note that in such situation the equilibrium Eq. (14.14) depends only on the increment of displacement Δu_n as unknown variable. Moreover, one may rewrite it to the corresponding residual of a nonlinear type:

$$r := \langle\!\langle A[q_n; \Delta u_n], v \rangle\!\rangle - \ell_n(f, v) = 0, \quad \forall v \in \mathcal{U} \tag{14.18}$$

where the nonlinear operator A depends on the parameter set q and the displacement u. Finally, following the residual one may define the measurement operator as

$$z_f = Y(u_n, q_n) = Y(q_n). \tag{14.19}$$

For notational simplicity the index n is dropped from the further text.

Before solving Eq. (14.18) one may notice that the spaces $\mathcal{U}$ and $\mathcal{Z}$ are infinite dimensional, as is the space $\mathcal{S} = L_2(\Omega)$. Thus, for further analysis their finite approximation has to be introduced. In an analogous fashion to Sect. 3.1, let us choose an N-dimensional subspace $\mathcal{U}_N = \mathrm{span}\{v_j : j = 1, \ldots, N\} \subset \mathcal{U}$ with the piecewise linear basis $\{v_j\}_{j=1}^N$. Then an element of $\mathcal{U}_N$ can be represented by the vector $\boldsymbol{u} = (u^1, \ldots, uj^N)^T \in \mathbb{R}^N$ such that $\sum_{j=1}^N u^j v_j \in \mathcal{U}_N$. Similarly, the spaces $\mathcal{P}$ and $\mathcal{C}$ are discretised by piecewise constant functions such that $\mathcal{Z}_N = \mathcal{U}_N \times \mathcal{P}_N \times \mathcal{C}_N$ is appropriate subspace of $\mathcal{Z}$. By inserting those ansatzes into Eq. (14.18) the residual becomes:

$$\boldsymbol{r}(\boldsymbol{u}) := \langle\!\langle A[\boldsymbol{q}; \Delta\boldsymbol{u}], \boldsymbol{v} \rangle\!\rangle - \ell(\boldsymbol{f}, \boldsymbol{v}) = \boldsymbol{0}, \quad \forall \boldsymbol{v} \in \boldsymbol{\mathcal{U}}_N := \mathcal{U}_N \otimes \mathcal{S} \tag{14.20}$$

and the measurement operator:

$$\boldsymbol{z}_f = \boldsymbol{Y}(\boldsymbol{u}, \boldsymbol{q}) = \boldsymbol{Y}(\boldsymbol{q}). \tag{14.21}$$

However, Eq. (14.20), and subsequently Eq. (14.21), are only semi-discretised due to the dependence on the uncertain parameter ω. For MCMC or any other Monte Carlo method [10, 12], Eq. (14.20) has to be solved for each sample point ω_z to obtain $\boldsymbol{u}(\omega_z)$. This then can be used to predict the measurement $\boldsymbol{z}_f(\omega_z) = \boldsymbol{Y}(\boldsymbol{u}(\omega_z), \boldsymbol{q}(\omega_z))$, which may be computationally quite costly. Thus we take another approach by assuming the PCE ansatz for the solution in a form:

$$\boldsymbol{u}(\omega) = \sum_{\alpha \in \mathcal{J}} \boldsymbol{u}^{(\alpha)} H_\alpha\big(\boldsymbol{\theta}(\omega)\big) \tag{14.22}$$

and projecting the residual in a Galerkin manner onto the finite dimensional subspace $\mathcal{S}_J$ according to:

$$\mathbf{r}(\mathbf{u}) = \left[\ldots, \mathbb{E}\left(H_\alpha(\cdot)\boldsymbol{r}(\cdot)\left[\sum_\alpha \boldsymbol{u}_\beta H_\beta\right]\right), \ldots\right] = \mathbf{0}, \tag{14.23}$$

where the block-version of the residual is denoted as $\mathbf{r}(\mathbf{u}) = (\ldots, \boldsymbol{r}^{(\alpha)}(\mathbf{u})^T, \ldots)^T$. In this manner the process of solving the residual Eq. (14.20) reduces to the evaluation of a possibly high-dimensional integral.

The integration of $\mathbb{E}(H_\alpha(\cdot)\boldsymbol{r}(\cdot)[\sum_\alpha \boldsymbol{u}_\beta H_\beta])$ can be done directly via PCE algebra as presented in [21] (Galerkin method) or numerically via high-dimensional integration (pseudo-Galerkin method). Whether we use the first or second approach the computational time of integration drastically reduces in comparison to the direct integration techniques. As the direct algebraic approach is already considered in [21] in this work we choose the pseudo-Galerkin (collocation) approach and compute:

$$\int_\Omega H_\alpha \boldsymbol{r}(\omega)\left[\sum_\beta \boldsymbol{u}^{(\beta)} H_\beta\right] \mathrm{d}\mathbb{P}(\omega) \approx \sum_{z=1}^{L} w_z H_\alpha(\boldsymbol{\theta}_z)\boldsymbol{r}(\boldsymbol{\theta}_z)\left[\sum_\beta \boldsymbol{u}^{(\beta)} H_\beta(\boldsymbol{\theta}_z)\right], \tag{14.24}$$

via the set of integration points $\Theta = \{\boldsymbol{\theta}_z, 1 \leq z \leq L\}$, $\boldsymbol{\theta} = \{\theta_1, \ldots, \theta_M\}$ with the corresponding weights $\boldsymbol{w} := \{w_z\}_{z=1}^L$. Note that the evaluation of the integral requires L evaluations of the residual, $\boldsymbol{r}(\boldsymbol{\theta}_z)$, $z = 1, \ldots, L$, each corresponding to the numerical integration over the spatial domain $\mathcal{G} \subset \mathbb{R}^d$ done in a classical FEM way. This could be seen as an advantage compared to the intrusive Galerkin method [21], as the FEM code is used in a black-box fashion. On other side, the number of calls of the deterministic software increases drastically with the stochastic dimension which may lead to the expensive or almost impractical procedures.

Once the solution of Eq. (14.20) is found the following procedure collapses to the (iterative) solution of a convex mathematical programming problem, which has for a goal to find the closest distance in the energy norm of a trial state to a convex set $\mathcal{K}$, known as a closest point projection. In other words, one search for the solution of:

$$\Sigma(\omega) = \arg\min_{\Sigma(\omega)\in\mathcal{K}} \Phi(\omega) \tag{14.25}$$

where

$$\Phi := \arg\min_{\Sigma(\omega)\in\mathcal{K}} \left[\frac{1}{2}\langle\!\langle \Sigma^{trial} - \Sigma, \boldsymbol{A}^{-1} : (\Sigma^{trial} - \Sigma)\rangle\!\rangle\right] \tag{14.26}$$

and

$$\mathcal{K} := \left\{\Sigma := \big(\sigma(\omega), \chi(\omega)\big) \in \mathcal{R} \times \mathcal{C} \mid \varphi(\Sigma) \leq 0 \text{ a.s. in } \Omega\right\}. \tag{14.27}$$

Here $\Sigma := (\sigma(\omega), \chi(\omega))$ denotes the so-called generalised stress and $\Sigma^{trial} = \Sigma_n + \boldsymbol{A} : \Delta\varepsilon$ its corresponding trial stress in time n. Note that the minimisation is done over the convex set $\mathcal{K}$ described by a von Mises yield function φ which must be non-positive almost surely. However, in order to perform the numerical computation one has first to spatially discretise the problem via $\mathcal{W}_N := \mathcal{R}_N \times \mathcal{C}_N$ and then to relax the almost sure condition by introduction of the discretised substitute of $\mathcal{K}$:

$$\mathcal{K}^\star = \{\Sigma \in \mathcal{W}_N \otimes \mathcal{S}_J \mid \varphi(\Sigma(\boldsymbol{\theta}_z)) \leq 0, \forall \boldsymbol{\theta}_z \in \Theta\}. \quad (14.28)$$

Here the set $\mathcal{K}^\star$ is taken as a set of "deterministic" constraints on a finite number of the integration points Θ. Such construction allows the decoupling of the problem in Eq. (14.25) into L smaller problems, which may be solved independently. Note that each of them corresponds to the normal deterministic optimization problem as presented in [25], for which the closest point projection consists of two steps called the predictor and the corrector step respectively.

Predictor Step The predictor step evaluates the displacement $\boldsymbol{u}_n(\boldsymbol{\theta}_z)$ by solving the equilibrium equation Eq. (14.20) [14, 19] and the strain increment $\Delta\boldsymbol{\varepsilon}_n(\boldsymbol{\theta}_z) = \nabla_s \Delta\boldsymbol{u}_n(\boldsymbol{\theta}_z)$ via linear symmetric mapping ∇_s. Once the increment of strain is computed one may define the trial stress $\boldsymbol{\Sigma}_n^{trial}(\boldsymbol{\theta}_z)$ assuming $\Delta\boldsymbol{\varepsilon}_n(\boldsymbol{\theta}_z)$ to be purely elastic. By checking if the stress $\boldsymbol{\Sigma}_n^{trial}(\boldsymbol{\theta}_z)$ lies outside of the admissible region $\mathcal{K}^*(\boldsymbol{\theta}_z)$ we proceed with the corrector step. Otherwise, $\boldsymbol{\Sigma}_n(\boldsymbol{\theta}_z) = \boldsymbol{\Sigma}_n^{trial}(\boldsymbol{\theta}_z)$ is our solution and we may proceed to the next step.

Corrector Step The purpose of the corrector step is to project the stress outside of admissible region back onto a point in $\mathcal{K}^*(\boldsymbol{\theta}_z)$. To do this, we define the corresponding Lagrangian to a minimisation problem Eq. (14.25):

$$\mathcal{L}(\boldsymbol{\theta}_z) = \Phi_n(\boldsymbol{\theta}_z) + \boldsymbol{\lambda}_n(\boldsymbol{\theta}_z)\varphi_n(\boldsymbol{\Sigma})(\boldsymbol{\theta}_z), \quad (14.29)$$

with $\varphi_n(\Sigma)(\boldsymbol{\theta}_z)$ being the yield function describing the convex set. The solution $\boldsymbol{\lambda}_n(\boldsymbol{\theta}_z)$ is then simply found by taking the derivative

$$\mathbf{0} \in \partial_\Sigma \mathcal{L} = \partial_\Sigma \Phi_n(\boldsymbol{\theta}_z) + \boldsymbol{\lambda}_n(\boldsymbol{\theta}_z)\partial_\Sigma \boldsymbol{\varphi}(\boldsymbol{\theta}_z) \quad (14.30)$$

and solving the corresponding system of equations. Once $\boldsymbol{\lambda}_n(\boldsymbol{\theta}_z)$ is known one may compute the update of the stress- and strain-like variables.

5 Numerical Results

The method considered in this paper is numerically tested on the example of a flat plate containing a circular hole. The plate is constrained on the left edge, and subjected to uniform tension f on the right edge as shown in Fig. 14.1. The material properties describing the elastoplastic behaviour are taken in two different scenarios: homogeneous and heterogeneous case. The homogeneous random quantities q

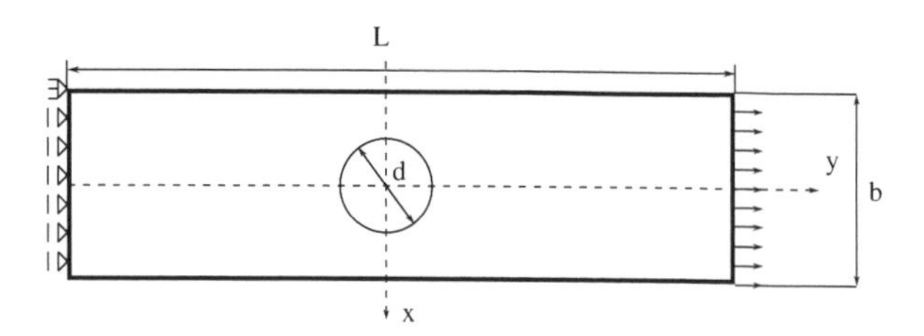

Fig. 14.1 Experimental set up. Here $b = 20$, $L = 56$ and $d = 10$

are modelled as a random variables of lognormal type, more precisely a modified lognormal random variable:

$$\kappa = \kappa_0 + \kappa_1 \exp(\mu + \sigma\theta), \tag{14.31}$$

where κ_0 and κ_1 are constants and θ the standard Gaussian random variable. On other side the heterogeneous case is specified by properties modelled as a lognormal random field (positive-definite):

$$\kappa(x, \omega) = \exp\big(\mu(x) + \sigma(x)\gamma(x, \omega)\big) \tag{14.32}$$

where $\gamma(x, \omega)$ represents the standard Gaussian random field with zero mean and unit variance. In case that one chooses $\mu(x)$ and $\sigma(x)$ as appropriate constants the random field becomes homogeneous in a mean sense, but locally in each realisation heterogeneous.

Due to the lack of the measurement data, the reality is simulated via computer by assuming the true values for q and running the corresponding finite element analysis. For simplicity reasons the truth is taken to be deterministic even though one realisation of some positive definite random field would be more appropriate. Additionally the obtained response is polluted by a Gaussian noise with zero mean and the standard deviation σ_ε.

5.1 Random Variable Update

In this particular example the prior elastoplastic behaviour is described by the random yield stress σ_y and bulk modulus K. Due to their positive definiteness, the mentioned properties are modelled according to Eq. (14.31) by taking for σ_{yf}: $\kappa_0 = 0.1$, $\kappa_1 = 0.25$, $\mu = 0$ and $\sigma = 0.3$ and for K_f: $\kappa_0 = 10$, $\kappa_1 = 15$, $\mu = 1$ and $\sigma = 0.3$. For such chosen probability distributions the corresponding forward response is computed with the help of the pseudo-Galerkin method (and Gauss–Hermite sparse grid) with the polynomial chaos expansion of the maximum order equal to four. More than that is not necessary to take as the input can be already accurately described by polynomial order 3. On other side, the first order polynomial expansion (i.e. Gaussian distribution) is not considered due to the violence of positive-definite requirements on K_f and σ_{yf}.

Table 14.1 The relative mean error ε_m [%] and relative variance ε_v [%] as a function of the PCE order and the number of the measurement points. The measurement is the first stress component

	Bulk modulus							Yield stress						
	p	2	3	4	5	6	7	p	2	3	4	5	6	7
ε_m	2	6.70	6.70	4.81	4.02	2.50	2.50	2	30.62	30.76	20.54	16.32	0.50	0.27
	3	6.70	6.70	4.79	3.92	2.47	2.43	3	30.73	30.79	20.60	16.42	0.78	0.11
	4	6.69	6.69	4.78	3.92	2.46	2.41	4	30.73	30.54	20.62	16.36	0.76	0.10
ε_v	2	14.68	14.31	11.92	9.65	7.35	7.38	2	99.42	97.66	87.65	82.37	48.32	44.06
	3	14.81	14.63	12.23	9.95	7.10	7.14	3	99.37	97.37	87.16	82.12	48.24	44.10
	4	14.80	14.72	12.25	9.96	7.09	7.07	4	99.37	99.37	87.12	82.08	48.12	44.03

In order to simulate the virtual measurement the true values $\sigma_y = 0.5$ and $K = 80$ are adopted together with the uniform tension $f = 10 \cdot t$ in three equal time steps $h = 1$. Further, the finite element analysis is performed by discretising the domain with the different number of elements than the one corresponding to the forward problem in order to escape the violation of the inverse law. The obtained response is then polluted by a Gaussian noise with σ_ε taking the values in a set $\{0.1, 0.01, 0.001\}$. As any kind of response (stress, strain, etc.) can be declared as the measurement quantity, in this work we choose the stress components (as a more abstract than real experiment) or the displacement (corresponding to the experiments performed in reality). In each of these cases the response is measured in 2 up to 7 measurement points mostly concentrated around the hole (where the measurements in reality are expected to be performed).

With respect to the previous description the results after the update are plotted in Table 14.1 for the measurement of the first stress component and noise $\sigma_\varepsilon = 0.01$. As the error estimates we adopt the relative mean error of posterior compared to the truth $\varepsilon_m = 100 \cdot |\mathbb{E}(\kappa_a) - \mathbb{E}(\kappa_t)|/|\mathbb{E}(\kappa_t))$ and the reduction of the variance compared to the prior $\varepsilon_v = 100(\operatorname{var}\kappa_a / \operatorname{var}\kappa_f)$. In Table 14.1 clearly is visible that the bulk modulus K approaches the truth in circa 6% of the mean error already with only two measurement points. This continues to drop to 2% with the number of the measurement points. Similar is valid for the reduction of variance ε_v. In addition to this, the slight decrease of ε_m and ε_v can be observed in the direction of the polynomial order increase. However, this improvement is not very drastic as the second order approximation already accurately describes the prior. Similar results are characterizing the update of the stress variable σ_y, though the errors drop much faster than in a case of K. The reason for this are the placements of the measurement points. Namely, the more points are lying in the plastic area the more informative data for σ_y enter the update process. It is interesting to notice that the mean error becomes smaller than 1%, while the variance more slowly reduces. For this one need more measurement points or possibly nonlinear approximation of Bayesian estimate. Namely, due to the nonlinear relationship between the parameter and data, the linear update as presented in Sect. 3 is not optimal for σ_y.

Table 14.2 The relative mean error ε_m [%] with respect to the number of the measurements in different experiments

Bulk modulus							Yield stress						
z	2	3	4	5	6	7	z	2	3	4	5	6	7
σ_{xx}	6.79	6.73	4.88	4.01	2.57	2.51	σ_{xx}	30.62	30.76	20.54	16.32	0.50	0.27
σ_{yy}	8.70	8.71	5.02	3.73	3.69	2.91	σ_{yy}	31.74	32.10	16.26	8.49	9.19	3.02
σ_{xy}	7.88	7.75	4.68	3.81	3.18	2.80	σ_{xy}	31.21	30.81	15.18	11.29	6.81	2.84
u	39.94	7.59	8.48	7.93	7.85	7.83	u	5.27	2.43	1.98	2.12	2.13	2.13

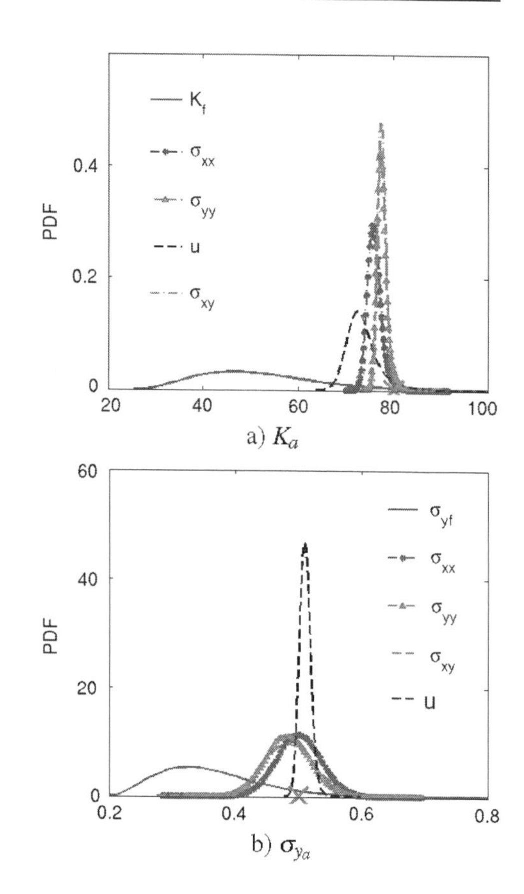

Fig. 14.2 (**a**) The posterior probability density function of bulk modulus in different experiments. (**b**) The posterior density function of yield stress in different experiments. The truth is denoted with *red X*

Following previous discussion in Table 14.2 are studied the results obtained by measuring different quantities such as σ_{xx}, σ_{yy} and σ_{xy} stress components as well as the displacement u (see Fig. 14.2). Here it is interesting to note that the stress measurement brings smaller errors in posterior for K than for yield stress, in both ε_m and ε_v (the later one gathered in Table 14.3). In contrast to this the displacement is more suitable measurement for the value of σ_y. More importantly the higher reduction in the variance is observed (see Table 14.3).

Table 14.3 The relative variance ε_v [%] with respect to the number of the measurements in different experiments

Bulk modulus							Yield stress						
z	2	3	4	5	6	7	z	2	3	4	5	6	7
σ_{xx}	14.68	14.31	11.92	9.65	7.35	7.38	σ_{xx}	99.42	97.66	87.65	82.37	48.32	44.06
σ_{yy}	14.64	18.41	13.47	10.11	9.94	8.79	σ_{yy}	98.80	95.87	80.19	60.23	59.77	45.46
σ_{xy}	17.06	16.67	12.33	9.68	8.14	7.86	σ_{xy}	99.10	98.66	78.22	70.9	63.20	52.42
u	96.92	43.54	33.34	29.35	29.34	18.10	u	28.31	18.55	12.32	11.28	11.27	11.27

Table 14.4 The relative mean ε_m [%] and variance ε_v [%] with respect to the measurement noise

Mean							Variance						
σ_ε	2	3	4	5	6	7	σ_ε	2	3	4	5	6	7
1e-1	9.98	9.95	6.74	5.8	5.65	5.54	1e-1	36.99	36.89	21.74	16.34	16.23	15.91
1e-2	6.79	6.73	4.88	4.01	2.57	2.51	1e-2	14.68	14.31	11.92	9.65	7.35	7.38
1e-3	6.75	6.79	0.95	1.14	1.98	1.31	1e-3	14.23	11.72	6.94	6.65	6.94	6.61

Besides the influence of the polynomial order on the update procedure, one may investigate the influence of the corresponding measurement error. As shown in Table 14.4 the mean error ε_m and the variance reduction ε_v for the bulk modulus are decreasing with smaller values of σ_ε as expected. The smaller measurement error is, the more we are certain about the experimental data (and hence the truth) and thus the better is update.

5.2 Random Field Update

The previous example was rather simple as the number of random variables representing the problem is relatively small. In order to properly investigate the update of material properties in this example we consider the identification of shear modulus G priory modelled as a lognormal random field with $\mu = 3.50 \cdot 10^4$ and $\sigma = 0.1\mu$ according to Eq. (14.32). The field has exponential correlation function with the correlation lengths equal to $l_c = 10$. The same geometrical problem (see Fig. 14.1) as before is considered with slightly different loading conditions, i.e. $f = 25$. In addition both, the purely elastic and nonlinear response are studied due to the comparison purposes. Due to the lack of the measurement data, the reality is simulated by modelling the shear modulus $G_t = 2.8 \cdot 10^4$ as a point-wise constant function, and measuring the values of the shear stress σ_{xy} in 30% of nodal points (including boundary conditions). The collected data are then disturbed by a central Gaussian noise with the diagonal covariance $\sigma_\varepsilon^2 \boldsymbol{I}$, where σ_ε is approximately equal to 1% of the measured value.

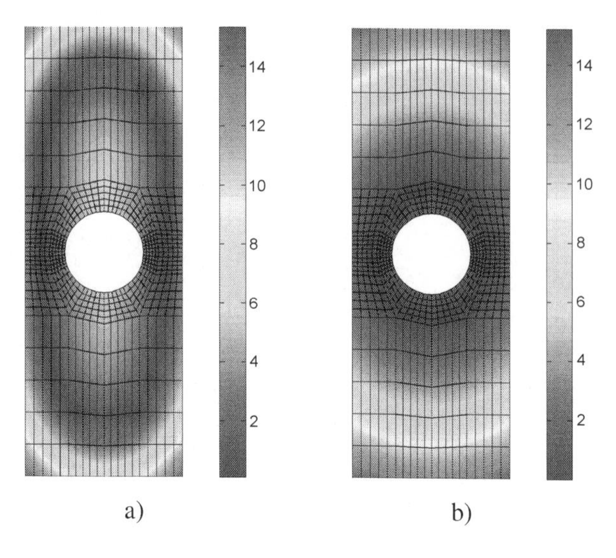

Fig. 14.3 The relative error ε_a [%] of updated shear modulus G via (**a**) linear model (elasticity) (**b**) nonlinear model (elasto-plasticity). For PCE is used order $p = 3$ in $M = 10$ random variables

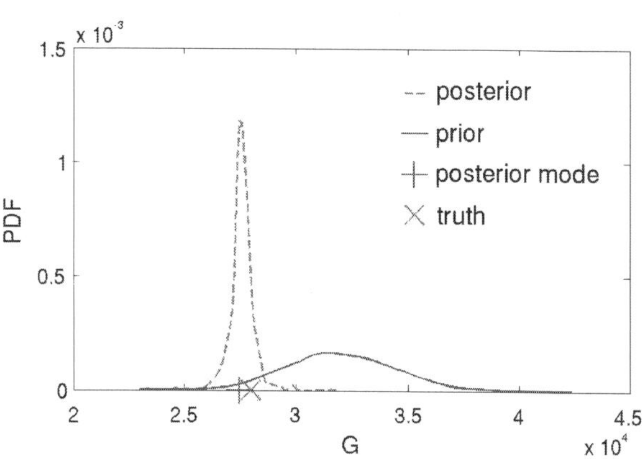

Fig. 14.4 Posterior probability distribution compared to the prior for the nonlinear model. The update is obtained by linear Bayesian method with third order PCE and $M = 10$ random variables

As the plots of the relative root mean square error

$$\varepsilon_a = 100 \cdot \frac{\|G_a - G_t\|_{L_2(\Omega)}}{\|G_f - G_t\|_{L_2(\Omega)}} \tag{14.33}$$

in Fig. 14.3 show the direct linear update performs better in a case of linear than the nonlinear model as expected. The 2% error region E_r in linear case spreads from the central part to the boundary resulting in much wider region than in nonlinear case. In contrast to this, the nonlinear model produces reduced E_r region strictly in the central plastifying zone. For the point in this domain the update performs well, i.e. the variance reduces, the mean moves in the direction of the truth and the truth is almost coinciding with the mode, see Fig. 14.4. However, in other nodes outside of E_r this may not be the case. This behaviour is expected as the linear Bayesian approximation is not optimal for the nonlinear models.

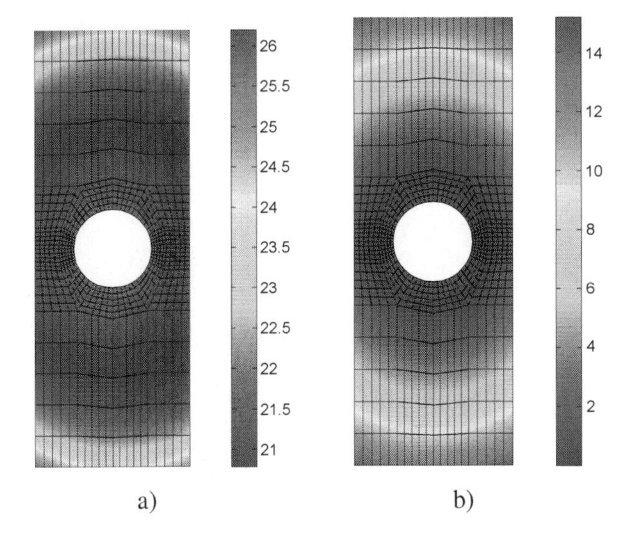

Fig. 14.5 Shear modulus G via elasto-plastic response, the relative error ε_a [%]: (**a**) EnKF result (**b**) direct PCE result. For PCE is used $p = 3$ and $M = 10$ random variables, while for EnKF 100 ensemble members

Besides the direct PCE update procedure the identification of the shear modulus for the nonlinear model is also done with the help of ensemble Kalman filter (EnKF) method (see [17, 18]) with 100 ensemble members. The comparison of results in Fig. 14.5 shows that the direct update produces much smaller value of the relative error ε_a. This can be explained by relatively small number of ensemble members. On other side the EnKF identifies G in a more unified way, i.e. the region of the minimal error is covering almost the whole computational area in contrast to PCE where it is placed around the hole edge.

In previous experiments the update results are influenced by the values of the different quantities such as the order of PCE, number of terms in truncated KLEs, etc. However, until now we did not consider the influence of the measured quantity on the update process. We assumed that the shear stress σ_{xy} is the most appropriate measurement. In order to investigate this, we substituted σ_{xy} in previous experiment by a stress σ_{yy}. This change significantly influences the update results by increasing the relative root mean square error three times as shown in Fig. 14.6 for nonlinear model.

6 Conclusion

In this paper is studied the problem of identifying parameters or quantities in elasto-plastic computational model by comparison with virtual reality models (e.g. more refined models). The introduced Bayesian approach starts from the idea that the choice of parameters should be such as to minimise a certain error functional. In other words, the update setting embeds the unknown quantity in a probability distribution, where the spread of the probability distribution should reflect the uncertainty about that quantity. Reformulating the classical Bayesian approach via conditional

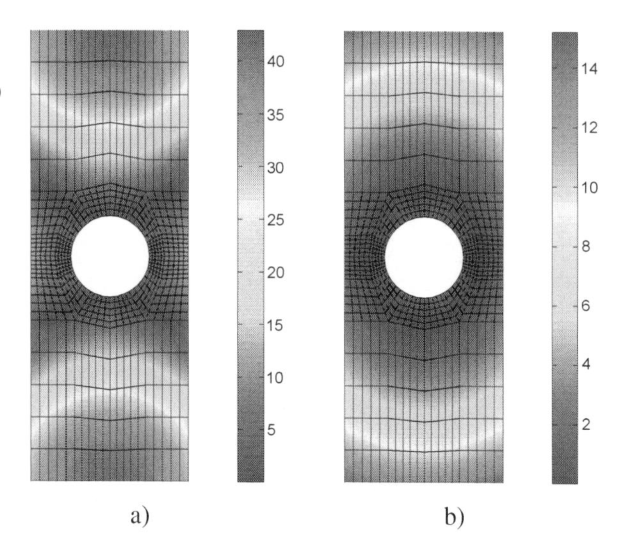

Fig. 14.6 The relative error ε_a [%] of updated shear modulus G (nonlinear model) with the help of (**a**) σ_{yy} measurement (**b**) σ_{xy} measurement. For PCE is used $p = 3$ and $M = 10$

expectation and the minimisation of variance as its background the Bayesian update reduces to the simple linear formula, containing the well-known Kalman filter as a special case.

The identification problem here considered is a very difficult one, an elasto-plastic system, or mathematically speaking a variational inequality of a second kind. The non-smoothness inherent in such problems makes the update procedure more complex and difficult to perform. However, as presented in numerical results the PCE based methods still succeed to preform well. Regarding this the PCE methods show great promise for the future parameter identification in nonlinear problems.

Acknowledgements The authors would like to acknowledge the financial support of Technical University Braunschweig, DAAD and DFG (Deutsche Forschungsgemeinschaft).

References

1. Blanchard, E.D., Sandu, A., Sandu, C.: A polynomial chaos-based Kalman filter approach for parameter estimation of mechanical systems. J. Dyn. Syst. Meas. Control **132**(6), 061404 (2010)
2. Bobrowski, A.: Functional Analysis for Probability and Stochastic Processes: An Introduction. Wiley Series in Probability and Statistics. Cambridge University Press, Cambridge (2005)
3. Gamerman, D., Lopes, H.F.: Markov Chain Monte Carlo: Stochastic Simulation for Bayesian Inference. Chapman & Hall, London (2006)
4. Ghanem, R.G., Spanos, P.D.: Stochastic Finite Elements: A Spectral Approach. Springer, New York (1991)
5. Hida, T., Kuo, H.-H., Potthoff, J., Streit, L.: White Noise Analysis—An Infinite Dimensional Calculus. Kluwer, Dordrecht (1993)
6. Higdon, D., Lee, H., Holloman, C.: Markov chain Monte Carlo-based approaches for inference in computationally intensive inverse problems. Bayesian Stat. **7**, 181–197 (2003)

7. Holden, H., Øksendal, B., Ubøe, J., Zhang, T.-S.: Stochastic Partial Differential Equations. Birkhäuser, Basel (1996)
8. Kalman, R.E.: A new approach to linear filtering and prediction problems. J. Basic Eng. **82**, 35–45 (1960)
9. Kolmogorov, A.: Foundations of the Theory of Probability, 2nd edn. Chelsea, New York (1956)
10. Kučerová, A., Matthies, H.G.: Uncertainty updating in the description of heterogeneous materials. Tech. Mech. **30**(1–3), 211–226 (2010)
11. Luenberger, D.G.: Optimization by Vector Space Methods. Wiley, New York (1969)
12. Marzouk, Y.M., Xiu, D.: A stochastic collocation approach to Bayesian inference in inverse problems. Commun. Comput. Phys. **6**(4), 826–847 (2009)
13. Matthies, H.G.: Stochastic finite elements: computational approaches to stochastic partial differential equations. Z. Angew. Math. Mech. **88**(11), 849–873 (2008)
14. Matthies, H.G., Rosić, B.: Inelastic media under uncertainty: stochastic models and computational approaches. In: Reddy, D. (ed.) IUTAM Symposium on Theoretical, Computational and Modelling Aspects of Inelastic Media. IUTAM Bookseries, vol. 11, pp. 185–194 (2008)
15. Matthies, H.G., Litvinenko, A., Pajonk, O., Rosić, B., Zander, E.: Parametric and uncertainty computations with tensor product representations. Technical report, Institut für Wissenschaftliches Rechnen, TU Braunschweig (2011)
16. El Moselhy, T., Marzouk, Y.M.: Bayesian inference with optimal maps. J. Comput. Phys. (2011). doi:10.1016/j.jcp.2012.07.022
17. Pajonk, O., Rosić, B.V., Litvinenko, A., Matthies, H.G.: A deterministic filter for non-Gaussian Bayesian estimation—Applications to dynamical system estimation with noisy measurements. Physica D **241**(7), 775–788 (2012)
18. Pajonk, O., Rosić, B.V., Matthies, H.G.: Deterministic linear Bayesian updating of state and model parameters for a chaotic model. Informatikbericht 2012-01, Institut für Wissenschaftliches Rechnen, Technische Universität. Braunschweig (2012)
19. Rosić, B., Matthies, H.G.: Computational approaches to inelastic media with uncertain parameters. J. Serbian Soc. Comput. Mech. **2**, 28–43 (2008)
20. Rosić, B.V., Matthies, H.G.: Stochastic plasticity—a variational inequality formulation and functional approximation approach I: the linear case. Technical Report 2012-02, Institut für Wissenschaftliches Rechnen (2012)
21. Rosić, B.V., Matthies, H.G., Živković, M.: Uncertainty quantification of infinitesimal elastoplasticity. Sci. Tech. Rev. **61**(2), 3–9 (2011)
22. Rosić, B.V., Pajonk, O., Litvinenko, A., Matthies, H.G.: Sampling-free linear Bayesian update of polynomial chaos representations. J. Comput. Phys. **231**(17), 5761–5787 (2012)
23. Rosić, B.V., Kučerová, A., Sýkora, J., Pajonk, O., Litvinenko, A., Matthies, H.G.: Parametric identification in a probabilistic setting. Technical report, Institut für Wissenschaftliches Rechnen, TU Braunschweig (2012)
24. Saad, G., Ghanem, R.: Characterization of reservoir simulation models using a polynomial chaos-based ensemble Kalman filter. Water Resour. Res. **45**, W04417 (2009)
25. Simo, J.C., Hughes, T.J.R.: Computational Inelasticity. Springer, New York (1998)
26. Soize, C.: Maximum entropy approach for modeling random uncertainties in transient elastodynamics. J. Acoust. Soc. Am. **109**(5I), 1979–1996 (2001)
27. Tarantola, A.: Popper, Bayes and the inverse problem. Nat. Phys. **2**(8), 492–494 (2006)
28. Wiener, N.: The homogeneous chaos. Am. J. Math. **60**, 1936–1938 (1938)

Chapter 15
SH Surface Waves in a Half Space with Random Heterogeneities

Chaoliang Du and Xianyue Su

Abstract Horizontally polarized shear waves (SH waves) do not exist in a homogeneous half space according to the traditional elastic wave theory. However, in this study, we proved both theoretically and numerically that there will be surface waves in a half space which has small, random density, but the mean value of the density is homogeneous. Historically, this type of half space is often treated as a homogeneous one with deterministic methods. In this investigation, a closed-form dispersion equation was derived stochastically, and the frequency spectrum, dispersion equation, phase/group velocity were plotted numerically to study how the random inhomogeneities will affect the dispersion properties of the half space with random density. This research may find its application in seismology, non-destructive test/evaluation, etc.

1 Introduction

In this study, the dispersion and attenuation properties of waves propagating in a half space (see Fig. 15.1) with random heterogeneities are investigated.

Shear horizontal surface waves (SHSW) are the most destructive waves in an earth quake and they can propagate through a very long distance without much loss of its energy. But, scientists have proved long ago that there is no SHSW in a homogeneous isotropic linearly elastic half-space [1]. However, in 1911, love predicted

C. Du
Department of Mechanics and Aerospace Engineering, College of Engineering, Peking University, Beijing, 100871, P.R. China

C. Du (✉)
Beijing Aeronautical Science and Technology Research Institute, COMAC, The Olympic Building, No.267 Middle Section of North 4th Ring Road, Beijing, 100083, P.R. China
e-mail: duchaoliang@comac.cc

X. Su
LTCS and Department of Mechanics and Aerospace Engineering, College of Engineering, Peking University, Beijing, 100871, P.R. China
e-mail: xyswsk@pku.edu.cn

M. Papadrakakis et al. (eds.), *Computational Methods in Stochastic Dynamics*, Computational Methods in Applied Sciences 26,
DOI 10.1007/978-94-007-5134-7_15,

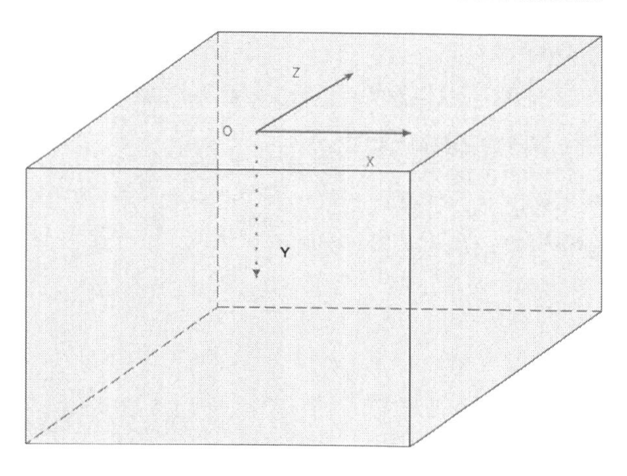

Fig. 15.1 Coordinate system of the half space

mathematically that SHSW could exist if the half-space is covered by a layer of a different material.

Since then, SHSW in a half space was mostly explained theoretically by Love's theory or its variant theories. But we know that the earth's surface is very complex. It is a mixture of many kinds of rocks, sands, soil, water, etc., and more complicatedly, these materials do not often distribute in deterministic ways, but distribute randomly. So do SHSW exist in such a complex, random half space?

Similar problems have been explored by some scientists. B. Collet et al. [2] studied SHSW in a Functionally Graded Material of which some material constants share the same depth-dependent function , and derived some of the depth-dependent functions which could be solved exactly. Using their solutions, they studied the influence of different inhomogeneity functions on the properties of SHSW. J. Achenbach et al. [3] studied SHSW in a purely elastic half-space whose shear modulus and mass density depend arbitrarily on the depth and gave a general solution that is quite exact for high frequencies. T.C.T. Ting [4] recently investigated SHSW in a half space of which C_{44} and ρ have the same function form, and C_{55}, C_{45} are correlated. Here, $C_{\alpha\beta}$ is the elastic stiffness in the contracted notation. Ting also got an asymptotic solution of general graded materials for large wave number k. Anti-plane shear waves for anisotropic graded materials have been considered for periodic half-spaces by A. Shuvalov et al. [5] and for a single plate by A. Shuvalov et al. [6]. Shear horizontal waves in functionally graded piezoelectric materials are also greatly studied by Tianjian Lu et al. [7–9].

But these researches haven't given an explicit solution of dispersion and attenuation of SHSW in a half space with random density in the depth direction by strict stochastic methods. In this study, we get the explicit dispersion equation by the first order smoothing approximation (FOSA) method. And we then analyze the dispersion and attenuation properties using the dispersion equation.

In this study we proved mathematically and numerically that SHSW could exist in a stochastically homogeneous half space. Some interesting properties of dispersion and attenuation found in this study could promote our understanding of waves propagating in a half space with random heterogeneities, e.g. earth's upper crust, alloys or composites. It will also help us to do the inverse problems, for example, to use seismic waves to detect the earth's crust structure, and to use ultrasonic waves to evaluate a structure with randomly distributed micro-cracks or heterogeneities.

2 Modeling and Mathematical Analysis

The fundamental dynamic equation system for statistically homogeneous, isotropic, linearly elastic solid is

$$\tau_{ij,j} + \rho f_i = \rho \ddot{u}_i \tag{15.1}$$

$$\tau_{ij} = \lambda \varepsilon_{kk} \delta_{ij} + 2\mu \varepsilon_{ij} \tag{15.2}$$

$$\varepsilon_{ij} = \frac{1}{2}(u_{i,j} + u_{j,i}) \tag{15.3}$$

To account for the random heterogeneities, the constants ρ, μ, λ in the equation system are changed to random processes of space.

Consider SH waves propagating in x direction in a half space (see Fig. 15.1).

It is known that for anti-plane waves that $u_x = u_y = 0$ and $\partial/\partial z = 0$. And if we assume that there is no body force, the equation system reduces to

$$\tau_{zj,j} = \rho \ddot{u}_z \tag{15.4}$$

$$\tau_{zj} = \mu u_{z,j} \tag{15.5}$$

in which, $j = x, y$.

So the dynamic equation for SH waves in a random half space is

$$(\mu u_{z,j})_{,j} = \rho \ddot{u}_z \tag{15.6}$$

And the boundary condition is

$$\tau_{zy}|_{y=0} = 0 \quad \text{i.e.} \tag{15.7}$$

$$\mu u_{z,y}|_{y=0} = 0 \tag{15.8}$$

Assume here that there is randomness only in the y direction. Consider an harmonic wave motion of the form

$$u_z = f(y)\exp\left[i(k_1 x - \omega t)\right] \tag{15.9}$$

in which, $f(y)$ is a random process. To study the surface shear wave, we assume the averaged $f(y)$ to be

$$\langle f(y) \rangle = A e^{-by}, \tag{15.10}$$

in which $b > 0$. Thus the mean wave motion $\langle u_z \rangle$ could be written as

$$\langle u_z \rangle = A e^{-by} e^{i(k_1 x - \omega t)} \tag{15.11}$$

If there is no random heterogeneities in the solid, a solution of Eq. (15.6) would be of the form [1]

$$u_z = Ae^{-by}e^{i(k_1x-\omega t)} \tag{15.12}$$

Substituting Eq. (15.12) into Eq. (15.6), we find

$$\frac{\omega^2}{C_s^2} - k_1^2 + b^2 = 0 \tag{15.13}$$

In which, C_s is the shear velocity of the homogeneous material without random heterogeneities,

$$C_s = \sqrt{\frac{\mu_0}{\rho_0}} \tag{15.14}$$

For a free surface, the boundary condition at $y = 0$ is

$$\frac{\mathrm{d}u_z}{\mathrm{d}y} = 0 \tag{15.15}$$

The boundary condition Eq. (15.15) can be satisfied only if either $A = 0$ or $b = 0$. Therefore, there is no surface SH wave in an homogeneous, isotropic, linearly elastic half space.

Firstly, we consider that random heterogeneities are only on the surface (as a practical example, the roughness of the earth surface could be viewed as a half space but with random heterogeneities on the surface). Under this circumstance, the boundary condition will be

$$\mu\frac{\partial u_z}{\partial y}\bigg|_{y=0} = 0 \quad \Rightarrow \quad (\mu_0 + \varepsilon\mu_1)\frac{\partial(\langle u_z\rangle + \varepsilon u_{z1})}{\partial y}\bigg|_{y=0} = 0 \tag{15.16}$$

in which, $\varepsilon\mu_1$ and εu_{z1} represent the surface roughness. Averaging both sides of Eq. (15.16), we get

$$\left\{\mu_0\frac{\mathrm{d}\langle u_z\rangle}{\mathrm{d}y} + \varepsilon^2\left\langle\mu_1\frac{\partial u_{z1}}{\partial y}\right\rangle\right\}\bigg|_{y=0} = 0 \tag{15.17}$$

ε is an averaging measure of how the properties of random heterogeneities deviate from the averaged properties, and we assume it to be small and take it as the small parameter. The randomness of the surface takes effect through the term $\varepsilon^2\langle\mu_1\frac{\partial u_{z1}}{\partial y}\rangle$. We assume here that

$$\varepsilon^2\left\langle\mu_1\frac{\partial u_{z1}}{\partial y}\right\rangle\bigg|_{y=0} = \mu_0 A\beta e^{i(k_1x-\omega t)} \tag{15.18}$$

in which, β is the surface parameter. Substituting Eqs. (15.18) and (15.11) into Eq. (15.17), we get

$$b = \beta \tag{15.19}$$

Considering Eq. (15.13), the dispersion equation for SH waves in a half space with random heterogeneities only on the surface is

$$\frac{\omega^2}{C_s^2} - k_1^2 + \beta^2 = 0 \tag{15.20}$$

Next, we will investigate the problem of the half space with random heterogeneities in the whole depth direction. Substituting Eq. (15.9) in Eq. (15.6) gives

$$\left(\rho\omega^2 - \mu k_1^2\right) f + (\mu f_{,y})_{,y} = 0 \tag{15.21}$$

Assuming that ρ, μ differ slightly from the mean value of them, ρ, μ can be written as

$$\rho(y) = \rho_0 + \varepsilon\rho_1(y) \qquad \mu(y) = \mu_0 + \varepsilon\mu_1(y) \tag{15.22}$$

where, ε is a small parameter, and

$$\langle \rho_1 \rangle = \langle \mu_1 \rangle = 0 \tag{15.23}$$

Substituting Eq. (15.22) in Eq. (15.21), we have

$$\left(\rho_0\omega^2 - \mu_0 k_1^2\right) f + \mu_0 f_{,yy} + \varepsilon\left(\left(\rho_1\omega^2 - \mu_1 k_1^2\right) f + (\mu_1 f_{,y})_{,y}\right) = 0 \tag{15.24}$$

According to FOSA theory, the deterministic operator of Eq. (15.21) is

$$L_0(y) = \mu_0\left(k_0^2 + \frac{\partial^2}{\partial y^2}\right) \tag{15.25}$$

in which,

$$k_0^2 = \frac{\omega^2}{C_s^2} - k_1^2 \tag{15.26}$$

And the first order random operator of Eq. (15.21) is

$$L_1(y) = P(y) + \mu_1(y)_{,y}\frac{\partial}{\partial y} + \mu_1(y)\frac{\partial^2}{\partial y^2} \tag{15.27}$$

in which,

$$P(y) = \rho_1(y)\omega^2 - \mu_1(y)k_1^2 \tag{15.28}$$

Considering Eq. (15.23), we can see that $\langle L_1 \rangle = 0$.

For steady waves, the Green function G_0 of the deterministic operator L_0 is

$$G_0(y_1, y_2) = -\frac{1}{2k_0\mu_0}\sin\left(k_0|y_1 - y_2|\right) \tag{15.29}$$

According to stochastic theory, the FOSA equation is

$$L_0\langle f(y_1)\rangle - \varepsilon^2\left\langle L_1(y_1)\int G_0(y_1,y_2)L_1(y_2)\langle f(y_2)\rangle \mathrm{d}y_2\right\rangle = 0 \tag{15.30}$$

To solve Eq. (15.30), let's calculate $L_1(y_1)G_0(y_1,y_2)$ first,

$$L_1(y_1)G_0(y_1,y_2) = -\left(P(y_1) + \mu_1(y_1)_{,y_1}\frac{\partial}{\partial y_1} + \mu_1(y_1)\frac{\partial^2}{\partial y_1^2}\right) \times \frac{1}{2k_0\mu_0}\sin\bigl(k_0|y_1-y_2|\bigr) \tag{15.31}$$

When $y_2 < y_1$

$$L_1(y_1)G_0(y_1,y_2) = Q_1\sin\bigl(k_0(y_1-y_2)\bigr) + Q_2\cos\bigl(k_0(y_1-y_2)\bigr) = M(y_1,y_2) \tag{15.32}$$

in which,

$$Q_1 = \left(\frac{\mu_1(y_1)k_0}{2\mu_0} - \frac{P(y_1)}{2k_0\mu_0}\right) \tag{15.33}$$

$$Q_2 = -\frac{\mu_1(y_1),y_1}{2\mu_0} \tag{15.34}$$

and, when $y_2 > y_1$

$$L_1(y_1)G_0(y_1,y_2) = -M(y_1,y_2) \tag{15.35}$$

Then, using Eq. (15.10), $L_1(y_2)\langle f(y_2)\rangle$ can be expressed as

$$L_1(y_2)\langle f(y_2)\rangle = \left(P(y_2) + \mu_1(y_2)_{,y_2}\frac{\partial}{\partial y_2} + \mu_1(y_2)\frac{\partial^2}{\partial y_2^2}\right)Ae^{-by_2} = A\bigl(P(y_2) - \mu_1(y_2)_{,y_2}b + \mu_1(y_2)b^2\bigr)e^{-by_2} = N(y_2) \tag{15.36}$$

If we assume that $\mu_1 = 0$, we could study the influence of the randomness of the density on the dispersion properties of the half space.

The random process $\rho_1(y_1;\gamma)$ is taken as Uhlenbeck–Ornstein process [10]. Although its correlation function is not mean-square differentiable, this process has been used in a number of investigations because it fits experimental data the best [11]. This process is a centered and stationary random process [10] and its correlation function is

$$R_{\rho_1(y_1;\gamma)\rho_1(y_2;\gamma)} = \int \rho_1(y_1;\gamma)\rho_1(y_2;\gamma)\,\mathrm{d}\gamma = \zeta^2 e^{-|y_1-y_2|/R_c} = R(y_1-y_2) \tag{15.37}$$

In which, $\zeta = \sqrt{\langle \rho_1{}^2\rangle}$ and it is the standard deviation of the random density function; γ is a random variable. And R_c is the integral radius (the correlation length) of the correlation function, which physically means the scale of heterogeneity [12], and it should be positive.

From Eq. (15.10), we have

$$L_0\langle f(y_1)\rangle = \mu_0\left(k_0^2 + \frac{\partial^2}{\partial y^2}\right)Ae^{-by_1} = \mu_0\left(k_0^2 + b^2\right)Ae^{-by_1} \tag{15.38}$$

Substituting Eqs. (15.31), (15.36) and (15.38) into Eq. (15.30), we get the dispersion equation,

$$k_0^2 + b^2 - \frac{\omega^4\zeta^2\varepsilon^2 b}{2k_0\mu_0^2}\left(\frac{1}{(b+1/R_c)^2 + b^2} + \frac{1}{(b-1/R_c)^2 + b^2}\right) = 0 \tag{15.39}$$

It could be seen from the dispersion equation Eq. (15.39) that if there is no random fluctuation, i.e. $\varepsilon = 0$ or $\zeta = 0$ then $k_0^2 + b^2 = 0$—the equation becomes the dispersion equation without random heterogeneities;

Considering the surface condition Eq. (15.19), the dispersion equation could be written as,

$$k_0^2 + \beta^2 - \frac{\omega^4\zeta^2\varepsilon^2 \beta}{2k_0\mu_0^2}\left(\frac{1}{(\beta+1/R_c)^2 + \beta^2} + \frac{1}{(\beta-1/R_c)^2 + \beta^2}\right) = 0 \tag{15.40}$$

If β is zero, then ε will be zero too according to Eq. (15.18), that is, there will be no heterogeneities in the half space. So if β is zero, the half space will be homogeneous, and there will be no SH surface waves according to Eq. (15.40). In the future, we may consider another model that, if β is zero, ε will not be zero, to study the interesting case which has randomness only below the surface.

To conveniently evaluate numerically the effect of random heterogeneities, the dispersion equation Eq. (15.40) is transformed into a dimensionless equation in the following. h is a symbol for the characteristic length and it could be the correlation length of the random heterogeneities. Introduce new dimensionless variables as,

$$\begin{aligned} \overline{\omega} &= \frac{2h\omega}{\pi C_s} & \overline{k} &= \frac{2hk_1}{\pi} \\ \overline{R_c} &= \frac{\pi R_c}{2h} & \overline{\zeta} &= \frac{\varepsilon\zeta}{\rho_0} \\ \overline{\mu_0} &= \frac{\mu_0}{\rho_0 C_s^2} = 1 & \overline{\beta} &= \frac{2h\beta}{\pi} \end{aligned} \tag{15.41}$$

From Eqs. (15.26) and (15.41), we get

$$k_0^2 = \frac{\omega^2}{C_s^2} - k_1^2 = \left(\frac{\pi}{2h}\right)^2\left(\overline{\omega}^2 - \overline{k}^2\right) \tag{15.42}$$

so the dimensionless k_0 is defined as,

$$\overline{k_0}^2 = \overline{\omega}^2 - \overline{k}^2 \tag{15.43}$$

Using Eqs. (15.41), we could get the dimensionless dispersion equation from Eq. (15.20),

$$\overline{\omega}^2 - \overline{k_1}^2 + \overline{\beta}^2 = 0 \tag{15.44}$$

Using Eqs. (15.41) and (15.43), the dimensionless dispersion equation of Eq. (15.40) is,

$$\overline{k_0}^2 + \overline{\beta}^2 - \Lambda = 0 \tag{15.45}$$

Λ denote the random term,

$$\Lambda = \frac{\overline{\omega}^4 \overline{\zeta}^2 \overline{\beta}}{2\overline{k_0}} \left(\frac{1}{(\overline{\beta} + 1/\overline{R_c})^2 + \overline{\beta}^2} + \frac{1}{(\overline{\beta} - 1/\overline{R_c})^2 + \overline{\beta}^2} \right) \tag{15.46}$$

3 Numerical Results and Analysis

The SH surface waves propagating in a half space with random densities is further studied numerically. The dimensionless dispersion equation Eq. (15.45) is used to compute the curves. The numerical results are explained and discussed in the following.

3.1 Random Heterogeneities Only on the Surface

The geomorphy of the earth's surface is always very complex. The reason for this complexness can come from both natural and man-made actions. In this study, we model the complex geomorphy by giving a surface parameter β. So in this section, we will study the dispersion properties for half spaces with random heterogeneities only on the surface. The dispersion curves are plotted according to Eq. (15.20).

From Fig. 15.2, it can be seen that the phase velocity will grow to 1 slowly, but for $\overline{k} < 2$, the phase velocity will be 0, i.e. the waves become standing waves in this circumstance.

From Fig. 15.3, it can be seen that, given a wave number, the phase velocity will decrease to 0 as the surface parameter $\overline{\beta}$ grows, i.e. the waves propagate more and more slowly when the surface becomes more and more rough, and all the waves will be blocked when $\overline{\beta}$ is large enough.

3.2 Frequency Spectrum Analysis

In the following, we will study the dispersion properties for half spaces with random heterogeneities not only on the surface but also in the whole half space. The related parameters are set to $\varepsilon = 0.1$, $\overline{\zeta} = 2$, $\overline{R_c} = 0.4$, $\overline{\beta} = 2$ respectively.

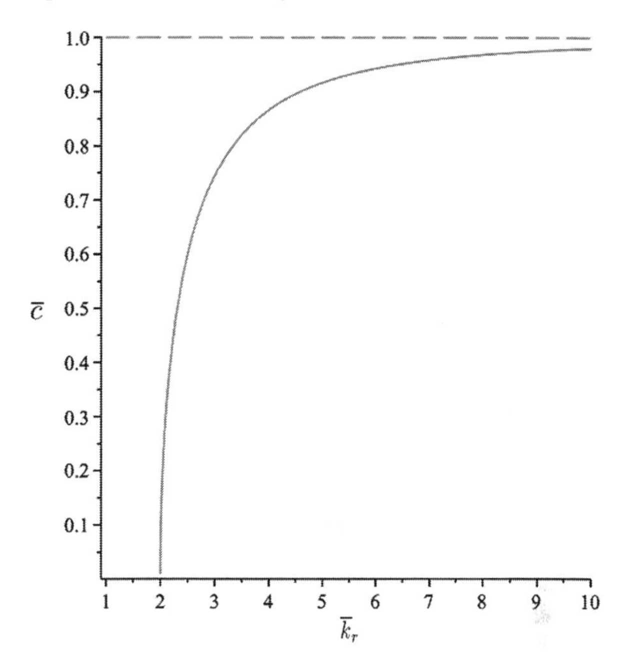

Fig. 15.2 Normalized phase velocity—normalized wave number. $\overline{\beta} = 2$

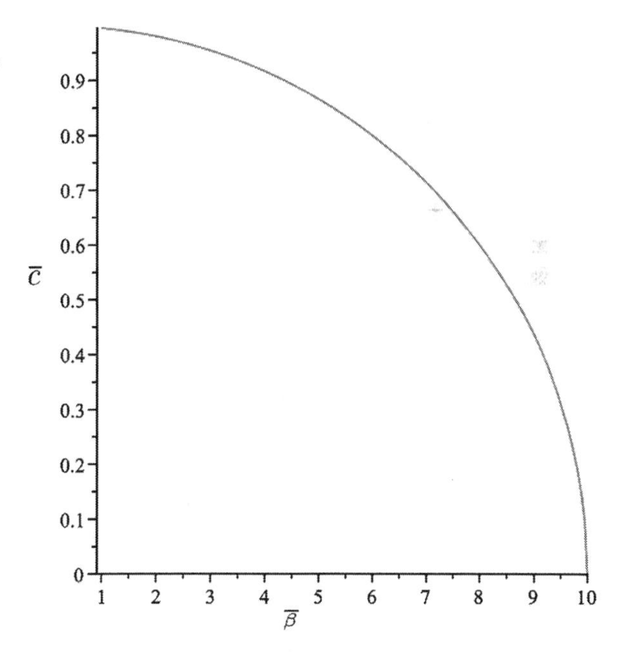

Fig. 15.3 Normalized phase velocity—normalized surface parameter. $\overline{k} = 10$

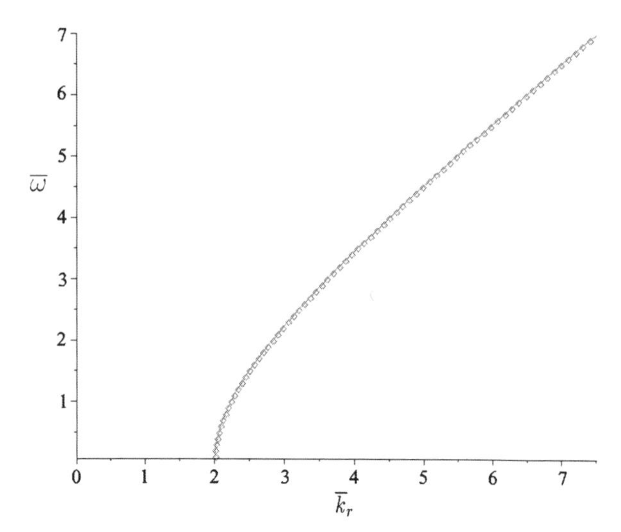

Fig. 15.4 Normalized circular frequency—normalized wave number

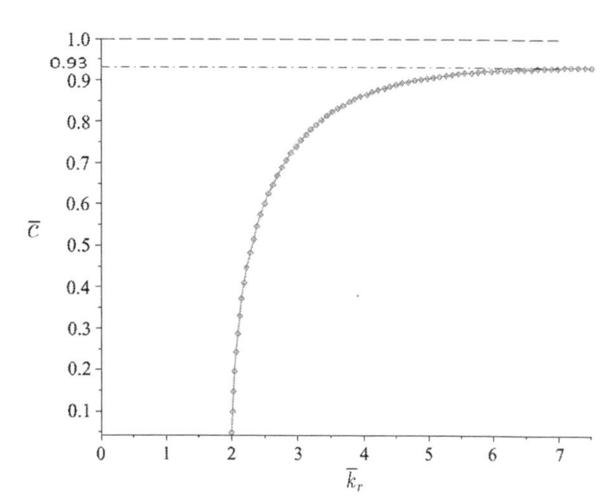

Fig. 15.5 Normalized phase velocity—normalized wave number

From Figs. 15.4, 15.5 and 15.6, we can see that

1. As the wave number grows, the velocity will grow to a value—approximately 0.93 in this case. The reason that it can not reach to 1 could be that the waves are reflected and scattered by the random heterogeneities.
2. The wave number does not start from 0, but 2. We can call this value the cut-off wave number. 2 is also the value of $\overline{\beta}$. From Eq. (15.45), we can see that the cut-off wave number equals the surface parameter.

Also, from Fig. 15.5, it can be seen that the phase velocity will decrease to 0 when the wave number decreases. This phenomenon agrees with the common knowledge that when the wave number decreases (the wave length increases), the effect of the random heterogeneities will be averaged out gradually, that is, the stochastically

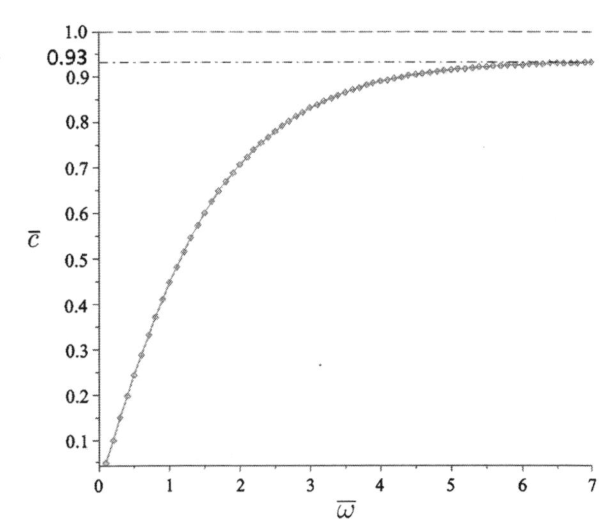

Fig. 15.6 Normalized phase velocity—normalized circular frequency

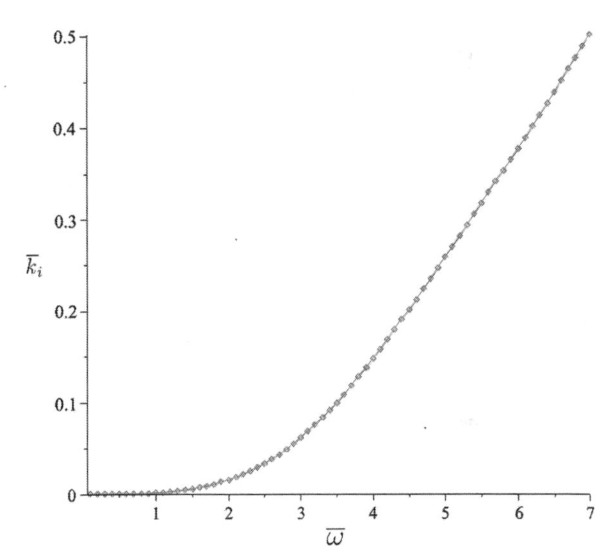

Fig. 15.7 Normalized imaginary wave number—normalized circular frequency

homogeneous half space will be more and more like a homogeneous half space, and we know that SHSW could not exist in a homogeneous half space, therefore, the phase velocity will decrease gradually to 0.

The imaginary wave number represents the attenuation rate. Therefore, we know from Fig. 15.7 that the bigger the circular frequency is, the faster the wave attenuates. This phenomenon should be caused by reflection and scattering. And from Fig. 15.4 we see that the wave length will decrease as the circular frequency grows. It is known that the smaller the wave length is, the easier the waves can be reflected or scattered by the random heterogeneities. Thus the wave should attenuates more fast as the frequency grows.

4 Conclusion

In this study, we proved that SHSW could exist in a stochastically homogeneous half space. The dispersion properties of SHSW in an half space with random density in the depth direction or only near the surface have been investigated both theoretically and numerically. The first order smoothing approximation method is used to solve the random differential equation. The dimensionless dispersion equation is obtained. And the dispersion properties is further studied numerically. The phase velocity is found increasing to an asymptotic value when the wave number is bigger than a critical value—the cut-off wave number, below which the phase velocity is 0. The interesting properties of dispersion and attenuation found here will help us understanding properties of waves in a half space with random heterogeneities, e.g. the earth's crust. It will also help us to do the inverse problems, for example, to use seismic waves to detect the earth's upper crust structure, and to extract information more exactly from the acoustic testing results.

Acknowledgements This study is supported by the National Natural Science Foundation of China under Grant no. 90916007.

References

1. Achenbach, J.: Wave Propagation in Elastic Solids. North-Holland, Amsterdam (1973)
2. Collet, B., Destrade, M., Maugin, G.: Bleustein–Gulyaev waves in some functionally graded materials. Eur. J. Mech. A, Solids **25**(5), 695 (2006)
3. Achenbach, J.D., Balogun, O.: Anti-plane surface waves on a half-space with depth dependent properties. Wave Motion **47**(1), 59 (2010)
4. Ting, T.: Existence of anti-plane shear surface waves in anisotropic elastic half-space with depth-dependent material properties. Wave Motion **47**(6), 350 (2010)
5. Shuvalov, A.L., Poncelet, O., Golkin, S.V.: Existence and spectral properties of shear horizontal surface acoustic waves in vertically periodic half-spaces. Proc. R. Soc., Math. Phys. Eng. Sci. **465**(2105), 1489 (2009)
6. Shuvalov, A., Poncelet, O., Kiselev, A.: Shear horizontal waves in transversely inhomogeneous plates. Wave Motion **45**(5), 605 (2008)
7. Qian, Z., Jin, F., Lu, T., Kishimoto, K.: Transverse surface waves in functionally graded piezoelectric materials with exponential variation. Smart Mater. Struct. **17**, 065005 (2008)
8. Eskandari, M., Shodja, H.: Love waves propagation in functionally graded piezoelectric materials with quadratic variation. J. Sound Vib. **313**(1–2), 195 (2008)
9. Cao, X., Jin, F., Wang, Z., Lu, T.: Bleustein–Gulyaev waves in a functionally graded piezoelectric material layered structure. Sci. China Ser. G, Phys. Mech. Astron. **52**(4), 613 (2009)
10. Frisch, U.: Wave Propagation in Random Media. Probabilistic Methods in Applied Mathematics. Academic Press, San Diego (1970)
11. Chen, K., Soong, T.: Covariance properties of waves propagating in a random medium. J. Acoust. Soc. Am. **49**, 1639 (1971)
12. Belyaev, A., Ziegler, F.: Uniaxial waves in randomly heterogeneous elastic media. Probab. Eng. Mech. **13**(1), 27 (1998)

Chapter 16
Structural Seismic Fragility Analysis of RC Frame with a New Family of Rayleigh Damping Models

Pierre Jehel, Pierre Léger, and Adnan Ibrahimbegovic

Abstract Structural seismic vulnerability assessment is one of the key steps in a seismic risk management process. Structural vulnerability can be assessed using the concept of fragility. Structural fragility is the probability for a structure to sustain a given damage level for a given input ground motion intensity, which is represented by so-called fragility curves or surfaces. In this work, we consider a moment-resisting reinforced concrete frame structure in the area of the Cascadia subduction zone, that is in the South-West of Canada and the North-West of the USA. According to shaking table tests, we first validate the capability of an inelastic fiber beam/column element, using a recently developed concrete constitutive law, for representing the seismic behavior of the tested frame coupled to either a commonly used Rayleigh damping model or a proposed new model. Then, for each of these two damping models, we proceed to a structural fragility analysis and investigate the amount of uncertainty to be induced by damping models.

Keywords Damping · Inelastic time-history analysis · Structural fragility · Fiber beam element · Reinforced concrete frame structure · Earthquake

---

P. Jehel (✉)
MSSMat, École Centrale Paris/CNRS, Grande Voie des Vignes, 92295 Châtenay-Malabry Cedex, France
e-mail: pierre.jehel@ecp.fr

P. Léger
Department of Civil Engineering, École Polytechnique of Montreal, University of Montreal Campus, P.O. Box 6079, Station CV, Montreal, QC, Canada H3C 3A7
e-mail: pierre.leger@polymtl.ca

A. Ibrahimbegovic
LMT-Cachan, ENS Cachan/CNRS/UPMC/PRES UniverSud Paris, 61 avenue du Président Wilson, 94235 Cachan Cedex, France
e-mail: ai@lmt.ens-cachan.fr

M. Papadrakakis et al. (eds.), *Computational Methods in Stochastic Dynamics*, Computational Methods in Applied Sciences 26,
DOI 10.1007/978-94-007-5134-7_16,

1 Introduction

Decision makers are interested in seismic risk analyses for predicting the post-earthquake situation in a given geographical region, so as to anticipate the human, social and economical impact of a major earthquake. For building structures, seismic risk assessment requires three main steps. (i) A seismic hazard analysis has first to be performed. It can be either deterministic or probabilistic. In the latter case, the seismic hazard is often expressed as an intensity measure—often the peak ground acceleration (*PGA*)—with a certain probability of being exceeded in a certain number of years. (ii) Then, the seismic fragility of the building considered has to be estimated: it corresponds to the conditional probability $P_{ij} = P[DI \geq DI_i | IM = IM_j]$ of the building to sustain a given damage index DI_i for a unique—or a set of—given input ground motion intensity measures IM_j. The probability to attain a damage index DI_i can then be computed as

$$P_i = \sum_j P[DI \geq DI_i | IM = IM_j] \cdot P[IM = IM_j]. \quad (16.1)$$

These damage indices have to be related to building performance requirements. (iii) Finally, the exposure of the buildings and populations has to be determined.

Structural seismic fragility analysis thus is a key step in the overall earthquake risk management process. This task is commonly achieved by constructing fragility curves from inelastic time-history analyses that take into account the variability in the seismic input motion alone or in both the input motion and the structural model. The main ingredients for fragility analyses are: (i) A set of seismic time-history records representative of the seismic hazard in the geographical region of interest for the project; (ii) An inelastic structural model along with a damping model; (iii) A mapping between damage indices and structural performance levels; and (iv) Statistical tools to analyze fragility curves.

In [1], Hwang and Huo present a methodology for constructing fragility curves accounting for uncertainties in the seismic, site, and structural parameters. On the one hand, 8 scenario earthquakes corresponding to different *PGA*, annual exceedance probabilities, magnitudes, and source-to-site distances are considered; for each scenario, 50 samples of ground motion time histories are generated using a seismological model that takes into account uncertainties in seismic and soils parameters. On the other hand, 50 samples of each of 6 random structural parameters are generated and then combined using the Latin Hypercube sampling technique to eventually generate 50 samples of the inelastic structural model. Then, for each earthquake scenario, the 50 ground motion samples are combined with the 50 structural samples to establish 50 earthquake-site-structure samples. Finally, for each earthquake scenario, 50 values of the damage index are computed from inelastic analyses and fragility curves are constructed.

The preceding approach defines a fully probabilistic approach in the sense that it takes into account uncertainty sources in both the seismic input motions and the inelastic structural behavior. It can however also be worth considering only uncertainties in ground motions. To that purpose, there exists, following Jalayer and Beck

[2], an alternative to the *IM-based approach* considered in this work: the *probabilistic ground motion time history approach*. This latter method is based on a stochastic ground motion model pertaining to seismic source parameters, which has to circumvent the most difficult drawback to be tackled in the *IM*-based approach, namely whether the selected set of *IM* thoroughly represent the input signal characteristics. On another hand, the recent work of Rosić *et al.* [3] considers the uncertain structural response of inelastic media and deterministic loading to provide maps of the probability for a component of the stress tensor to reach a given value at a given point.

The concept of fragility curve reduces the vulnerability analysis to the consideration of a unique intensity measure. This limitation has been pointed out in research work where the concept of fragility surfaces emerged. In a study dedicated to the analyze of the limitations of commonly used intensity measures for fragility analysis of single-degree-of-freedom linear and nonlinear systems [4], Kafali and Grigoriu propose to construct fragility surfaces instead of curves for assessing the seismic performance of nonlinear systems. For a given state of damage in the structure, the proposed fragility surface is the graphical representation of the relationship between the failure probability and the set of intensity parameters (m, r) constituted of the moment magnitude m and source-to-site distance r. This concept of fragility surface is also used in [5] by Seyedi *et al.* who extend it to other intensity measures. Indeed, they construct fragility surfaces that provide the probability for an inelastic reinforced concrete structure to sustain a given inter-story drift ratio, according to the spectral displacement at both eigenperiod T_1 and T_2. They finally conclude that when dealing with uncertainties propagation, fragility surfaces allow for estimating the variability of structural fragility due to a second *IM*, which should lead to more accurate seismic risk analyses.

Fragility curves have been used as a baseline to deal with a wide range of issues pertaining to earthquake engineering. In [6], Sáez *et al.* study the effect of considering inelastic dynamic soil-structure interaction on the seismic vulnerability analysis. They construct fragility curves from a very large number of artificially generated input earthquakes. They also introduce the so-called Fisher information concept which allows for measuring the amount of information contained in the seismic ground motions and thus provide a tool for the statistical analysis of fragility curves. In [7], Saxena *et al.* address the issue of assuming identical support ground motion in the analysis of the seismic response of long, multi-span, reinforced concrete bridges. Analyzing fragility curves, they show that considering spatial variation of earthquake ground motions is of first importance. In [8], Popescu *et al.* construct fragility curves to present the results of deterministic and both 2D and 3D stochastic analyses of the seismic liquefaction potential of saturated soil deposits. In the context of design rules assessments, Lagaros [9] computes fragility curves to analyze the seismic performance of multi-story RC buildings designed according to Greek and European building codes. The fragility curves are plotted from 10,000 simulations based on Monte Carlo techniques to take into account both uncertainty in the seismic signal and in key structural parameters for assessing structural stability.

There is a likely source of uncertainty in inelastic seismic time history analyses which is only rarely considered in fragility analyses, namely damping added to the inelastic structural model so as to introduce in the simulations an amount of energy dissipation coming from inelastic mechanisms that are not explicitly accounted for in the structural model. In [1], the critical viscous damping ratio added for the seismic analyses of a reinforced concrete frame building is described by the uniform distribution restricted to the range 2%–4%. In [10], the structural damping is described by a lognormal distribution with a mean of 2% and a coefficient of variation of 0.62 for the purpose of steel frame fragility analysis. In both these works, Rayleigh damping is added and the uncertainty pertaining to additional damping thus stems from the critical damping ratio, not from the damping model type—Rayleigh in this case. Because it has been shown that it can be difficult to control the amount of damping generated by common Rayleigh damping models throughout inelastic time history analyses [11–13], the main purpose of the work presented in this chapter is to investigate the likely amount of uncertainty introduced by the damping model, in the context of fragility analysis.

To that aim, we proceed as follows. In the next section, we first present a reinforced concrete moment-resisting frame structure—simply referred to as "RC frame" throughout this chapter—which was tested on a shaking table. We use the corresponding experimental data as a reference to validate the developments that we present all along this chapter. Then, we detail in Sect. 3 the numerical model that we use to perform inelastic seismic time-history analyses of the RC frame. The issue of modeling damping is discussed and a new family of Rayleigh damping models is proposed. Results from simulations performed with both a "classical" Rayleigh damping model and the proposed new damping model are compared to the shaking table test results so as to validate the capability of the proposed combination of hysteretic with additional damping models for representing the behavior of the RC frame. The proposed new family of Rayleigh damping models can rely on a physical background which often lacks to commonly used Rayleigh damping models. In Sect. 4, we proceed to the selection of a set of real seismic ground motion records compatible with the seismic hazard in the Cascadia subduction zone. A seismic fragility analysis of the RC frame in this geographical region is then carried out in Sect. 5: fragility curves, along with their statistical analysis, as detailed in [6], are constructed for every damping models considered so as to investigate the amount of uncertainty these latter could bring in structural fragility analyses.

2 RC Frame Tested on a Shaking Table

The test structure considered throughout this chapter is represented in Fig. 16.1. It was designed at a reduced scale of $1/2$ according to the provisions of the National Building Code of Canada [14] and of the Canadian concrete standard [15]. The structure was assumed to have a nominal ductility, which corresponds to a force reduction factor $R = 2$ to compute the design base shear. The various assumptions and parameters used in the design of the this structure can be found in [16]. Four inverted U-shape concrete blocks attached in each span of the beams were used to

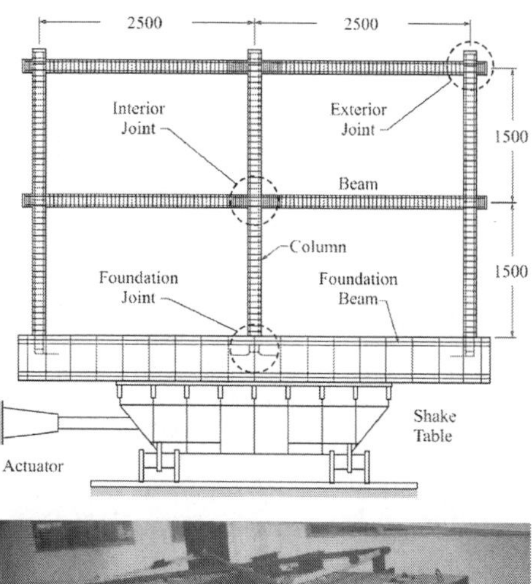

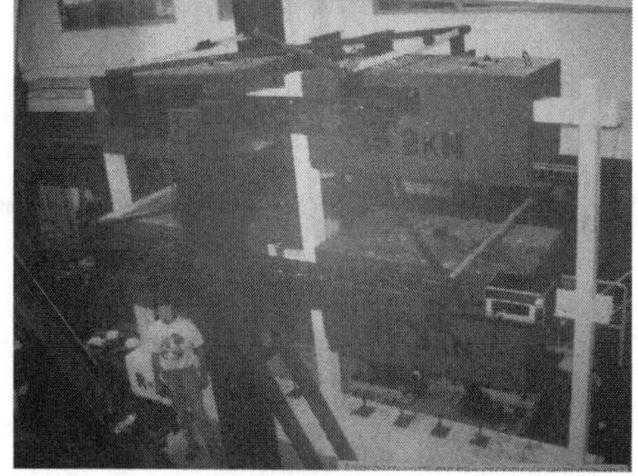

Fig. 16.1 RC frame structure tested on the shaking table at École Polytechnique in Montreal. Dimensions are in [mm]

simulate concentrated gravity loads from framing joints. The centers of gravity of the added masses were computed such that they coincide with the center of gravity of the beams. Service cracks were induced by these added masses. The total weight of the frame was 95 kN. The fundamental period T_1 of the structure with added masses was measured at 0.36 s in a free-vibration test. Mode 1 excites 91% of the total mass of the structure and when mode 2 is also considered approximately all the mass is excited.

The structure was assumed to be located in a seismic zone 4 in Canada, as depicted in the 1985 seismic zoning map of the National Building Code of Canada. The seismic hazard in this zone is such that peak horizontal ground acceleration between 0.16 g and 0.23 g is likely to be observed with 10% probability of exceedance in 50 years. Such seismic zones can be found in Western, Eastern and Northern Canada. The ground motion record that was selected for the test program corresponds to the N04W component of the accelerogram recorded in Olympia, Washington on April 13, 1949. Figure 16.2 presents the feedback record measured during the test initially scaled to a peak ground acceleration $PGA = 0.21\ g$ as well as the corresponding elastic response spectrum with 5% viscous damping ratio.

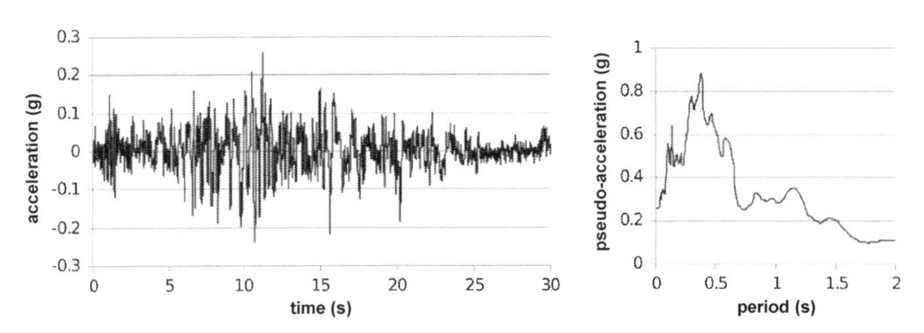

Fig. 16.2 Acceleration time history recorded on the shaking table during the test and corresponding elastic response spectrum with a critical viscous damping ratio of 5%

3 Seismic Inelastic Time History Analysis of the RC Frame

The set of equations of motion for the discretized structure is written as:

$$\mathbf{M}\ddot{\mathbf{d}}(t) + \mathbf{C}(t)\dot{\mathbf{d}}(t) + \mathbf{F}_R(t) = -\mathbf{M}\boldsymbol{\Delta}\ddot{u}_g(t) \tag{16.2}$$

where $\mathbf{d}(t)$ is the vector containing the nodal displacements, $\mathbf{M}$ is the mass matrix, $\mathbf{C}(t)$ is the damping matrix, $\mathbf{F}_R(t)$ is the inelastic resisting forces vector, $\boldsymbol{\Delta}\ddot{u}_g(t)$ is the vector of the rigid body acceleration induced by the ground displacement $u_g(t)$. In this section, we first present the inelastic structural model used to compute $\mathbf{F}_R(t)$, then we define two damping models: a "classical" Rayleigh model $\mathbf{C}_1(t)$ and a new model $\mathbf{C}_2(t)$, and we finally validate the capability of the two models for representing the behavior of the RC frame presented in the previous section.

3.1 Fiber Beam/Column Element

The inelastic structural model is based on a fiber frame element suitably implemented in the framework of a displacement-based formulation so that it can integrate the uniaxial concrete behavior law recently developed by the authors [17] and briefly presented in this section. This constitutive model is capable of representing the main energy dissipative phenomena likely to occur in concrete: appearance of permanent deformation, strain hardening and softening, stiffness degradation, local hysteresis loops, appearance of cracks. Its theoretical development and numerical implementation are based on thermodynamics with internal variables [18, 19] and on the finite element method with embedded strong discontinuities [20–22]. FEAP [23] is the finite element program used for the numerical implementation of the developments presented in this section.

Enhanced Kinematics The first ingredient of this model is the definition of an enhanced kinematics that takes strong discontinuities into account. This is done, as depicted in Fig. 16.3 by writing the displacement field $u(\mathbf{x},t)$ as the sum of

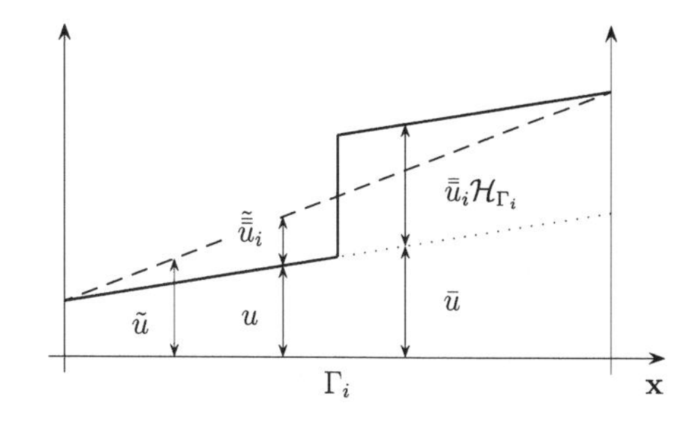

Fig. 16.3 Construction of an enhanced displacement field $u(\mathbf{x}, t)$ as the sum of a continuous displacement $\bar{u}(\mathbf{x}, t)$ and of a displacement jump $\bar{\bar{u}}_i(t)\mathcal{H}_{\Gamma_i}(\mathbf{x})$ pertaining to discontinuity Γ_i

a continuous displacement $\bar{u}(\mathbf{x}, t)$—that is the displacement as it would be in the absence of strong discontinuity—and of displacement jumps $\bar{\bar{u}}_i(t)$ in sections Γ_i of the solid domain Ω:

$$u(\mathbf{x}, t) = \bar{u}(\mathbf{x}, t) + \sum_{i=1}^{n_{dis}} \bar{\bar{u}}_i(t)\mathcal{H}_{\Gamma_i}(\mathbf{x}) \tag{16.3}$$

where $\mathcal{H}_{\Gamma_i}(\mathbf{x})$ is the Heaviside's function which, for a left-to-right oriented domain, is null on the left side of the discontinuity Γ_i and unity on its right side.

With the hypothesis of small transformation, we have the following expression for the normal strain field:

$$\epsilon_{xx}(\bar{u}, \bar{\bar{u}}_i, t) = \frac{\partial \bar{u}(\mathbf{x}, t)}{\partial \mathbf{x}} + \sum_{i=1}^{n_{dis}} \bar{\bar{u}}_i(t)\delta_{\Gamma_i}(\mathbf{x}) \tag{16.4}$$

where $\delta_{\Gamma_i}(\mathbf{x})$ is the Dirac's function.

Stored Energy Function When the Lagrange's variational principle is for instance chosen to derive the governing equations of the system (see Eq. (16.2) for the corresponding discretized form), the internal potential energy U^{int} as to be written:

$$\begin{aligned} U^{int}(\bar{u}, \boldsymbol{\alpha}, t) &= \int_{\Omega} \psi(\bar{u}, \boldsymbol{\alpha}, t)\, d\Omega \\ &= \sum_{f=1}^{n^c_{fib}} \int_{\Omega^c_f} \psi^c(\bar{u}, \boldsymbol{\alpha}^c, t)\, d\Omega^c_f + \sum_{f=1}^{n^s_{fib}} \int_{\Omega^s_f} \psi^s(\bar{u}, \boldsymbol{\alpha}^s, t)\, d\Omega^s_f \end{aligned} \tag{16.5}$$

where $n^{c,s}_{fib}$ is the total number of concrete or steel fibers, $\Omega^{c,s}_f$ is the volume of the fiber, $\psi^{c,s}$ is the stored energy function for concrete or steel which depends on the continuous displacement field $\bar{u}(\mathbf{x}, t)$ and on the set of internal variables $\boldsymbol{\alpha}^{c,s}$. Normal stresses are computed from these functions as

$$\sigma_{xx} = \frac{\partial \psi}{\partial \epsilon_{xx}}. \tag{16.6}$$

Set of Internal Variables The set of internal variables $\boldsymbol{\alpha}$ is defined to characterize the evolution of the main energy dissipative—inelastic—mechanisms which develop in the system. These internal variables are the memory of the system. The

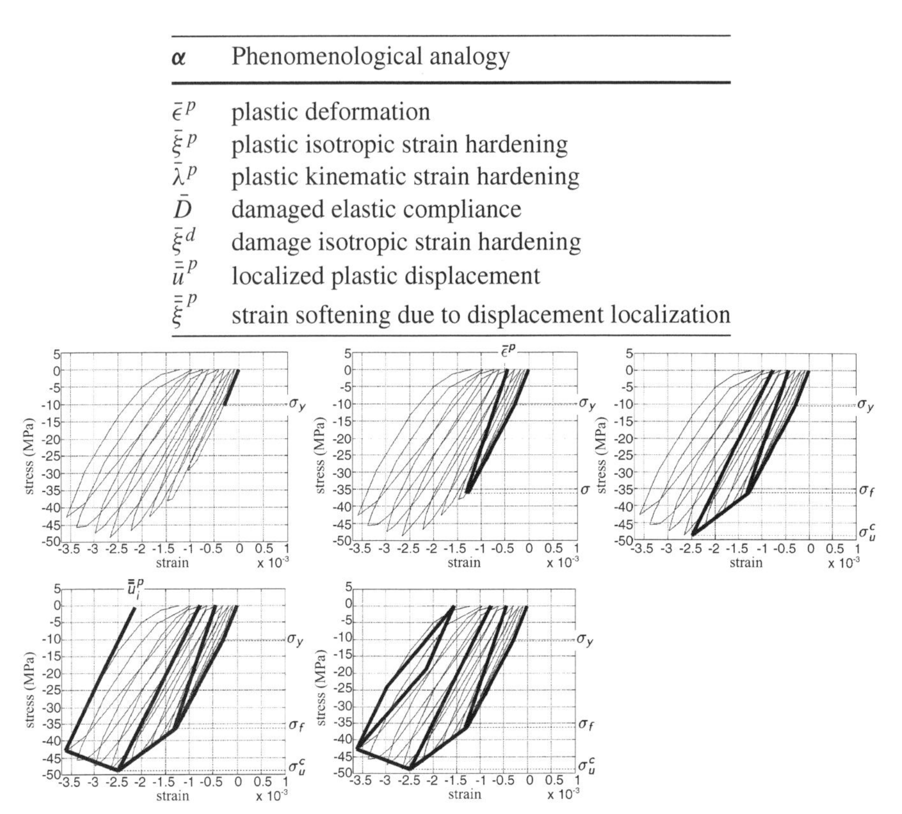

$\boldsymbol{\alpha}$	Phenomenological analogy
$\bar{\epsilon}^p$	plastic deformation
$\bar{\xi}^p$	plastic isotropic strain hardening
$\bar{\lambda}^p$	plastic kinematic strain hardening
$\bar{D}$	damaged elastic compliance
$\bar{\xi}^d$	damage isotropic strain hardening
$\bar{\bar{u}}^p$	localized plastic displacement
$\bar{\bar{\xi}}^p$	strain softening due to displacement localization

Fig. 16.4 Internal variables for compression and their phenomenological interpretation. [*top, left*] Elastic response until yield stress σ_y is reached. [*top, centre*] In the plastic domain, plastic deformation ($\bar{\epsilon}^p$) and strain hardening ($\bar{\xi}^p$) develop. [*top, right*] Once the limit stress σ_f is reached, damage mechanisms are activated too, leading to a degradation of stiffness ($\bar{D}$) and a change in the strain hardening evolution ($\bar{\xi}^d$). [*bottom, left*] Once the ultimate stress σ_u is reached, deformation localizes ($\bar{\bar{u}}^p$) and strain-softening is observed ($\bar{\bar{\xi}}^p$). [*bottom, center*] Local hysteresis loops are represented with kinematic hardening in the plastic domain ($\bar{\lambda}^p$). Note that each parameter of this model has a clear interpretation regarding the constitutive law to be identified

physical interpretation of each of them is provided in Fig. 16.4. Note that the constitutive law used here can handle different behavior in compression and tension, and can also reproduce a visco-elastic response (see [17] for a full description). Viscosity is not considered in this work and, for the sake of conciseness, we only focus on the compressive part of the behavior law in Fig. 16.4. The local admissible state of the system is expressed according to criteria functions in the stress-like domain of the set of variables dual to $\boldsymbol{\alpha}$. When irreversible mechanisms are activated in the structure, internal variables have to be updated and their evolution is governed by the principle of maximum dissipation. From the computational point of view, because we only consider linear hardening and softening laws, there is no need for

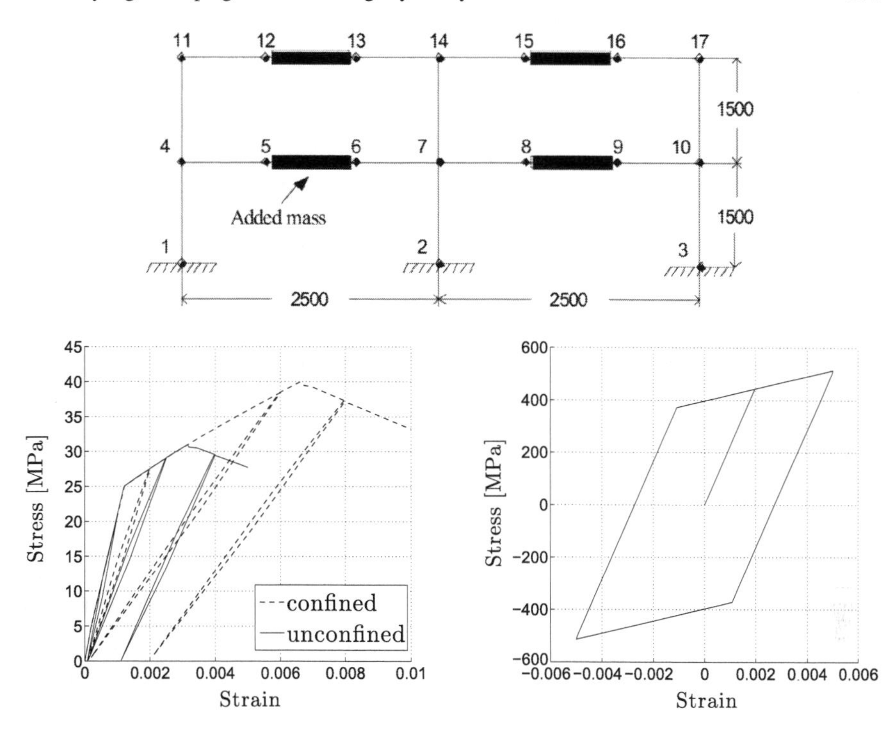

Fig. 16.5 Finite element mesh (dimensions in [mm]) and material constitutive laws for the inelastic structural modeling. [*bottom left*] Confined and unconfined concrete behavior laws. [*bottom right*] Steel constitutive law

local iteration when internal variables are updated, except for transitions between hardening and softening regimes, which leads to an efficient resolution procedure.

3.2 Inelastic Structural Model

The finite element mesh and the uniaxial constitutive laws for steel and both confined and unconfined concrete fibers used for the inelastic structural model of the RC frame are shown in Fig. 16.5. Material behavior laws have been identified to fit experimental monotonic ($\sigma_{xx} - \epsilon_{xx}$) curves. The structure is assumed fixed at its base. Rigid end zones are defined to model the beam-to-column connections and rebar slip in surrounding concrete is not represented. These later hypotheses are questionable because the connections exhibit inelastic behavior during the test. This obvious limitations of the structural model has to be reminded when damping model is added in the simulation.

The loading time history consists in two successive phases: (i) Static dead load is first applied step by step and then kept constant; (ii) The seismic loading is applied. A first validation check of the inelastic structural model is carried out by simulating a free vibration test. Before dead load is applied, the elastic fundamental period of the structure is computed as $T_1^{ela} = 0.28$ s; then, due to the inelastic behavior of the structure, the elongated period is evaluated as $T_1^{ini} = 0.36$ s when dead load is completely applied. Both T_1^{ela} and T_1^{ini} coincide with the experimental values.

3.3 Proposition of a New Family of Rayleigh Damping Models

Basic Definitions In the context of inelastic time history numerical analysis, the definition of damping might differ according to the reference cited. On the one hand, in [24], damping consists in both (i) *inherent damping* resulting from the dissipation of energy by inelastic structural elements, and (ii) *additional viscous damping* added in the simulation to take into account inherent energy dissipation sources not otherwise explicitly considered in the inelastic structural model. On the other hand, in [25], damping is defined as "the portion of energy dissipation that is not captured in the hysteretic response of components that have been included in the model", and it is then suggested in [25] to use "un-modeled energy dissipation" as a more appropriate terminology for damping.

Because, in experimental investigations, measured damping results from all the energy dissipative phenomena, we herein decide to define damping as the combination of both *hysteretic damping* due to the energy dissipated by all the inelastic phenomena explicitly accounted for in the structural model and *additional viscous damping* that should be consistent with the inelastic structural model namely, that does not introduce energy dissipation already accounted for in the inelastic structural model.

Problems Encountered with Rayleigh Damping Controlling the amount of additional viscous damping energy dissipated in inelastic time history analyses is a very challenging task [11–13]. This is especially the case for commonly used Rayleigh proportional damping models, that is when the damping matrix is computed, in its most general form, as

$$\mathbf{C}(t) = \alpha(t)\mathbf{M} + \beta(t)\mathbf{K}(t), \tag{16.7}$$

where $\mathbf{K}(t)$ is the tangent stiffness matrix. Several researchers have provided insight in the comprehension of Rayleigh damping regarding the inelastic structural model it is coupled to, have highlighted limitations, and have eventually provided recommendations to cope with them [11–13]. Nevertheless, adding damping and controlling its consistency with the inelastic structural model still remains an issue to be addressed.

Three Common Phases in Seismic Response We now discuss in a qualitative way the notion of *consistency* for additional viscous damping. To that purpose, we start by stating that seismic structural response is composed by three main consecutive phases, as illustrated in Fig. 16.6. Both inelastic structural model and additional damping model must then be capable of representing the salient phenomena

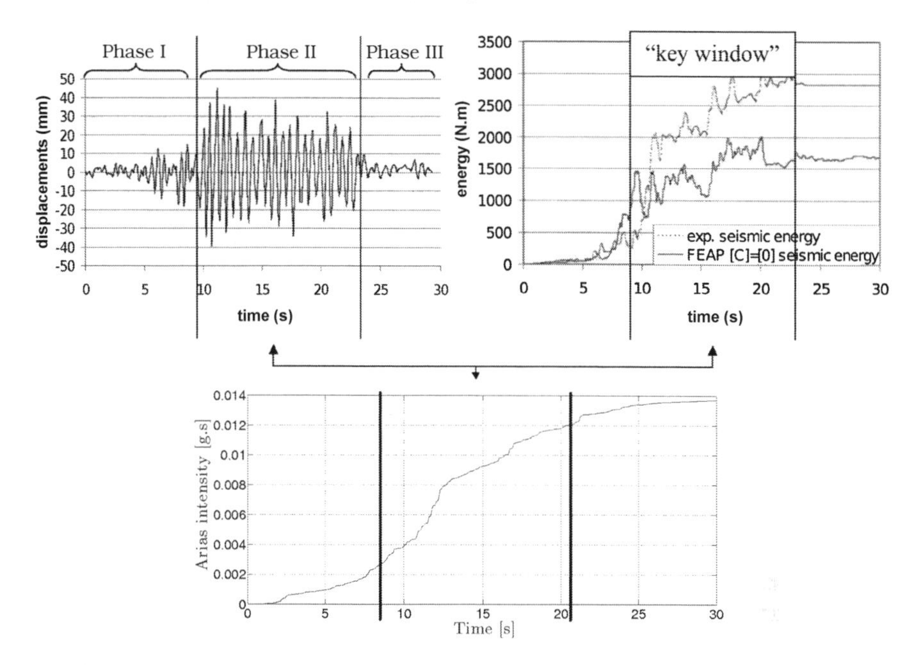

Fig. 16.6 Three common phases in the seismic response and the concept of "key window". [*top left*] Structural relative displacement time-history. [*top right*] Total relative seismic energy time-history in the structure. [*bottom*] Arias intensity of the seismic signal: $AI(t) = \frac{\pi}{2g} \int_0^t \ddot{u}_g^2(\tau)\, d\tau$

corresponding to each of these three phases. Foremost has to be properly modeled what we call here the "key window", namely the time interval within which the major inelastic modifications for structural performance assessment develop. For instance, key mechanisms that control near-collapse structural behavior are listed in [26]: degradation of strength and stiffness, and structure P-delta effects. From experimental results, we know that strain rate is another major issue.

A consistent additional damping model should be adapted to each of these three phases as follows:

- Phase 1: None or only few incursions in the inelastic domain occur. Energy dissipation in phase 1 thus comes from the friction in the cracks that appeared when applying dead load and from many other mechanisms always present in mechanical systems. When used, visco-elasticity and constitutive laws with local hysteresis [17, 27] in the structural model could account for these energy dissipation sources but a small amount of additional damping usually has to be added. For the RC frame considered in this work, a free-vibration test was carried out after dead load had been applied and a first modal damping ratio of 3.3% was measured [16].
- Phase 2: As the ground motion becomes stronger (at around 8 s in Fig. 16.6), an important amount of seismic energy is imparted to the structure and some parts of the structure then exhibit inelastic behavior. Inelastic structural models

are designed to explicitly model part of the numerous inherent nonlinear energy dissipative mechanisms involved in the structural response. The energy dissipation due to the mechanisms not explicitly accounted for in the inelastic structural model has to be introduced with the additional damping model.
- Phase 3: The structure has suffered irreversible degradations that modified its dynamic properties. Thus, even if the seismic demand is again as low as in phase 1, the energy dissipative mechanisms are different because of frictions in the cracks that appeared within phase 2 or at degraded bound between steel and concrete. Here again, visco-elasticity and behavior laws with local hysteresis [17, 27] in the structural model could account for these damping sources, but it generally has to be completed by additional damping.

Proposition of a New Family of Rayleigh Damping Models In the following, two damping models will be used:

- A commonly used Rayleigh model based on tangent stiffness matrix and with two constant coefficients

$$\mathbf{C}_1(t) = \alpha\mathbf{M} + \beta\mathbf{K}(t); \tag{16.8}$$

- We propose a new family of models that is directly dependent on both the two key notions in the definition of the three phases introduced above: the capacity of the inelastic structural model to absorb energy and the seismic demand. The model is based on Rayleigh damping with tangent stiffness matrix and with coefficients adapted to each of the three phases:

$$\mathbf{C}_2(t) = \alpha(t)\mathbf{M} + \beta(t)\mathbf{K}(t) \tag{16.9}$$

The idea of adapting Rayleigh damping to the capabilities of the inelastic structural model for dissipating energy is present in the use of the tangent stiffness rather than the initial one: it is expected that the choice of tangent stiffness dependent damping will have the main advantage of providing the significant additional source of damping only in the domains/modes that are not accounted for by inelastic model. Such a choice allows to provide the physically based damping phenomena interpretation, which leads to damping coefficients that are easier to identify. The same idea has been further exploited in [28] where 1% viscous damping is added to an inelastic dam model before cracking and 10% after cracking to represent localized high dissipation by friction between crack lips. Another instance is the work presented in [29] where viscous damping is added only in the shear wall zones which remain elastic, while no damping is added in the inelastic zones where the structural model is let alone to dissipate the seismic energy.

In spite of its stronger physical background, implementing damping model $\mathbf{C}_2(t)$ is not as straightforward as damping model $\mathbf{C}_1(t)$. First, three sets of Rayleigh coefficients $(\alpha_p, \beta_p)_{p=1,2,3}$ corresponding to each of the three phases p have to be identified to define the appropriate damping ratios ξ. Second, the instants which delimit the three phases have to be determined, which can be automatically accomplished by the computer program that is capable of detecting the activation of significant inelastic behavior.

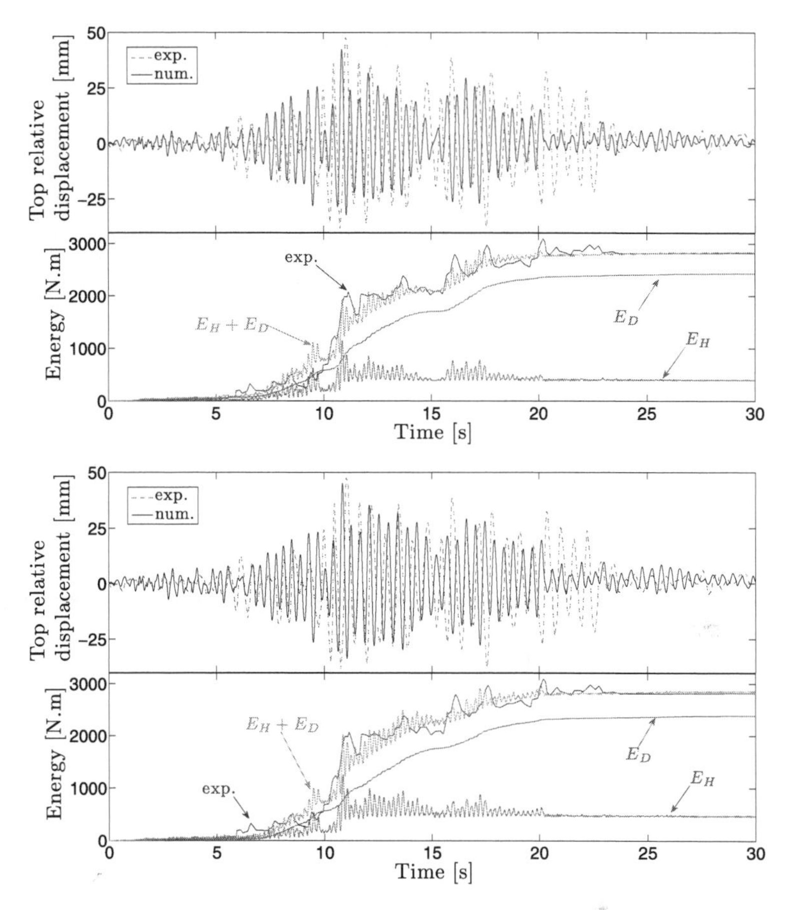

Fig. 16.7 Experimental and simulated top-displacement; simulated hysteretic plus elastic energies (E_H), simulated additional damping energy (E_D), and both experimental and simulated total internal work. [*top*] With common added damping model $\mathbf{C}_1(t)$; [*bottom*] With the proposed new family of Rayleigh damping models $\mathbf{C}_2(t)$. The structural responses shown here for damping models $\mathbf{C}_1(t)$ and $\mathbf{C}_2(t)$ looks very similar because both models have been calibrated to experimental data; however, model $\mathbf{C}_2(t)$ has more capability for representing transient evolution of added damping

3.4 Calibration and Validation of the Models

Seismic inelastic time history analyses of the RC frame have been carried out with the inelastic structural model presented above coupled to either additional damping model $\mathbf{C}_1(t)$ or $\mathbf{C}_2(t)$. The implicit Newmark integration scheme with parameters $\beta = 0.25$ and $\gamma = 0.5$ is used with a time step of 0.005 s. Figure 16.7 shows a comparison between the simulated top-displacement and energies time histories and the respective experimental results reproduced from [30]. Good agreement between simulated and experimental data can be observed. Moreover, there is very good agreement between the hysteretic plus elastic (E_H) and damping (E_D) energy quantities computed with the models proposed here and an analogous Perform3D [31] simulation we carried out for comparison purpose, namely $E_H \approx 550$ N.m and $E_D \approx 2250$ N.m, corresponding to approximately 20% and 80% of the total work done by the structure during seismic motions. Last, the fundamental period of the RC frame in post-earthquake conditions is estimated by a free-vibration test performed at the end of the seismic signal and it is observed that simulated value is $T_1^{sim} = 0.45$ s whereas the experimental measure comes to $T_1^{exp} = 0.55$ s. This shows, as expected by regarding the limitations of the structural model used (elastic beam to column connections and no rebar slip), that not all structural stiffness degradation mechanisms are always well represented by the inelastic structural model. In particular, the joints and supports often need special attention and more elaborate models.

For damping model $\mathbf{C}_1(t)$, the good results shown in Fig. 16.7 have been obtained with α and β computed so that $\xi_1 = \xi_2 = 3.3\%$. For damping model $\mathbf{C}_2(t)$, curves plotted in Fig. 16.7 have been obtained with the following parameters identified so as to obtain good match between experimental and simulated responses:

- Phase 1: from $0 \le t \le t_1$, $\xi_1 = \xi_2 = 1.0\%$. t_1 is defined such that two conditions are satisfied. First, the hysteretic energy which is dissipated by the inelastic response of the structural model has to reach—for the frame considered in this work—150 N.m. Then, the seismic demand must be such that the increase in Arias intensity *AI* [32] within the time range $[t_1; t_1 + 10 \times T_1^{ini}]$ is larger than 0.0025 g.s; T_1^{ini} is the fundamental period of the structure after dead load has been applied and before the earthquake ($T_1^{ini} = 0.36$ s for the structure considered here).
- Phase 2: from $t_1 \le t \le t_2$, $\xi_1 = \xi_2 = 4.0\%$. t_2 is defined as $t_2 = t_1 + 10 \times T_1^{ini}$.
- Phase 3: from $t_2 \le t \le \bar{T}$, $\xi_1 = \xi_2 = 2.5\%$, where $\bar{T}$ is the duration of the seismic signal.

It might happen that one of these two criteria is never satisfied within the earthquake duration. In such a case, only phase 1 is effectively active throughout the analysis.

In the rest of this chapter, we focus only on the new model with changing coefficients, for it is very likely to deliver better prediction from the standard Rayleigh damping models given its more sound physical basis. The damping parameters defined above are expected to be suitable to model the response of the frame structure for every seismic signal that will be used in the next sections, as we usually proceed with damping model $\mathbf{C}_1(t)$. In the next sections, we then investigate likely consequences of using damping model $\mathbf{C}_2(t)$ instead of $\mathbf{C}_1(t)$ when it comes to structural fragility analysis.

4 Selection of a Set of Real Ground Motions

Following [33], there are three basic options available for obtaining accelerograms for inelastic time-history analysis: (i) to use spectrum-compatible synthetic accelerograms with realistic energy, duration and frequency content; (ii) to use synthetic accelerograms generated from seismological source model and accounting for path and site effects; and (iii) to use real accelerograms recorded during earthquakes. We chose this latter option and present in the following how we proceed to select ground motion time histories in the PEER ground motion database [34].

4.1 Likely Earthquake Scenarios in the Cascadia Subduction Zone

Ground motion time history recorded from the Olympia, Washington 1949 earthquake has been used for the shaking table test at École Polytechnique of Montreal [16]. It is then assumed that the frame structure considered in this investigation is located in the Cascadia subduction zone. The Western margin of the North American plate—from the North of California state up to Vancouver island, British Columbia, Canada—is characterized by the so-called Cascadia subduction zone, where it is subducted by the Juan de Fuca plate beneath the Pacific ocean.

The seismic activity in the Cascadia subduction zone has been investigated for several decades [35–38]. Three types of earthquakes are produced in this zone:

- *Shallow crustal earthquakes* are associated to surface faults in the American continental plate with magnitude M_w larger than 7.0 and hypocenter depth less than 30 km;
- *Thrust interplate or interface earthquakes* are due to differential motion in the interface between the Juan de Fuca and the North American plates. They happen offshore with surface hypocenter, generally with depth less than 30 km. The Cascadia subduction zone has the potential to produce a large event of $M_w = 8.3 \pm 0.5$;
- *Intraplate or intraslab earthquakes* occur deep within the Cascadia subduction zone (depth > 40 km) beneath the Puget Sound of Western Washington state. These types of earthquakes have occurred frequently including in 1949 Olympia ($M_w = 6.8$), 1965 Seattle–Tacoma ($M_w = 6.8$), 1999 Satsop ($M_w = 5.9$) and 2001 Nisqually ($M_w = 6.8$);
- No seismicity has been observed for depth larger than 100 km.

4.2 Search for a Real Ground Motions Dataset

To that purpose, we use the PEER ground motion database with its Web application [34]. Currently, this database is limited to recorded time series from shallow crustal earthquakes only. A basic criterion used by the Web application to select a representative acceleration time history is that its elastic response spectrum provides a good match to a user target spectrum over a range of periods of interest. We define the target spectrum as the elastic response spectrum corresponding to the feedback accelerogram recorded on the shaking table during the test (see

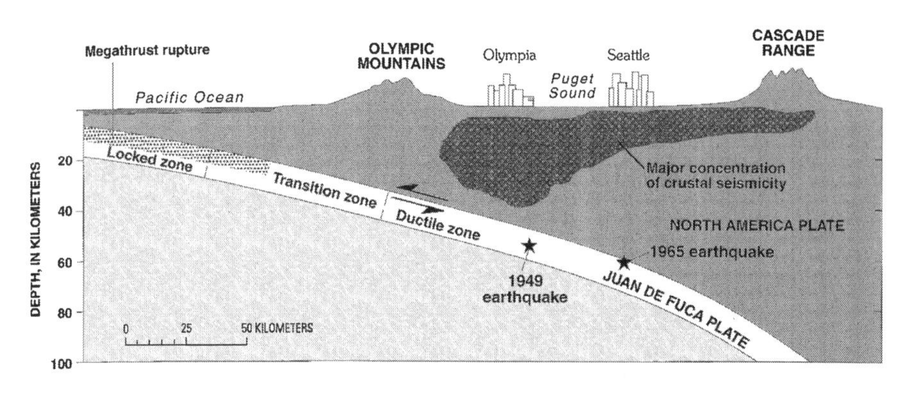

Fig. 16.8 Cross section of the Cascadia subduction zone (adapted from [36])

Table 16.1 Geologic profile for the Olympic Highway Test Lab strong-motion recording site in Washington [36]. v_s is the shear wave velocity and ρ the density

Depth [m]	Geology	Description	v_s [m/s]	ρ [kg/m^3]
0–3	Fill	Loose sand	165	1500
3–12	Deposits	Medium dense fine to medium sand	220	1500
12–20	Deposits	Interbedded very stiff to hard sandy silt and very dense silty fine to medium sand	270	1500
20–41	Deposits	(*same as layer just above*)	330	1500

Fig. 16.2). The Web application allows for assigning different weights to different period ranges so that the matching process is guided by the period ranges with the higher weights. In our case, the response of the structure is governed by its first eigenmode and the fundamental period was experimentally measured to vary within the range $0.36\text{ s} \leq T_1 \leq 0.55\text{ s}$ corresponding to the pre- and post-seismic states. We consequently favor the target and actual spectra to match within this period range.

Then, several parameters—or acceptance criteria—can be input into the Web application to characterize the likely earthquake scenarios of interest for the study. According to (i) what was stated in the previous Sect. 4.1, (ii) the cross section of the Cascadia subduction zone depicted in Fig. 16.8, and (iii) the geological profile of the Puget Sound region described in Table 16.1, we define suitable ranges for these parameters as summarized in Table 16.2. Practically, due to the above-mentioned current limitation of the PEER database, we only look for ground motion records corresponding to shallow crustal earthquakes.

Because a first search with these criteria led to a selection of around 20 earthquakes only, we proceeded to a second search with the range of allowed moment magnitudes extended to $6.0 \leq M_w \leq 9.0$. Then, we only retained 48 earthquakes which had the best fitting coefficients with the target response spectrum and finally multiply each of them by a scale factor of 5.0. Note that PEER ground motion database provides the fault normal (FN) and fault parallel (FP) components of the

Table 16.2 Acceptance criteria for initial ground motions search with the PEER database Web application [34]. M_w is the moment magnitude, R_{JB} is the Joyner–Boore distance, R_{rup} is the closest distance to rupture plane and v_{s30} is the average shear wave velocity in the top 30 meters of the site. All fault types are considered

Earthquake:	crustal	interface	intraslab
M_w	[7.0, 9.0]	[7.8, 8.8]	[5.5, 7.0]
R_{JB} [km]	[0, 150]	[30, 200]	[30, 200]
R_{rup} [km]	[0, 150]	[30, 100]	[30, 100]
v_{s30} [m/s]	[0, 200]	[0, 200]	[0, 200]

seismic signal and that the 48 records we selected either corresponds to the FN or FP component pertaining to 48 different earthquakes.

5 Seismic Fragility Analysis

In this section, we focus on the vulnerability analysis of the test RC frame structure presented in Sect. 2. Uncertainty is only considered in the seismic loading: the selected time-history ground motions that are likely to occur in the Cascadia subduction zone are used as inputs of inelastic time-history deterministic simulations to compute fragility curves. The damping model—either $\mathbf{C}_1(t)$ or $\mathbf{C}_2(t)$—is the only variable considered in the RC frame model.

5.1 Theoretical Background

Structural vulnerability analysis is evaluated here by computing fragility curves which provide the conditional probability for a structure to sustain a given damage level for a given earthquake intensity. Following [39], it is assumed that the fragility curve can be expressed in the form of a two-parameter lognormal distribution function. The estimation of these two parameters is then performed with the maximum likelihood method.

Let consider a sample of n independent observations x_j that can be classified in two classes as "success" ($x_j = 1$) and "failure" ($x_j = 0$). Let also consider that each realization x_j has a relative frequency of success equal to $F(IM_j, \theta)$, that is a function that depends on the intensity measure IM_j associated to each realization x_j and of a set of parameters θ. Then, the probability of observing a set of realizations $\mathbf{x} = (x_1, \ldots, x_n)$ composed of p successes $x_1, \ldots, x_p$ and $n - p$ failures $x_{p+1}, \ldots, x_n$, whatever the order, is ([40], p. 77):[1]

[1]Henri Poincaré (1854–1912) is a French mathematician, physician and philosopher. This year is the hundredth anniversary of his death.

$$f(\mathbf{x}|\theta) = \frac{n!}{p!(n-p)!}\big(F(IM_j,\theta)\big)^p \cdot \big(1 - F(IM_j,\theta)\big)^{n-p} \qquad (16.10)$$

The problem one has to solve can be expressed as: *Given the observed data* $\mathbf{x}$*, find the set of parameters* θ *that is most likely to have produced these observed data.* To solve this inverse problem, we define the likelihood function as a function of θ given $\mathbf{x}$:

$$\mathcal{L}(\theta|\mathbf{x}) = f(\mathbf{x}|\theta). \qquad (16.11)$$

The principle of maximum likelihood estimation states that: given the data $\mathbf{x}$ actually observed, the set of parameters θ looked for is the one that makes $\mathbf{x}$ the most likely data to be observed. θ can thus be identified by maximizing the likelihood function $\mathcal{L}$. For computational convenience, the log-likelihood $\ln\mathcal{L}(\theta|\mathbf{x})$ is introduced and maximized, which provides the same estimators because $\ln x$ is a monotonic function. The problem one has to solve thus reads:

$$\theta_e = \arg\max_{\theta} \ln\mathcal{L}(\theta|\mathbf{x}) \qquad (16.12)$$

Under the lognormal assumption, the fragility curve for a particular damage index DI_i—defining what is "success" and "failure"—is defined as:

$$F(IM_j,\theta_e) = \phi\left(\frac{1}{\zeta_e}\ln\frac{IM_j}{c_e}\right) = P[DI \geq DI_i|IM_j] \qquad (16.13)$$

where $\theta_e = \{\zeta_e, c_e\}$ is the set of estimated parameters and $\phi(\cdot)$ is the standardized normal distribution function.

Constructing fragility curves in such a framework raises issues concerning their statistical significance. In [39], Shinozuka *et al.* provide tools to test the goodness of fit between the inferred fragility curve and the realization of the random variable X_j following Bernoulli distribution: $X_j = 1$ when the damage index is reached and $X_j = 0$ otherwise. They also present a Monte Carlo technique they use to demonstrate the extent of the statistical variations in the estimators θ. Another very important contribution for the statistical analysis of fragility curves is the work of Sáez *et al.* [6]. They show how to compute the amount of Fisher information about the set of parameters θ—the terms of the Fisher information matrix being by definition $\mathcal{F}_{ij}(\theta) = \mathrm{cov}(\frac{\partial \ln\mathcal{L}}{\partial\theta_i}; \frac{\partial \ln\mathcal{L}}{\partial\theta_j})$, where $\mathrm{cov}(\cdot)$ denotes the covariance—which is provided by the ground motions used to construct fragility curves, and then how to compute a lower bound for the standard deviation of the elements of θ_e. This method thus provides a way to somehow measure the ability of the data to estimate θ.

5.2 Intensity Measures and Structural Damage Indices

To characterize the seismic ground motion time-histories used for structural vulnerability analysis, they are assigned intensity measures (*IM*). 18 of them are reviewed in [41] in the context of the issue of selecting earthquakes for incremental dynamic analysis of inelastic steel frame structures. 44 *IM* are reviewed or proposed for masonry structures in [42]. Among the most common *IM*, one finds: the peak ground acceleration (*PGA*); the acceleration at the fundamental period $S_A(T_1, 5\%)$; the Arias intensity (*AI*) which is effectively a measure of the total energy in the ground motion and computed as

$$AI(t) = \frac{\pi}{2g} \int_0^t \ddot{u}_g^2(\tau)\,d\tau; \tag{16.14}$$

the significant duration D_{5-95}, which the time needed to build up between 5% and 95% of the total Arias intensity. Other measures are: the root mean square of acceleration (RMSA) computed as [41]:

$$RMSA = \sqrt{\frac{1}{\tau_d}\frac{2g}{\pi}AI}, \tag{16.15}$$

where τ_d is an effective duration of the record, taken here as D_{5-95}; spectrum intensities such as [42]:

$$SI_V = \int_{T_a}^{T_b} S_V(T, \xi)\,dT, \tag{16.16}$$

where S_V is the spectral velocity at period T and with damping ξ. Here, we take $T_A = 0.36$ s and $T_B = 0.55$ s which are the fundamental periods in pre- and post-earthquake conditions. For fragility surfaces, Seyedi *et al.* [5] chose the spectral displacement $S_D(T_1)$ and $S_D(T_2)$ where T_1 and T_2 are the two main eigenperiods in the direction along which the ground motions are applied.

Then, to characterize the structural response, damage indices (*DI*) are used. One can distinguish between three categories: (i) *DI* computed from energy quantities such as the Park–Ang–Wen damage accumulation model [43] or the normalized hysteretic energy used in [44] or [41]; (ii) Other *DI* based on quantities directly related to the structural inelastic response such as the ductility demand [41], the cumulative ductility index as defined in [44] or the maximum strength degradation ratio [42]; (iii) Other quantities not necessarily pertaining to—but affected by—the inelastic structural behavior such as the maximum relative roof displacement (*MRD*) or the maximum inter-story drift ratio (*MISDR*). To these two latter indices, because particular attention is paid in this work on the energy dissipated by both the damping and the inelastic structural models, we also compute the amount of hysteretic energy (E_H) dissipated by the inelastic mechanisms explicitly accounted for in the structural model and the amount of damping energy (E_D), as well as their respective ratio $E_{H,D}R = E_{H,D}/(E_H + E_D)$.

5.3 *Fragility Curves*

We first investigate in Table 16.3 the correlation between the intensity measures and damage indices considered in this work, when a linear model is used to predict a

Table 16.3 Square of the coefficient of correlation R^2 (in %) between actual *DI* and *DI* obtained from a linear model between *IM* and *DI*, for added damping models $\mathbf{C}_1(t)/\mathbf{C}_2(t)$

	MRD	*MISDR*	$E_{D/H}R$	E_D	E_H
PGA	73/71	71/69	3.5/0.6	51/52	36/37
AI	68/68	68/68	9.1/5.9	86/83	61/62
D_{5-95}	8/7	8/7	1.3/1.0	1/1	2/2
RMSA	75/73	74/73	8.2/4.5	57/55	43/45
$S_{A,V,D}(T_1, 5\%)$	85/82	84/82	5.7/2.6	68/68	53/55
$SI_{A,V,D}(5\%)$	77/74	77/75	8.5/4.6	57/57	48/51

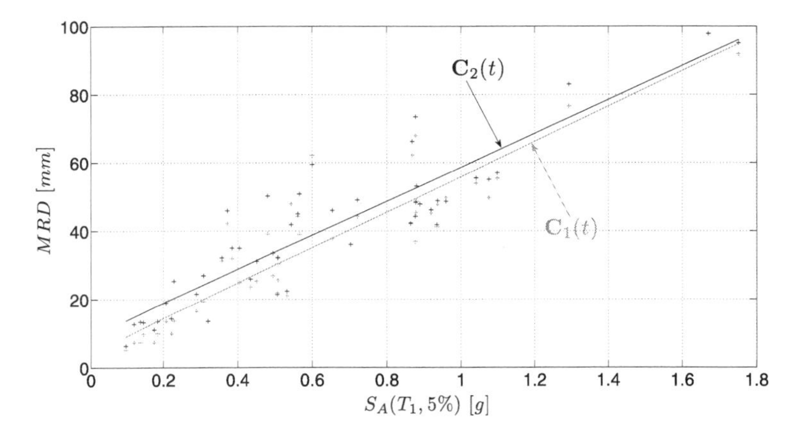

Fig. 16.9 Linear model between $S_A(T_1, 5\%)$ and *MRD* for both damping models $\mathbf{C}_1(t)$ and $\mathbf{C}_2(t)$

DI from an *IM*. Correlation is the lowest for intensity measure D_{5-95} because there is no explicit influence of the earthquake duration in the various *DI* considered. The correlation for the energy ratios is poor for all the *IM* considered, which is not the case for the dissipated energy quantities where, in particular, correlation is good with the energy contained in the seismic signal (*AI*). The very good correlated maximum roof displacement and spectral displacement at fundamental period T_1 with 5% viscous damping—as depicted in Fig. 16.9—will be used in the following to construct fragility curves.

In Fig. 16.9, one can see an obvious tendency to obtain larger maximum roof displacement with damping model $\mathbf{C}_2(t)$. Constructing fragility curves for a given *MRD* level along with a proper statistical analysis provides another way to infer some likely consequences of using a damping model instead of another. To that purpose, Fig. 16.10 shows the fragility curves (solid lines) pertaining to *DI* *MRD* ≥ 40 mm obtained with both damping models $\mathbf{C}_1(t)$ and $\mathbf{C}_2(t)$. Dashed lines approximately represent the smallest area the fragility curves would describe when its parameters $\theta = \{\zeta, c\}$ varies around $\theta_e = \{\zeta_e, c_e\}$ plus or minus their standard de-

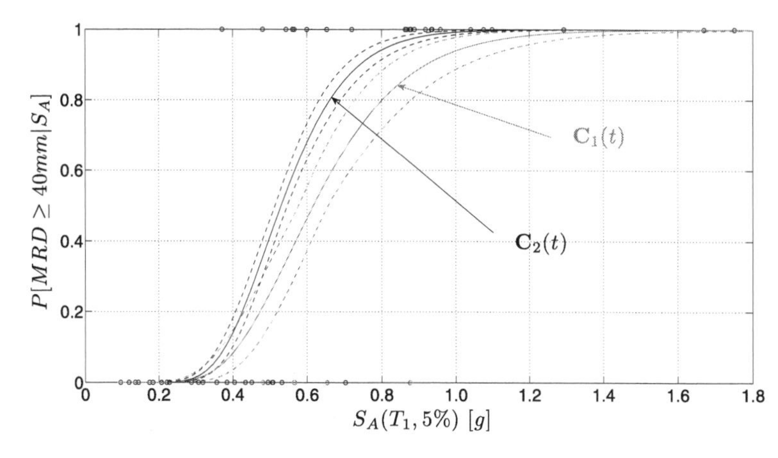

Fig. 16.10 Structural fragility curves (*solid lines*) for both damping models $\mathbf{C}_1(t)$ and $\mathbf{C}_2(t)$. *Dashed lines* are plotted according to the method presented in [6] and show that, for the *DI* level considered here, the fragility curves are constructed from a number of ground motions which is sufficient for drawing pertinent conclusions from this comparative study. *Circles* represent the realizations of the Bernoulli random variable $X_i = 1$ when $RMD_i \geq 40$ mm and $X_i = 0$ otherwise, $i \in [1; 48]$

viation. Dashed lines are constructed as in [6], relying on the inverse of the Fisher information matrix provided by the selected ground motion about θ.

From Fig. 16.10, one can infer that the additional damping model entails very significant uncertainty in structural fragility analysis of inelastic RC frame structures. Figure 16.11 is shown to provide better insight into the discrepancies one can expect to observe in the structural response when either damping model $\mathbf{C}_1(t)$ or $\mathbf{C}_2(t)$ is used. Both seismic signals considered have the same intensity measure $S_A(T_1, 5\%) = 0.51$ g. For the analysis in concern on the left part of Fig. 16.11, either the maximum capacity of the structure to store hysteretic energy $E_H = 150$ N.m or the minimum seismic demand $0.0025/(t_2 - t_1)$ is not reached, so that, for damping model $\mathbf{C}_2(t)$, only the phase 1 with $\xi_1 = \xi_2 = 1\%$ is effectively active throughout the analysis; this makes both damping models analogous but with critical damping ratios in sharp contrast—3.3% against 1%—leading to very different structural responses. For the analysis in concern on the right part of Fig. 16.11, phase 2 begins at $t_1 = 15.4$ s until $t_2 = 19$ s. Both models predict almost identical *MRD* but oscillation amplitude is more rapidly attenuated for model $\mathbf{C}_1(t)$.

6 Conclusions and Perspectives

The inelastic structural response in seismic loading results from the combination of an inelastic structural model with an added damping model. The damping model has to be consistent with the inelastic structural model, which implies that it has, and only has, to model the energy dissipation sources not otherwise explicitly accounted

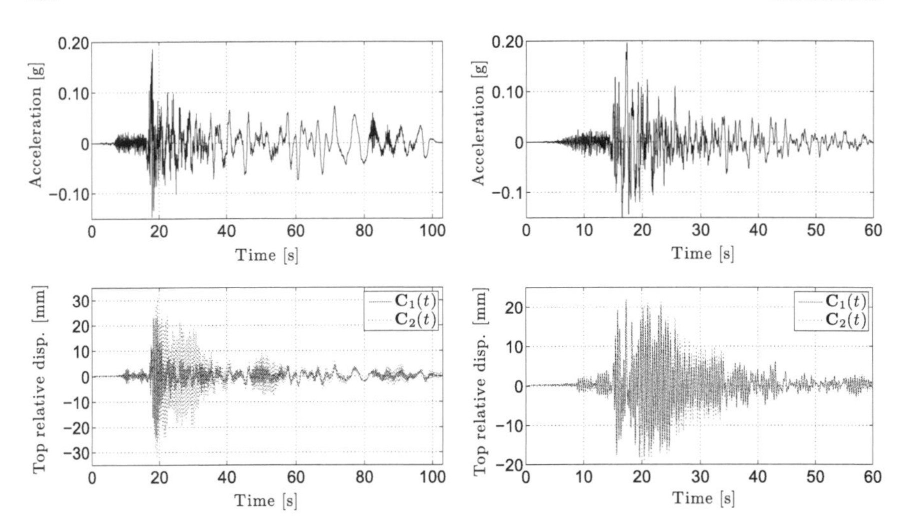

Fig. 16.11 [*left*] Analysis where only phase 1 is activated for damping model $\mathbf{C}_2(t)$. [*right*] Analysis where three phases are accounted for in damping model $\mathbf{C}_2(t)$. [*top*] Ground motion time-history. [*bottom*] Top relative displacement time-history; Model $\mathbf{C}_1(t)$ with $\xi_1 = \xi_2 = 3.3\%$ all along the simulation; [*bottom left*] Model $\mathbf{C}_2(t)$ with $\xi_1 = \xi_2 = 1\%$ all along the simulation; [*bottom right*] Model $\mathbf{C}_2(t)$ with $\xi_1 = \xi_2 = 1\%$ for $0 \leq t \leq 15.4$ s, $\xi_1 = \xi_2 = 4\%$ for $15.4\ \text{s} \leq t \leq 19$ s and $\xi_1 = \xi_2 = 2.5\%$ for $19\ \text{s} \leq t \leq 60$ s

for in the inelastic structural model, nothing more or less. Rayleigh damping models are the most commonly used for earthquake engineering applications, although it is well established that controlling the amount of energy dissipation these models introduce throughout inelastic time history simulations is difficult to achieve.

On another hand, inelastic time history analyses are widely used for structural fragility assessment in seismic loading. Uncertainties arising from the seismic signal likely to excite a building in a given geographical region as well as from the lack of knowledge on the structural parameters which characterize the inelastic structural model have both been considered in structural fragility analyses. On the opposite, the damping model is scarcely considered as a source of uncertainty; moreover, when it is the case, the damping ratio is assigned a probabilistic distribution but whether commonly used added damping models are suitable or not is an issue that is not regarded. This is the issue addressed in the work presented above.

In this chapter, we propose a new family of Rayleigh damping models that relies both on the capacity of the inelastic structural model to absorb energy and on the seismic demand and compare it to a commonly used Rayleigh model in the context of the fragility analysis of a RC moment-resisting frame. Its inelastic response is modeled by fiber elements using a constitutive law recently developed by the authors [17]. The structure is supposed to be built in the Cascadia subduction zone, a seismically active zone in the South-West of Canada and the North-West of the USA. From this comparative analysis, it can be inferred that the added damping model entails very significant uncertainty in structural fragility analysis of inelastic RC frame structures.

As a further development, we seek the proposed new family of Rayleigh damping models to be confronted to other experimental evidence, criticized and improved. Albeit constructed on a stronger physical basis than commonly used damping models, there is yet no clear guarantee that this model remains consistent with the inelastic structural model it is coupled to, throughout inelastic time history analysis. Finally, such a proposition for a damping model should not eclipse the need for further improvement of the inelastic structural model, which would, for the case treated in this work, at least involve explicitly accounting for the inelastic response of the beam to column connections.

Acknowledgements The authors thank Pr. André Filiatrault for providing the data from the shaking table tests used in this work. The first author benefited from partial funding from Électricité de France (EDF) within the research project "MARS" ("Modèles Avancés pour le Risque Sismique").

References

1. Howard, H.H.M., Huo, J.-R.: Generation of hazard-consistent fragility curves. Soil Dyn. Earthq. Eng. **13**(5), 345–354 (1994)
2. Jalayer, F., Beck, J.L.: Effects of two alternative representations of ground-motion uncertainty on probabilistic seismic demand assessment of structures. Earthquake Eng. Struct. Dyn. **37**, 61–79 (2008)
3. Rosic, B.V., Matthies, H.G., Zikovic, M., Ibrahimbegovic, A.: Formulation and computational application of inelastic media with uncertain parameters. In: Oñate, E., Owen, D.R.J. (eds.) Proceedings of the 10th International Conference on Computational Plasticity (COMPLAS X), CIMNE, Barcelona (2009)
4. Kafali, C., Grigoriu, M.: Seismic fragility analysis: application to simple linear and nonlinear systems. Earthquake Eng. Struct. Dyn. **36**, 1885–1900 (2007)
5. Seyedi, D.M., Gehl, P., Douglas, J., Davenne, L., Mezher, N., Ghavamian, S.: Development of seismic fragility surfaces for reinforced concrete buildings by means of nonlinear time-history analysis. Earthquake Eng. Struct. Dyn. **39**, 91–108 (2010)
6. Sáez, E., Lopez-Caballero, F., Modaressi-Farahmand-Razavi, A.: Effect of the inelastic dynamic soil-structure interaction on the seismic vulnerability assessment. Struct. Saf. **33**(1), 51–63 (2011)
7. Saxena, V., Deodatis, G., Shinozuka, M., Feng, M.Q.: Development of fragility curves for multi-span reinforced concrete bridges. In: Proceedings of International Conference on Monte Carlo Simulation, Monte-Carlo, Monaco (2000)
8. Popescu, R., Prevost, J.H., Deodatis, G.: 3d effects in seismic liquefaction of stochastically variable soil deposits. Géotechnique **55**(1), 21–31 (2005)
9. Lagaros, N.D.: Probabilistic fragility analysis: a tool for assessing design rules of rc buildings. Earthq. Eng. Eng. Vib. **7**(1), 45–56 (2008)
10. Ellingwood, B.R.: Earthquake risk assessment of building structures. Reliab. Eng. Syst. Saf. **74**, 251–262 (2001)
11. Léger, P., Dussault, S.: Seismic-energy dissipation in MDOF structures. J. Struct. Eng. **118**(6), 1251–1267 (1992)
12. Hall, J.F.: Problems encountered from the use (or misuse) of Rayleigh damping. Earthquake Eng. Struct. Dyn. **35**, 525–545 (2006)
13. Charney, F.A.: Unintended consequences of modeling damping in structures. J. Struct. Eng. **134**(4), 581–592 (2008)

14. Associate Committee on the National Building Code: National building code of Canada. Technical report, National Research Council of Canada, Ottawa, Ontario, Canada (1995)
15. Design of concrete structures for buildings. Standard CAN-A23.3-94, Canadian Standards Association, Rexdale, Ontario, Canada (1994)
16. Filiatrault, A., Lachapelle, É., Lamontagne, P.: Seismic performance of ductile and nominally ductile reinforced concrete moment resisting frames. I. Experimental study. Can. J. Civ. Eng. **25**, 331–341 (1998)
17. Jehel, P., Davenne, L., Ibrahimbegovic, A., Léger, P.: Towards robust viscoelastic-plastic-damage material model with different hardenings/softenings capable of representing salient phenomena in seismic loading applications. Comput. Concr. **7**(4), 365–386 (2010)
18. Germain, P., Nguyen, Q.S., Suquet, P.: Continuum thermodynamics. J. Appl. Mech. **50**, 1010–1020 (1983)
19. Maugin, G.A.: The Thermodynamics of Nonlinear Irreversible Behaviors—An Introduction. World Scientific, Singapore (1999)
20. Garikipati, K., Hughes, T.J.R.: A study of strain localization in a multiple scale framework—the one-dimensional problem. Comput. Methods Appl. Mech. Eng. **159**, 193–222 (1998)
21. Ibrahimbegovic, A., Brancherie, D.: Combined hardening and softening constitutive model of plasticity: precursor to shear slip line failure. Comput. Mech. **31**, 89–100 (2003)
22. Oliver, J., Huespe, A.E.: Theoretical and computational issues in modelling material failure in strong discontinuity scenarios. Comput. Methods Appl. Mech. Eng. **193**, 2987–3014 (2004)
23. Taylor, R.L.: FEAP: A Finite Element Analysis Program, User Manual and Programmer Manual, Version 7.4. University of California Press, Berkeley (2005)
24. Applied Technology Council: Quantification of building seismic performance factors. Technical Report FEMA P695, Federal Emergency Management Agency, Washington, DC (June 2009)
25. Applied Technology Council: Modeling and acceptance criteria for seismic design and analysis of tall buildings. Technical Report PEER 2010/111 or PEER/ATC-72-1, Pacific Earthquake Engineering Research Center, Richmond, CA (October 2010)
26. Krawinkler, H.: Importance of good nonlinear analysis. Struct. Des. Tall Spec. Build. **15**, 515–531 (2006)
27. Ragueneau, F., La Borderie, C., Mazars, J.: Damage model for concrete-like materials coupling cracking and friction, contribution towards structural damping: first uniaxial applications. Mech. Cohes.-Frict. Mater. **5**, 607–625 (2000)
28. Tinawi, R., Léger, P., Leclerc, M., Cipolla, G.: Seismic safety of gravity dams: from shake table experiments to numerical analyses. J. Struct. Eng. **126**(4), 518–529 (2000)
29. Luu, H., Ghorbanirenani, I., Léger, P., Tremblay, R.: Structural dynamics of slender ductile reinforced concrete shear walls. In: EURODYN 2011 (2011)
30. Filiatrault, A., Lachapelle, É., Lamontagne, P.: Seismic performance of ductile and nominally ductile reinforced concrete moment resisting frames. II. Analytical study. Can. J. Civ. Eng. **25**, 342–351 (1998)
31. CSI: Perform3D User's manual. Technical report, California (2007)
32. Arias, A.: A measure of earthquake intensity. In: Seismic Design for Nuclear Power Plants, pp. 438–483. MIT Press, Cambridge (1970)
33. Bommer, J.J., Acevedo, A.B.: The use of real earthquake accelerograms as input to dynamic analysis. J. Earthq. Eng. **8**(S1), 43–91 (2004)
34. Users manual for the PEER ground motion database web application. Technical report, Pacific Earthquake Engineering Research Center (November 2011)
35. Baker, G.E., Langston, C.: Source parameters of the 1949 magnitude 7.1 South Puget Sound, Washington, earthquake as determined from long-period body waves and strong ground motion. Bull. Seismol. Soc. Am. **77**, 1530–1557 (1987)
36. Silva, W.J., Wong, I.G., Darragh, R.B.: Engineering characterization of earthquake strong motions in the Pacific Northwest. In: Assessing Earthquake Hazards and Reducing Risk in the Pacific Northwest, vol. 2. U.S. Geological Survey Professional Paper 1560, pp. 313–324. United States Government Printing Office, Washington (1998)

37. Saragoni, G.R., Concha, P.: Damaging capacity of Cascadia subduction earthquakes compared with Chilean subduction. In: 13th World Conference on Earthquake Engineering, Vancouver, BC, Canada, August 1–6 (2004)
38. Wiest, K.R., Doser, D.I., Velasco, A.A., Zollweg, J.: Source investigation and comparison of the 1939, 1946, 1949 and 1965 earthquakes, Cascadia subduction zone, western Washington. Pure Appl. Geophys. **164**, 1905–1919 (2007)
39. Shinozuka, M., Feng, M.Q., Jongheon, L., Naganuma, T.: Statistical analysis of fragility curves. J. Eng. Mech. **126**(12), 1224–1231 (2000)
40. Poincaré, H.: Calcul des Probabilités. Cours de la Faculté des Sciences de Paris—Cours de Physique mathématique, 2nd edn. Gauthier-Villars, Paris (1912) (in French)
41. Léger, P., Kervégant, G., Tremblay, R.: Incremental dynamic analysis of nonlinear structures: selection of input ground motions. In: Proceedings of the 9th U.S. National and 10th Canadian Conference on Earthquake Engineering, Toronto, Ontario, Canada, July 25–29 (2010)
42. Bommer, J., Magenes, G., Hancock, J., Penazzo, P.: The influence of strong-motion duration on the seismic response of masonry structures. Bull. Earthq. Eng. **2**, 1–26 (2004). doi:10.1023/B:BEEE.0000038948.95616.bf
43. Park, Y.J., Ang, A.H.S., Wen, Y.K.: Seismic damage analysis of reinforced concrete buildings. J. Struct. Eng. **111**(4), 740–757 (1985)
44. Castiglioni, C.A., Pucinotti, R.: Failure criteria and cumulative damage models for steel components under cyclic loading. J. Constr. Steel Res. **65**, 751–765 (2009)

Chapter 17
Incremental Dynamic Analysis and Pushover Analysis of Buildings. A Probabilistic Comparison

Yeudy F. Vargas, Luis G. Pujades, Alex H. Barbat, and Jorge E. Hurtado

Abstract Capacity-spectrum-based-methods are also used for assessing the vulnerability and risk of existing buildings. Capacity curves are usually obtained by means of nonlinear static analysis. Incremental Dynamic Analysis is another powerful tool based on nonlinear dynamic analysis. This method is similar to the pushover analysis as the input is increasingly enlarged but it is different as it is based on dynamic analysis. Moreover, it is well known that the randomness associated to the structural response can be significant, because of the uncertainties involved in the mechanical properties of the materials, among other uncertainty sources, and because the expected seismic actions are also highly stochastic. Selected mechanical properties are considered as random variables and the seismic hazard is considered in a probabilistic way. A number of accelerograms of actual European seismic events have been selected in such a way that their response spectra fit well the response spectra provided by the seismic codes for the zone where the target building is constructed. In this work a fully probabilistic approach is tackled by means of Monte Carlo simulation. The method is applied to a detailed study of the seismic response of a reinforced concrete building. The building is representative for office buildings in Spain but the procedures used and the results obtained can be extended to other types of buildings. The main purposes of this work are (1) to analyze the differences when static and dynamic techniques are used and (2) to obtain a measure of the uncertainties involved in the assessment of the vulnerability of structures. The results show that static based procedures are somehow conservative and that uncertainties increase with the severity of the seismic actions and with the damage. Low dam-

Y.F. Vargas (✉) · L.G. Pujades
Department of Geotechnical Engineering and Geosciences, Technical University of Catalonia (BarnaTech), Jordi Girona 1-3, Building D2, Campus Norte UPC, 08034 Barcelona, Spain
e-mail: yeudy.felipe.vargas@upc.edu

A.H. Barbat
Structural Mechanics Department, Technical University of Catalonia (BarnaTech), Jordi Girona 1-3, Building C1, Campus Norte UPC, 08034 Barcelona, Spain

J.E. Hurtado
National University of Colombia, Apartado 127, Manizales, Colombia

M. Papadrakakis et al. (eds.), *Computational Methods in Stochastic Dynamics*, Computational Methods in Applied Sciences 26,
DOI 10.1007/978-94-007-5134-7_17,

age state fragility curves have little uncertainty while high damage grades fragility curves show great scattering.

1 Introduction

To prevent the seismic risk, it is necessary to assess the vulnerability of existing structures. To do that, several methods have been proposed, starting from different approaches. One is the vulnerability index method in which the action is defined by EMS-98 macroseismic intensities and structural behaviour through a vulnerability index [1, 2]. Another highly used method is based on the capacity spectrum. In this, the seismic action is defined by means of the 5% damped elastic response spectra and the vulnerability or fragility of the building by using the capacity curve. Capacity curves are calculated from an incremental nonlinear static analysis, commonly known as "Pushover Analysis" (PA) [3–5]. Another tool used to evaluate the performance of structures against seismic actions is the Incremental Dynamic Analysis (IDA) proposed by Vamvatsikos & Cornell [6]. The purpose of IDA is to obtain a measure of damage in the structure by increasing the intensity of the action record, in this case the peak ground acceleration. Vamvatsikos & Cornell makes an interesting analogy between PA and IDA, as both procedures are based on incremental increases of the loads on the structure and on the measure of its response in terms of a control variable which usually is the maximum displacement at the roof or the maximum inter storey drift, among others. Furthermore IDA allows obtaining the dynamic response of a structure for increasing seismic actions. On the other hand, most of the parameters involved in the structural response are random variables. In this work only the randomness due to the mechanical properties of the materials and the seismic action is considered. The randomness expected in the vulnerability and fragility of the building is analysed by means of Monte Carlo techniques. Therefore, a probabilistic comparison between the PA and the IDA is performed when calculating the fragility and expected damage of an existing reinforced concrete building. The main result of this work is the quantitative assessment of the expected randomness of the structural response, defined by its capacity curve, as well as of the fragility curves and the expected damage, which can be given in terms of mean values and standard errors. The damage assessment through nonlinear static procedures is tested against the results of fully nonlinear dynamic analyses. One of the main conclusions of this work is the importance of measuring the vulnerability of structures taking into account that the variables involved are random. Furthermore, this approach incorporates detailed information about the building and uses powerful tools to analyze the structure such as the PA and the IDA, providing valuable key information that can hardly be obtained with other simplified methods in which the building and the seismic actions are defined by only one parameter.

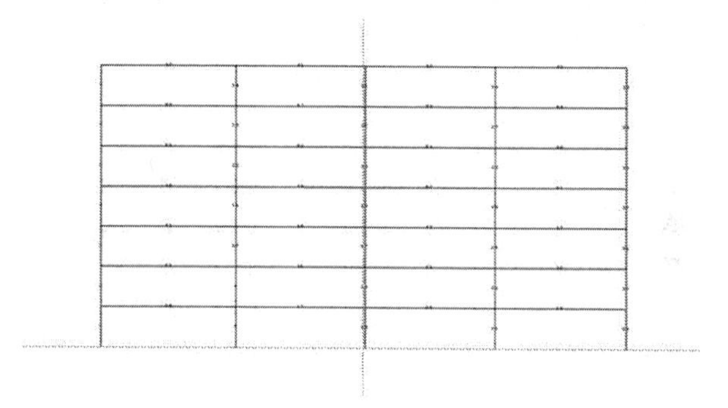

Fig. 17.1 Picture of the block of buildings omega located in the Technical University of Catalonia (BarnaTech) (*above*) and sketch of the 2D structural model (*below*)

2 Building Description

This paper analyzes a reinforced concrete structure, consisting of columns and waffle slabs, which is part of the North Campus of the Technical University of Catalonia in Barcelona, Spain. It has 7 stories and 4 spans, the height is 24.35 m and the width is 22.05 m. Figure 17.1 shows a block of four buildings as the analyzed one. In the first building 5 levels can be clearly seen; the other two stories are under the ground. The fundamental period of the building is 0.97 seconds. This value is higher when compared to that of conventional reinforced concrete buildings, because in the numerical model, the waffle slabs are approximated with beams of equivalent inertia and, therefore, are structural elements wide and flat leading to a reduction of the lateral stiffness of the structure. In the calculation model, the structural elements (equivalent beams and columns) follow an elasto-plastic constitutive law, which does not take into account either hardening or softening. Yielding surfaces are defined by the bending moment-axial load interaction diagram in columns and by the bending moment-angular deformation interaction diagram in beams.

Table 17.1 Parameters defining the Gaussian random variables considered in this work

	Mean Value (kPa)	Standard deviation (kPa)	Coefficient of variation
fc	25000	2500	0.1
fy	500000	50000	0.1

3 Damage Index Based on Pushover Analysis

Pushover analysis is the tool more often used to evaluate the behaviour of the structures against seismic loads. This numerical tool consists in applying horizontal loads to the structure, according to a certain pattern of forces and increasing its value until the structural collapse is reached. The result is a relationship between the displacement at the roof of the building and the base shear, called capacity curve. In this article, due to the probabilistic approach, the PA is performed repeatedly, therefore, it is appropriate to apply a procedure for obtaining automatically the horizontal load limit. To do that, Satyarno [7] proposes the adaptive incremental nonlinear analysis that establishes the horizontal load limit as a function of the tangent fundamental frequency, i.e. the frequency associated with the first mode of vibration, which is being calculated for each load increment. Therefore, the first mode of vibration to determine the shape of the load in height is calculated in each step. A detailed description of this procedure is found in the manuals of the program Ruaumoko [8] used here for calculating the static and dynamic nonlinear structural response. As mentioned above, the mechanical properties of the materials are considered as random variables. The impact of epistemic uncertainties in the structural response has been treated by Crowley et al. in [9] by considering the variation of the ground floor storey height, column depth and beam length, among others. The aim of that article is to generalize the results for a structural typology. In the present study, the aim is to obtain a measure of the uncertainties in the structural response for one building and, for this reason, we consider only the epistemic uncertainties associated to the compressive strength of concrete and the yield strength of steel. Thus, the values used in the structural design for concrete compressive strength *fc*, and the tensile strength associated with steel yielding strength *fy*, are considered as random variables assuming they follow a Gaussian probability function whose parameters are shown in Table 17.1. For the Monte Carlo analysis 1000 random samples are generated by means of the inversion method of the cumulative probability distribution curve. This method warranties the homogeneous distribution of the samples. Figure 17.2 shows the capacity curves obtained by means of the PA analysis.

The capacity curves shown in Fig. 17.2 are transformed into capacity spectra, which relate the spectral displacement to spectral acceleration by means of the following equations [10]:

$$sd_i = \frac{\delta_i}{PF_1}; \qquad sa_i = \frac{V_i/W}{\alpha_i} \tag{17.1}$$

The subscript i in Eq. (17.1) refers to the applied load increments on the structure during the PA; sd_i is the spectral displacement; δ_i is the displacement at the roof of

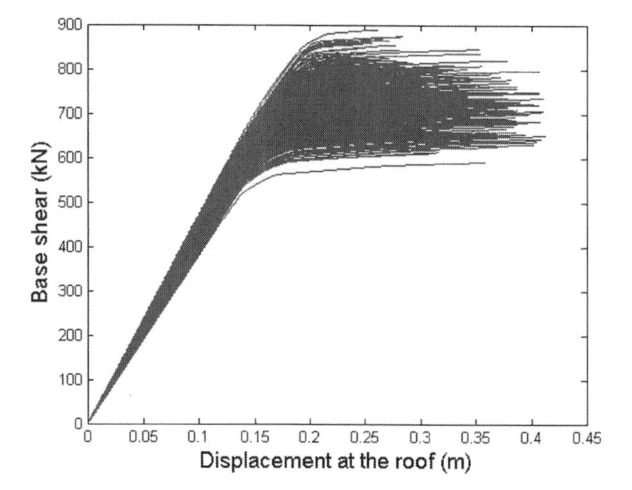

Fig. 17.2 Capacity curves obtained from the PA, taking into account the uncertainty in the mechanical properties of materials

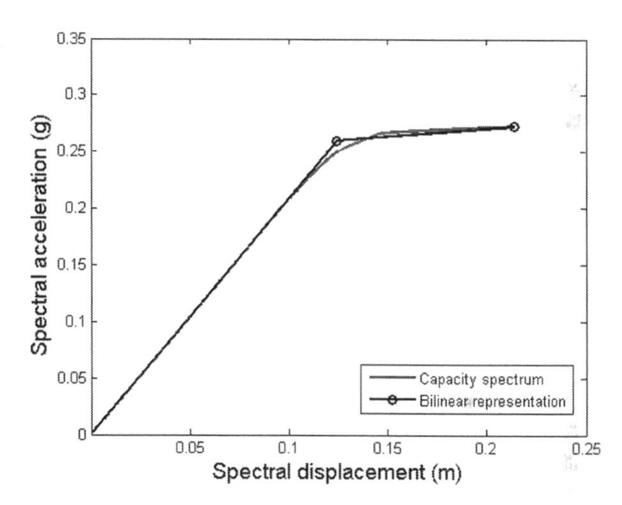

Fig. 17.3 Capacity spectrum and the bilinear representation

the building; PF_i is the modal participation factor of the first mode of vibration; sa_i is the spectral acceleration; V_i is the base shear; W is the weight of the building and α_i is the modal mass coefficient of the first mode of vibration.

On the other hand, the capacity spectrum can be represented in a bilinear form, which is defined completely by the yielding (Dy, Ay) and ultimate (Du, Au) capacity points. As we will see later on, this simplified form is useful for defining damage state thresholds in a straightforward manner; see also [5]. Assumptions to build the bilinear capacity spectrum are: (1) the area under the bilinear curve must be equal to the area of the original curve; (2) the coordinates of the point of maximum displacement must be the same in both curves; (3) the slope of the initial branch should be equal in both curves. Figure 17.3 shows an example of the bilinear representation of the capacity spectrum. Different studies have been proposed to calculate the damage of the structure from the definition of damage states (ds), which are a description of

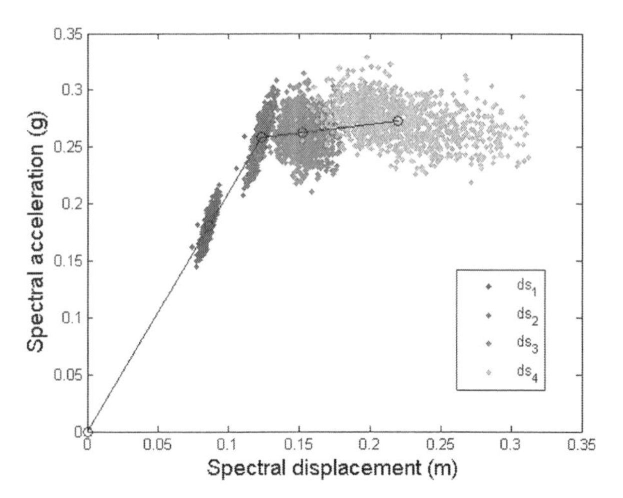

Fig. 17.4 Damage states as random variables

the damage in the structure for a given spectral displacement. For example, FEMA [11] and Risk UE [12], define 4 *ds*, namely *slight*, *moderate*, *extensive* and *complete*. Description of the damage states depends on the type of structure. For instance, according to FEMA [11], in the case of reinforced concrete structures, the *ds slight* is described as: "*beginning of cracking due to bending moment or shear in beams and columns*". *Collapse* state considers that the structure reaches an imminent risk of collapse. Risk UE defines the damage states in simplified form, starting from the capacity spectrum in its bilinear representation.

Based on the values (Dy, Ay) and (Du, Au), the spectral displacements for the four damage states threshold ds_i are obtained according to the following equations:

$$\begin{aligned} ds_1 &= 0.7 * Dy \\ ds_2 &= Dy \\ ds_3 &= Dy + 0.25 * (Du - Dy) \\ ds_4 &= Du \end{aligned} \tag{17.2}$$

Therefore, after calculating the capacity spectrum in bilinear representation and applying Eq. (17.2), it is possible to obtain the damage states thresholds as random variables, as is shown in Fig. 17.4. The mean, standard deviation and coefficient of variation of the damage states are shown in Table 17.2. It is worth noting how the co-

Table 17.2 Mean value, standard deviation and coefficient of variation of the damage states

	ds_1 (cm)	ds_2 (cm)	ds_3 (cm)	ds_4 (cm)
μ_{ds}	8.6	12.3	15.2	21.9
σ_{ds}	0.27	0.38	1.00	3.25
c.o.v.	0.03	0.03	0.06	0.15

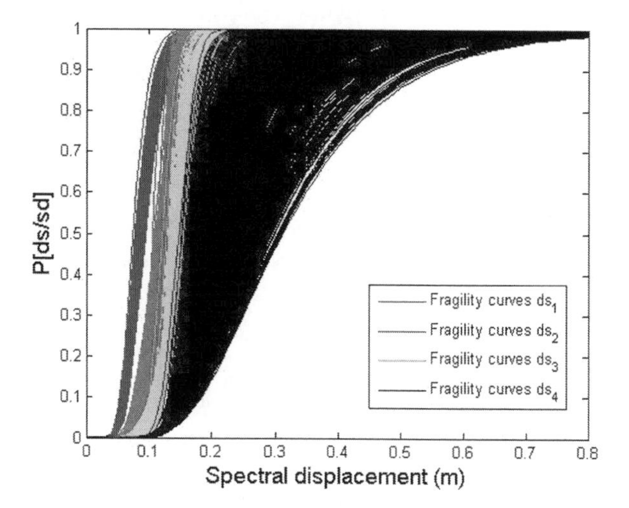

Fig. 17.5 Fragility curves as random variables

efficient of variation of the damage state 4 is greater than those of the input variables. This effect is due to the fact that this type of systems is not robust, mainly because of the nonlinearity of the problem. In addition, these results show the importance of the probabilistic approach in this type of analysis as the expected uncertainties in the output can be greater than those of the input variables. After obtaining the damage states as random variables it is also possible to calculate the fragility curves which, for each damage state, represent the probability of reaching or exceeding the corresponding damage state. Fragility curves are represented as a function of a parameter representing the seismic action, for instance spectral displacement, PGA, etc.

The following simplified assumptions to construct fragility curves from damage states thresholds are considered: (1) the probability that the spectral displacements in each damage state threshold, ds_i, equals or exceeds the damage state is 50%; (2) for each damage state ds_i, the corresponding fragility curve, follows a lognormal cumulative probability function described by the following equation:

$$P[ds_i/sd] = \phi\left[\frac{1}{\beta_{ds_i}}\,\mathrm{Ln}\left(\frac{sd}{ds_i}\right)\right] \tag{17.3}$$

where sd is the spectral displacement and β_{ds_i} is the standard deviation of natural logarithm of the damage state ds_i; (3) the expected seismic damage in buildings follows a binomial probability distribution. Figure 17.5 shows all fragility curves calculated after applying the described procedure.

Since the probabilities of occurrence of each damage state are easily obtained from the fragility curves, one can calculate the expected damage index, *DI*, which is the normalized mean damage state. *DI* can be interpreted as a measure of the overall expected damage in the structure.

$$DI = \frac{1}{n}\sum_{i=0}^{n} i\,P(ds_i) \tag{17.4}$$

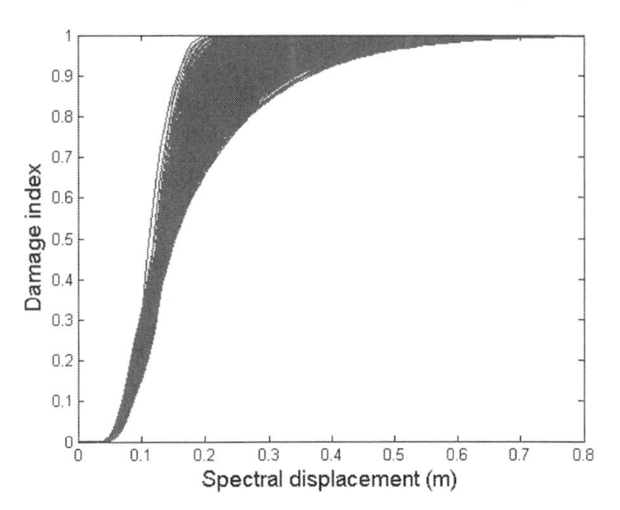

Fig. 17.6 Damage index curves obtained starting from the PA, considering the mechanical properties of the materials as random variables

where n is the number of damage states considered, in this case 5 (four non-null) and $P(ds_i)$ is the probability of occurrence of ds_i. Figure 17.6 shows the DI calculated from the fragility curves of Fig. 17.5. The curves of Fig. 17.6 must be interpreted as random vulnerability curves. These curves are an important result of this work as they allow linking PA and IDA procedures by comparing the obtained results.

4 Damage Index Based on the Incremental Dynamic Analysis

Dynamic analysis allows obtaining the time history of the response of a structure to an earthquake action. In IDA, the earthquake is scaled to various PGA, allowing obtaining the maximum response as a function, for instance, of PGA. As mentioned above, the main purpose of this article is to compare the results obtained with PA and IDA. An important element of the uncertainty related to the seismic response of structures is the random variability in the ground-motion prediction, whose influence has been studied in [13]. According to the probabilistic approach it is necessary to obtain the seismic action as a random variable. To do that, 20 earthquakes have been selected from two databases, one from Spain and the other from Europe [14], whose elastic response spectra are compatible with elastic response spectrum taken from EuroCode 8 (EC8) [15]. In this case, the elastic spectrum type 1 and soil D is selected. This spectrum corresponding to great earthquakes and soft soils has been chosen in order to submit the structure to strong enough seismic actions to obtain significant damage. Figure 17.7 shows the spectra of the selected earthquakes, their median value, and the spectrum type 1 soil D, taken from EC8. After selecting the accelerograms, the dynamic response of the structure is calculated for different PGA increasing at intervals of 0.04 g, until the value that causes the collapse. This value is 0.8 g. In each run of the nonlinear dynamic analysis, the damage index proposed by Park & Ang [16] and the maximum displacement at the roof of the building are calculated, allowing comparing these results with those of the PA analysis. Fig-

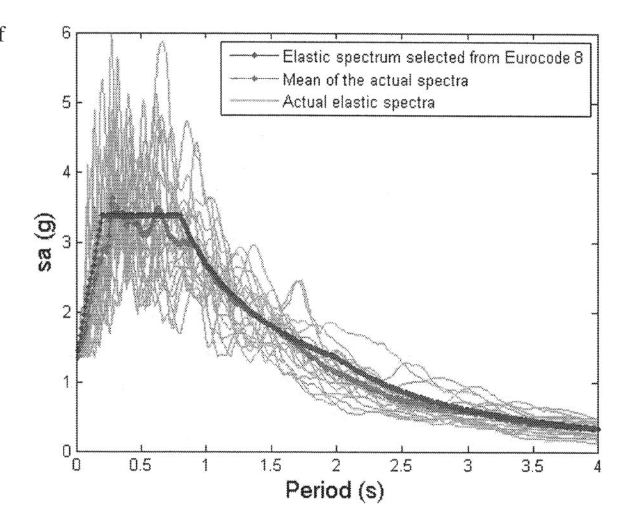

Fig. 17.7 Selected spectra of the accelerograms that are compatible with spectrum type 1 soil D of Eurocode 8

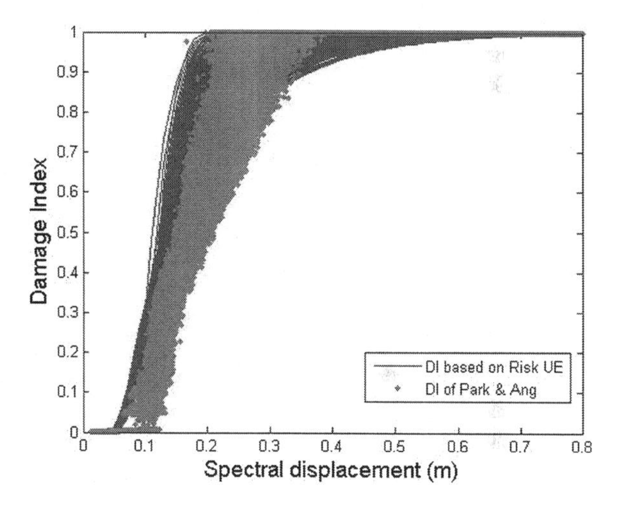

Fig. 17.8 Damage index obtained with static and dynamic procedures

ure 17.8 shows the results obtained. It is important to note the large scatter in both cases, showing the importance of assessing the vulnerability of structures from a probabilistic perspective, whichever procedure is used.

Figure 17.8 shows that the damage index obtained with the procedure based on the PA is conservative. However, for extreme cases when the damage index is close to 0 and 1, which correspond to the *null* and *collapse* damage states, similar values are obtained with both procedures. On the other hand, it can be seen that the curves obtained with the PA procedure are somehow conservative, as the structural damage begins for a smaller spectral displacement. PA based curves are shifted with respect to the IDA based curves. This behavior can be attributed to the fact that the damage state thresholds ds_1 and ds_3 in Eq. (17.2) are based on expert opinion. A little change in these values would avoid this shift. The use of constant coefficients, namely of 0.7

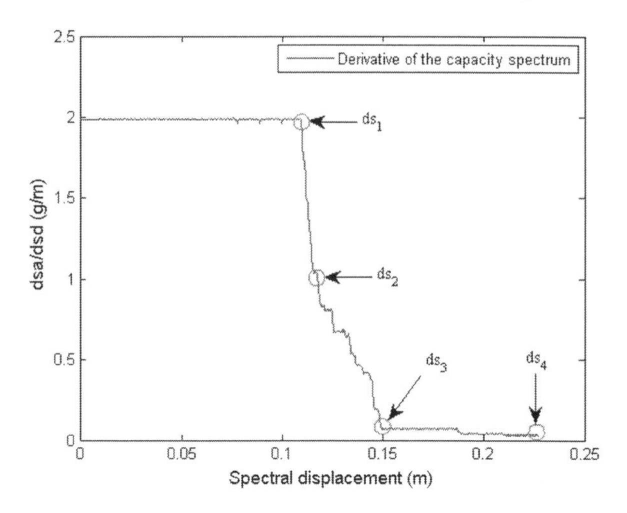

Fig. 17.9 Derivative of a capacity spectrum and new damage states

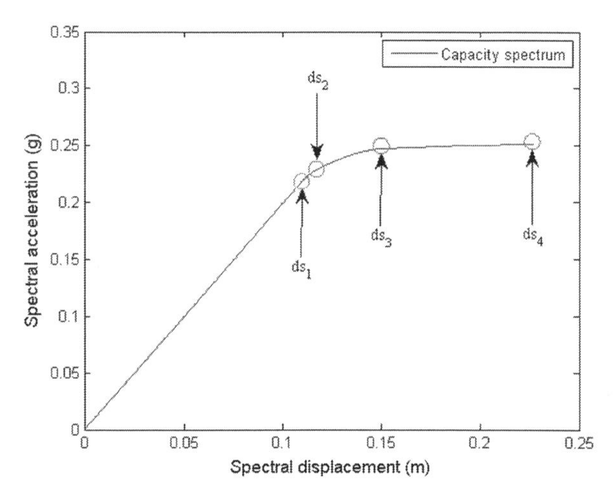

Fig. 17.10 Capacity spectrum and new damage states

and 0.25, in these equations are useful for massive large scale assessments. In this type of studies [5] a great amount of buildings is evaluated based on the use of simplified structural typologies owing to the difficulty to obtain specific capacity curves and coefficients for each building. This approach leads to reasonably good results in average sense. A new method for estimating the damage state thresholds is proposed here. This method is based on an accurate analysis of the variation of the slope of the capacity curve, namely of its derivative. It is worth noting too that, as we will see below, the new procedure of assessing the ds_1 and ds_3 thresholds avoids the shifting between PA and IDA based damage curves of Fig. 17.8.

Figure 17.9 shows an example of the derivative function of the capacity spectrum plotted in Fig. 17.10. In both figures the new damage states thresholds are shown.

In fact, the derivative function is related to the degradation of the stiffness as it gives the actual stiffness of the structure as a function of the spectral displacement

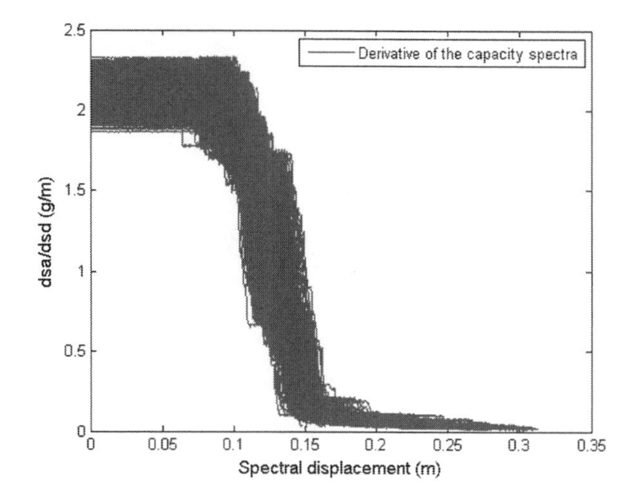

Fig. 17.11 Derivative functions of all capacity spectra

caused by lateral load increases in the pushover analysis. Then, ds_1 is defined by the spectral displacement where the lateral stiffness start to decrease; in other words, the point where the damage starts to increase. At this stage of the method, ds_2 has been defined as the spectral displacement corresponding to a reduction of 50% of the initial stiffness. ds_3 is defined by the spectral displacement where the derivative tends to be constant, indicating the end of the degradation of the stiffness which remains almost constant till the displacement of collapse. Finally, ds_4 is maintained as the spectral displacement corresponding to the ultimate point. It is worth noting that the shapes, but not the values, of the derivative functions are very similar for all the 1000 capacity spectra analyzed. See Fig. 17.11. Therefore, the new damage states based on the stiffness degradation and the damage states calculated via Risk UE approach, which will be called $ds_{i\text{-}S}$ and $ds_{i\text{-}R}$ respectively, are compared. In order to characterize the statistical properties of the distribution of the old (see Fig. 17.4) and new defined damage states, the Kolmogorov–Smirnov test [17] has shown that the damage states calculated with both approaches follow a Gaussian distribution. Figure 17.12 shows the comparison between both probability density functions. For the damage states different to *extensive* and *collapse*, the mean values and the standard deviations of the $ds_{i\text{-}S}$ are higher than those of $ds_{i\text{-}R}$. Then, the procedure described above for obtaining the fragility curves and damage indices was applied again by using the new damage states. Figure 17.13 shows the obtained results. For comparison purposes, the damage indices obtained by means of the dynamic analyses are also plotted in this figure. This figure allows comparing new and old damage index functions as well as each of these functions with the results of the dynamic analyses.

Concerning PA results, blue and black colour curves, a clear shift towards increasing spectral displacements of the new damage functions can be seen, indicating that the Risk UE choice is somehow conservative. Furthermore, new black curves fit better the IDA results (red points).

In order to quantitatively improve this comparison, Fig. 17.14 and Fig. 17.15 compare respectively the first and second moments of these distributions. These sta-

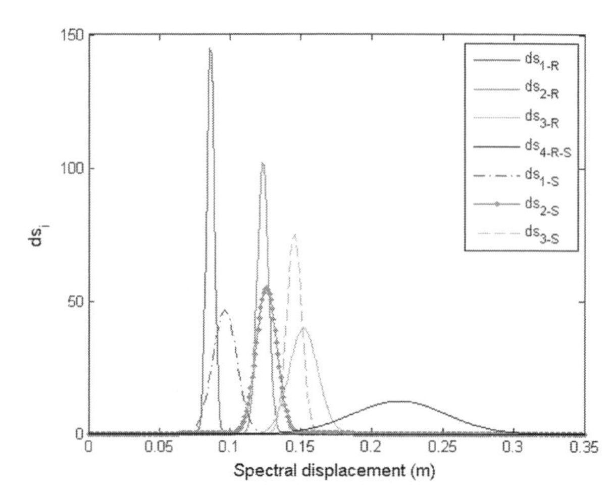

Fig. 17.12 Comparison between damage states based on Risk UE and stiffness degradation approach

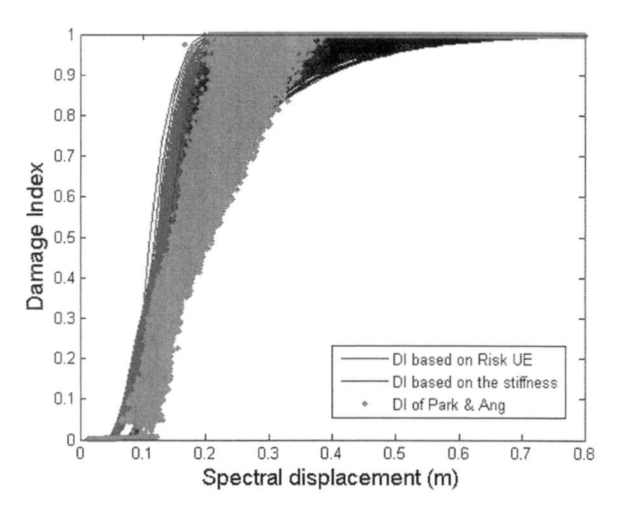

Fig. 17.13 Comparison between damage indices obtained with all methodologies

tistical properties, namely the mean values and standard deviations are computed for each spectral displacement by using the corresponding random ordinates. It can be clearly seen in Fig. 17.14 how the mean of the random variable obtained with the derivative approach fits quite well the mean of the damage index obtained via non-linear dynamic analysis. Figure 17.15 shows that, for spectral displacements in the range 0.1 to 0.3 m, the standard deviation corresponding to PA results is lower than one corresponding to IDA results. This effect is attributed to the fact that PA results do not consider the seismic actions leading to lower uncertainties. To consider the uncertainties of the seismic action, we use a simplified method allowing obtaining the expected displacement as a function of PGA for a given seismic input, represented by the 5% damped elastic response spectrum. Obviously, the building in this analysis is defined by its capacity spectrum. In this procedure, the elastic response spectrum is reduced based on the ductility of the building which is calculated from

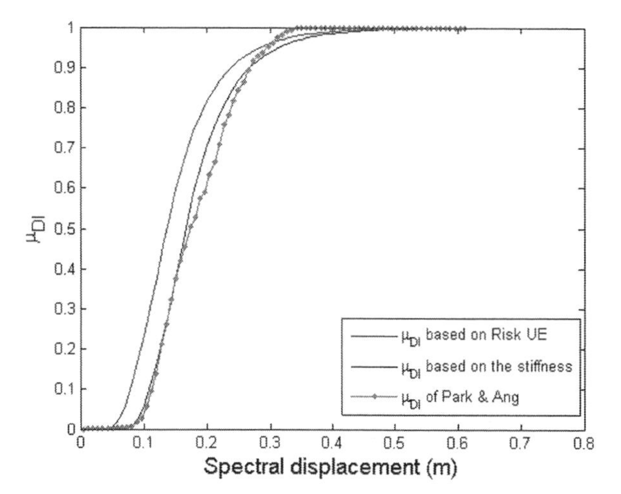

Fig. 17.14 Mean of the damage indices

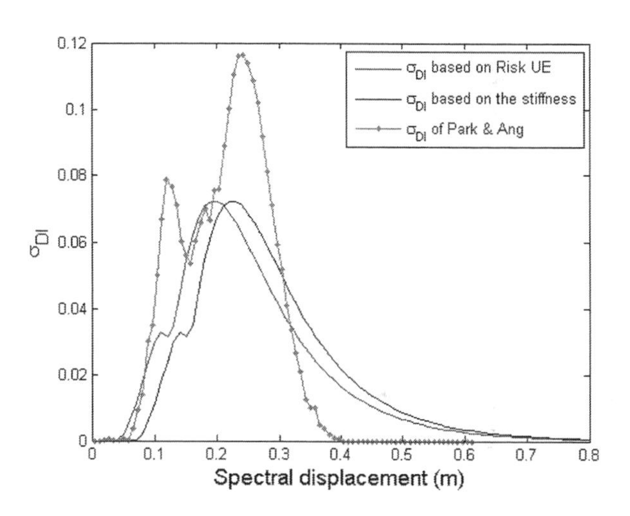

Fig. 17.15 Standard deviations of the damage indices

the capacity spectrum as the ratio between the spectral displacements of the ultimate capacity point (Du) and that of the yielding point (Dy) (see Fig. 17.3). An extended explanation of this technique can be found in [9] and has been also used in [18], it has been initially proposed by [19] and its development has been reviewed in [20]. In this way, increasing the PGA at intervals of 0.04 g between 0.04 and 0.8 g, as in the IDA, a relation between the PGA and the spectral displacement, sd, is obtained for each spectrum corresponding to each of the 20 accelerograms used and for each of the 1000 capacity spectra. Therefore, a total of 20000 relations between sd and PGA are obtained.

Figure 17.16 and Fig. 17.17 show the mean and the standard deviation curves of the damage indices as a function of PGA for PA, by using the new defined damage states, and IDA results.

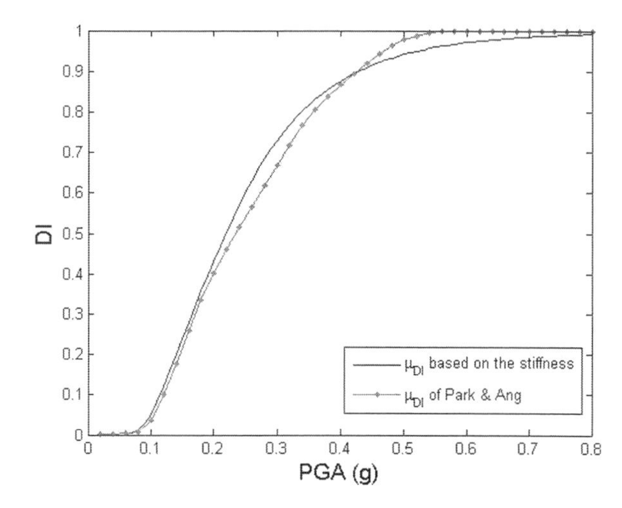

Fig. 17.16 Mean of the damage indices as a function of the PGA

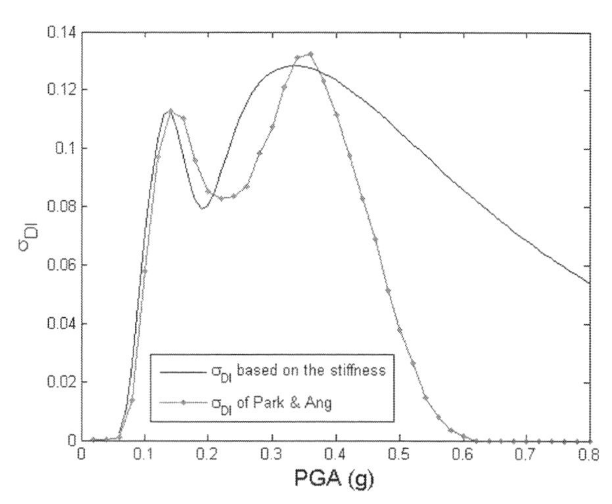

Fig. 17.17 Standard deviation of the damage indices as a function of the PGA

It can be seen how, in the range between 0 and 0.4 g, the mean values and the standard deviations show a good agreement. Note that now the uncertainties in the seismic actions are included in both curves. For greater values, standard deviations in the new PA approach are larger than for the IDA approach but both decrease because damage indices greater than one were not allowed. The fact of the better agreement between the PA and IDA results, when using the new damage states thresholds, indicates that this proposal based on the stiffness degradation, obtained from the derivatives of the capacity curves, should be preferred to the expert-opinion based one as proposed in the Risk UE approach. Furthermore, the damage index calculated in this way is able to represent, not only the expected damage obtained via nonlinear dynamic analysis, but also the uncertainties associated to the mechanical properties of the materials and the seismic action. Finally, it is important to note that in the case study building analysed here the Risk UE approach is a little conservative as the damage appears before the new approach. This is so because the new damage states thresholds are greater. Obviously the spectral displacements of the damage states thresholds can coincide but if the new defined grades are smaller, the Risk UE approach may underestimate the expected damage. In any case, the new approach to determine damage state thresholds capture better the degradation of the buildings strength as indicated by the agreement with the IDA results.

5 Conclusions

In this work, the vulnerability of a real reinforced concrete structure, with columns and waffle slabs has been assessed, taking into account that the input variables are random. Only the randomness of the concrete compressive and the steel yielding strengths has been taking into account but the seismic action has been also considered in a stochastic way. Two approaches to evaluate the expected damage of the building have been used. The first one is based on the pushover analysis and the second one is based on the incremental dynamic analysis. An important conclusion is that, despite working with advanced structural analysis, these procedures show significant uncertainties when taking into account the randomness of the variables associated with the problem. It should be emphasized that in this work relatively small coefficients of variation for input variables have been considered taking into account the uncertainties that may exist in older structures that did not have quality control and have not been designed according to the earthquake-resistant criteria. The results obtained give support to the idea that static procedures are conservative when compared with the dynamic analysis. Furthermore, for expected damage analysis, a new procedure has been proposed to define the damage states thresholds. The technique is based on the degradation of the stiffness which can be observed in the derivative function of the capacity curve. The results using this new approach show a better agreement with the dynamic analysis than the obtained ones when using damage states thresholds based on expert-opinion.

Probably one of the most relevant conclusions of this work is that whichever procedure is used to evaluate the expected seismic damage of a structure, the input parameters of the structural problem to be treated, must be considered as random variables. We have seen how the probabilistic consideration of a few of these parameters produces significant uncertainties in the seismic response. Simplified deterministic procedures based on characteristic values usually lead to conservative results but some abridged assumptions on the definition of the seismic actions and on the estimation of the seismic damage states and thresholds can lead also to underestimate the real damage that can occur in a structure.

Acknowledgements This work was partially funded by the Geologic Institute of Catalonia (IGC), by the ministry of science and innovation of Spain and by the European Commission through research projects CGL-2005-04541-C03-02/BTE, CGL2008-00869/BTE, CGL2011-23621 INTERREG: POCTEFA 2007-2013/ 73/08 y MOVE—FT7-ENV-2007-1-211590. The first author has a scholarship funded by a bilateral agreement between the IGC and the Polytechnic University of Catalonia (BarnaTech).

References

1. Barbat, A.H., Yépez Moya, F., Canas, J.A.: Damage scenarios simulation for risk assessment in urban zones. Earthq. Spectra **2**(3), 371–394 (1996)
2. Barbat, A.H., Mena, U., Yépez, F.: Evaluación probabilista del riesgo sísmico en zonas urbanas. Revista internacional de métodos numéricos para cálculo y diseño en ingeniería **14**(2), 247–268 (1998)
3. Borzi, B., Phino, R., Crowley, H.: Simplified pushover analysis for large-scale assessment of RC buildings. Eng. Struct. **30**, 804–820 (2008)
4. Barbat, A.H., Pujades, L.G., Lantada, N., Moreno, R.: Seismic damage evaluation in urban areas using the capacity spectrum method: application to Barcelona. Soil Dyn. Earthq. Eng. **28**, 851–865 (2008)
5. Lantada, N., Pujades, L.G., Barbat, A.H.: Vulnerability index and capacity spectrum based methods for urban seismic risk evaluation. A comparison. Nat. Hazards **51**, 501–524 (2009)
6. Vamvatsikos, D., Cornell, C.A.: The incremental dynamic analysis. Earthquake Eng. Struct. Dyn. **31**(3), 491–514 (2002)
7. Satyarno, I.: Pushover analysis for the seismic assessment of reinforced concrete buildings. Dissertation, University of Canterbury (1999)
8. Carr, A.J.: Ruaumoko—inelastic dynamic analisys program. Dept. of Civil Engineering, Univ. of Canterbury, Christchurch, New Zealand (2000)
9. Crowley, H., Bommer, J.J., Pinho, R., Bird, J.F.: The impact of epistemic uncertainty on an earthquake loss model. Earthquake Eng. Struct. Dyn. **34**(14), 1635–1685 (2005)
10. ATC-40: Seismic evaluation and retrofit of concrete buildings. Applied Technology Council, Redwood City, California (1996)
11. FEMA: HAZUS99 technical manual. Federal Emergency Management Agency, Washington, DC, USA (1999)
12. RISK-UE: An advanced approach to earthquake risk scenarios with applications to different European towns. Project of the European Commission (2004)
13. Bommer, J.J., Crowley, H.: The influence of ground motion variability in earthquake loss modelling. Bull. Earthq. Eng. **4**(3), 231–248 (2006)
14. Ambraseys, N., Smit, P., Sigbjornsson, R., Suhadolc, P., Margaris, B.: Internet-site for European strong-motion data. European Commission, Research-Directorate General, Environment and Climate Programme. http://www.isesd.hi.is/ESD_Local/frameset.htm. Accesed 17 Apr 2011
15. Eurocode 8: Design of structures for earthquake resistance. Part 1: general rules, seismic actions and rules for building (2002)
16. Park, Y.J., Ang, A.H.S.: Mechanistic seismic damage model for reinforced concrete. J. Struct. Eng. **111**(4), 722–757 (1985)
17. Lilliefors, H.W.: On the Kolmogorov–Smirnov test for normality with mean and variance unknown. J. Am. Stat. Assoc. **318**, 399–402 (1967)
18. Vargas, Y.F., Pujades, L.G., Barbat, A.H., Hurtado, J.E.: Evaluación probabilista de la capacidad, fragilidad y daño sísmico en edificios de hormigón armado. Revista internacional de métodos numéricos para cálculo y diseño en ingeniería **29**(1) (2013, to appear)
19. Freeman, S.A., Nicoletti, J.P., Tyrell, J.V.: Evaluation of existing buildings for seismic risk—a case study of Puget Sound Naval Shipyard, Bremerton, Washington. In: Proceedings of U.S. National Conference on Earthquake Engineering, Berkeley, USA, pp. 113–122 (1975)
20. Freeman, S.A.: Review of the development of the capacity spectrum method. ISET J. Earthq. Technol. **41**, 1–13 (2004)

Chapter 18
Stochastic Analysis of the Risk of Seismic Pounding Between Adjacent Buildings

Enrico Tubaldi and Michele Barbato

Abstract Seismic pounding can induce severe damage and losses in buildings. The corresponding risk is particularly relevant in densely inhabited metropolitan areas, due to the inadequate clearance between buildings. In order to mitigate the seismic pounding risk, building codes provide simplified procedures for determining the minimum separation distance between adjacent buildings. The level of safety corresponding to the use of these procedures is not known a priori and needs to be investigated. The present study proposes a reliability-based procedure for assessing the level of safety corresponding to a given value of the separation distance between adjacent buildings exhibiting linear elastic behaviour. The seismic input is modelled as a nonstationary random process, and the first-passage reliability problem corresponding to the pounding event is solved employing analytical techniques involving the determination of specific statistics of the response processes. The proposed procedure is applied to estimate the probability of pounding between linear single-degree-of-freedom systems and to evaluate the reliability of simplified design code formulae used to determine building separation distances. Furthermore, the capability of the proposed method to deal with complex systems is demonstrated by assessing the effectiveness of the use of viscous dampers, according to different retrofit schemes, in reducing the probability of pounding between adjacent buildings modelled as multi-degree-of-freedom systems.

1 Introduction

Earthquake ground motion excitation can induce pounding in adjacent buildings with inadequate separation distance. The corresponding risk is particularly relevant

E. Tubaldi (✉)
Department of Architecture, Construction and Structures, Marche Polytechnic University, Via Brecce Bianche, 60131, Ancona, Italy
e-mail: etubaldi@libero.it

M. Barbato
Department of Civil & Environmental Engineering, Louisiana State University and A&M College, Nicholson Extension, Baton Rouge, LA 70803, USA
e-mail: mbarbato@lsu.edu

M. Papadrakakis et al. (eds.), *Computational Methods in Stochastic Dynamics*, Computational Methods in Applied Sciences 26,
DOI 10.1007/978-94-007-5134-7_18,

in densely inhabited metropolitan areas, due to the need of maximizing the land use and the consequent limited separation distance between adjacent buildings.

The problem of seismic pounding has been investigated by several researchers in the last two decades. A significant number of early studies focused on the definition of simplified rules, such as the Double Difference Combination (DDC) rule, for determining the peak relative displacement response of adjacent buildings at the potential pounding locations [1–3]. A critical separation distance (CSD) was defined and set equal to the mean peak relative displacement between adjacent buildings, by neglecting the associated probability of pounding. In the same context, considerable research effort was devoted to the assessment of the accuracy of code rules (e.g., the absolute sum (ABS) and square-root-of-the-sums-squared (SRSS) rules [4, 5]) in determining the mean peak relative displacement response (i.e., the CSD) of adjacent buildings.

More recent studies have proposed a probabilistic approach for the assessment of the seismic pounding risk. In Lin [6], a method was developed to estimate the first two statistical moments of the random variables describing the peak relative displacement response between linear elastic structures subjected to stationary base excitation. In Lin and Weng [7], a numerical simulation approach was suggested to evaluate the pounding probability, over a 50-year design lifetime, of adjacent buildings separated by the code-specified CSD. The latter study considered both the uncertainty affecting the seismic input intensity (by using a proper hazard model) and the record-to-record variability (by using artificially generated spectrum-compatible ground acceleration time histories as input loading). The buildings were modelled as multi-degree of freedom systems with inelastic behaviour and deterministic properties. In Hong et al. [8], a procedure was developed to assess the fractiles of the CSD between linear elastic systems with deterministic and uncertain structural properties subjected to stationary base excitation. The previous study was later extended by Wang and Hong [9] to include non-stationary seismic input.

Despite the numerous studies available in the literature on seismic pounding, to the best of the authors' knowledge, a reliability-based methodology for the evaluation of the safety levels associated with specified CSDs is still needed. In addition, the gradual progress of seismic design codes from a prescriptive to a performance-based design philosophy generates a significant need for new advanced, accurate, and computationally efficient reliability-based methodologies for the assessment and mitigation of seismic pounding risk.

This paper presents a fully probabilistic methodology for assessing the seismic pounding risk between adjacent buildings with linear behaviour. This methodology is consistent with and can be easily incorporated into a performance-based earthquake engineering (PBEE) paradigm such as the Pacific Earthquake Engineering Research centre (PEER) framework [10, 11]. The presented methodology considers the uncertainty affecting both the seismic input (i.e., site hazard and record-to-record variability) and the parameters used to describe the structural systems of interest (i.e., material properties, geometry, and damping properties). The seismic input is modelled as a nonstationary random process. The seismic pounding risk is computed from the solution of a first-passage reliability problem. While the approach proposed is general, the methodology presented here is specialized to linear elastic systems subjected to Gaussian loading. Under these assumptions, approximate analytical solutions and efficient simulation techniques can be used to solve the relevant first-passage reliability problem. Thus, this methodology is appropriate for structural systems that remain in their linear elastic behaviour range before pounding (which is a very common condition for low values of the CSDs and, thus, high seismic pounding risk), although it can be extended to account for nonlinear behaviour of the considered structural systems.

2 PBEE Framework for Seismic Pounding Risk Assessment

The PEER PBEE framework is a general probabilistic methodology, based on the total probability theorem, for risk assessment and design of structures subjected to seismic hazard [10, 11]. The PEER PBEE methodology involves four probabilistic analysis components: (1) probabilistic seismic hazard analysis (PSHA), (2) probabilistic seismic demand analysis (PSDA), (3) probabilistic seismic capacity analysis (PSCA), and (4) probabilistic seismic loss analysis (PSLA). PSHA provides the probabilistic description of an appropriate ground motion intensity measure (*IM*), usually expressed as mean annual frequency (MAF), $\nu_{IM}(im)$, of exceedance of a specific value *im*. PSDA provides the statistical description of structural response parameters of interest, usually referred to as engineering demand parameters (*EDP*s), conditional to the value of the seismic intensity *IM*. PSCA consists in computing the probability of exceeding a specified physical limit-state, defined by structure-specific damage measures (*DM*s), and conditional to the values of the *EDP*s. Finally, PSLA provides the probabilistic description of a decision variable (*DV*), which is a measurable attribute of a specific structural performance and can be defined in terms of cost/benefit for the users and/or the society.

The reliability-based procedure developed in this paper consists in computing the MAF of pounding, ν_p, between two adjacent buildings. This procedure is a specialization for the seismic pounding problem of the first three steps of the general PEER PBEE framework (i.e., PSLA is beyond the scope of this paper). It is noteworthy that the proposed approach is conceptually very different from the computation of the CSD, which does not explicitly provide the probability of pounding associated with a given separation distance. The computation of the MAF of pounding can be expressed as

$$\nu_p = \int_{edp}\int_{im} G_{DM|EDP}(dm|edp) \cdot \left|\mathrm{d}G_{EDP|IM}(edp|im)\right| \cdot \left|\mathrm{d}\nu_{IM}(im)\right| \qquad (18.1)$$

in which, $G_{DM|EDP}(dm|edp)$ = complementary cumulative distribution function of variable *DM* conditional to $EDP = edp$, and $G_{EDP|IM}(edp|im)$ = complementary cumulative distribution function of variable *EDP* conditional to $IM = im$, where upper case symbols indicate random variables and lower case symbols denote specific values assumed by the corresponding random variable. The *IM* must be selected so that it can be readily related to the stochastic description of an appropriate random process model for the input ground motion. This selection must also account for sufficiency and efficiency of the *IM* in describing the effects of the ground motion excitation on the structural response [12]. However, an exhaustive selection of appropriate *IM*s for different types of structures and structural performances is beyond the scope of this paper.

The maximum value $U_{rel,\max}$ of the relative distance $U_{rel}(t)$ between the adjacent buildings observed during the seismic event (i.e., for $t \in [0, t_{\max}]$, with $t =$ time and $t_{\max} =$ duration of the seismic event) is assumed here as *EDP*. The probabilistic distribution of $U_{rel,\max}$ reflects the record-to-record variability of the ground motions expected to occur at the site for a given intensity, as well as the effects of the uncertainty in the parameters used to describe the structural model. Finally, the pounding event is assumed as the controlling limit-state in PSCA, by using the following limit-state function, g:

$$g = \Xi - U_{rel,\max} \tag{18.2}$$

in which $\Xi =$ random variable describing the building separation distance, and the pounding event corresponds to $g \leq 0$. Thus, $G_{EDP|IM}(epd|im) = P[U_{rel,\max} \geq u|IM = im]$ and $G_{DM|EDP}(dm|edp) = P[g < 0|U_{rel,\max} = u]$. An important intermediate result of the procedure is the convolution of PSCA and PSDA, also called fragility analysis, which yields a fragility curve. Fragility curves describe the probability $P_{p|IM}$ of pounding conditional on the seismic intensity, i.e.,

$$P_{p|IM} = \int_{edp} G_{DM|EDP}(dm|edp) \cdot \left|\mathrm{d}G_{EDP|IM}(edp|im)\right| \tag{18.3}$$

The MAF of pounding, ν_p, can be used to compute the MAF of exceeding a specified value of *DV*, e.g., the MAF of repair cost due to pounding damage. The computation of the latter quantity requires the definition of a realistic loss model, based on appropriate structural response models (e.g., dynamic impact between adjacent systems) and damage models (e.g., damage produced by floor-to-floor and floor-to-column pounding). Structural response and damage models involve the definition of other *EDP*s and *DM*s, respectively, in addition to those already employed in this paper for assessing the pounding risk. Several structural response and damage models available in the literature could be employed to define an appropriate loss model [13–16].

In addition, ν_p can be directly used to determine the pounding risk, $P_p(t_L)$, for a given structure over its design life, t_L (e.g., 50 years). Assuming that the occurrence of a pounding event can be described by a Poisson process and that the buildings are immediately restored to their original condition after pounding occurs, $P_p(t_L)$ can be easily computed as

$$P_p(t_L) = 1 - e^{-\nu_p \cdot t_L} \tag{18.4}$$

3 Seismic Pounding Risk Assessment Methodology

Fragility analysis is the most computationally challenging component of the probabilistic PBEE framework. A simple and general approach for fragility analysis in seismic pounding assessment is provided by Monte Carlo simulation (MCS) [5, 7]. For any given value of IM, MCS-based fragility analysis requires (1) the definition of a set of ground motions that are selected from an appropriate database of real records or generated from an appropriate random process, (2) the sampling of the structural parameters that define the structural systems and of their separation distances, (3) the numerical simulation of the structural response for each ground motion time history and each set of structural parameters and separation distances, and (4) the evaluation of $P_{p|IM}$ as the ratio between the number of failures and the number of samples. However, the computational cost associated with MCS can be very high and even prohibitive when small failure probabilities need to be estimated by numerically simulating the time history response of complex multi-degree-of-freedom (MDOF) systems.

In this paper, an efficient combination of analytical and simulation techniques is proposed for the calculation of $P_{p|IM}$ under the assumptions of linear elastic behaviour for the buildings and of Gaussian input ground motion. The methodology is described first for linear elastic systems with deterministic structural properties and separation distance, and then generalized to stochastic linear systems.

It is noteworthy that, for low values of the building separation distance, ξ, the buildings are expected to behave elastically before pounding occurs, while the assumption of linear behaviour of the buildings before pounding becomes less realistic for larger values of ξ. If the buildings are expected to enter their nonlinear behaviour before pounding, the methodology described in the remainder of this paper needs to be extended to nonlinear systems, e.g., by using statistical linearization techniques [17] or subset simulation [18]. This extension is beyond the scope of this paper.

3.1 Linear Systems with Deterministic Structural Properties

The computation of the conditional failure probability $P_{p|IM}$ can be expressed in the form of a single-barrier first-passage reliability problem as [5, 9]

$$P_{p|IM} = P\left\{\max_{0\leq t\leq t_{\max}}\left[U_{rel}(t)\right] \geq \xi \,|IM = im\right\} \tag{18.5}$$

in which $U_{rel}(t) = U_{\mathrm{A}}(t) - U_{\mathrm{B}}(t)$, $U_{\mathrm{A}}(t)$ and $U_{\mathrm{B}}(t)$ = displacement response of the adjacent buildings A and B at the (most likely) pounding location, and ξ = deterministic value of the building separation distance (Fig. 18.1).

Under the hypotheses of deterministic linear elastic systems subjected to Gaussian loading processes and deterministic threshold, several analytical approximations of $P_{p|IM}$ exist in the literature [19–22]. These analytical approximations require computing the following statistics of the relative displacement process $U_{rel}(t)$ for a given $IM = im$: $\sigma^2_{U_{rel}}(t)$ = variance of $U_{rel}(t)$, $\sigma^2_{\dot{U}_{rel}}(t)$ = variance of the relative velocity process $\dot{U}_{rel}(t)$, $\rho_{U_{rel}\dot{U}_{rel}}(t)$ = correlation coefficient between $U_{rel}(t)$ and $\dot{U}_{rel}(t)$, and $q_{U_{rel}}(t)$ = bandwidth parameter of $U_{rel}(t)$.

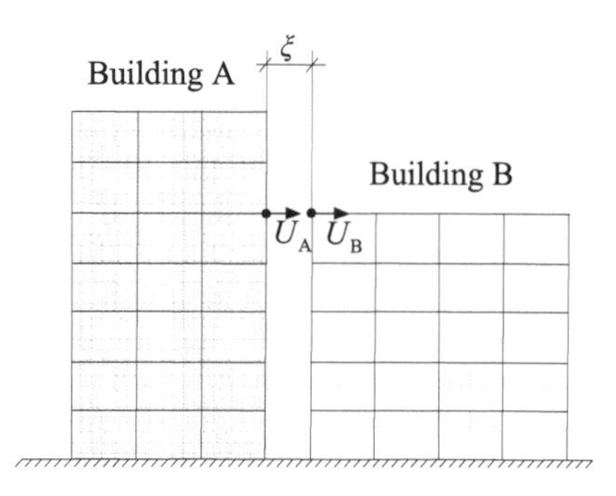

Fig. 18.1 Geometric description of the pounding problem between adjacent buildings

These statistics can be obtained from the spectral characteristics of order zero to two of process $U_{rel}(t)$ [23–25].

Following the methodology described in Barbato and Conte [24], a state-space formulation of the equations of motion for the two buildings is employed to compute exactly and in closed-form the required spectral characteristics. The seismic input is modelled as a time-modulated Gaussian colored noise process. For this specific input ground motion process, the spectral characteristics of the displacement processes (and of any response process obtained as a linear combination of the displacement processes) are available in exact closed-form for single-degree-of-freedom (SDOF) systems and both classically and non-classically damped MDOF systems [25].

The equations of motion for the linear system constituted by two non-connected adjacent buildings can be expressed as follows:

$$\mathbf{m} \cdot \ddot{\mathbf{U}}(t) + \mathbf{c} \cdot \dot{\mathbf{U}}(t) + \mathbf{k} \cdot \mathbf{U}(t) = \mathbf{p} \cdot F(t) \tag{18.6}$$

in which $\mathbf{m} = \begin{pmatrix} \mathbf{m}_A & \mathbf{0} \\ \mathbf{0} & \mathbf{m}_B \end{pmatrix}$, $\mathbf{c} = \begin{pmatrix} \mathbf{c}_A & \mathbf{0} \\ \mathbf{0} & \mathbf{c}_B \end{pmatrix}$, $\mathbf{k} = \begin{pmatrix} \mathbf{k}_A & \mathbf{0} \\ \mathbf{0} & \mathbf{k}_B \end{pmatrix}$, $\mathbf{U} = \begin{pmatrix} \mathbf{U}_A \\ \mathbf{U}_B \end{pmatrix}$, $\mathbf{m}_i$, $\mathbf{k}_i$, $\mathbf{c}_i$ and $\mathbf{U}_i$ = mass matrix, damping matrix, stiffness matrix, and vector of nodal displacements of building i, respectively (i = A, B), $\mathbf{p}$ = load distribution vector, $F(t)$ = scalar function describing the time-history of the external loading (input random process), and a superposed dot denotes differentiation with respect to time. It is noteworthy that connections between the two buildings (e.g., damping devices interposed between the building to mitigate seismic pounding risk) can be easily modelled by introducing the appropriate terms in matrix $\mathbf{c}$. The response process of interest $U_{rel}(t)$ can be related to the displacement response vector $\mathbf{U}(t)$ by means of a linear operator $\mathbf{b}$ as $U_{rel}(t) = \mathbf{b} \cdot \mathbf{U}(t)$.

The probability of pounding conditional on $IM = im$ is given by

$$P_{p|IM} = 1 - P\big[U_{rel}(t=0) < \xi \,|IM = im\big] \times \exp\left\{-\int_0^{t_{\max}} h_{U_{rel}|IM}(\xi,\tau)\cdot \mathrm{d}\tau\right\} \tag{18.7}$$

in which $P[U_{rel}(t=0) < \xi\,|IM = im]$ = probability that the random process $U_{rel}(t)$ is below the threshold ξ at time $t = 0$, and $h_{U_{rel}|IM}(\xi,t)$ = time-variant hazard function (i.e., up-crossing rate of threshold ξ conditioned on zero up-crossings before time t) conditional on $IM = im$. For systems with at rest initial conditions, $P[U_{rel}(t=0) < \xi | IM = im] = 1$.

To date, no exact closed-form expressions exist for the time-variant hazard function $h_{U_{rel}|IM}(\xi, t)$. However, several approximate solutions are available in the literature, e.g., Poisson's (P), $h^{(P)}_{U_{rel}|IM}(\xi, t) = \nu_{U_{rel}|IM}(\xi, t)$, classical Vanmarcke's (cVM), $h^{(cVM)}_{U_{rel}|IM}(\xi, t)$, and modified Vanmarcke's (mVM), $h^{(mVM)}_{U_{rel}|IM}(\xi, t)$, approximations [22, 26]. These analytical approximations can be readily computed based on the closed-form expressions of the spectral characteristics of process $U_{rel}(t)$, as shown in Barbato and Vasta [25]. In addition, for linear elastic systems subjected to Gaussian loading, $P_{p|IM}$ can be efficiently and accurately estimated by using the Importance Sampling using Elementary Events (ISEE) method [27].

3.2 Linear Systems with Uncertain Structural Properties and Separation Distance

In addition to the uncertainty in the seismic input, significant uncertainty can be found in geometrical, mechanical, and material properties characterizing the structural systems and their models. Hereinafter, the uncertainty in geometrical, mechanical, and material properties of the structural models, as well as in their separation distance, Ξ, is referred to as model parameter uncertainty (MPU). MPU can significantly modify the structural performance and, thus, must be considered in the assessment of seismic pounding risk.

In order to include the effects of MPU, the total probability theorem is employed to compute the conditional probability of pounding as follows:

$$P_{p|IM} = \int_{\mathbf{X}} P_{p|IM,\mathbf{X}}(\mathbf{x}) \cdot f(\mathbf{x}) \cdot d\mathbf{x} = E_{\mathbf{X}}[P_{p|IM,\mathbf{X}}] \tag{18.8}$$

in which $\mathbf{X}$ = vector of uncertain model parameters (including the uncertain separation distance Ξ) with joint probability density function $f_{\mathbf{X}}(\mathbf{x})$, $P_{p|IM,\mathbf{X}}(\mathbf{x})$ = probability of pounding conditional on $\mathbf{X}$ and IM, and $E_{\mathbf{X}}[\ldots]$ = expectation operator with respect to vector $\mathbf{X}$.

MCS, or any variance reduction technique such as stratified sampling, can be employed to evaluate $P_{p|IM}$ in Eq. (18.8). For example, Latin hypercube sampling (LHS) can be employed for its computational efficiency [28]. The samples of vector $\mathbf{X}$ generated by using LHS can be used to define a set of deterministic linear elastic models with deterministic separation distance, for which the conditional probability of pounding can be computed as in Eq. (18.7).

4 Application Examples

In this section, the proposed methodology is applied to: (1) compute the pounding risk for SDOF systems with deterministic model parameters, (2) evaluate the re-

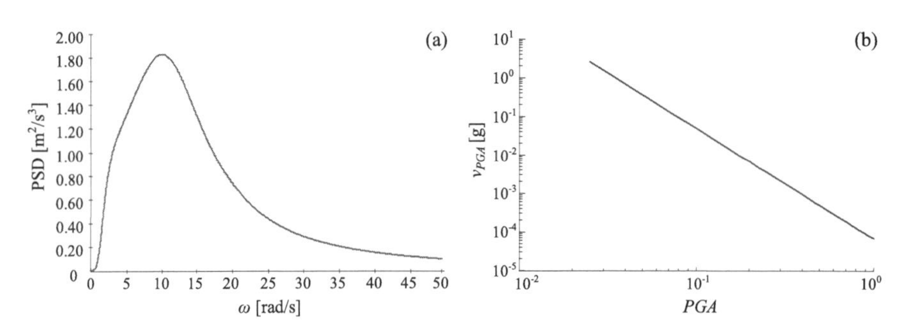

Fig. 18.2 Input ground motion: (**a**) PSD function of the embedded stationary process, and (**b**) site hazard curve

liability of simplified design code formulae used to determine building separation distance, and (3) to evaluate the effectiveness of different retrofit solutions using viscous dampers in reducing the pounding risk for deterministic MDOF models of multi-storey buildings. In all the application examples considered here, the input ground acceleration is modelled by a time-modulated Gaussian process. The time-modulating function, $I(t)$, is represented by the Shinozuka–Sato's function [29], i.e.,

$$I(t) = c \cdot \left(e^{-b_1 \cdot t} - e^{-b_2 \cdot t}\right) \cdot H(t) \tag{18.9}$$

in which $b_1 = 0.045\pi$ s^{-1}, $b_2 = 0.050\pi$ s^{-1}, $c = 25.812$, and $H(t) =$ unit step function. A duration $t_{\max} = 30$ s is considered for the seismic excitation.

The power spectral density (PSD) of the embedded stationary process is described by the widely-used Kanai–Tajimi model, as modified by Clough and Penzien [30], i.e.,

$$S_{CP}(\omega) = S_0 \frac{\omega_g^4 + 4\xi_g^2\omega^2\omega_g^2}{[\omega_g^2 - \omega^2]^2 + 4\xi_g^2\omega^2\omega_g^2} \cdot \frac{\omega^4}{[\omega_f^2 - \omega^2]^2 + 4\xi_f^2\omega^2\omega_f^2} \tag{18.10}$$

in which $S_0 =$ spectral amplitude of the bedrock excitation (considered to be a white noise process), ω_g and $\xi_g =$ fundamental circular frequency and damping factor of the soil, respectively, and ω_f and $\xi_f =$ parameters describing the Clough–Penzien filter. The values of the parameters employed for all the applications are $\omega_g = 12.5$ rad/s, $\xi_g = 0.6$, $\omega_f = 2$ rad/s, and $\xi_f = 0.7$. The PSD function in Eq. (18.10) is shown in Fig. 18.2(a) for $S_0 = 1$.

The peak ground acceleration, *PGA*, is assumed as *IM*. In order to derive the fragility curves in terms of the selected *IM*, the relationship between the parameter S_0 of the Kanai–Tajimi spectrum and the *PGA* at the site is assessed empirically. A set of 500 synthetic stationary ground motion records are generated using the spectral representation method [31] based on the PSD function given in Eq. (18.10) with $S_0 = 1$. Each ground motion realization is then modulated in time using the function defined in Eq. (18.9). The peak ground acceleration corresponding to $S_0 = 1$, $PGA_{S_0=1}$, is estimated as the mean of the *PGA*s of the sampled ground

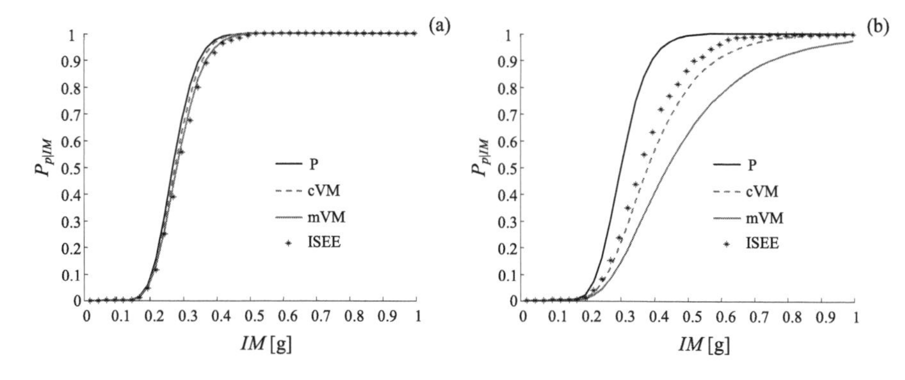

Fig. 18.3 Fragility curves for $\xi = 0.1$ m: (**a**) $t_A = 1.0$ s and $t_B = 0.5$ s, and (**b**) $t_A = 1.0$ s and $t_B = 0.9$ s

motion time histories. The values of S_0 corresponding to different values of PGA are obtained as follows:

$$S_0 = \left(\frac{PGA}{PGA_{S_0=1}} \right)^2 \tag{18.11}$$

In this study, the site hazard curve is expressed in the approximate form used in Cornell et al. [32], i.e.,

$$\nu_{IM}(im) = P[IM \geq im|1yr] = k_0 \cdot im^{-k_1} \tag{18.12}$$

in which k_0 and k_1 = parameters obtained by fitting a straight line through two known points of the site hazard curve in logarithmic scale. The site hazard curve is taken from Eurocode 8-Part 2 [33], assuming that, for the site of interest, $PGA =$ 0.3 g corresponds to a return period of 475 years. Using $k_1 = 2.857$ [34], the site hazard curve becomes (see Fig. 18.2(b)).

$$\nu_{PGA}(pga) = 6.734 \cdot 10^{-5} \cdot pga^{-2.857} \tag{18.13}$$

4.1 Pounding Risk for Linear SDOF Systems with Deterministic Model Parameters

The first application example consists in the assessment of the pounding risk between two adjacent buildings modelled as deterministic linear elastic SDOF systems with periods t_A and t_B, and damping ratios $\zeta_A = \zeta_B = 5\%$. The conditional probability of pounding $P_{p|IM}$ is calculated using the approximate analytical hazard functions $h^{(P)}_{U_{rel}|IM}(\xi, t)$, $h^{(cVM)}_{U_{rel}|IM}(\xi, t)$, and $h^{(mVM)}_{U_{rel}|IM}(\xi, t)$, for a deterministic distance between the buildings $\xi = 0.1$ m and for two different combinations of natural periods of the two systems, i.e., (1) $t_A = 1.0$ s and $t_B = 0.5$ s, referred to as well separated natural periods (Fig. 18.3(a)), and (2) $t_A = 1.0$ s and $t_B = 0.9$ s, referred

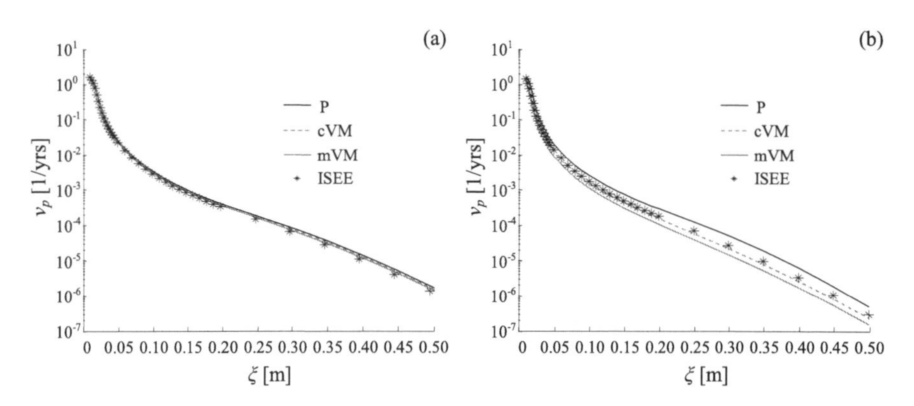

Fig. 18.4 MAF of pounding for varying separation distance: (**a**) $t_A = 1.0$ s and $t_B = 0.5$ s, and (**b**) $t_A = 1.0$ s and $t_B = 0.9$ s

to as close natural periods (Fig. 18.3(b)). The obtained conditional probabilities are presented in Fig. 18.3 as fragility curves and compared with the corresponding results obtained using ISEE method [27], which are assumed as reference solution.

In the case of well separated natural periods for the structures (Fig. 18.3(a)), the fragility curves estimated using the P, cVM, and mVM approximations are very similar and close to the fragility curves obtained using the ISEE method. In the case of close natural periods (Fig. 18.3(b)), the fragility curves estimated with the approximate analytical methods show significant differences, and only the cVM approximation provides results that are close to the fragility curves estimated using the ISEE method. The observed result can be explained by recognizing that the relative displacement process $U_{rel}(t)$ can be interpreted as a response process of a two-degree-of-freedom system. This multi-modal characteristic of $U_{rel}(t)$ can significantly affect the accuracy of the different approximations of the time-variant hazard function $h_{U_{rel}|IM}(\xi, t)$ [35]. In the case of well separated natural periods, the contribution of the higher period vibration mode to $U_{rel}(t)$ is significantly larger than the contribution of the lower period vibration mode. By contrast, in the case of close natural periods, both vibration modes provide a significant contribution to the response process.

Figure 18.4 shows the MAF of pounding, ν_p, as a function of the building separation distance ξ (in semi-logarithmic scale) for the cases of well separated natural periods (Fig. 18.4(a)) and of close natural periods (Fig. 18.4(b)), respectively. The estimates of the MAF of pounding obtained using the analytical approximations (P, cVM, and mVM) of the hazard function are compared to the corresponding estimate obtained using the ISEE method. Figure 18.5 plots (in semi-logarithmic scale) the pounding risk for a design lifetime of 50 years, evaluated according to Eq. (18.4), for the same two cases of well separated and close natural periods. Considerations similar to the ones made for the fragility curves can be made also for the MAF of pounding and the 50-year pounding risk, i.e., the analytical approximations provide very accurate results for the case of well separated natural periods and less accu-

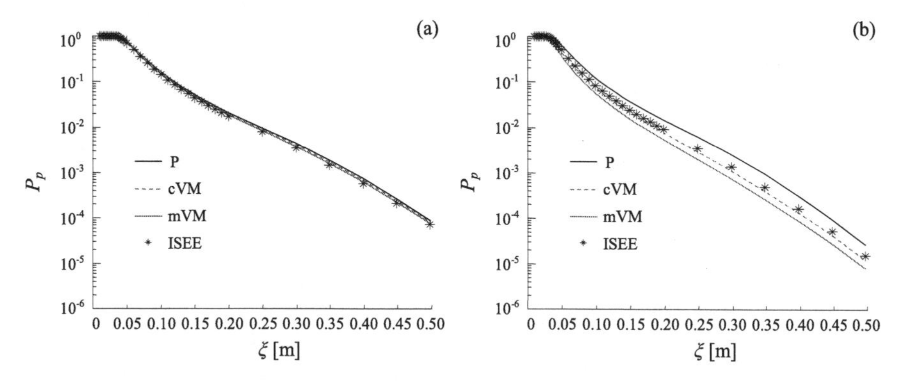

Fig. 18.5 50-year pounding risk for varying separation distance: (**a**) $t_A = 1.0$ s and $t_B = 0.5$ s, and (**b**) $t_A = 1.0$ s and $t_B = 0.9$ s

rate results for the case of close natural periods, with the exception of the cVM approximation, which is accurate in both cases.

It is observed that the P approximation of the time-variant hazard function always yields conservative results, while the mVM approximation underestimates the risk computed using the ISEE method for the case of close natural periods. Similar results have been documented for the first-passage reliability problem of SDOF and MDOF systems subjected to time-modulated white and colored noise excitations [26].

4.2 Reliability of Formulae Used in Seismic Design Codes

The proposed methodology is applied here to evaluate the pounding risk corresponding to the separation distance prescribed by anti-seismic design codes. In order to avoid pounding between new adjacent buildings, current seismic design codes (e.g., [4, 33]) prescribe a minimum clearance to be provided between the structures. This minimum clearance between two adjacent buildings is assumed equal to the expected value of the peak relative displacement, for a given site-specific earthquake action and a value of the seismic intensity (hazard level) corresponding to a given probability of exceedance. Given the seismic input, the peak relative displacement is obtained by combining (using simplified combination rules) the values of the peak displacements of the two adjacent structural systems, which are computed using (deterministic) structural analysis. The most commonly employed rules are the ABS method or the slightly more accurate SRSS method. The major limit of these approximate rules is that they neglect the response phase differences between the adjacent structures. In order to overcome this drawback, the use of the Double Difference Combination rule for determining the CSD has been proposed and investigated by several researchers [1–3].

In the application presented here, the values of the CSD according to the ABS, SRSS, and DDC rules are calculated following the procedure described in [5]. This

Table 18.1 Critical separation distance and corresponding 50-year pounding risk using different combination rules

	$t_A = 1.0$ s and $t_B = 0.5$ s				$t_A = 1.0$ s and $t_B = 0.9$ s		
	ABS	SRSS	DDC		ABS	SRSS	DDC
ξ [m]	0.1379	0.1049	0.1042	ξ [m]	0.1832	0.1298	0.0946
P_p	0.0620	0.1351	0.1376	P_p	0.0106	0.0334	0.0857

procedure involves (1) generating a set of 500 samples of input ground motion time histories for the reference value of the peak ground acceleration (i.e., $PGA = 0.3$ g, corresponding to a probability of exceedance of 10% in 50 years), (2) computing the corresponding 500 peak displacement responses of systems A and B ($U_{A,\max}$ and $U_{B,\max}$), (3) computing the sample means $\bar{U}_{A,\max}$ and $\bar{U}_{B,\max}$ of $U_{A,\max}$ and $U_{B,\max}$, respectively, and (4) combining $\bar{U}_{A,\max}$ and $\bar{U}_{B,\max}$ using the ABS, the SRSS, and the DDC rule to derive estimate of the peak relative displacement $\bar{U}_{rel,\max}$.

Table 18.1 shows the values of the separation distance computed according to different combination rules and the corresponding 50-year probability of failure, computed based on the cVM approximation of the time-variant hazard function. It is observed that the CSDs obtained using simplified combination rules yield inconsistent values of the failure probability, which are also strongly dependent on the natural periods of the two adjacent buildings. Thus, it is concluded that a methodology is still needed to determine the CSD between adjacent buildings corresponding to consistent safety levels for different combinations of the buildings' natural periods and location's seismic hazard.

4.3 MDOF Models of Multi-storey Buildings Retrofitted by Means of Viscous Dampers

As a third application, the proposed methodology is employed to assess the risk of pounding between two adjacent multi-storey buildings modelled as linear MDOF systems, before and after retrofit with viscous dampers (Fig. 18.6(a)). Different retrofit solutions are considered and their effectiveness in reducing the seismic pounding risk is compared (Fig. 18.6(b)). The considered buildings are steel moment-resisting frames with shear-type behaviour. The properties of the buildings are taken from Lin [36]. Building A is a six-story building with story stiffness $k_A = 548{,}183$ kN/m (equal for every story) and floor mass $m_A = 454.545$ tons (equal for each floor). Building B is a four-story building with story stiffness $k_B = 470{,}840$ kN/m and floor mass $m_B = 454.545$ tons. A Rayleigh-type damping matrix $\mathbf{c}_R$ is used to model the inherent buildings' damping and is built by considering a damping ratio $\zeta_R = 2\%$ for the first two vibration modes of each system. MPU is not considered in this application. The fundamental vibration periods of building A and B are $t_A = 0.751$ s and $t_B = 0.562$ s, respectively.

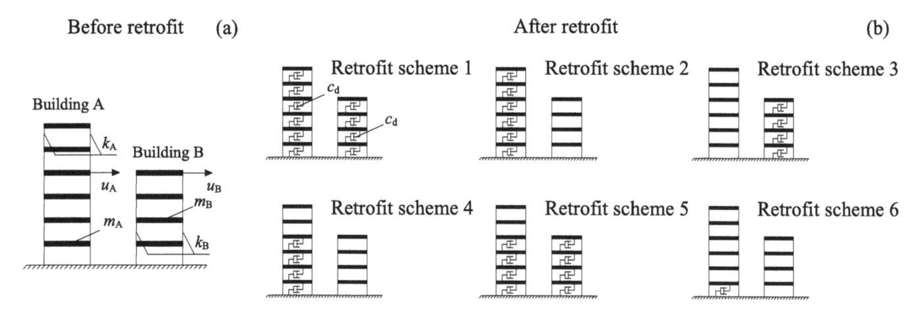

Fig. 18.6 Pounding between adjacent multi-storey buildings: (**a**) building A and B before retrofit, and (**b**) different retrofit schemes considered in this study

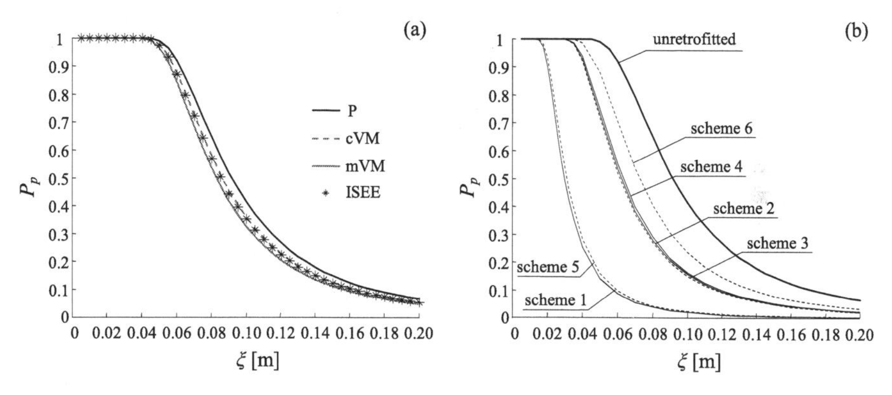

Fig. 18.7 Pounding risk between multi-storey buildings A and B: (**a**) comparison of different analytical solution and ISEE results, and (**b**) comparison of different retrofit schemes

The following six different retrofit solutions, based on the use of braces with purely viscous behaviour [37], are considered: (1) braces located at each story of both buildings (retrofit scheme 1), (2) braces located at all stories of the tall building only (retrofit scheme 2), (3) braces located at all stories of the short building only (retrofit scheme 3), (4) braces located at the lower four stories of the tall building only (retrofit scheme 4), (5) braces located at the lower four stories of both buildings, and (6) a single brace located at the first story of the tall building only. The two buildings before retrofit are shown in Fig. 18.6(a), while the six retrofit schemes are shown in Fig. 18.6(b). The viscous braces provide an additional source of damping, modelled by means of a damping matrix $\mathbf{c_V}$. The total damping matrix for the two buildings' systems becomes $\mathbf{c} = \mathbf{c_R} + \mathbf{c_V}$. The damping coefficient corresponding to the dampers at each floor of buildings A and B is $c_d = 10{,}000$ kN·s/m. The systems corresponding to retrofit schemes 4, 5, and 6 are non-classically damped and their analysis requires the use of the complex modal analysis technique [25].

Figure 18.7(a) shows three different analytical estimates (P, cVM, and mVM approximations) of the 50-year probability of pounding between the two un-retrofitted buildings, for different values of the separation distance. Figure 18.7 also reports the

50-year probability of pounding obtained using the ISEE method, which is considered as reference solution. The analytical estimates provide a very good approximation of the pounding risk for a wide range of separation distances. In this particular case, the results obtained using the cVM hazard function give the best approximation of the ISEE results.

Figure 18.7(b) compares the 50-year probability of pounding of the un-retrofitted buildings and of the buildings retrofitted following the six different retrofit solutions considered in this application example. The results presented in Fig. 18.7(b) are obtained using the cVM approximation of the hazard function.

It is observed that the use of viscous dampers can be very effective in reducing the risk of pounding between the two buildings. Furthermore, the introduction of viscous braces according to scheme 3, scheme 5, and scheme 6 (corresponding to the dotted lines in Fig. 18.7(b)) is a very efficient retrofit solution, since it obtains a significant reduction of the pounding risk at a significantly lower retrofit cost when compared with other retrofit schemes. In particular, retrofit scheme 3 appears to achieve a very good compromise between retrofit cost and reduction of pounding risk.

5 Possible Applications and Future Work

The innovative performance-based approach proposed in this paper for estimating the risk of pounding between adjacent buildings under earthquake excitation presents several practical applications. A first application is the evaluation of the safety levels corresponding to current seismic code provisions for building separation distance under different design conditions. The current code provisions are affected by several limitations: (1) they do not provide an explicit control on the performance of the structures [38]; (2) they do not account for uncertainties in the model properties of the structures; (3) they consider only a single value of the seismic event intensity instead of a continuous representation of the site seismic hazard [39]; and (4) they are based on simplified rules for combining the peak responses of the buildings [5], the accuracy of which is limited. Although these limitations have already been pointed out in many studies (including the present paper), an extensive parametric study involving a wide range of building properties and seismic input models is required to assess their effects on the performance and reliability of adjacent buildings subjected to seismic pounding hazard.

As an additional application, the proposed assessment methodology can be used as a first step toward improved code provisions and/or a performance-based design methodology for the separation distances between adjacent buildings. A performance-based design methodology should aim to obtain a target (sufficiently small and consistent for different design conditions) probability of pounding between two adjacent systems. The computation of the separation distance corresponding to the target pounding probability can be regarded as an inverse reliability problem, i.e., a problem in which one is seeking to determine the values to assign to design parameters such that target reliability levels are attained for the limit state considered. Possible design parameters that can be varied in the design are the separation distance between the two buildings, for newly design buildings, or the properties of viscous or viscoelastic dampers in the buildings or between the buildings in the case of existing buildings with a given insufficient separation distance.

An efficient solution to this design problem can be sought by recasting the inverse reliability problem as a zero-finding problem:

$$\mathbf{y}^* = \text{Zero}\big[P_p(\mathbf{y}, t_L) - \bar{P}_p\big] \tag{18.14}$$

where $\mathbf{y} =$ vector collecting the design parameters, the functional expression Zero[...] = zero of the function in the parentheses, and $P_p(\mathbf{x}, t_L) =$ probability of pounding in the design life-time t_L. This zero-finding problem can be solved efficiently using classical iterative constrained optimization algorithms.

6 Conclusions

This paper presents a fully probabilistic performance-based methodology for assessment of the seismic pounding risk between adjacent buildings. This methodology, which is consistent with the PEER PBEE framework, is able to account for all pertinent sources of uncertainty that can affect the pounding risk, e.g., uncertainty in the seismic input (i.e., site hazard and record-to-record variability) and in the parameters used to describe the structural systems of interest (i.e., material properties, geometry, damping properties, separation distance).

An efficient combination of analytical and simulation techniques is proposed for the calculation of the pounding risk under the assumptions of linear elastic behaviour for the buildings and of non-stationary Gaussian input ground motion. The pounding problem is recast as a first-passage reliability problem, which is solved analytically by using the spectral characteristics (up to the second order) of the non-stationary stochastic process representing the relative displacement between the buildings. Three different analytical approximations of the time-variant hazard function are used: (1) the Poisson's approximation, (2) the classical Vanmarcke's approximation, and (3) the modified Vanmarcke's approximation. Results obtained by employing the importance sampling using elementary events method are assumed as reference solutions to evaluate the absolute and relative accuracy of the three analytical approximations considered here. The effects of uncertainty in the model parameters are efficiently included by means of the total probability theorem and the Latin hypercube sampling technique.

The proposed methodology is applied in this paper to investigate the risk of pounding between SDOF systems with deterministic properties. With reference to this specific application example, the following observations are made:

(1) The proposed combination of analytical and simulation techniques provides sufficiently accurate estimates of the pounding risk when the classical Vanmarcke's approximation is used to estimate the time-variant hazard function.

(2) The accuracy of the analytical approximations of the time-variant hazard function depends on the ratio between the natural periods of the adjacent buildings. Higher accuracy is reached when the natural periods of the two buildings are well separated.
(3) The Poisson's approximation of the time-variant hazard function yields always conservative estimates of the risk.
(4) The simplified combination rules suggested in modern seismic design codes for calculating the critical separation distance yield inconsistent values of the pounding probability, which are also strongly dependent on the natural periods of the adjacent buildings.

In addition, the capabilities of the proposed method are demonstrated in a second application example by assessing the effectiveness of the use of viscous dampers, according to different retrofit schemes, in reducing the pounding probability of two adjacent multi-story buildings modelled as linear elastic multi-degree-of-freedom systems. Based on the results presented, the following considerations are made:

(1) The analytical approximations provide very accurate estimates of the pounding risk, due to the fact that the fundamental periods of the two buildings are well separated.
(2) The use of viscous dampers can dramatically reduce the risk of pounding between the two systems for any given separation distance.
(3) The use of viscous braces in the lower levels of the taller building is a very efficient and cost-effective technique for minimizing the pounding risk.

Acknowledgements The authors gratefully acknowledge support of this research by (1) the Louisiana Board of Regents (LA BoR) through the Pilot Funding for New Research (Pfund) Program of the National Science Foundation (NSF) Experimental Program to Stimulate Competitive Research (EPSCoR) under Award No. LEQSF(2011)-PFUND-225; (2) the LA BoR through the Louisiana Board of Regents Research and Development Program, Research Competitiveness (RCS) subprogram, under Award No. LESQSF(2010-13)-RD-A-01; (3) the Longwell's Family Foundation through the Fund for Innovation in Engineering Research (FIER) Program; and (4) the LSU Council on Research through the 2009–2010 Faculty Research Grant Program. Any opinions, findings, conclusions or recommendations expressed in this publication are those of the authors and do not necessarily reflect the views of the sponsors.

References

1. Jeng, V., Kasai, K., Maison, B.F.: A spectral difference method to estimate building separations to avoid pounding. Earthq. Spectra **8**, 201–223 (1992)
2. Kasai, K., Jagiasi, R.A., Jeng, V.: Inelastic vibration phase theory for seismic pounding. J. Struct. Eng. **122**, 1136–1146 (1996)
3. Penzien, J.: Evaluation of building separation distance required to prevent pounding during strong earthquake. Earthquake Eng. Struct. Dyn. **26**, 849–858 (1997)
4. International Conference of Building Officials (ICBO): Uniform Building Code, Whittier, California (1997)
5. Lopez-Garcia, D., Soong, T.T.: Assessment of the separation necessary to prevent seismic pounding between linear structural systems. Probab. Eng. Mech. **24**, 210–223 (2009)

6. Lin, J.H.: Separation distance to avoid seismic pounding of adjacent buildings. Earthquake Eng. Struct. Dyn. **26**, 395–403 (1997)
7. Lin, J.H., Weng, C.C.: Probability analysis of seismic pounding of adjacent buildings. Earthquake Eng. Struct. Dyn. **30**, 1539–1557 (2001)
8. Hong, H.P., Wang, S.S., Hong, P.: Critical building separation distance in reducing pounding risk under earthquake excitation. Struct. Saf. **25**, 287–303 (2003)
9. Wang, S.S., Hong, H.P.: Quantiles of critical separation distance for nonstationary seismic excitations. Eng. Struct. **28**, 985–991 (2006)
10. Porter, K.A.: An overview of PEER's performance-based earthquake engineering methodology. In: 9th International Conference on Application of Statistics and Probability in Civil Engineering (ICASP9), San Francisco, California (2003)
11. Zhang, Y., Acero, G., Conte, J., Yang, Z., Elgamal, A.: Seismic reliability assessment of a bridge ground system. In: 13th World Conference on Earthquake Engineering, Vancouver, Canada (2004)
12. Luco, N., Cornell, C.A.: Structure-specific scalar intensity measures for near-source and ordinary earthquake ground motions. Earthq. Spectra **23**, 357–392 (2007)
13. Karayannis, C.G., Favvata, M.J.: Earthquake-induced interaction between adjacent reinforced concrete structures with non-equal heights. Earthquake Eng. Struct. Dyn. **34**, 1–20 (2005)
14. Jankowski, R.: Non-linear viscoelastic modelling of earthquake-induced structural pounding. Earthquake Eng. Struct. Dyn. **34**, 595–611 (2005)
15. Muthukumar, S., DesRoches, R.: A Hertz contact model with nonlinear damping for pounding simulation. Earthquake Eng. Struct. Dyn. **35**, 811–828 (2006)
16. Cole, G., Dhakal, R., Carr, A., Bull, D.: An investigation of the effects of mass distribution on pounding structures. Earthquake Eng. Struct. Dyn. **40**, 641–659 (2010)
17. Roberts, J.B., Spanos, P.D.: Random Vibrations and Statistical Linearization. Dover, New York (1993)
18. Au, S., Beck, J.L.: Estimation of small failure probabilities in high dimensions by subset simulation. Probab. Eng. Mech. **16**, 263–277 (2001)
19. Crandall, S.H.: First-crossing probabilities of the linear oscillator. J. Sound Vib. **12**, 285–299 (1970)
20. Wen, Y.K.: Approximate methods for nonlinear time-variant reliability analysis. J. Eng. Mech. **113**, 1826–1839 (1987)
21. Vanmarcke, E.H.: On the distribution of the first-passage time for normal stationary random processes. J. Appl. Mech. **42**, 215–220 (1975)
22. Barbato, M.: Use of time-variant spectral characteristics of nonstationary random processes in structural reliability and earthquake engineering applications. In: Computational Methods in Stochastic Dynamics. Springer, Berlin (2011)
23. Michaelov, G., Sarkani, S., Lutes, L.D.: Spectral characteristics of nonstationary random processes—a critical review. Struct. Saf. **21**, 223–244 (1999)
24. Barbato, M., Conte, J.P.: Spectral characteristics of non-stationary random processes: theory and applications to linear structural models. Probab. Eng. Mech. **23**, 416–426 (2008)
25. Barbato, M., Vasta, M.: Closed-form solutions for the time-variant spectral characteristics of non-stationary random processes. Probab. Eng. Mech. **25**, 9–17 (2010)
26. Barbato, M., Conte, J.P.: Structural reliability applications of spectral characteristics of non-stationary random processes. J. Eng. Mech. **137**, 371–382 (2011)
27. Au, S.K., Beck, J.L.: First excursion probabilities for linear systems by very efficient importance sampling. Probab. Eng. Mech. **16**, 193–207 (2001)
28. Iman, R.L., Conover, W.J.: Small sample sensitivity analysis techniques for computer models, with an application to risk assessment. Commun. Stat. **A9**, 1749–1842 (1980)
29. Shinozuka, M., Sato, Y.: Simulation of nonstationary random processes. J. Eng. Mech. Div. **93**(EM1), 11–40 (1967)
30. Clough, R.W., Penzien, J.: Dynamics of Structures. McGraw-Hill, New York (1993)
31. Shinozuka, M., Deodatis, G.: Simulation of stochastic processes by spectral representation. Appl. Mech. Rev. **44**, 191–203 (1991)

32. Cornell, A.C., Jalayer, F., Hamburger, R.O.: Probabilistic basis for 2000 SAC federal emergency management agency steel moment frame guidelines. J. Struct. Eng. **128**, 526–532 (2002)
33. European Committee for Standardization (ECS): Eurocode 8—design of structures for earthquake resistance, EN1998. Brussels, Belgium (2005)
34. Lubkowski, Z.A.: Deriving the seismic action for alternative return periods according to Eurocode 8. In: 14th European Conference on Earthquake Engineering, Ohrid, Macedonia (2010)
35. Toro, G.R., Cornell, C.A.: Extremes of Gaussian processes with bimodal spectra. J. Eng. Mech. **112**, 465–484 (1986)
36. Lin, J.H.: Evaluation of seismic pounding risk of buildings in Taiwan. J. Chin. Inst. Eng. **28**, 867–872 (2005)
37. Occhiuzzi, A.: Additional viscous dampers for civil structures: analysis of design methods based on effective evaluation of modal damping ratios. Eng. Struct. **31**, 1093–1101 (2009)
38. Collins, K.R., Wen, Y.K., Foutch, D.A.: Dual-level seismic design: a reliability-based methodology. Earthquake Eng. Struct. Dyn. **25**, 1433–1467 (1996)
39. Krawinkler, H., Miranda, E.: Performance-based earthquake engineering. In: Bozorgnia, Y., Bertero, V.V. (eds.) Earthquake Engineering: From Engineering Seismology to Performance-Based Engineering. CRC Press, Boca Raton (2004)

Chapter 19
Intensity Parameters as Damage Potential Descriptors of Earthquakes

Anaxagoras Elenas

Abstract This paper provides a methodology to quantify the interrelationship between the seismic intensity parameters and the structural damage. First, a computer-supported elaboration of the accelerograms provides several peak, spectral and energy seismic parameters. After that, nonlinear dynamic analyses are carried out to provide the structural response for a set of seismic excitations. Among the several response characteristics, the overall structure damage indices after Park/Ang and the maximum inter-storey drift ratio are selected to represent the structural response. Correlation coefficients are evaluated to express the grade of interrelation between seismic acceleration parameters and the structural damage. The presented methodology is applied to a six-story reinforced concrete frame building, designed according to the rules of the recent Eurocodes. As seismic input for the nonlinear dynamic analysis, a set of spectrum-compatible synthetic accelerograms has been used. As the numerical results have shown, the spectral and energy parameters provide strong correlation to the damage indices. Due to this reason, spectral and energy related parameters are better qualified to be used for the characterization of the seismic damage potential.

1 Introduction

It is well-known that seismic accelerograms are ground acceleration time-histories that cannot be described analytically. Several seismic parameters have been presented in the literature during the last decades. These can be used to express the intensity of the seismic excitations and to simplify its description. Post-seismic field observations and numerical investigations have indicated the interdependency between the seismic parameters and the damage status of buildings after earthquakes [1, 2]. The latter can be expressed by proper damage indices, while the interdependency between the considered quantities can be quantified numerically by appropri-

A. Elenas (✉)
Institute of Structural Mechanics and Earthquake Engineering, Democritus University of Thrace, 67100 Xanthi, Greece
e-mail: elenas@civil.duth.gr

M. Papadrakakis et al. (eds.), *Computational Methods in Stochastic Dynamics*, Computational Methods in Applied Sciences 26,
DOI 10.1007/978-94-007-5134-7_19,

ate correlation coefficients. Their values deliver the correlation grade (low, medium or high) between the examined quantities.

This paper provides a method for quantifying the interrelationship between the seismic parameters and global damage indices. First, a computer analysis of the accelerograms provided several peak ground motion, spectral and energy seismic parameters. After that, nonlinear dynamic analyses were carried out to provide the structural response for a set of seismic excitations and a given reinforced concrete frame structure. Keeping in mind that most of the seismic loading parameters are characterized by a single numerical value, single-value damage indicators have also been selected to represent the structural response. Thus, the overall structural damage index (OSDI) after Park/Ang ($DI_{G,PA}$) and the maximum inter-storey drift ratio (MISDR) are selected to represent the structural response. Finally, correlation coefficients are evaluated to express the grade of interdependency between seismic acceleration parameters and the used damage indices. The presented methodology is applied to a six-story reinforced concrete frame building subjected to several artificial accelerograms.

2 Seismic Intensity Parameters

In general, the intensity parameters can be classified with peak, spectral and energy parameters. In this work the following parameters have been selected to represent the seismic intensity: peak ground acceleration PGA, peak ground velocity PGV, the term PGA/PGV, spectral acceleration (SA), spectral velocity (SV), spectral displacement (SD), central period (CP), absolute seismic input energy (E_{inp}), Arias intensity (I_A), strong motion duration after Trifunac/Brady (SMD_{TB}), seismic power ($P_{0.90}$), root mean square acceleration (RMS_a), intensity after Fajfar/Vidic/Fischinger (I_{FVF}), spectral intensities after Housner (SI_H), after Kappos (SI_K) and after Martinez-Rueda (SI_{MR}). They have been chosen from all three of the seismic parameter categories. Table 19.1 provides an overview of the used parameters and their literature references, respectively. The definition of each parameter is presented in the mentioned literature.

3 Seismic Acceleration Time Histories

The seismic excitations used for the dynamic analyses in this study are based on artificial accelerograms created to be compatible with the design spectra of the current Greek antiseismic code (2004). The reason for choosing this approach rather than relying on natural accelerograms was dictated by the need to have a sufficiently large database for statistical reasons. For the creation of the aforementioned artificial accelerograms the program SIMQKE [14] has been utilized. As artificial accelerogram creation parameters the PGA, the total duration (TD) and the design spectra for all three Greek seismic regions (nominal PGA equal to 0.16 g, 0.24 g and 0.36 g) have

Table 19.1 Seismic parameters

No	Seismic parameters	Reference	No	Seismic parameters	Reference
1	PGA	[3]	9	I_A	[7]
2	PGV	[3]	10	SMD_{TB}	[8]
3	PGA/PGV	[3]	11	$P_{0.90}$	[9]
4	SA	[4]	12	RMS_a	[3]
5	SV	[4]	13	I_{FVF}	[10]
6	SD	[4]	14	SI_H	[11]
7	CP	[5]	15	SI_K	[12]
8	E_{inp}	[6]	16	SI_{MR}	[13]

been used. All created for subsoil category B, as described in Eurocode 8 (EC8) [15] and the Greek Antiseismic Code [16]. This subsoil category belongs to deep deposits of medium dense sand or over-consolidated clay at least 70 m thick. In order to cover most types of Greek region seismic activity, an artificial accelerogram creation procedure has been devised comprising the creation of 5 random artificial accelerograms for each of the 15 preselected PGA values that were assigned for the three different Greek seismic regions. Thus, 75 different synthetic accelerograms have been compiled, which ensures that the overall structural damages of the examined structure will cover all the possible damage grades, from low to severe, in order to cover statistical demands as well.

4 Global Damage Indices

As explained previously, attention is focused on damage indicators that consolidate all member damage into one single value that can be easily and accurately be used for the statistical exploration of the interrelation with the also single-value seismic parameters in question. Thus, in the OSDI model after Park/Ang [17] the global damage is obtained as a weighted average of the local damage at the ends of each element. The local damage index is a linear combination of the damage caused by excessive deformation and that contributed by the repeated cyclic loading effect that occurs during seismic excitation. Thus, the local DI is given by the following relation:

$$DI_{L,PA} = \frac{\theta_m - \theta_r}{\theta_u - \theta_r} + \frac{\beta}{M_y \theta_u} E_T \quad (19.1)$$

where, $DI_{L,PA}$ is the local damage index, θ_m the maximum rotation attained during the load history, θ_u the ultimate rotation capacity of the section, θ_r the recoverable rotation at unloading, β a strength degrading parameter, M_y the yield moment of the section and E_T the dissipated hysteretic energy. The Park/Ang damage index is a linear combination of the maximum ductility and the hysteretic energy dissipation demand imposed by the earthquake on the structure.

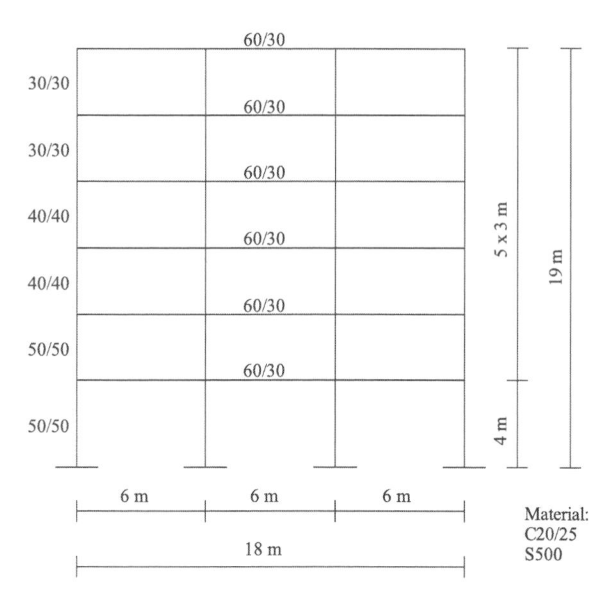

Fig. 19.1 Reinforced concrete frame structure

The global damage index after Park/Ang [17] takes into account the local damages of all elements of the examined structure (e.g. beams and columns of a frame). Thus, it depends both, the distribution and the severity of the localized damage and is given by the following relation:

$$DI_{G,PA} = \frac{\sum_{i=1}^{n} DI_{L,PA} E_i}{\sum_{i=1}^{n} E_i} \tag{19.2}$$

where, $DI_{G,PA}$ is the global damage index, $DI_{L,PA}$ the local damage index after Park/Ang, E_i the energy dissipated at location i and n the number of locations at which the local damage is computed.

The MISDR [18, 19] is a simple overall structural damage index that describes satisfactorily various forms of damages after an earthquake. The post-seismic damage degree can be classified according to this index. Equation (19.3) defines the maximum inter-storey drift ratio (MISDR) as the ratio of the maximum absolute inter-storey drift $|u|_{max}$ to the inter-storey height h:

$$MISDR = \frac{|u|_{max}}{h} 100\,[\%] \tag{19.3}$$

5 Application

The six-storey reinforced concrete frame structure shown in Fig. 19.1 has been detailed in agreement with the rules of the recent Eurocodes for structural concrete and antiseismic structures, Eurocode 2 (EC2) and Eurocode 8 (EC8) [15, 20], respectively. According to the EC8 Eurocode, the structure shown in Fig. 19.1, has been considered as an "importance class II, ductility class M"-structure with a subsoil category B (deep deposits of medium dense sand or over-consolidated clay at least 70 m thick). The building belongs on a seismic zone with nominal seismic design acceleration equal to 0.16 g. The cross-sections of the beams are considered as T-beams with 30 cm width, 20 cm slab thickness, 60 cm total beam height and 1.45 m effective slab width. The distances between each frame of the structure is equal to 6 m while the ground floor has a 4 m height and all subsequent floors 3 m. The eigenperiod of the frame was 1.0 s. In addition to the seismic load, live, snow and wind loads have also been taken into account as well as the eccentricity of structural element from verticality. The numerical values of loads, safety factors as well as load combinations have been chosen in accordance with the current design codes (Eurocodes).

After the design procedure of the reinforced concrete frame structure, a nonlinear dynamic analysis evaluates the structural seismic response, using the computer program IDARC [21]. A three-parameter Park model specifies the hysteretic behavior of beams and columns at both ends of each member. This hysteretic model incorporates stiffness degradation, strength deterioration, slip-lock and a tri-linear monotonic envelope. Experimental results of cyclic force-deformation characteristics of typical components of the studied structure, specifies the parameter values of the above degrading parameters. This study uses the nominal parameter for stiffness degradation. Among the several response parameters, the focus is on the overall structural damage indices (OSDI) described in the previous section.

6 Results

The first step was the creation of the aforementioned set of 75 synthetic accelerograms using the SIMQKE program. This program generates baseline corrected acceleration-time histories. The next step was a computer supported evaluation of 16 seismic parameters as presented in Table 19.1. Nonlinear dynamic analyses has been performed for the reinforced concrete frame building under question, including all artificial acceleration-time histories, in order to obtain the structural damage indices after Park/Ang and the MISDR. Statistical procedures provide the correlation coefficients after Pearson and Spearman [22], between all the evaluated seismic parameters and damage indices. The Pearson correlation shows how well the data fit a linear relationship, while the Spearman correlation shows how close the examined data are to monotone ranking. The latter coefficient is more important in the present study. Table 19.2 summarizes the results of the correlation study.

It is supposed that correlation coefficients up to 0.5 means low correlation, coefficients between 0.5 and 0.8 means medium correlations, while coefficients greater than 0.8 means strong correlation between the two variables. Table 19.2 presents the correlation coefficients after Pearson and the rank correlation coefficients after Spearman among all the examined seismic parameters presented and the examined the damage indices. Thus, the results show low Pearson and Spearman correlation

Table 19.2 Correlation coefficients between the seismic parameters and the OSDIs

Seismic parameters	Pearson correlation		Spearman rank correlation	
	$DI_{G,PA}$	MISDR	$DI_{G,PA}$	MISDR
PGA	0.568	0.523	0.635	0.631
PGV	0.657	0.659	0.788	0.795
PGA/PGV	−0.355	−0.367	−0.393	−0.394
SA	0.711	0.678	0.803	0.806
SV	0.724	0.696	0.804	0.804
SD	0.738	0.706	0.849	0.845
CP	−0.342	−0.326	−0.351	−0.332
E_{inp}	0.668	0.667	0.812	0.821
I_A	0.682	0.659	0.824	0.821
SMD_{TB}	0.103	0.086	0.155	0.145
$P_{0.90}$	0.685	0.662	0.823	0.820
RMS_a	0.713	0.677	0.824	0.821
I_{FVF}	0.655	0.656	0.789	0.796
SI_H	0.703	0.664	0.796	0.795
SI_K	0.702	0.670	0.802	0.806
SI_{MR}	0.614	0.558	0.725	0.725

between the term PGA/PGV, CP, SMD_{TB} and the examined damage indices. All the remaining seismic parameters provided medium Pearson correlation with the examined damage indices. On the other hand, high rank correlation is observed between SA, SV, SD, E_{inp}, I_A, $P_{0.90}$, RMS_a, SI_K and the used damage indices. In addition, medium rank correlation is observed between PGA, PGV, I_{FVF}, SI_H, SI_{MR} and the damage indices.

Thus, spectral (SA, SV, SD, SI_K) and energy (E_{inp}, I_A, $P_{0.90}$, RMS_a) seismic intensity parameters provided high correlation with the examined overall structural damage indices. These parameters are appropriate descriptors of the damage potential of a seismic excitation. Finally, the seismic parameters show the same correlation grade with the global damage index of Park/Ang ($DI_{G,PA}$) and with the maximum inter-storey drift ratio (MISDR) in all the cases. All the seismic parameters show the same correlation grade for both, Pearson and Spearman correlation, with exception the cases with high rank correlation. There, the Pearson correlation grade is medium.

7 Conclusions

In this paper a methodology for the value estimation of the interdependence between seismic acceleration intensity parameters and damage indices has been presented.

Peak, spectral and energy parameters have been considered. The global damage index after Park/Ang and the MISDR represented the post-seismic structural damage status. The degree of the interrelationship between seismic parameters and damage indices has been expressed by the Pearson correlation coefficient and by the Spearman rank correlation coefficient.

The results show low Pearson and Spearman correlation between the term PGA/PGV, CP, SMD_{TB} and the examined damage indices. Medium correlation is observed between PGA, PGV, I_{FVF}, SI_H, SI_{MR} and the damage indices, in all the cases. High rank correlation is observed between SA, SV, SD, E_{inp}, I_A, $P_{0.90}$, RMS_a, SI_K and the damage indices. In all these cases, the corresponding Pearson correlation grade was medium. The seismic parameters show the same correlation grade with $DI_{G,PA}$ with MISDR in all the cases.

All these results lead to conclude that the spectral and energy seismic parameters are reliable descriptors of the seismic damage potential and to recommend them as appropriate descriptors of the seismic damage potential.

References

1. Elenas, A.: Correlation between seismic acceleration parameters and overall structural damage indices of buildings. Soil Dyn. Earthq. Eng. **20**, 93–100 (2000)
2. Elenas, A., Meskouris, K.: Correlation study between seismic acceleration parameters and damage indices of structures. Eng. Struct. **23**, 698–704 (2001)
3. Meskouris, K.: Structural Dynamics. Ernst & Sohn, Berlin (2000)
4. Chopra, A.K.: Dynamics of Structures. Prentice Hall, New York (1996)
5. Vanmarcke, E.H., Lai, S.S.P.: Strong-motion duration and RMS amplitude of earthquake records. Bull. Seismol. Soc. Am. **70**, 1293–1307 (1980)
6. Uang, C.M., Bertero, V.V.: Evaluation of seismic energy in structures. Earthquake Eng. Struct. Dyn. **19**, 77–90 (1990)
7. Arias, A.: A measure of earthquake intensity. In: Hansen, R.J. (ed.) Seismic Design for Nuclear Power Plants, pp. 438–483. MIT Press, Cambridge (1970)
8. Trifunac, M.D., Brady, A.G.: A study on the duration of strong earthquake ground motion. Bull. Seismol. Soc. Am. **65**, 581–626 (1975)
9. Jennings, P.C.: Engineering seismology. In: Kanamori, H., Boschi, E. (eds.) Earthquakes: Observation, Theory and Interpretation, pp. 138–173. Italian Physical Society, Varenna (1982)
10. Fajfar, P., Vidic, T., Fischinger, M.: A measure of earthquake motion capacity to damage medium-period structures. Soil Dyn. Earthq. Eng. **9**, 236–242 (1990)
11. Housner, G.W.: Spectrum intensities of strong motion earthquakes. In: Proceedings of Symposium on Earthquake and Blast Effects on Structures, pp. 20–36. EERI, Oakland (1952)
12. Kappos, A.J.: Sensitivity of calculated inelastic seismic response to input motion characteristics. In: Proceedings of the 4th U.S. National Conference on Earthquake Engineering, pp. 25–34. EERI, Oakland (1990)
13. Martinez-Rueda, J.E.: Definition of spectrum intensity for the scaling and simplified damage potential evaluation of earthquake records. In: Proceedings of the 11th European Conference on Earthquake Engineering, CD-ROM. Balkema, Rotterdam (1998)
14. Gasparini, D.A., Vanmarcke, E.H.: SIMQKE, a Program for Artificial Motion Generation, User's Manual and Documentation. Publication R76-4. MIT Press, Cambridge (1976),
15. EC8: Eurocode 8: design of structures for earthquake resistance—part 1: general rules, seismic actions, and rules for buildings. European Committee for Standardization, Brussels, Belgium (2004)

16. EAK: National Greek antiseismic code. Earthquake Planning and Protection Organization (OASP) Publication, Athens (2003)
17. Park, Y.J., Ang, A.H.S.: Mechanistic seismic damage model for reinforced concrete. J. Struct. Eng. **111**, 722–739 (1985)
18. Structural Engineers Association of California (SEAOC): Vision 2000: performance based seismic engineering of buildings. Sacramento, California (1995)
19. Rodriguez-Gomez, S., Cakmak, A.S.: Evaluation of seismic damage indices for reinforced concrete structures. Technical Report NCEER-90-0022, State University of New York, Buffalo (1990)
20. EC2: Eurocode 2: design of concrete structures—part 1: general rules and rules for buildings. European Committee for Standardization, Brussels, Belgium (2000)
21. Valles, R.E., Reinhorn, A.M., Kunnath, S.K., Li, C., Madan, A.: IDARC 2D version 4.0: a program for inelastic damage analysis of buildings. Technical Report NCEER-96-0010, State University of New York, Buffalo (1996)
22. Ryan, T.P.: Modern Engineering Statistics. Wiley, Hoboken (2007)

Chapter 20
Classification of Seismic Damages in Buildings Using Fuzzy Logic Procedures

Anaxagoras Elenas, Eleni Vrochidou, Petros Alvanitopoulos, and Ioannis Andreadis

Abstract It is well-known that damage observations on buildings after severe earthquakes exhibit interdependence with the seismic intensity parameters. Numerical elaboration of structural systems quantified the interrelation degree by correlation coefficients. Further, the seismic response of buildings is directly depended on the ground excitation. Consequently, the seismic response of buildings is directly depended on the used accelerogram and its intensity parameters. Among the several response quantities, the focus is on the overall damage. Thus, the Maximum Inter-Storey Drift Ratio and the damage index of Park/Ang are used. Intervals for the values of the damage indices are defined to classify the damage degree in low, medium, large and total. This paper presents an Adaptive Neuro-Fuzzy Inference System for the damage classification. The seismic excitations are simulated by artificial accelerograms. Their intensity is described by seismic parameters. The proposed system was trained and tested on a reinforced concrete structure. The results have shown that the proposed fuzzy technique contributes to the development of an efficient blind prediction of seismic damages. The recognition scheme achieves correct classification rates over 90%.

A. Elenas (✉)
Institute of Structural Mechanics and Earthquake Engineering, Democritus University of Thrace, 67100 Xanthi, Greece
e-mail: elenas@civil.duth.gr

E. Vrochidou · P. Alvanitopoulos · I. Andreadis
Laboratory of Electronics, Department of Electrical and Computer Engineering, Democritus University of Thrace, 67100 Xanthi, Greece

E. Vrochidou
e-mail: evrochid@ee.duth.gr

P. Alvanitopoulos
e-mail: palvanit@ee.duth.gr

I. Andreadis
e-mail: iandread@ee.duth.gr

M. Papadrakakis et al. (eds.), *Computational Methods in Stochastic Dynamics*, Computational Methods in Applied Sciences 26,
DOI 10.1007/978-94-007-5134-7_20,

1 Introduction

Seismic accelerograms are records of ground acceleration versus time during earthquakes that cannot be described analytically. However, several seismic parameters have been presented in the literature during the last decades that can be used to express the intensity of a seismic excitation and to simplify its description. Post-seismic field observations and numerical investigations have indicated the interdependency between the seismic parameters and the damage status of buildings after earthquakes [1, 2]. The latter can be expressed by proper damage indices (DIs). The Maximum Inter-Storey Drift Ratio (MISDR) and the global damage index as defined by Park/Ang ($DI_{G,PA}$) characterize effectively the structural damage caused to buildings during earthquakes and thus, are used as metrics to classify the damage degree into 4 categories, low, medium, large and total. In this context, the damage degrees denote undamaged or minor damage-repairable damage-irreparable damage-partial or total collapse of the building, respectively.

This paper suggests a technique based on an Adaptive Neuro-Fuzzy Inference System (ANFIS) for seismic structural damage classification. A total set of 200 artificial accelerograms has been used and were correctively assigned to one of the above four categories with performances up to 90% and 87% of accuracy, for MISDR and $DI_{G,PA}$, respectively. High classification rates indicate that the proposed methodology is suitable for adaptive predictive control of the behavior of the concrete construction used, for any unknown seismic signal. The proposed method is applied to an eight-story reinforced concrete frame building, designed after the rules of the recent Eurocodes.

2 Damage Indices

MISDR is an overall structural damage index (OSDI) that can define the level of post-seismic corruption in a building [3, 4] and can be evaluated by Eq. (20.1):

$$\mathrm{MISDR} = \frac{|u|_{\max}}{h} 100[\%] \tag{20.1}$$

where $|u|_{\max}$ is the absolute maximum inter-storey drift and h the inter-storey height.

Additionally, the OSDI after Park/Ang ($DI_{L,PA}$) is used to describe the structural damage [5]. First, the local damage index according to Park/Ang is calculated. The local damage index is a linear combination of the damage caused by excessive deformation and that contributed by the repeated cyclic loading effect that happens during an earthquake. The local DI is given by the relation:

$$DI_{L,PA} = \frac{\theta_m - \theta_r}{\theta_u - \theta_r} + \frac{\beta}{M_y \theta_u} E_T \tag{20.2}$$

where θ_m is the maximum rotation during the load history, θ_u is the ultimate rotation capacity of the section, θ_r is the recoverable rotation at unloading, β is a strength

Table 20.1 Structural damage classification according to MISDR and $DI_{G,PA}$

Structural Damage Indices	Structural Damage Degree			
	Low	Medium	Large	Total
MISDR	≤ 0.5	$0.5 < \text{MISDR} \leq 1.5$	$1.5 < \text{MISDR} \leq 2.5$	> 2.5
$DI_{G,PA}$	≤ 0.3	$0.3 < DI_{G,PA} \leq 0.6$	$0.6 < DI_{G,PA} \leq 0.8$	> 0.8

degrading parameter (0.1–0.15), M_y is the yield moment of the section and E_T is the dissipated hysteretic energy.

The global damage index after Park/Ang is a combination of the maximum ductility and the hysteretic energy dissipation demand forced by the earthquake on the structure. Thus, the global damage index after Park/Ang ($DI_{G,PA}$) is given by:

$$DI_{G,PA} = \frac{\sum_{i=0}^{n} DI_L E_i}{\sum_{i=0}^{n} E_i} \tag{20.3}$$

where E_i is the energy dissipated at location i and n is the number of locations at which the local damage is calculated.

The two used DIs are utilized extensively in earthquake engineering, as they are experimentally proved to express the behavior of structures [5–12]. In Table 20.1, intervals for the values of the DIs are defined to classify the damage degree in low, medium, large and total [11]. These categories refer to minor, reparable damage, irreparable damage and severe damage or collapse of buildings, respectively.

3 Seismic Intensity Parameters

It is well-known that seismic intensity parameters are simple descriptors of the complex seismic accelerogram and they exhibit interdependency with observed post-seismic damages. Correlation studies manifested the interrelation degree between seismic intensity parameters and the damage indicators [1, 2]. Therefore, the following parameters are evaluated: peak ground acceleration PGA, peak ground velocity PGV, the term PGA/PGV, spectral acceleration (SA), spectral velocity (SV), spectral displacement (SD), central period (CP), absolute seismic input energy (E_{inp}), Arias intensity (I_A), strong motion duration after Trifunac/Brady (SMD_{TB}), seismic power ($P_{0.90}$), root mean square acceleration (RMS_a), intensity after Fajfar/Vidic/Fischinger (I_{FVF}), spectral intensities after Housner (SI_H), after Kappos (SI_K) and after Martinez-Rueda (SI_{MR}), effective peak acceleration (EPA), maximum EPA (EPA_{max}), cumulative absolute velocity (CAV) and destructiveness potential after Araya/Saragoni (DP_{AS}). Table 20.2 presents the examined intensity parameters and their literature references, respectively.

Table 20.2 Seismic intensity parameters

No	Seismic Intensity Parameter	References
1	Peak Ground Acceleration (PGA)	[13, 14]
2	Peak Ground Velocity (PGV)	[13, 14]
3	PGA to PGV ratio (PGA/PGV)	[13, 14]
4	Spectral Velocity (SV)	[13, 14]
5	Spectral Acceleration (SA)	[13, 14]
6	Spectral Displacement (SD)	[13, 14]
7	Central Period (CP)	[15]
8	Seismic Energy Input (E_{inp})	[16]
9	Arias Intensity (I_A)	[17]
10	Strong Motion Duration after Trifunac/Brady (SMD_{TB})	[18]
11	Power ($P_{0.90}$)	[19]
12	Root Mean Square Acceleration (RMS_a)	[13]
13	Seismic Intensity after Fajfar/Vidic/Fischinger (I_{FVF})	[20]
14	Spectrum Intensity after Housner (SI_H)	[21]
15	Spectrum Intensity after Kappos (SI_K)	[22]
16	Spectrum Intensity after Martinez-Rueda (SI_{MR})	[23]
17	Effective Peak Acceleration (EPA)	[24, 25]
18	Cumulative Absolute Velocity (CAV)	[26]
19	Maximum EPA (EPA_{max})	[24, 25]
20	Destructiveness Potential after Araya/Saragoni (DP_{AS})	[27]

4 Structural Model

Figure 20.1 presents the examined reinforced concrete structure. The eigenfrequency of the frame is 0.85 Hz. The design of the 8-storey building is based on the recent Eurocode rules EC2 and EC8 [28, 29]. The cross-sections of the beams are T-beams with 40 cm width, 20 cm slab thickness, 60 cm total beam height and 1.45 m effective slab width. The distance between the frames of the structure is 6 m. The structure has been characterized as an "importance class II-ductility class medium" structure according to the EC8 Eurocode. The subsoil is of type C and the region seismicity of category 2 after the EC8 Eurocode (design around acceleration value equal to 0.24 g). External loads are taken under consideration and are incorporated into load combinations due to the rules of EC2 and EC8. With the help of the IDARC software, the characteristics of the building are inserted into the program and a dynamic analysis is taking place, so as to estimate the structural behaviour of the building [7].

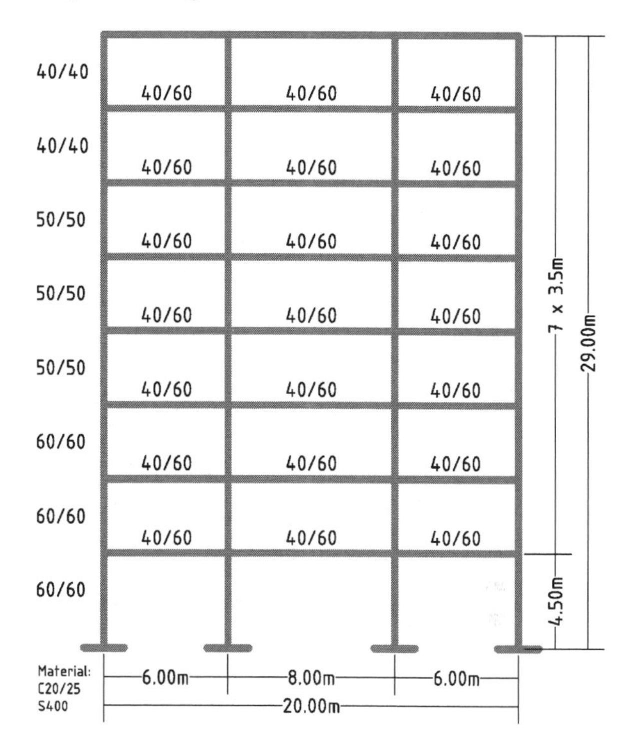

Fig. 20.1 Reinforced concrete frame structure

5 ANFIS Algorithm

ANFIS was introduced in 1993. ANFIS is able to extract a set of fuzzy "if-then" rules and define the membership functions in order to establish the association between inputs and outputs. Its structure is shown in Fig. 20.2. Basically, ANFIS suggests a method that, through the training procedure, can estimate the membership function parameters that serve the fuzzy inference system (FIS) to consequently specify the desired output for a certain given input [30].

ANFIS creates a fuzzy inference system in order to relate a certain input to the appropriate output. FIS interprets inputs into a set of fuzzy membership values and similarly the output membership functions to outputs. During the learning process, all parameters which define the membership functions will change. In order to optimize the model, these parameters are evaluated. Usually a gradient vector is used and an optimization routine could be applied in order to tune the parameters, so as to lead the model to a better generalization performance.

In this work, 20 seismic parameters are used as input data to describe the damage caused by one seismic event, and a total of 200 seismic events are used to train the system. All 20 seismic features have been normalized to belong in the interval [0, 1]. The 200 seismic events are distributed equally to all four damage categories in order to create a uniform data set.

First, inputs are related to membership functions (MFs) (Fig. 20.3 shows the initial MF for one of the seismic parameters), to rules to outputs MFs, by using

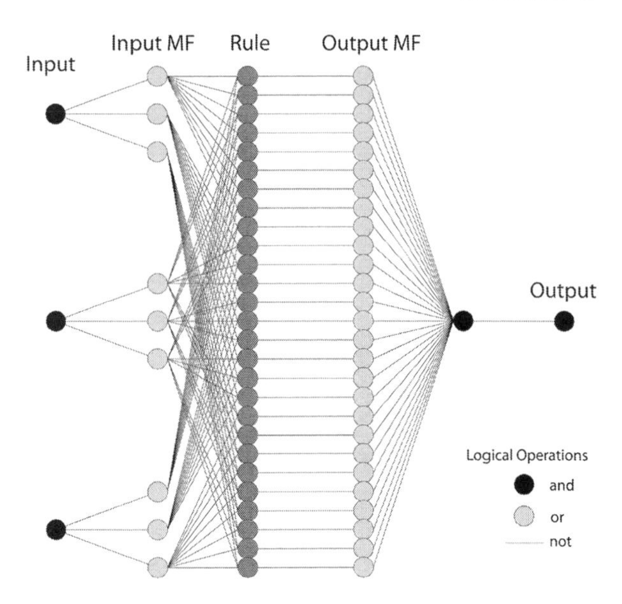

Fig. 20.2 ANFIS structure

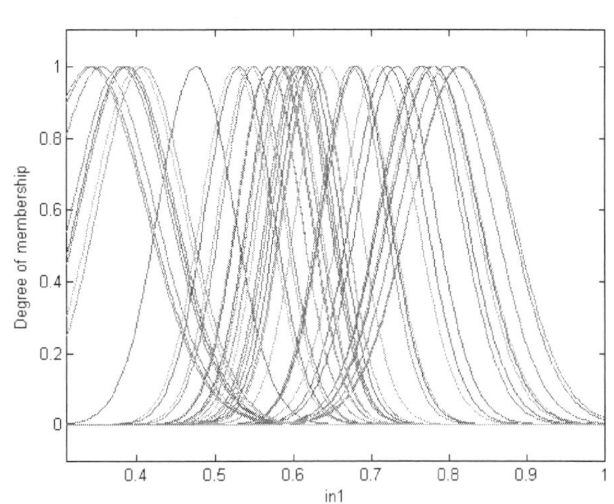

Fig. 20.3 Initial membership function on input 1

Fuzzy C-Means (FCM) technique [31, 32], which is analyzed later in this section. Next, the input/output data, which is a uniform set of 100 accelerograms, is used for training the model. The membership function parameters are tuned through the training process.

After the training, a model validation procedure is performed. During this procedure, an unknown input data set is presented to the trained fuzzy model for simulation. Thus, it can be evaluated the efficiency of the model. When a checking data set is presented to ANFIS, the fuzzy model selects the appropriate parameters associated with the minimum checking data model error. One crucial point with model validation, is selecting a suitable data set. This set must be representative of the

Table 20.3 Classification results based on the structural damage indices MISDR and $DI_{G,PA}$

Structural Damage Index	MISDR	$DI_{G,PA}$
Correct Classification Percentage (%)	90%	87%

data that the model is trying to simulate, and at the same time distinguishable from the training data. If a large amount of samples is collected, then all possible cases are contained and thus, the training set is more representative. In our case, a total number of 200 seismic excitations are considered as the data set.

FCM is a wildly used data clustering technique. Each data point is assigned to a cluster with a membership grade that is specified by a membership grade. It provides a method that shows how to group data points that populate some multidimensional space into a specific number of different clusters. The purpose of data clustering is to discover similarities between input patterns from a large data set, in order to design an effective classification system. At first; the FCM algorithm selects randomly the cluster centers. This initial choice for these centers is not always the appropriate. Furthermore, the variation of the cluster centers leads to different membership grades for each one of the clusters. Through the iteration process of the FCM algorithm, the cluster centers are gradually moved towards to their proper location. This is achieved by minimizing the weighted distance between any data point and the cluster centre. Finally, FCM function defines the cluster centers and the membership grades for every data point.

6 Results

The results are summarized in Table 20.3. The structural damage is presented by means of the two used DIs, MISDR and $DI_{G,PA}$, and the algorithm was tested for both DIs. The results indicate that the MISDR leads to higher performance, up to 90%, compared with the results when using $DI_{G,PA}$ which rates up to 87%.

In Figs. 20.4 and 20.5, blue circles represent the seismic signals that have been misclassified with ANFIS algorithm using MISDR and $DI_{G,PA}$ respectively.

7 Conclusions

This paper presents an efficient algorithm based on ANFIS techniques for seismic signal classification. A number of 20 seismic parameters and a set of 200 artificial accelerograms with known damage effects were used. For each seismic excitation the induced structural damage of the examined building is estimated and quantified according to two widely used damage indices, MISDR and $DI_{G,PA}$. The structural damage is expressed in the form of 4 damage categories. The 4 damage categories (classes) are defined through threshold values of the used damage indices.

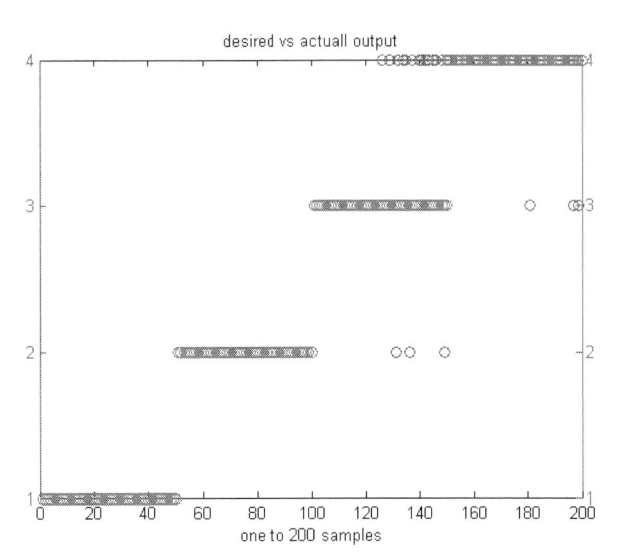

Fig. 20.4 Classification of 200 seismic signals into 4 damage classes with MISDR as metric. Correct classification percentage: 90%

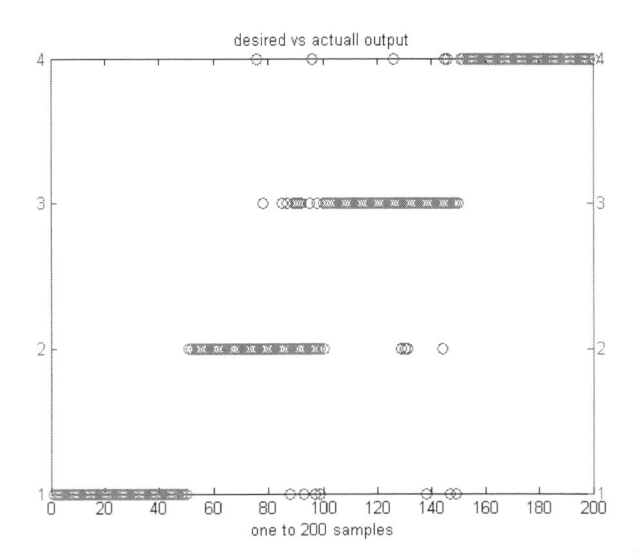

Fig. 20.5 Classification of 200 seismic signals into 4 damage classes with $DI_{G,PA}$ as metric. Correct classification percentage: 87%

An ANFIS model is trained and tested. The classification results reveal the effectiveness of the proposed system to estimate the earthquake's impact (damage category) on the examined structure. Classification rates up to 90% in the case of MISDR and 87% in the case of $DI_{G,PA}$ are achieved. The high percentage of correct classification in both cases, prove the efficiency of the method and show that the fuzzy technique that is implemented, contributes to the development of a competent blind prediction of the seismic damage potential that an accelerogram possesses.

References

1. Elenas, A.: Correlation between seismic acceleration parameters and overall structural damage indices of buildings. Soil Dynamics and Earthquake. Engineering **20**, 93–100 (2000)
2. Elenas, A., Meskouris, K.: Correlation study between seismic acceleration parameters and damage indices of structures. Eng. Struct. **23**, 698–704 (2001)
3. Structural Engineers Association of California (SEAOC): Vision 2000: performance based seismic engineering of buildings. Sacramento, California (1995)
4. Rodriguez-Gomez, S., Cakmak, A.S.: Evaluation of seismic damage indices for reinforced concrete structures. Technical Report NCEER-90-0022, State University of New York, Buffalo (1990)
5. Park, Y.J., Ang, A.H.S.: Mechanistic seismic damage model for reinforced concrete. J. Struct. Eng. **111**, 722–739 (1985)
6. Altoontash, A.: Simulation and damage models for performance assessment of reinforced concrete beam-column joints. Dissertation, Stanford University, Stanford (2004)
7. Valles, R.E., Reinhorn, A.M., Kunnath, S.K., Li, C., Madan, A.: IDARC 2D version 4.0: a program for inelastic damage analysis of buildings. Technical Report NCEER-96-0010, State University of New York, Buffalo (1996)
8. Freeman, S.A.: Drift limits: are they realistic. Earthq. Spectra **1**, 355–362 (1985)
9. CEB-FIP: Displacement-based design of reinforced concrete buildings. State-of-Art report, Fédération Internationale du Béton, Lausanne (2003)
10. Toussi, S., Yao, J.T.P.: Assessment of structural damage using the theory of evidence. Struct. Saf. **1**, 107–121 (1982)
11. Gunturi, S.K.V., Shah, H.C.: Building specific damage estimation. In: Proceedings of the 10th World Conference on Earthquake Engineering, pp. 6001–6006. Balkema, Rotterdam (1992)
12. Andreadis, I., Tsiftzis, Y., Elenas, A.: Intelligent seismic acceleration signal processing for structural damage classification. IEEE Trans. Instrum. Meas. **56**, 1555–1564 (2007)
13. Meskouris, K.: Structural Dynamics. Ernst & Sohn, Berlin (2000)
14. Chopra, A.K.: Dynamics of Structures. Prentice Hall, New York (1996)
15. Vanmarcke, E.H., Lai, S.S.P.: Strong-motion duration and RMS amplitude of earthquake records. Bull. Seismol. Soc. Am. **70**, 1293–1307 (1980)
16. Uang, C.M., Bertero, V.V.: Evaluation of seismic energy in structures. Earthquake Eng. Struct. Dyn. **19**, 77–90 (1990)
17. Arias, A.: A measure of earthquake intensity. In: Hansen, R.J. (ed.) Seismic Design for Nuclear Power Plants, pp. 438–483. MIT Press, Cambridge (1970)
18. Trifunac, M.D., Brady, A.G.: A study on the duration of strong earthquake ground motion. Bull. Seismol. Soc. Am. **65**, 581–626 (1975)
19. Jennings, P.C.: Engineering seismology. In: Kanamori, H., Boschi, E. (eds.) Earthquakes: Observation, Theory and Interpretation, pp. 138–173. Italian Physical Society, Varenna (1982)
20. Fajfar, P., Vidic, T., Fischinger, M.: A measure of earthquake motion capacity to damage medium-period structures. Soil Dyn. Earthq. Eng. **9**, 236–242 (1990)
21. Housner, G.W.: Spectrum intensities of strong motion earthquakes. In: Proceedings of Symposium on Earthquake and Blast Effects on Structures, pp. 20–36. EERI, Oakland (1952)
22. Kappos, A.J.: Sensitivity of calculated inelastic seismic response to input motion characteristics. In: Proceedings of the 4th U.S. National Conference on Earthquake Engineering, pp. 25–34. EERI, Oakland (1990)
23. Martinez-Rueda, J.E.: Definition of spectrum intensity for the scaling and simplified damage potential evaluation of earthquake records. In: Proceedings of the 11th European Conference on Earthquake Engineering, CD-ROM. Balkema, Rotterdam (1998)
24. ATC 3-06 Publication: Tentative provisions for the development of seismic regulations for buildings. Applied Technology Council, Redwood City, CA (1978)
25. Lungu, D., Aldea, A., Zaicenco, A., Cornea, T.: PSHA and GIS technology tools for seismic hazard macrozonation in Eastern Europe. In: Proceedings of the 11th European Conference on Earthquake Engineering, CD-ROM. Balkema, Rotterdam (1998)

26. Cabanas, L., Benito, B., Herraiz, M.: An approach to the measurement of the potential structural damage of earthquake ground motions. Earthquake Eng. Struct. Dyn. **26**, 79–92 (1997)
27. Araya, R., Saragoni, G.R.: Earthquake accelerograms destructiveness potential factor. In: Proceedings of the 8th World Conference on Earthquake Engineering, pp. 835–842. EERI, San Francisco (1984)
28. EC2: Eurocode 2: design of concrete structures—part 1: general rules and rules for buildings. European Committee for Standardization, Brussels, Belgium (2000)
29. EC8: Eurocode 8: design of structures for earthquake resistance—part 1: general rules, seismic actions, and rules for buildings. European Committee for Standardization, Brussels, Belgium (2004)
30. Duda, R.O., Hart, P.E., Stock, D.G.: Pattern Classification. Wiley, New York (2001)
31. Kurian, C.P., George, V.I., Bhat, J., Aithal, R.S.: ANFIS model for time series prediction of interior daylight illuminance. ICGST Int. J. Artif. Intell. Mach. Learn. (Online) **6**, 35–40 (2006)
32. Theodoridis, S., Koutroumbas, K.: Pattern Recognition. Academic Press, Kidlington (2009)

Chapter 21
Damage Identification of Masonry Structures Under Seismic Excitation

G. De Matteis, F. Campitiello, M.G. Masciotta, and M. Vasta

Abstract In the present paper, a spectral based damage identification technique is addressed on historical masonry structures. The seismic behavior of a physical 1:5.5 scaled model of the church of the Fossanova Abbey (Italy) is investigated by means of numerical and experimental analyses. Aiming at defining the seismic vulnerability of such a structural typology a wide experimental campaign was carried out. The achieved experimental results lead to the definition of a refined FE model reproducing the dynamic behavior of the whole structural complex. Then, the central transversal three-central bays of the church, as it mostly influences the seismic vulnerability of the Abbey, was investigated in a more detail by means of a shaking table test on a 1:5.5 scaled physical model in the Laboratory of the Institute for Earthquake Engineering and Engineering Seismology in Skopje. In the present paper a brief review of the numerical activity related to the prediction of the shaking table test response of the model is first proposed. Then, the identification of frequency decay during collapse is performed through decomposition of the measured power spectral density matrix. Finally, the localization and evolution of damage in the structure is analyzed. The obtained results shown that a very good agreement is achieved between the experimental data and the predictive/interpretative numerical analyses.

Keywords Masonry structures · Earthquake engineering · Shaking table tests · Power spectral matrix decomposition · Dynamic damage identification

1 Introduction

Gothic architecture spread as from the 12th Century and broke out during the Middle Ages in the cultural and religious area of the Christianity of Western Europe, with some trespasses in the Middle East and in the Slavic–Byzantine Europe. Many important abbeys were built in those areas, providing a key impulse to the regional

G. De Matteis · F. Campitiello · M.G. Masciotta · M. Vasta (✉)
Engineering Department, University "G. D'Annunzio" of Chieti-Pescara, Viale Pindaro 42, 65123 Pescara, Italy
e-mail: mvasta@unich.it

M. Papadrakakis et al. (eds.), *Computational Methods in Stochastic Dynamics*, Computational Methods in Applied Sciences 26,
DOI 10.1007/978-94-007-5134-7_21,

economy and contributing to a general social, economic and cultural development. The most interested areas sprawl from the northern Countries (England) to those facing the Mediterranean Basin (Italy), but also spread out from the Western (Portugal) to the Eastern Countries, as Poland and Hungary. Monastic orders and in particular the Cistercian one, with its monasteries, had an important role for broaden the new architectonic message, adapting to the local traditions the technical and formal heritage received by the Gothic style [1, 2] and [3].

Gothic cathedrals may result particularly sensitive to earthquake loading. Therefore, within the European research project "Earthquake Protection of Historical Buildings by Reversible Mixed Technologies" (PROHITECH), this structural typology has been investigated by means of shaking table tests on large scale models [4]. Based on a preliminary study devoted to define typological schemes and geometry which could be assumed as representative of many cases largely present in the seismic prone Mediterranean Countries, the Fossanova cathedral, which belongs to the Cistercian abbatial complex built in a small village in the central part of Italy, close to the city of Priverno (LT), has been selected as an interesting and reference example of pre-Gothic style church [5]. In order to assess the vulnerability of the church against seismic actions a wide numerical and experimental activity was developed. Firstly, the identification of the geometry of the main constructional parts as well as of the mechanical features of the constituting materials of the cathedral was carried out. Then, Ambient Vibration Tests were performed in order to characterize the dynamic behavior of the church and to calibrate refined FE models developed using the ABAQUS code [5]. To this purpose elastic FEM analyses were performed to predict the behavior of the three-central bays of the church, which were detected as the key-part of the structural complex [6, 7]. The recognized resistant unit about transversal direction was designed in length scale 1:5.5 according to "true replica" modeling principles and tested on the shaking table in the IZIIS Laboratory in Skopje [8]. The physical model was tested and the as-built model was loaded until heavy damage occurred. The structural response of the tested physical model has been deeply investigated by means of non-linear numerical analyses that has shown good agreements with experimental measurements [8].

In this paper, the identification of frequency decay during collapse is performed through decomposition of the measured power spectral density matrix. Finally, the localization and evolution of damage in the structure is analyzed. The obtained results has shown that a very good agreement is achieved between the experimental data and the predictive/interpretative numerical analyses.

2 The Fossanova Abbey: Model and Experimental Test

The Fossanova Abbey (Fig. 21.1) was built in the XII century and opened in 1208. The architectural complex presents three rectangular aisles with seven bays, a transept and a rectangular apse. Between the main bay and the transept raises the skylight turret with a bell tower. The main dimensions are 69.85 m (length), 20.05 m

Fig. 21.1 The Fossanova church

Fig. 21.2 The vaulted system of the Fossanova church

(height), and 23.20 m (width). The nave, the aisles, the transept and the apse are covered by ogival cross vaults. Detailed information on the main dimensions of the bays are provided in De Matteis et al. [6].

The previously mentioned vaulted system does not present ribs, but only ogival arches transversally oriented with respect to the span and ogival arches placed on the confining walls (Fig. 21.2). The ridge-poles of the covering wood structure is supported by masonry columns placed on the boss of the transversal arches of the nave and apse. The crossing between the main bay and the transept is covered by a wide ogival cross vault with diagonal ribs sustained by four cross shaped columns delimiting a span with the dimensions of 9.15×8.85 m.

The main structural elements constituting the central nave and the aisles are four longitudinal walls (West–East direction). The walls delimiting the nave are sustained by seven couples of cross-shaped piers (with dimensions of 1.80×1.80 m) with small columns laying on them and linked to the arches. The bays are delimited inside the church by columns with adjacent elements having a capital at the top. The columns-capital system supports the transversal arches of the nave. The external of the clearstory walls are delimited by the presence of buttresses with a hat on the top that reaches the height of 17.90 m. The walls of the clearstory present large splayed windows and oval openings that give access to the garret of the aisles. Also the walls that close the aisles present seven coupled column-buttresses systems reaching the height of 6.87 m and further splayed windows.

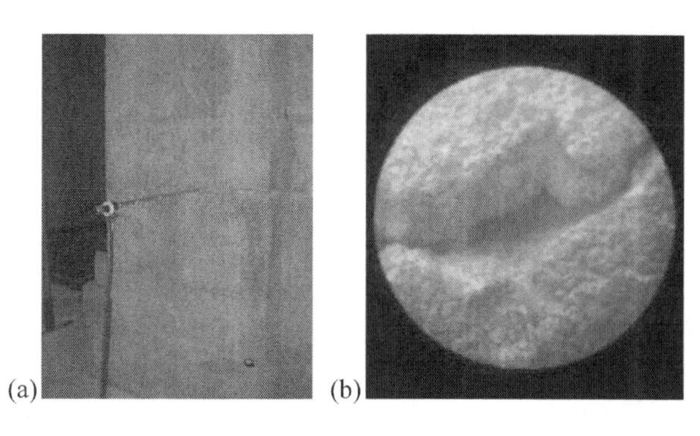

Fig. 21.3 Endoscope tests

During the centuries, the complex suffered some aesthetic modifications: the main prospect was modified since the narthex was eliminated installing an elaborate portal with a large rose-window; a part of the roof and of the lantern were rebuilt, introducing a Baroque style skylight turret; additional modifications on the roofing of the church were applied, with the reduction of the slope of pitches and with the restoration of the same slope as in the original form.

In order to determine the actual geometry and the mechanical features of the main constructional elements, an accurate experimental activity has been developed. In particular, both in situ inspection and laboratory tests have been carried out [5, 6].

It has been determined that the basic material constituting the constructional elements of the church is a very compact sedimentary limestone. In particular, columns and buttresses are made of plain stones with fine mortar joints (thickness less than 1 cm). The lateral walls (total thickness 120 cm) consist of two outer skins of good coursed ashlar (the skins being 30 cm thick) with an internal cavity with random rubble and mortar mixture fill.

In order to inspect the hidden parts of the constituting structural elements, endoscope tests have been executed on the right and left columns of the first bay, on the third buttress of the right aisle, on the wall of the main prospect and at the end on the filling of the vault covering the fourth bay of the nave. The test on the columns (Fig. 21.3a) allowed the exploration of the internal nucleus of the pier, relieving a total lack of internal vacuum, with the predominant presence of limestone connected with continuum joints of mortar (Fig. 21.3b). The test on the buttress was performed at the level of 143 cm, reaching the centre of the internal wall. The presence of regular stone blocks having different dimensions and connected to each other with mortar joints without any significant vacuum was detected. The tests on the wall put into evidence the presence of a two skins and rubble fill. The test made on the extrados of the vault, with a drilling depth of 100 cm, allowed a first layer of 7 cm made of light concrete and then a filling layer of irregular stones and mortar with the average thickness of 10 cm to be identified.

In order to define the mechanical features of the material, original blocks of stone were taken from the cathedral and submitted to compression tests (Fig. 21.4a). In total, 10 different specimens having different sizes have been tested, giving rise to an average ultimate strength of about 140 MPa and an average density $\gamma = 1700\ \mathrm{kg/m^3}$.

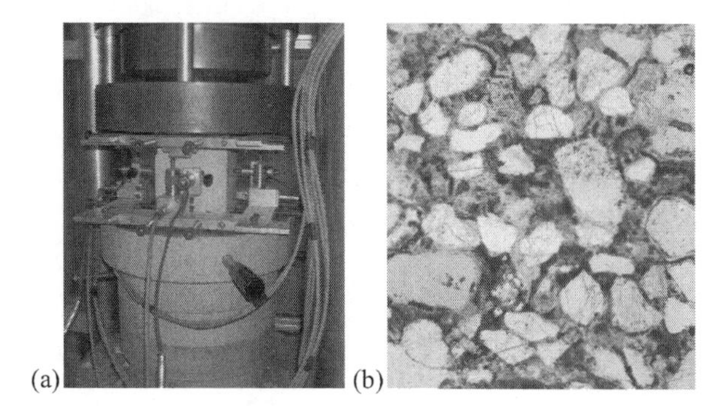

Fig. 21.4 Compression tests on limestone (**a**) and microscope analysis on mortar (**b**)

Besides, based on the results obtained for three different specimens, a Young's modulus equal to 42.600 MPa has been assessed, while a Poisson's ratio equal to 0.35 has been estimated.

Also, mortar specimens were extracted from the first column placed on the left of the first bay, from the wall of the aisle on the right and from the wall on the northern side of the transept. The specimens were catalogued as belonging to either the external joints (external mortar) or to the filling material (internal mortar). Compression tests have been carried out according to the Italian provisions (UNI EN 1926:2001), relieving a noticeable reduction of the average compressive strength for the specimens belonging to the external mortar (3.33 MPa) with respect to the internal ones (10.30 MPa). Besides, the Young's modulus has been determined on three different mortar specimens, according to the UNI EN 1015-11:2001 provisions, providing values ranging from 8.33 MPa to 12.16 MPa.

Chemical and petrography analyses have been also performed on the mortar specimens. In particular, chemical tests were made by X rays diffractometer analysis, according to the UNI 11088:2003 provisions. The prevalence of three material, namely, quartz crystal SiO_2, crystallized calcium carbonate $CaCO_3$ and some traces of felspate, was noticed. Also, a petrography study on thin sections of mortar specimens have been done by using two electronic microscopes, according to the UNI EN 932-3:1998 provisions (Fig. 21.4b). The analysis relieved the presence of quartz crystal sand end felspate, without any significant presence of crystallized calcium carbonate. The binding was quantified with a percentage of 60% of the total volume.

A FE model of the entire Abbey was calibrated on the basis of the in-situ experimental activity. The seismic analyses on such a model revealed that the more important structural part of the structural complex was to be recognized in the three-central bays of the main nave shown in Fig. 21.5 [7]. For the above reason a physical model of the key-structural part was designed and constructed in the IZIIS Laboratory in Skopje (Fig. 21.6). The model was executed in a 1:5.5 scale ratio (length) which was the maximum value compatible with the capacity of the shaking table.

The Buckingham's theorem was followed to define all the physical parameters needed to the construction of the model, according to the "true replica" modeling principles. All the involved quantities was scaled on the base of the three main parameters Length ($L_r = 1/5.5$), Mass Density ($\rho_r = 1$) and Acceleration ($a_r = 1$)

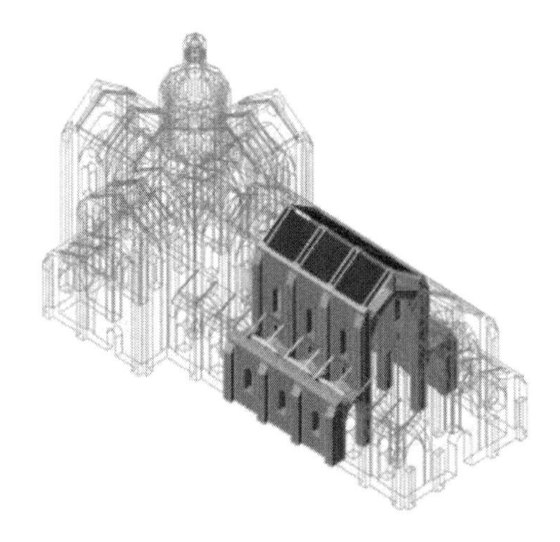

Fig. 21.5 Recognized seismic resistant unit in transversal direction

Fig. 21.6 Fossanova physical model (UPM) in scale 1:5.5 tested on shaking table

so that for the stresses a scaling ratio $\sigma_r = 0.18$ is obtained. The dimensions of the model were 3.97×4.44 m at the base, 3.67 m was the maximum height.

Some simplifications were adopted in the construction of the prototype: free edges was left in longitudinal direction, in fact no boundary restrains was applied on the fronts, neglecting the longitudinal continuity as it is in the reality. Then, the wooden roof structure wasn't realized because it wasn't considered as an active element in the evaluation of the seismic vulnerability [7].

The input signal of the test was assumed to be the scaled natural Calitri record (North–South direction) of Irpinia (Italy) 1980 earthquake record. The main feature of the selected earthquake are a maximum acceleration of 0.155 g (compatible with seismic hazard of the site), a quite long duration time (80 s), a high input energy for the relevant frequency (0.5 Hz–10 Hz) and typical two peak accelerations (or two strong motions). The record and the derived elastic spectra (with damping ratio $\zeta = 5\%$) are shown in Fig. 21.7a, b. The shaking table test was performed by considering three phases: phase 1, phase 2A and phase 2B. In the first phase the as-built unreinforced physical model (UPM) was tested and heavy damaged at the end. In the second phase (2A) the model was repaired and reinforced with carbon fiber ties.

Fig. 21.7 The accelerometric record (Irpinia, 1980)

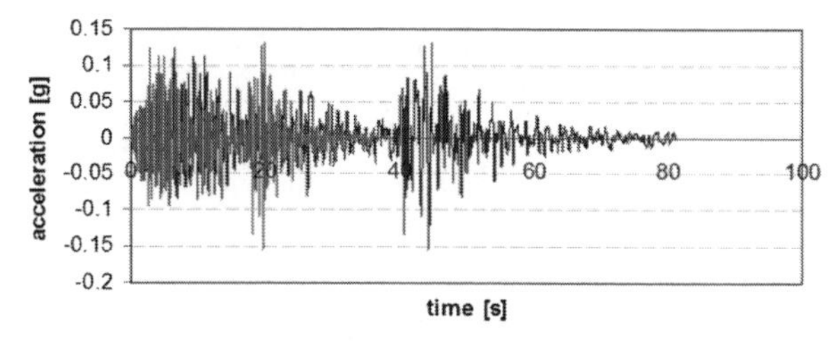

a. Calitri North-South record (scaled and not scaled)

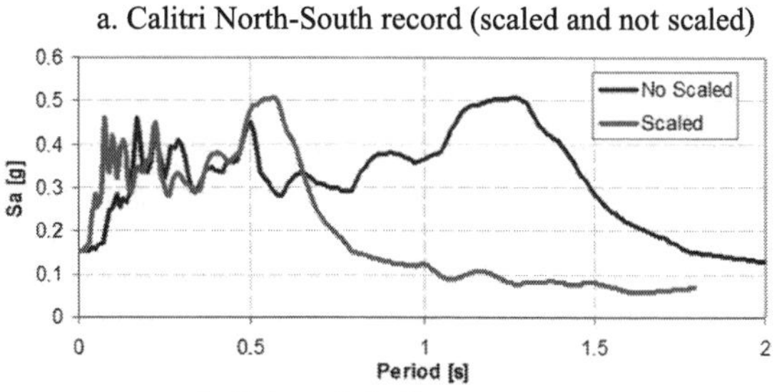

b. Relevant elastic response spectra

Fig. 21.8 Equivalent capacity curve of the tested model

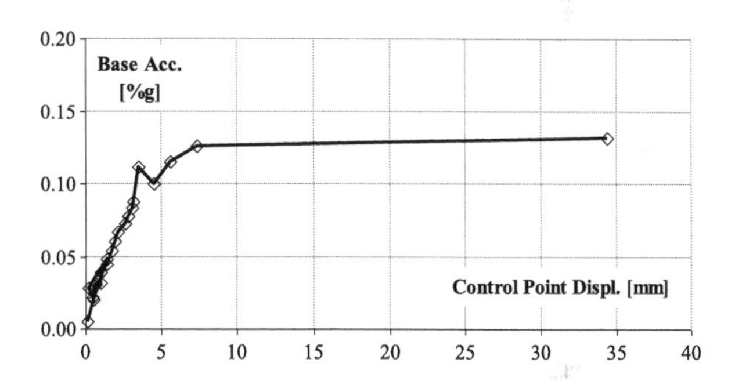

Finally, in the phase 2B, the reinforcing system was modified and the model was loaded until failure. Even though the examination of the reinforced systems is not the object of the present paper, in the following the level of input intensity which provoked serious damage to the model for every phase is listed:

- 0.14 g for the original model (phase 1);
- 0.28 g for the strengthened model (phase 2A);
- 0.40 g for the strengthened model (phase 2B);

In particular, the maximum acceleration measured at the base of the shaking table, by means of the accelerometer "CH1" [9], versus the maximum absolute displacement measured at the top of the buttresses at each step of the phase 1 (unreinforced physical model) is shown in Fig. 21.8. The curve can be assumed as an equivalent capacity curve for the tested UPM [8].

3 Dynamic Damage Identification

Assuming in first instance linearity under earthquake condition, the following equations of motion holds for the structure

$$\boldsymbol{M}\ddot{\boldsymbol{X}} + \boldsymbol{C}\dot{\boldsymbol{X}} + \boldsymbol{K}\boldsymbol{X} = a_g(t)\boldsymbol{V} \tag{21.1}$$

where $\boldsymbol{X}$(t) is the nodal structural response process, $\boldsymbol{M}$, $\boldsymbol{C}$ and $\boldsymbol{K}$ are $n \times n$ mass, damping and stiffness matrices respectively, $\boldsymbol{V}$ a $n \times 1$ vector while $a_g(t)$ represent the ground acceleration.

A more suitable representation of the structural response may be achieved by means of the decomposition in fully coherent independent vectors. Despite the non Gaussianity of the structural response vector $\mathbf{X}(t) = [X_i(t)]$ $(i = 1, \ldots, n)$, collecting the nodal response processes, its main characteristics can be represented by the knowledge of the second order spectral properties. Let us consider the power spectral density (PSD) matrix of $\boldsymbol{X}(t)$

$$\mathbf{S}_X(\omega) = \begin{bmatrix} S_{X_1}(\omega) & S_{X_1X_2}(\omega) & \cdots & S_{X_1X_n}(\omega) \\ S_{X_2X_1}(\omega) & S_{X_2}(\omega) & \cdots & S_{X_2X_n}(\omega) \\ \vdots & & \ddots & \vdots \\ S_{X_1X_n}(\omega) & \cdots & \cdots & S_{X_n}(\omega) \end{bmatrix} \tag{21.2}$$

The elements of $\mathbf{S}_X(\omega)$ are the direct and cross power spectral densities, defined as the Fourier transform of the correlation components

$$S_{X_1X_j}(\omega) = \frac{1}{2\pi}\int_{-\infty}^{+\infty} R_{X_1X_j}(\tau)e^{-j\omega\tau}\,d\tau \tag{21.3}$$

For finite length measurements of the process $\mathbf{X}(t)$ the elements of the spectral matrix becomes

$$S_{X_1X_j}(\omega, T) = \frac{E[X_i(\omega, T)X_j^*(\omega, T)]}{2\pi T} \tag{21.4}$$

where $X_i(\omega, T)$ denotes the Fourier transform of $X_i(t)$ over the observation time T

$$X_i(\omega, T) = \int_0^T X_i(t)e^{-j\omega t}\,dt \tag{21.5}$$

The input–output spectral relationship can be written as

$$\boldsymbol{S}_X(\omega, T) = \mathrm{S}_a(\omega, T)\left[\boldsymbol{H}^{*T}(\omega)\boldsymbol{V}^T\boldsymbol{V}\boldsymbol{H}(\omega)\right] \tag{21.6}$$

where $\mathrm{S}_a(\omega, T)$ represent the input ground acceleration PSD while $\boldsymbol{H}(\omega)$ is the transfer matrix

$$\boldsymbol{H}(\omega) = \left[\boldsymbol{K} + i\omega\boldsymbol{C} - \omega^2\boldsymbol{M}\right]^{-1} \tag{21.7}$$

Dividing both members of Eq. (21.6) by $\mathrm{S}_a(\omega, T)$

$$\boldsymbol{S}(\omega) = \frac{\boldsymbol{S}_X(\omega, T)}{\mathrm{S}_a(\omega, T)} = \boldsymbol{H}^{*T}(\omega)\boldsymbol{V}^T\boldsymbol{V}\boldsymbol{H}(\omega) \tag{21.8}$$

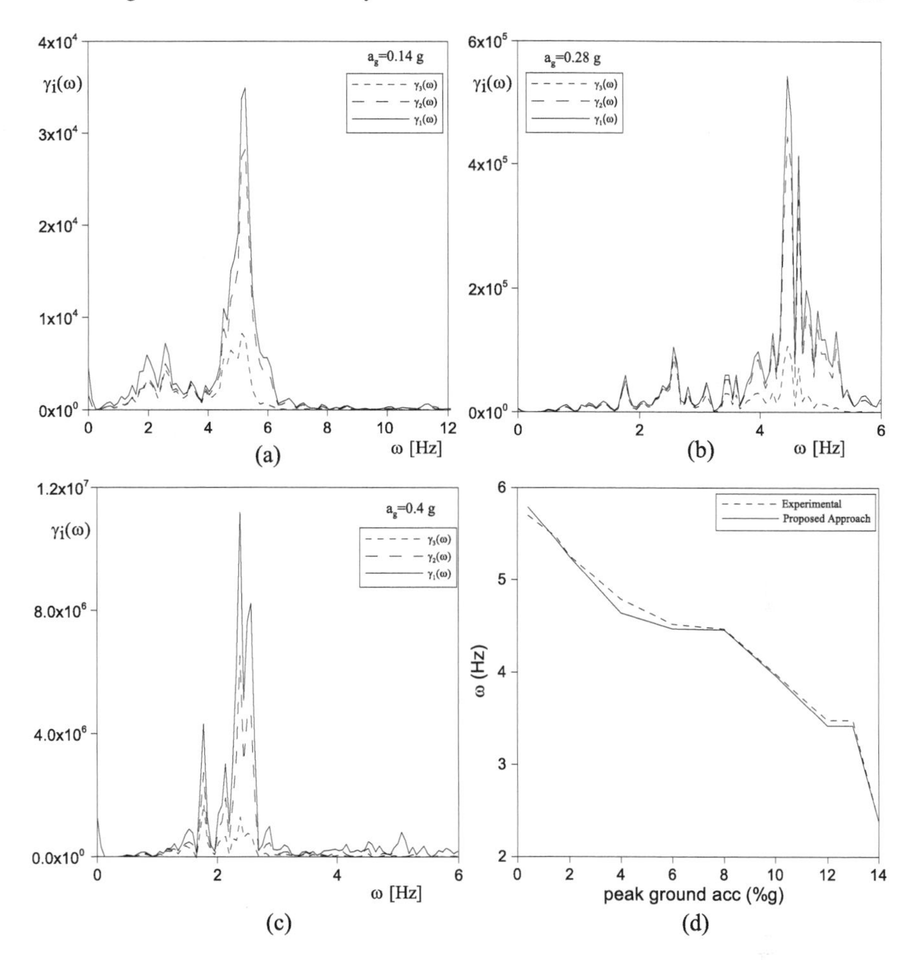

Fig. 21.9 Response spectral eigenvalues $\gamma_i(\omega)$ increasing peak ground acceleration: (**a**) $a_g = 0.14$ g; (**b**) $a_g = 0.28$ g; (**c**) $a_g = 0.4$ g; (**d**): Comparison between experimental and identified frequency decay curves for different values of ground peak acceleration

we observe that the matrix $\boldsymbol{S}(\omega)$ does not depend on the observation time T and, as well as the response PDF matrix $\boldsymbol{S}_X(\omega, T)$, is Hermitian and non-negative definite, thus its eigenvalues $\boldsymbol{\Gamma}(\omega) = \mathrm{diag}(\gamma_1(\omega)\gamma_2(\omega)\cdots\gamma_n(\omega))$ are real and non-negative, with orthonormal complex eigenvectors $\boldsymbol{\Psi}(\omega) = [\boldsymbol{\psi}_1(\omega)\boldsymbol{\psi}_2(\omega)\cdots\boldsymbol{\psi}_n(\omega)]$

$$\boldsymbol{\Psi}(\omega)^{*T}\boldsymbol{\Psi}(\omega) = \boldsymbol{I}, \qquad \boldsymbol{\Psi}(\omega)^{*T}\boldsymbol{S}(\omega)\boldsymbol{\Psi}(\omega) = \boldsymbol{\Gamma}(\omega) \tag{21.9}$$

$$\boldsymbol{S}(\omega)\boldsymbol{\Psi}(\omega) = \boldsymbol{\Psi}(\omega)\boldsymbol{\Gamma}(\omega) \tag{21.10}$$

The eigenvalues $\gamma_i(\omega)$ $(i = 1, \ldots, n)$ may be sorted in decreasing order, and their importance in principal component analysis is usually limited to a reduced numbers.

Figures 21.9a–c show the response spectral eigenvalues for different intensities of ground peak acceleration $a_g = 0.14$ g, $a_g = 0.28$ g and $a_g = 0.4$ g. In the fre-

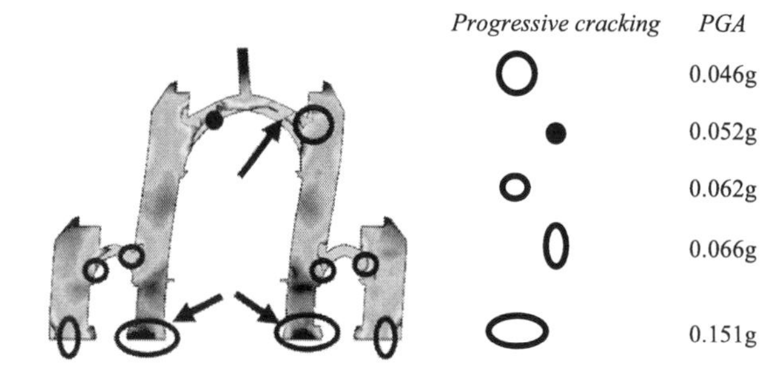

Fig. 21.10 Progressive damage of the model with increasing PGA

quency axis, the abscissa of the peak of the maximum eigenvalue $\gamma_1(\omega)$ allows for the identification of the frequency decay curve by varying a_g, as shown in Fig. 21.9d in comparison with experimental results.

Damage localization has been performed using the Parameter Method (PM) proposed by Dong et al. [10, 11] using a combination of frequency and mode shapes. The expression for the PM method is

$$\Delta\varphi = \sum_{j=1}^{n} \left[\phi_j^d \left(\frac{\omega_j^u}{\omega_j^d} \right) - \phi_j^u \right] \tag{21.11}$$

where ϕ is the structural mode, n the mode number while upper script u, d stands for undamaged and damage state respectively.

In Fig. 21.10 the progressive development of the cracks with increasing PGA is shown. Firstly, tensile stress is attained in both the arches for quite the same value of the PGA (0.046 g ÷ 0.062 g). Then, stress peaks were observed in the lateral buttresses for a PGA value equal to 0.066 g. Finally, the collapse mechanism was identified for a PGA of 0.151 g, when also the base sections of the internal piers exceed the adopted conventional limit state.

The experimental evidence of the collapse mode and crack evolution as shown in Fig. 21.10 can be identified by the PM method considering a simplified 2D model of the structure. In Fig. 21.11a the 2D finite element model is shown allowing the evaluation of the modes of the damaged and undamaged structure [12]. Once the evolutionary modes are known, damage location was performed by the PM method as shown in Fig. 21.11b–d were damage evolutionary localization is identified in the nodes of the 2D model.

4 Conclusions

The seismic behavior of a physical 1:5.5 scaled model of the church of the Fossanova Abbey has been investigated by means of numerical and experimental analyses. The achieved experimental results lead to the definition of a refined FE model reproducing the dynamic behavior of the whole structural complex. Then, the central transversal three-central bays of the church, as it mostly influences the seismic

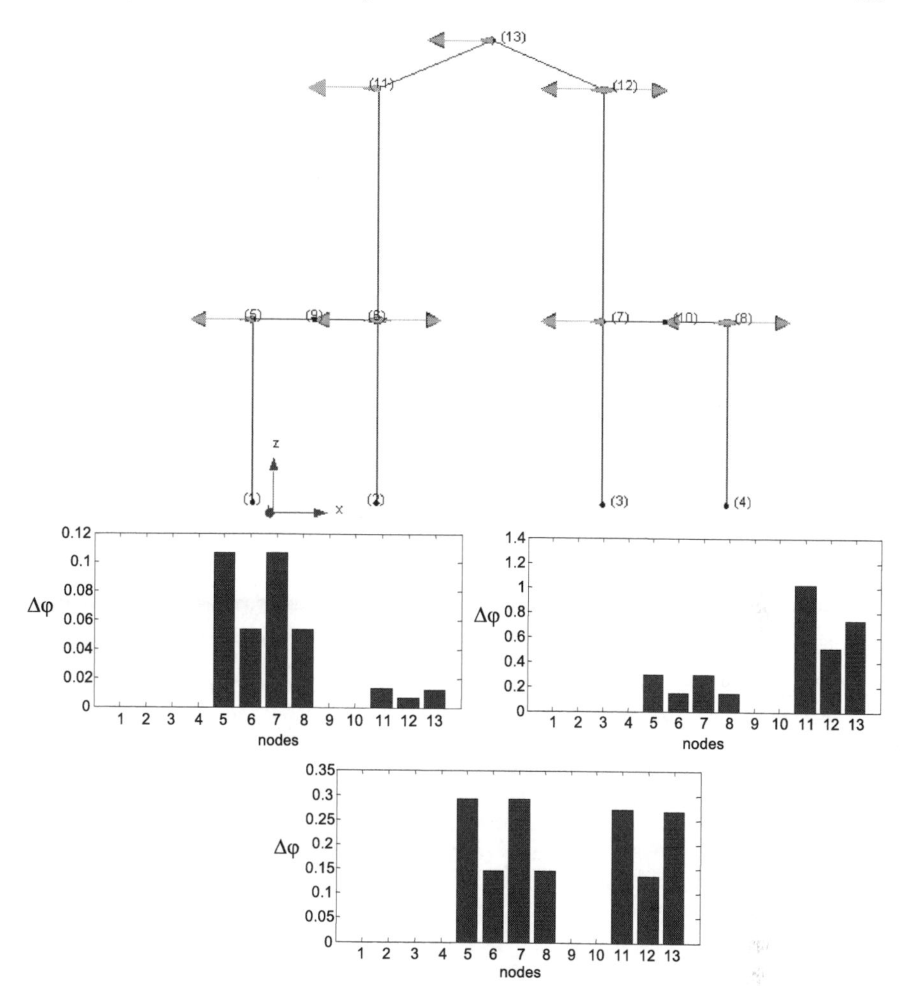

Fig. 21.11 (**a**) 2D finite element model, *arrows* symbol for measured displacement, (**b**) Progressive damage of the model with increasing PGA—PM method identification

vulnerability of the Abbey, was investigated in a more detail by means of a shaking table test on a 1:5.5 scaled physical model in the Laboratory of the Institute for Earthquake Engineering and Engineering Seismology in Skopje. In the present paper a brief review of the numerical activity related to the prediction of the shaking table test response of the model is first proposed. Then, the identification of frequency decay during collapse is performed through decomposition of the measured power spectral density matrix. Finally, the localization and evolution of damage in the structure is analyzed using a simplified 2D FEM model of the structure. The obtained results shown that a very good agreement is achieved between the experimental data and the predictive/interpretative numerical analyses.

References

1. Grodecki, L.: L'architettura gotica. Mondadori, Milan (1976) (in Italian)
2. Gimpel, J.: I costruttori di cattedrali. Jaca Book, Milan (1982) (in Italian)
3. De Longhi, F.: L'architettura delle chiese cistercensi Italiane. Ceschina, Milan (1958) (in Italian)
4. Mazzolani, F.M.: Earthquake protection of historical buildings by reversible mixed technologies. In: Proceedings of Behaviour of Steel Structures in Seismic Areas (STESSA 2006), pp. 11–24. Taylor & Francis, London (2006)
5. De Matteis, G., Colanzi, F., Mazzolani, F.M.: La chiesa abbaziale di Fossanova: indagini sperimentali per la valutazione della vulnerabilità sismica. In: Proceedings of the Workshop on Design for Rehabilitation of Masonry Structures (Wondermasonry 2), Ischia, Italy, 11–12 October (2007)
6. De Matteis, G., Langone, I., Colanzi, F., Mazzolani, F.M.: Experimental and numerical modal identification of the Fossanova Gothic cathedral. In: Proceedings of the 7th International Conference on Damage Assessment of Structures (DAMAS 2007), Torino, Italy, 25–27 June (2007)
7. De Matteis, G., Colanzi, F., Langone, I., Eboli, A., Mazzolani, F.M.: Numerical evaluation of the seismic response of the Fossanova cathedral based on experimental dynamic identification. In: Proceedings of the Third International Conference on Structural Engineering, Mechanics and Computation (SEMC 2007), Cape Town, South Africa, 10–12 September (2007)
8. De Matteis, G., Campitiello, F., Eboli, A., Mazzolani, F.M.: Analisi sismica della chiesa abbaziale di Fossanova mediante modelli numerici e sperimentali. In: L'Ingegneria Sismica in Italia, SM7.6, ANIDIS, Bologna (2009)
9. De Matteis, G., Mazzolani, F.M., Krstevska, L., Tashkov, Lj.: Seismic analysis and strengthening intervention of the Fossanova Gothic church: numerical and experimental activity. In: Urban Habitat Construction Under Catastrophic Events, COST C26, Malta, pp. 247–254 (2008)
10. Dong, C., Zhang, P.Q., Feng, W.Q., Huang, T.C.: The sensitivity study of the modal parameters of a cracked beam. In: Proceedings of the 12th International Modal Analysis Conference, pp. 98–104 (1994)
11. Ramos, L.F., De Roeck, G., Lourenco, P.B., Campos-Costa, A.: Damage identification on arched masonry structures using ambient and random impact vibrations. Eng. Struct. **32**, 146–162 (2010)
12. ARTeMIS version 5.2. Structural Vibration Solution A/S, NOVI Science Park

Printed by Publishers' Graphics LLC

Felix E. Browder

WORLD ECONOMIC DEVELOPMENT

INTERNATIONAL LABOUR OFFICE

WORLD ECONOMIC DEVELOPMENT

EFFECTS ON ADVANCED INDUSTRIAL COUNTRIES

by

Eugene STALEY

MONTREAL
1944

Series
LC 0
108847
International Labor Office

Studies and Reports, Series B (Economic Conditions), No. 36

PUBLISHED BY THE INTERNATIONAL LABOUR OFFICE,
3480 University Street, Montreal, Canada

Published in the United Kingdom for the INTERNATIONAL LABOUR OFFICE
by P. S. King & Staples, Ltd., London

Distributed in the United States by the INTERNATIONAL LABOUR OFFICE,
Washington Branch, 734 Jackson Place, Washington, D.C.

PREFACE

This volume appears on the eve of the Twenty-sixth Session of the International Labour Conference, at which particular attention is to be given to the social objectives of economic policy, and especially to the measures that will have to be taken, nationally and internationally, to ensure full employment, social security and rising standards of living. It is evident that development of industry in countries which have not hitherto been industrial will be actively sought and will be a most important feature of the economic situation in the near future. That development must necessarily affect world economic policy to a great extent and its social repercussions, in old as well as new industrial countries, obviously call for close study. The International Labour Office therefore feels that it is timely to publish this survey of the subject by so eminent an authority as Professor Eugene Staley.

The work was put into galley proofs several months ago and was circulated by the author to a number of persons who had an interest in, and expert knowledge of, its subject matter. Valuable criticisms and suggestions on all or portions of the text rendered it possible to undertake a rather extensive revision. The author desires to express his thanks particularly to Mr. Louis Bean, Dr. Percy W. Bidwell, Dr. Gerhardt Colm, Professor Allan G. B. Fisher, Dr. H. D. Fong, Mr. Hal Lary, Mr. August Maffry, Mr. William K. Miller, Professor R. J. Saulnier, Mr. Theodore Sumberg, Mrs. Maxine Sweezy, and Professor Jacob Viner. None of these friendly critics, of course, takes responsibility for any of the conclusions drawn by the author.

The International Economics and Statistics Unit of the United States Bureau of Foreign and Domestic Commerce (Dr. Amos Taylor, Director) rendered invaluable assistance in the preparation of much of the statistical and chart material, particularly for Chapter VIII. The interest manifested and the care taken by the staff of the Bureau made possible a better and earlier graphic presentation than could otherwise have been achieved.

The Rockefeller Foundation gave generous financial support several years ago to a research programme on "The Economics of Transition and Adaptation" at the Fletcher School of Law and

Diplomacy. The original plan of this study and the development of part of its basic ideas and materials date from this programme, but the present monograph was prepared specially for, and in collaboration with, the International Labour Office.

It should be added that the final proofs were not seen by the author, owing to his absence on an important mission to China.

The International Labour Office.

TABLE OF CONTENTS

INTRODUCTION AND SUMMARY

I. THE PROBLEM

Article V of the Atlantic Charter announces as a policy of the United Nations "the fullest collaboration" in the economic field "with the object of securing for all, improved labour standards, economic advancement and social security". Article VI looks forward to a world where "all the men in all the lands may live out their lives in freedom from fear and want". The Conference of the International Labour Organisation held at New York in November 1941 specifically endorsed both these articles in a resolution which also proclaimed once more the principle set forth in the I. L. O. Constitution, namely, that a lasting peace "can be established only if it is based on social justice". Furthermore, the Conference resolved that the close of the war must be followed by immediate action, previously planned and arranged,

> for the feeding of peoples in need, for the reconstruction of the devastated countries, for the provision and transportation of raw materials and capital equipment necessary for the restoration of economic activity, for the re-opening of trade outlets, for the resettlement of workers and their families under circumstances in which they can work in freedom and security and hope, for the changing over of industry to the needs of peace, for the maintenance of employment and for the raising of standards of living throughout the world.

In the network of Mutual Aid Agreements between the United States and Governments receiving lend-lease aid the signatories have pledged themselves to "agreed action . . . open to participation by all other countries of like mind, directed to the expansion, by appropriate international and domestic measures, of production, employment, and the exchange and consumption of goods, which are the material foundations of the liberty and welfare of peoples". Many other official and semi-official pronouncements of United Nations leaders could be cited in which the same hope is held out to the common people of the world: expansion of production and consumption, higher and more secure standards of living and of labour, "freedom from want".

Production and Freedom from Want

There is basically only one way in which "all the men in all the lands" can approach freedom from want, namely, through vastly increased production. It cannot be too often repeated that improvement in living standards depends fundamentally on improvement in the capacity of a people to produce. No programme of charity can abolish want. Immediately after the war, and during the war as occupied areas are liberated, the more fortunate of the United Nations will, of course, assist the people suffering from hunger and destitution by helping them with foods and medicines and the means of restarting production. But this can be no more than a palliative. The real, fundamental, permanent attack on want is through measures to multiply the efficiency of labour.

To some people, accustomed to think of the wealthier economies, it may seem that steadier use of existing productive capacity and more even distribution of its fruits would yield fairly satisfactory living standards. Yet when one looks at the unfilled material wants of the vast majority of the world's people it is quickly apparent that nothing but a radical advance in their *capacity to produce* will bring them within hailing distance of any modern conception of freedom from want. Even in the advanced countries, comparisons of actual consumption of large numbers of people in the lower income groups with goals of material well-being that seem modest enough to modern men reveal plenty of need for greater production if freedom from want is to be attained.

Of course, freedom from want is a very elastic goal. As peoples' standards of living rise they discover new wants. But the minimum definition of freedom from want, under modern conditions, would include these great essentials: enough food of the right kinds to maintain vigorous health, adequate clothing for comfort and cleanliness, houses safe and pleasant to live in, health care, and at least elementary education for all. There is still a long road to freedom from want, even on this minimum definition, for most of the people of the world. Production will have to be increased enormously in order to achieve it. If one were to go further and suppose that "all the men in all the lands" will some day be able to afford (that is, to produce) a standard of living that would be judged superior in our times—including, let us say, such new devices as radios, telephones, and automobiles to the same extent as the people of the United States use them today—the expansion in production of these and related goods and the host of raw material industries and technical services which feed into them would need to be fantastic.

Food. It is estimated that 60 per cent. of the world's gainfully occupied people are normally engaged in the production of food.

Yet the Mixed Committee on Nutrition of the League of Nations concluded in 1937 that about three quarters of the more than one thousand million inhabitants of Asia have a diet far below the standard for health, while even in the United States and western Europe much malnutrition exists among the lower income groups, extending to 20 or 30 per cent. of the entire population. The United Nations Conference on Food and Agriculture declared in 1943 that "there has never been enough food in the world for the people of the world to eat. To provide the food needs of mankind will require a vast increase in food production in every land."[1]

Clothing. If all people used as much cotton per person for wearing apparel and household furnishing as was used before the war in the United States of America, the cotton consumption of the world would be three times its highest point in the past[2], and textile output would have to be increased in proportion.

Shelter. Even the countries with the highest average living standards are discussing the need for great programmes of housing and rebuilding of slum areas after the war. The dwellings of most of the world's people certainly fall short of even low standards that might be set with regard to health, not to mention comfort and convenience.

Health Care. Soap is an elementary essential of sanitation. For the rest of the world to have one half as much per person as the average American, world production would have to be more than doubled.[3] Hospital beds are a rough index of the availability of modern medical science, including physicians, laboratories, vaccines, and other medicines. In the United States there were 10.7 hospital beds per 1000 persons before the war, and many rural areas were considered to be inadequately provided. But no Latin American country has one half as many in proportion to population, and experts are of the opinion that one hospital bed per 5000 persons would represent a significant achievement in the next stage of advance for such countries as China.[4]

Education. An enormous new investment would be required in equipment and in the training of personnel to make elementary

---

[1] UNITED NATIONS CONFERENCE ON FOOD AND AGRICULTURE, Hot Springs: *Final Act and Section Reports* (Washington, U. S. Government Printing Office, 1943), p. 36.

[2] *Memorandum on Consumption Levels and Requirements for Other Needed Agricultural Products* (submitted by the Delegation of the United States to the United Nations Conference on Food and Agriculture, Section 1 B, Hot Springs, Va., 18 May 1943).

[3] *Ibid.*

[4] A study prepared in the United States for the National Planning Association by George SOULE and associates, tentatively entitled *Latin America and Freedom from Want*, to be published in 1944 by Farrar and Rinehart; information supplied orally by a public health specialist.

education everywhere available and to provide the vocational training and advanced technical training necessary to make a modern economic system productive.

What about the new types of goods that no-one had a few decades ago? It would take some 600 million radios to bring the rest of the world up to the American consumption level. The number turned out by United States factories in 1940 was 11.5 million. For all the world to have as many telephones in proportion to population as the United States, which had 23.5 million at the beginning of 1942, would require nearly 350 million new instruments, with the vast amount of central station equipment and service organisation necessary to make them work. There were 32.6 million automobiles in use in the United States in 1941. For the rest of the world to have as many in proportion to population would have required 450 million additional vehicles, beyond the 45 million then in use outside the United States. At the highest annual production rate ever attained by the great American automobile industry in the past, it would take it more than seventy-five years to turn out that number of cars, and the need for production of rubber, metals, and service equipment would be staggering to the imagination. Highways (not to mention railways, waterways, and airlines) are fundamental in making possible a modern type of efficient production. The United States has three million miles of roads, one mile for each square mile of area. Sparsely populated Canada, with a 20 per cent. greater area including vast uninhabited frozen spaces and forest lands, has one mile of road for each 7.6 square miles of territory. For the entire world to have a network equal, not to that of the United States, but to the average for the United States and Canada combined, some 17.5 million miles of new road construction would be necessary.[1]

These calculations are not attempts to say what people ought to want, nor are they predictions of what they will attain in any near future. They are simply designed to show that for the attainment of a minimum level of "freedom from want" and still more for the satisfaction of some of the newer wants that arise where

---

[1] Rough calculations based on data from the following sources:

Radio: *Radio and Television Retailing* (Apr. 1941); U.S. DEPARTMENT OF COMMERCE: *Radio Today* (Jan. 1939).

Telephones: AMERICAN TELEPHONE AND TELEGRAPH COMPANY: *Telephone Statistics of the World* (1 Apr. 1942).

Automobiles: Automobile Manufacturers' Association, Detroit, and U.S. Department of Commerce.

Road mileage and areas: C. P. ROOT: *Highways of the World: Annual Statistical Survey of World Highways* (U.S. Bureau of Foreign and Domestic Commerce, Department of Commerce, Industrial Reference Service, No. 35, June 1941).

World population: LEAGUE OF NATIONS: *Statistical Year-Book, 1941-42*.

superior living standards are attainable, the level of production throughout the world would have to be raised enormously.

Economic Development

What is economic development? It is a combination of methods by which the capacity of a people to produce (and hence to consume) may be increased. It means introduction of better techniques; installing more and better capital equipment; raising the general level of education and the particular skills of labour and management; and expanding internal and external commerce in a manner to take better advantage of opportunities for specialisation.

Economic development is a broader term than "industrialisation"—if the latter is understood, as is generally the case, to stress the increase of manufacturing and other "secondary" production as compared with agriculture and other "primary" production. The greatest opportunities for raising productivity and income in many less developed areas will lie in modernisation of their agriculture, their forestry, their fisheries, etc., and not, at least at first, in the increase of manufacturing. Efficiency of production, adapted to the particular resources and circumstances of the country and to its best opportunities for specialisation in trade with the rest of the world, not mere imitation of the types of production that have expanded most dramatically in other countries, is the key to economic advancement.

A certain amount of "industrialisation", however, will practically always accompany substantial economic development, even in countries that, like Denmark and New Zealand, find much of the basis for their high living standards in intelligent agricultural specialisation. Increasing efficiency in agriculture, which is one of the strategic points for attack in the economic development of less developed countries, enables a smaller percentage of the occupied population to supply more food. Productive power is then available for non-agricultural pursuits. Also, as a community gets more income per family it spends a smaller proportion of its total income on food and a larger proportion on an increasing variety of other goods and services. It is a statistical fact that for every great region of the earth income levels are higher where the proportions of the working population engaged in agriculture are lower.[1]

[1] Louis H. Bean: *Industrialisation, the Universal Need for Occupational Adjustment out of Agriculture*, an unpublished manuscript made available by the author. Mr. Bean writes that China, India, many sections of Latin America, Africa, eastern Europe, and south-eastern United States are obviously over-agriculturalised, with 60 to 85 per cent. of their populations devoted to producing food and other farm products. On the basis of statistical comparisons with other areas, he concludes that the low per capita incomes of China and India could be doubled

(Footnote continued overleaf)

For the less developed countries as a whole, therefore, and for individual countries of large size and diversified resources, economic development will require and will bring about a considerable amount of industrialisation, in the sense of growth of "secondary" production. As incomes rise, an increasing proportion of workers will also be engaged in retail and wholesale trade, transportation and communication, administration, medical, educational and other professional services—that is, in so-called "tertiary" production, which is produced and consumed in much larger proportions in the wealthier than in the poorer communities.[1]

In the early stages of economic development industrialisation is likely to manifest itself in: (1) increased processing of local raw materials, including processing for export (*e.g.*, meat packing, canning, first stages in refining of minerals); (2) manufacture of simpler consumption goods (especially textiles and foodstuffs, but also furniture, soap, etc.); (3) assembly of products using imported parts (*e.g.*, automobiles and aircraft); (4) utilities and their maintenance (power plants, railway repair shops, etc.). In later stages the heavy, capital goods industries (metallurgical, metal working, machine tools, etc.), take on increased importance. This is the usual sequence, and the one likely to be easiest and also most immediately effective in raising living standards. The Soviet Union, however, has demonstrated that, at least under some circumstances, the heavy, capital goods industries can be pushed ahead of light industry, although this postpones the rise in living standards.

Development Programmes throughout the World

Throughout the world, programmes of post-war economic development, many of them very ambitious, some of them tremendous in scope, are being discussed.[2]

---

if, with more efficient use of human and natural resources, only 15 per cent. of their working populations were shifted from food production to other pursuits and that additional shifts of less than 10 per cent. would treble income.

> For the world as a whole, . . . there are about 800 million people classed as gainfully occupied and probably 60 per cent. of these (or about 500 million) are engaged in agriculture. If in the course of a reasonable period—say the first decade after the war—through appropriate regional development programmes, it were possible to alter the world's agricultural-industrial balance, so that 40 per cent. (instead of over 60 per cent.) were engaged in farming, the general gain in productivity and income and living standards would be enormous. . . From the standpoint of the number of working people involved in this broad objective of occupational adjustment, the bulk of reductions would be found in China (about 70 million), India (about 27 million), U.S.S.R. (about 13 million), Poland (about 2 million), Japan (about 2 million), and Latin America (about 2 million).

[1] Colin Clark: *Conditions of Economic Progress* (London, Macmillan, 1940), Chapter V.

[2] Cf. Lewis L. Lorwin: *Postwar Plans of the United Nations* (New York, Twentieth Century Fund, 1943), *passim.*

In China, development programmes today start with the vision of Dr. Sun Yat-sen, founder of the Chinese Republic. Shortly after the First World War he published a book in which he outlined a comprehensive scheme for installing modern transportation and communication and for developing industries and agriculture.[1] He proposed 100,000 miles of railway, one million miles of hard surfaced highway, improvement of existing canals and construction of new canals, regulation of rivers, construction of telegraph and telephone lines and radio stations, the development of three great ocean ports and various smaller harbours and docks, construction of public utilities and building of modern cities at all transportation centres, water power development, iron and steel works and cement works on a large scale in order to provide construction materials, mineral and agricultural development, great irrigation works, reafforestation in central and northern China, and colonisation in outlying regions.

Both official and unofficial groups have been at work in China during this war endeavouring to lay specific plans and to translate into technical data the materials and equipment and trained man-power needed for China's development over five, ten, twenty years and longer. More immediate proposals are being drawn up for the years following the war. One subject of controversy is the relative stress to be put on light industry and heavy industry. From the point of view of civilian economy and living standards light industry should come first, together with reform of agriculture. Military security, however, is more directly related to heavy industry. The decision on whether to bear the economic cost of heavy industry is likely to depend in large part on the nature of the peace.[2]

Some of the long-range goals which have been put forward by Chinese planners, for achievement over twenty years or more, include Dr. Sun's 100,000 miles of railway, half to be double-tracked, needing a total of 20 million tons of steel, 25,000 locomotives, 300,000 freight cars, and 30,000 passenger cars; half a million new automobiles a year for ten years, in order to achieve a total of three million at any given time; one million miles of highway; power plants capable of producing 20 million kilowatts; telephones

---

[1] *The International Development of China*, originally published in 1922. Second edition, with a Preface by Sun Fo, published in 1929. The book was reissued in 1943 by the Ministry of Information of the Republic of China, reprinted from the second edition.

[2] Theodore H. White, Chungking correspondent for *Time-Life-Fortune:* "China's Postwar Plans", in *Fortune*, Oct. 1943. For a discussion of China's resources, the directions that development might take, and problems connected with it, see H. D. Fong: *The Post-War Industrialisation of China* (National Planning Association, June 1942).

to the number of 80,000,000; one million new homes a year; modern furniture and sanitation industries; 320,000 cotton looms; 16,000 woollen looms; 94,000 silk looms; and, eventually, 10,000,000 tons of ocean-going shipping.[1]

Leaders of thought among the peoples of eastern and southeastern Europe stress the need for systematic economic development. The low standards of living in the area are based on an inefficient peasant agriculture. Competent students agree that, for its present methods of production, the area is heavily overpopulated. Persons engaged in agriculture actually work only a part of the year and yields are low. It has been estimated that about 25 per cent. of the agrarian population is either totally or partially unemployed—"disguised unemployment"—and industrial diversification is suggested as the way to make use of this excess labour.[2]

The area needs better transport, including railways and a network of local roads, better communications, power developments, and investment of capital in industries which would absorb some of the people whose capacity cannot be fully employed in agriculture. Suggestions have often been advanced to the effect that a comprehensive regional programme like that of the Tennessee Valley Authority is needed in this area—improvement of agricultural techniques, combined with pioneering work for the establishment of suitable industries.[3]

There is a great need and considerable desire throughout Latin America for programmes of economic improvement to raise living standards.[4] In seven Latin American countries national development corporations have come into existence since 1939. They have been engaged in wartime measures of stabilisation and in mobilisation of strategic resources, in co-operation with the Government of the United States, which has made credits available through the Export-Import Bank. Their work is also directed, however, towards the long-range expansion of production and is likely to continue after the war. Most of the other countries of Latin America are carrying on similar work under different forms of organisation. An Inter-American Development Commission was established in 1940 by the Inter-American Financial and Economic Advisory Committee (composed of representatives of the twenty-one American republics), and national commissions have been set up in

---

[1] Theodore H. WHITE, *op. cit.*

[2] P. N. ROSENSTEIN-RODAN: "Problems of Industrialisation of Eastern and Southeastern Europe", in *Economic Journal*, June-Sept. 1943, pp. 202-11.

[3] See, for example, Leon BARANSKI: *East and Central Europe* (published by New Europe, New York, 1943).

[4] See the detailed survey of the problem in the forthcoming study, *Latin America and Freedom from Want*, referred to in footnote 4 on p. 3.

each country to provide liaison with leaders in financial, industrial, engineering, and government circles.

Chile's prospects for economic development, which have recently been the subject of a competent first-hand appraisal[1], are indicative to some extent of those in other countries of Latin America. Extreme advocates of industrialisation have argued that the way to increase the country's production is to stimulate manufactures almost to the point of self-sufficiency. One Chilean industrialist and engineer wrote that "Whatever production replaces an import is and always will represent an increase in the national wealth, independently of its apparent cost in money value". Careful examination, however, leads to rejection of this view. A much more promising course for Chile would be a mixed development in which increased productivity in agriculture, now the chief industry, would play a prominent part.

Chile possesses more arable land, and more per capita, than California, Sweden, Switzerland, or New Zealand. It has good soil and climatic conditions. Yet, as a Chilean writer has pointed out, its per capita production of all agricultural commodities is less than half that of Sweden and Switzerland, one third that of California, and less than one seventh that of New Zealand. "Chile's agriculture, perhaps even more than her industry, offers great possibilities of development through modernisation. The chief obstacle appears to be the influence of the easy-going tradition of the *hacienda* . . . "[2]

With the aid of modern fishing boats, cold storage plants, and refrigerated cars, Chile could draw on its fishing resources to effect a great improvement in the diet of its people. At present fish is not regularly available even in Santiago and is as costly as meat, although there are rich fishing waters off the long Chilean coast. The country could possibly become one of the world's leading exporters of fresh, frozen, and canned fish; and it could perhaps export other types of canned and dehydrated food. Other promising lines of development include: exploitation of water power resources; possibly a small steel industry utilising cheap sea transport and cheap power; expansion of the small-scale copper fabricating industry that already exists; certain chemical industries based upon nitrates and other salts present in Chile; production of lumber for domestic use and for export, and for manufacture of paper, plywood, and rayon pulp. Better transportation and communication, adequate housing facilities, and improved conditions of public health

[1] Chapter VII, "Prospects for Economic Development", in *Chile: An Economy in Transition*, by P. T. ELLSWORTH, to be published in 1944 by Macmillan. Manuscript kindly made available by the author.
[2] *Ibid.*

are needed to integrate and to supplement industrial and agricultural progress.[1]

The Choice before the Industrially Advanced Countries

There are solid reasons of a social and political character which impel the peoples and governments of the advanced industrial countries to look sympathetically on the efforts of their less well equipped and less productive neighbours to achieve higher living standards through economic development.

There is the humanitarian wish to see people who suffer from chronic poverty acquire the material basis for better living. Furthermore, prominent spokesmen of the leading United Nations have pledged economic co-operation after the war to raise living standards for all people, not just for the more fortunate few. True, the pledges have been in general terms and have been expressed more often as hopes and aims than as legal commitments. But the reputation for reliability of the nations from which these pledges have come is to some extent at stake.

The industrially advanced countries have a great stake in future prospects for a more peaceful and orderly world. Poverty-stricken areas in an age when poverty is no longer necessary and in a world where some have great wealth may become festering trouble spots. Poverty-stricken areas lack the means of defending themselves and, by their weakness, invite conquest. It is generally recognised that there can be no really secure basis for a durable peace unless the peoples of the world are able to co-operate in some effective political organisation. Gross disparities in living standards do not make such co-operation easier, nor does the lack of educational opportunity which goes with low productivity. On a planet where aviation will soon have brought every place within less than forty-eight hours of every other place and where opposite sides of the globe are fractions of a second apart by radio communication, the advanced countries can no longer trust to the insulation of distance to protect themselves against the consequences of political discontent and disorder in other regions.

The same reasoning applies to physical disease as to political disorder. Planes will fly daily between the most out-of-the-way places, which may be incubation centres for epidemics, and the more fortunate lands, where pestilences are rare and immunities high. Can the advanced countries afford to leave great patches of the earth under conditions of poverty and malnutrition which breed disease?

---

[1] Chapter VII, "Prospects for Economic Development", in *Chile: An Economy in Transition,* by P. T. Ellsworth, to be published in 1944 by Macmillan.

If living standards remain low in other areas, there may be increasingly bitter complaints and increasing pressure against the immigration barriers of the more advanced and wealthy countries. The great constructive alternative to mass migration out of low-standard areas is a rapid flow of capital, modern equipment, and technology into those areas—in other words, economic development.

Finally, the advanced industrial countries will no doubt realise that they cannot prevent the modernisation of many of the less developed areas of the world, although obstruction or indifference on their part would delay the process. The Soviet Union has proved that a relatively "backward" economy, given sufficiently determined leadership, can modernise itself with little help from outside or even in the face of outside opposition. Capital can be accumulated internally by holding living standards down, even pushing them further down, while a large fraction of production is turned into new equipment. The process is a painful one. In countries which start from a low standard of living it can be carried through only at the cost of great misery, and probably only under a political dictatorship capable of restraining consumption drastically in order to accumulate capital. Also, other less developed countries lack the great size and variety of resources which helped make the achievement of the Soviet Union possible. Nevertheless, the leaders of at least some of the less developed countries are likely to urge that no sacrifice is too great in order to establish modern industry. For modern industry provides not merely higher living standards; it also provides the tools of modern warfare. This is a point which China, for example, after suffering so sadly from lack of equipment in its years of struggle against the invader, is unlikely to forget.

The process of modernising the production methods of the less developed countries will probably be pushed ahead regardless of the attitudes of the countries that hold industrial leadership today. Indeed, indifference or obstruction on the part of some of them would probably not be reflected in the policies of all; some help is likely to be forthcoming, in any case, to the countries undergoing modernisation. What the industrially advanced countries can influence by their attitudes is the degree and rapidity of economic progress, its timing, and to some extent, the direction that development takes. The real choice before them, therefore, is not "Shall there be economic development elsewhere or shall there not?" The choice is between positive co-operation in that development or indifference and antagonism towards it.

Economic Repercussions of Development Abroad: Hopes and Fears

It has been said above that there will be an insistent demand in many parts of the world for rapid progress in economic development after the war. Also, for social and political reasons, some of which have been suggested, the people of the industrially advanced countries will probably be inclined to encourage the process of development and to co-operate in speeding it. But what of the economic impact on them?

That is the question which it is the purpose of this study to explore. What are the effects—primarily the economic effects—which are likely to be felt in the advanced industrial countries of the world as a result of economic development elsewhere? Will the provision of capital, equipment, and technical and educational aid designed to increase production in the less developed areas of the world raise or lower the prosperity of the supplying countries? Will the introduction of modern methods into places where production has hitherto been carried on by more primitive means result in gain or loss to peoples that now have well established modern industries? Whatever may be the benefits that improved labour efficiency and higher living standards bring to the countries where economic development takes place, will other countries suffer from loss of markets and from new competition? Or will they find themselves enjoying new opportunities and higher real income as a result of the increased purchasing power of their customers and the more efficient production of the things they buy abroad? What may be the effects of economic development abroad upon employment, trade, business opportunities, and living standards?

The attitudes in the advanced industrial countries on these points express a mixture of hope and fear.

Some persons stress the hopeful side. They see in the growing prosperity of other countries the promise of expanding markets based on permanently increased purchasing power and higher living standards. They also make the point that the expansionary effects of investment for economic development throughout the world, together with domestic investment-fostering policy, will help to achieve peacetime full employment and high prosperity and will make the post-war problems of economic adjustment easier for the industrially advanced countries.

Dr. Sun Yat-sen gave vigorous and concrete expression to this view after the First World War in his book *The International Development of China*, referred to above. He was prompted to write the book, said Dr. Sun, "by the desire to contribute my humble part in the realisation of world peace". President Wilson had proposed a League of Nations to end military war. "I desire to propose to end the trade war by co-operation and mutual help in the development of China. This will root out probably the greatest cause of future wars." Dr. Sun pointed to the problem that would be created by the sudden collapse of the "war market" for billions of dollars, worth of munitions and supplies. Reconstruction, and the resumption of accustomed civilian demands for comforts and luxuries, would absorb some of the productive power so released. But at the same time improvements in technology and better economic organisation, amounting together to a "second industrial revolution" would increase the productive power of man many times over. "Where in this world can Europe and America look for a market to consume this enormous saving from the war? If the billions of dollars' worth of war industries can find no place in the *post-bellum* readjustment, then they will be a pure economic waste. The result will not only disturb the economic condition of the producing countries, but will also be a great loss to the world at large." China, continued Dr. Sun, if developed according to his programme, would require machinery in vast quantities for its agriculture, its mines, for the building of factories and for extensive transportation systems and public utilities. "The workshops that turn out cannon can easily be made to turn out steam rollers for the construction of roads in China. The workshops that turn out tanks can be made to turn out trucks for the transportation of the raw materials that are lying everywhere in China. And all sorts of warring machinery can be converted into peaceful tools for the general development of China's latent wealth."

The Chinese people, he said, will welcome this development, provided it can be so organised as to "ensure the mutual benefit of China and of the other countries co-operating with us". Some people in Europe and America might fear that the use of war machinery, war organisation and foreign technical experts to develop the latent human and material resources of China would create competition unfavourable to their industry. "I therefore propose a scheme", wrote Dr. Sun, "to develop a new market in China big enough both for her own products and for products from foreign countries."

Other persons take a pessimistic view of the effects upon advanced countries of economic development abroad or look upon it with fear, especially when it involves, as it often will, a substantial measure of industrialisation. They stress the new competition likely to come from improvements in the productive efficiency of other peoples. They argue that business men in the advanced countries will find their markets shrinking as former customers learn how to make imported goods for themselves. Cheap labour in low-standard areas, once it is equipped with modern machines, may turn out goods at such low cost that the higher-paid labour of the advanced countries will be unable to compete. They fear that the result will be loss of jobs and undermining of living standards. Capital and "know how", according to this view, had better stay at home to give employment to home labour and opportunities to home business firms, instead of going abroad to provide jobs for foreign workers and to build up foreign competitors.

In January 1942, *Harper's Magazine* published an article under the title "A Warning to the Peace-Planners: America's New Industrial Rivals".[1] It expounded the thesis that "outlets for manufactures" are being cut down over the world as one customer after another learns to make at home the goods that were previously imported.

> . . . a little bit shaved off in Cuba and a little bit shaved off in Ceylon is going to be, when added up, a whole lot shaved off the world demand. . . Whose steel trade, whose textile trade, will India cut into after the war? . . . Australia will be a hard competitor in post-war markets. . . Don't overlook New Zealand either. . . There is also Japan. . .
>
> If Asiatic industrialism fails to make organic connection with the "crowds of Asia" then it may be expected, first, to push the overseas suppliers hard in the markets of the fringe; and, second, to enter into competition with the older industrial countries in every price market of the world. These will be "unhealthy" results for America and Europe.
>
> As industrialism invades one country after another the basis of trading relations shifts. Failure to realise that this is true, or blind resistance to its implications, brought depressed areas to England between the wars. Failure to keep it in mind in the future will bring more and better depressed areas, not only in England, but also in other highly industrialised exporting countries. If every country turns to making as many things as possible at home, obviously those who formerly sold them these things must find some other way of disposing of them, preferably also at home, or get out of the business of making them altogether.

The same debate has been carried on in every country which, being relatively advanced in industrial techniques and prosperous for its time, has seen other countries adopting its methods or improving their own. Adam Smith wrote sarcastically of the eighteenth century mercantilist laws which restrained the export of machinery and tools and prohibited artificers from going abroad to practise or teach their trade: "The laudable motive of all these regulations is to extend our own manufactures, not by their own improvement, but by the depression of those of all our neighbours . . . "[2] In all countries which have been great suppliers

---

[1] C. Hartley GRATTAN in *Harper's Magazine*, Jan. 1942, pp. 126-133.

[2] Adam SMITH: *The Wealth of Nations*, Book IV, Chapter VIII.

of investment funds for development abroad serious misgivings have been expressed lest this "drain of capital"—in the tendentious phrase often used—impair the economic prosperity of the country and build up competitors.[1]

The Argument Today

Today there is much public interest in the advanced countries on the subject of the policies they should pursue towards post-war economic programmes of less developed countries, and the time-honoured divisions of opinion are finding their spokesmen. It may be useful to illustrate the course of the discussion in one country. The United States is chosen because its great capacity to supply capital, technical knowledge and equipment, together with its general position in world economics and politics, will make it particularly influential in advancing or retarding the pace of economic development in other regions.

---

[1] Jacob Viner: "Political Aspects of International Finance" in *Journal of Business of the University of Chicago*, Vol. I, Apr. 1928, p. 144. Mr. Viner cites the following instances, among others:

In 1813 Sir Henry Parnell explained a report of a British Parliamentary committee hostile to the export of capital as "influenced by no other motives than . . . the impolicy of sending our money to improve other countries while we have so much of our own lands that stand in need of the same kind of improvement". (Report from Select Committee on Corn Trade, 1812-13, *Hansard*, 1st series, Vol. XXV, Appendix.)

Palmerston, asserting in 1848 the right of the British Government to intervene to force debtor governments to meet the just claims of British creditors, explained that the Government had hitherto not exercised that right because "It has hitherto been thought by the successive Governments of Great Britain undesirable that British subjects should invest their capital in loans to foreign governments instead of employing it in profitable undertakings at home . . . " (Hall: *International Law*, 4th ed., Oxford, 1895, p. 205.)

In France there was opposition to the export of capital in the late nineteenth century and it became particularly vehement in the years immediately preceding the outbreak of the First World War. Statesmen, publicists and manufacturers argued that the great outward flow of French savings was a transfer to foreign industry of financial resources which were needed for the development of home industries. In Germany before the First World War the opposition to the export of capital on a great scale was especially vigorous on the part of the agrarians, who, as a debtor class, objected to anything which would raise interest rates at home. In America during the 1920's there were signs of similar concern as increasing amounts of capital went abroad. For example, Secretary of Labor Davis advocated that the sending of "American money abroad to develop foreign industries" be stopped in order to maintain American prosperity and to check unemployment. (*New York Times*, 17 Feb. 1928.)

Even economists with strong free-trade convictions have sometimes questioned the wisdom of large-scale investment abroad. Thus, Ricardo wrote early in the nineteenth century that he would be sorry to see weakened "those feelings which . . . induce most men of property to be satisfied with a low rate of profit in their own country, rather than seek a more advantageous employment for their wealth in foreign nations". (*The Principles of Political Economy and Taxation*, Everyman edition, p. 83.) Other economists answered, however, that if the exported capital is invested in productive enterprises abroad it serves to develop export markets, promotes the production of cheaper raw materials, foodstuffs or manufactures which can be purchased advantageously by the capital-exporting country and may open up territory for settlement. (For example, John Stuart Mill: *Principles of Political Economy*, Ashley edition, p. 739.)

President Roosevelt has expressed his views on this topic more than once. In an informal discussion at his press conference on 24 November 1942 he alluded to the establishment of a United States office to aid foreign relief and rehabilitation and to the fact that there were then in Washington certain official visitors from South American countries which were being assisted in developmental projects by United States loans. According to newspaper reports:

There would be rehabilitation abroad, Mr. Roosevelt declared, not only for humanitarian reasons but from the standpoint of America's own interest, for it would mean in the final analysis better purchasing power abroad for American products. It will mean safety in the future from attack and from war and it will encourage the development of democracy, he added.

The President recalled, in setting forth his views, that twenty years ago the South, particularly the rural areas, was so impoverished that there was a lack of buying power. Under the New Deal, he declared, this had changed, and the increased buying power gained, especially through agricultural rehabilitation, had led to the purchase of Northern goods. That in turn, he said, had provided more labour and so more wealth for the North.

Yet, he added, when the programme was undertaken there was much criticism of it in the North.

The same thing that was done for the South, he stressed, could be done for nations; and it would help them and us, he contended.[1]

Carrying his illustrations further, President Roosevelt recalled that in 1933 hardly any stores in rural Georgia were solvent, to say nothing of the banks. There were no Saturday afternoon customers.

The prospective buyer of a hat would be shown an eight-year-old model, because there was no turnover of stock. And because there was no turnover of stock the merchant was going deeper and deeper into the red.

Since that time, the President held, there has been an increase of Southern purchasing power, and Northern sellers are enjoying a Southern trade they never had before.

The same conception is involved in aid of the economy of the Republics of Central and South America, he went on. When the United States helps them it is helping itself, although a lot of people remain to be convinced of the advantages of putting other people on their feet.[2]

On his return from conference, at Cairo and Teheran, the President reported to Congress and to the people in a message which again stressed as one of the basic essentials of peace "a decent standard of living for all individual men and women and children in all nations. Freedom from fear is eternally linked with freedom from want." The President continued:

There are people who burrow through our nation like unseeing moles, and attempt to spread the suspicion that if other nations are encouraged to raise their standards of living, our own American standard of living must of necessity be depressed.

---

[1] *New York Times*, 25 Nov. 1942.
[2] *Christian Science Monitor*, 25 Nov. 1942.

The fact is the very contrary. It has been shown time and again that if the standard of living of any country goes up, so does its purchasing power—and that such a rise encourages a better standard of living in neighbouring countries with whom it trades. That is just plain common sense—and it is the kind of plain common sense that provided the basis for our discussions at Moscow, Cairo and Teheran.[1]

In the fall of 1943 the Secretary of the Treasury, Mr. Henry Morgenthau, Jr., wrote, in a foreword to proposals for a Bank for Reconstruction and Development prepared by United States Treasury experts:

It is imperative that we recognise that the investment of productive capital in undeveloped and in capital-needy countries means not only that those countries will be able to supply at lower costs more of the goods the world needs, but that they will at the same time become better markets for the world's goods. By investing in countries in need of capital, the lending countries, therefore, help themselves as well as the borrowing countries. If the capital made available to foreign countries would not otherwise have been currently employed, and if it is used for productive purposes, then the whole world is truly the gainer. Foreign trade everywhere will be increased; the real cost of producing the goods the world consumes will be lowered; and the economic well-being of the borrowing and lending countries will be raised.[2]

Mr. Wendell L. Willkie, Republican candidate for the Presidency in 1940, told an audience at St. Louis on 15 October 1943 that military power to repel aggression is not the final means of achieving peace. "The real foundation of peace and development must be economic":

The right way through is plain: expansion and development. Literally millions of people around this world are eager to work with us in co-operative economic effort. And if there is any one thing that we have learned in America it is that no man's prosperity needs to be had at the cost of another's. That well-being is a multiplying, not a dividing process.[3]

The Postwar Committee of the National Association of Manufacturers issued some "preliminary observations" in March 1943 which contained the following passage on "economic opportunity throughout the world":

It must always be remembered that the economic value of trade between the United States and other countries increases in proportion to the development of those countries with which we trade.

There is a widespread impression that if nations which formerly had little, if any, manufacturing activity should subsequently develop through their own efforts substantial manufacturing enterprise suited to their labour and resources they will thereby reduce the potential export markets of American industries and will reduce employment and living standards in this country. This, however, does

[1] Message to Congress of 11 Jan. 1944.

[2] *Preliminary Draft Outline of a Proposal for a Bank for Reconstruction and Development of the United and Associated Nations* (Washington, U. S. Treasury, 24 Nov. 1943), p. iii.

[3] *New York Times*,16 Oct. 1943.

not necessarily follow. Abundant statistics show that, as manufacturing increases, buying power also increases and so does the demand for imports.

The volume of trade depends much more upon the buying power per individual than upon the degree to which the products of one country are complementary to those of another.

It follows, therefore, that world-wide efforts to raise the standards of living of the underdeveloped peoples through the more intensive use of their natural resources are bound to be beneficial to the people of the United States, as well as to those whose opportunities are thus broadened.[1]

Vice-President Henry A. Wallace, in a widely discussed address in May 1942, recalled that he had once remarked, "half in fun and half seriously" that "the object of this war is to make sure that everybody in the world has the privilege of drinking a quart of milk a day". He continued:

The peace must mean a better standard of living for the common man, not merely in the United States and England, but also in India, Russia, China and Latin America—not merely in the United Nations, but also in Germany and Italy and Japan. . . Everywhere the common man must learn to build his own industry with his own hands in a practical way. Everywhere the common man must learn to increase his productivity so that he and his children can eventually pay to the world's community all that they have received. . . . [2]

At the next annual meeting of the National Association of Manufacturers, its outgoing president, Mr. William P. Witherow, took issue with the Vice-President:

Immediately after the war, government aid to war-torn countries is a foregone conclusion. But not the rehabilitation of their economy or the reforming of their lives. I am not fighting for a quart of milk for every Hottentot, or for a T.V.A. on the Danube, or for governmental handouts of free Utopia.

If the Government undertakes a "share-the-wealth" plan on an international basis, he said, "it may benefit those in foreign lands, but only by the impoverishment of the American people".[3]

A year later Mr. Witherow, now Chairman of the Board of the Association, again addressed its convention. Alluding to the "storm of concern" and the "lively exchange of ideas with so many alert minds" that his previous speech had provoked, he took as his theme "Every Hottentot a Capitalist!" He said:

Practical-minded Americans can see that the answer to the problem of the Hottentot is not to deliver a quart of milk to his doorstep. . . The real answer is to help him find the way to a better life; don't try to give it to him. . . Establish enterprise and trade, if you please. Then he can buy a cow of his own. . .

---

[1] *Jobs, Freedom, Opportunity, in the Postwar Years: Preliminary Observations by the Postwar Committee of the National Association of Manufacturers* (1943), pp. 37-8.

[2] "The Price of Victory", address before the Free World Association, 8 May 1942, New York City (printed in *Free World*, June 1942).

[3] *New York Times*, 3 Dec. 1942.

and instead of waiting for the international milkman—Uncle Sam—he can have not only a quart a day, but a gallon a day. . .

Helping the Hottentot to help himself is my recipe for establishing international economic security. He may need a few baskets in which to gather his coconuts. He may need a cow, or even a tractor, to cultivate his land and increase his crops. Most likely he'll need a little education not only on how to use tools but on the advantages of having milk, and the desirability of unparking himself and doing a bit of hustling in order to get it.

There is sense in our helping him to get his start. That would be a real humane investment and a good economic investment too if it brought us more coconuts in the long run. There is nothing foolish about investment—it is the very seed of betterment. . .

We must not fear the industrialisation of other nations. Not that we can stop it, but it has definite advantages for us. It may cause some shift in the price of some of our imports; but the industrialisation of South America, for example, should make more customers for American products, putting purchasing power in the pockets of millions of underprivileged who never were customers before. . . But we can't sell abroad unless we are willing to buy. . .

. . . We can invest our wealth in the rebuilding of the world. But I said "invest", not "dissipate" or "squander". We will be investing only if we can know that what we provide will be used productively— to *energise* not to *enervate* those who receive it. . . Our best assurance . . . will be to insist on a fair return on our investment and to supply capital only for purposes giving promise of such return. . . Finally, we can give the post-war world American ideas—industrial technique—management "know-how". These are as important in production as physical equipment—and only by production can the shattered world be rebuilt.[1]

Mr. Henry J. Kaiser, whose feats of shipbuilding have earned him industrial fame, has frequently stressed the positive view of world economic development. On the same occasion as Mr. Witherow's first speech, he told the National Association of Manufacturers:

Let it be said again that there will never be any significant prosperity in America as long as there are great hosts of people living on the margins of poverty anywhere on earth. This is the hour for action and now is the time to begin the heroic and magnificent task of reconstruction.[2]

Mr. Robert J. Watt, international representative of the American Federation of Labor, said on his return from a visit to South America:

Helping to raise the purchasing power of the Americas is not only good for workers, but also excellent for business. Workers earning fifty cents a day are poor customers. . . Industrialisation of the other Americas . . . is a healthy trend. We need not fear it. There is convincing experience to show that export markets develop with industry, that rising standards of living flow from industrialisation. . . The labour movement of our country wants to help the countries of South America to raise their standards of living, so that we can better maintain

[1] Mimeographed speech, press release of the National Association of Manufacturers, 9 Dec. 1943.

[2] *Christian Science Monitor*, 5 Dec. 1942.

our own. We try to be realistic in our friendly approach because we realise that cheap labour markets and exploited workers are a threat to the standards in our own nation.[1]

Mr. James B. Carey, Secretary-Treasurer of the Congress of Industrial Organizations, has expressed his views as follows:

Abundance cannot be achieved at home if misery and poverty prevail abroad. When the people of one land cannot buy, crops rot in their neighbours' fields. Walls of isolation cannot be built high enough to keep out airplanes with loads of either goods or bombs. Disease is no respecter of boundaries. Cycles of unemployment spread from nation to nation just as they do from State to State. We must be concerned with raising the levels of living in other nations both because we are interested in the human beings who live there and because if their living standards and wages are low, they will undercut us on the world market and drag our standards down towards theirs. Moreover, misery abroad will continue to result in unrest that will give aggressive demagogues a chance to seize the helm and steer to new wars. . . There must be international co-operation in applying modern technology and scientific knowledge to the maximum development of the world's resources. . . Organised labour is not interested in a new American imperialism. . . But we should use our influence to see that the peoples of other nations have an opportunity to develop their institutions. . . They cannot do it if they are kept in poverty and misery by out-worn economic practices that prevent development of national resources.[2]

These quotations give an idea of the way the discussion is proceeding. On the whole, there is probably less open expression at the moment of the negative attitudes and fears than of the positive, encouraging views of world economic development, such as those which appear in the Atlantic Charter and in many statements of United Nations leaders. Perhaps this is partly because co-operative, hopeful, forward-looking utterances are wartime morale-builders and hence get a good reception at present. Will negative attitudes, in which countries figure primarily as competitors of each other, gain the upper hand when military victory has been won?

Anyone who discusses the problems of world economic development with ordinary people can testify that, combined with considerable goodwill towards the people of countries on lower living standards, there is in the advanced industrial countries a strong latent fear. When the cheap labour of foreign countries has been equipped with modern tools, how shall we be able to compete? Will not our markets disappear? Are not the resources of the world limited, so that more for the Chinese workers means less for other workers? A question put to the writer by a member of an audience in a mid-western American town sums up very well the concern of many people in the industrially advanced

[1] Excerpts from speeches made in April and May 1943, supplied by Mr. Watt's office.

[2] Speech at Grinnell College, 16 June 1943, from mimeographed text.

nations: "Can we help to raise the living standards of these other countries without lowering our own?"

The object of this study is to find considered, objective answers to questions such as these.

II. MAIN CONCLUSIONS

It may be well at this point to state the fundamental assumptions upon which succeeding chapters will be based and then to outline in a preliminary way the main drift of the argument and the principal conclusions that are reached.

Throughout this study it is assumed that the United Nations will make use of their victory to establish some reasonably effective system of world political security. It is also assumed that economic development will not be prevented by civil wars or extreme factional bitterness in the countries where it would otherwise take place, or at least that there will be many areas where sufficient internal political stability exists for development to go forward. It is also assumed that international co-operation for promoting economic development will be forthcoming from the industrially advanced countries (in the form of technical assistance and capital loans) and will be acceptable to the developing countries. If any of these conditions are not present after the war, then the analysis given below of the effect of new economic development upon already established industrial areas may not be applicable or may be applicable only with considerable modification.

The economic analysis of the effects of modern development in new areas upon the established industrial areas will be presented in two main parts: first, effects arising out of *investment relationships* and second, effects arising out of those *changes in trade and production* which are set in motion by the increasing productivity of regions undergoing development. These two parts will be followed by a third which draws attention to certain cultural, political and politico-economic repercussions of new economic development. Specific attention will be given in all three parts to the question of policies and safeguards that might be adopted in order to bring the most beneficial effects and the least detrimental effects to the advanced economies and the newly developing economies alike.

General Thesis

The general thesis which emerges from this study is that economic development of new areas brings both opportunities and dangers to existing industrial areas, but that it is definitely possible, by policies of mutual co-operation and intelligent adaptation, to make the advantages far outweigh the disadvantages.

The equipping of industrially less advanced areas could help to provide a great new frontier of economic expansion. Particularly in the first difficult decade or two after the war, properly organised international co-operation in order to achieve a mutually beneficial *timing* and *direction* of equipment orders for the outfitting of new areas could help to stabilise employment and income in the equipment-supplying countries. In the absence of organised international co-operation for this purpose, however, the effect of new economic development both immediately and later might be much less mutually beneficial.

Economic development of new areas means an increase in the capacity of their peoples to produce and consume. This may be expected, on the basis of past experience, to bring a great increase in international trade, if political conditions are favourable. For established industrial areas the result will be a combination of new market opportunities and new competition. In part, the net effect of these developments upon any particular country of advanced industrialism will depend upon the accidents of geography and history—what lines of production the established area happens to depend upon, how specialised it is, what resources are available for new lines of production, and what specific things are being demanded and supplied by the developing areas. More important, however—and certainly more subject to control—is *the way in which the established industrial area reacts to the new situation.* The balance of advantage or disadvantage can be influenced decisively by the wisdom or unwisdom of the policy followed by its business enterprises, its organisations of workers, and its government. The key to the situation is *industrial adaptability.* Leading industrial countries can retain their lead and move on to still higher standards of living as other areas develop if they succeed in being adaptable, that is, if they shift labour and capital into lines of production where rising world income is bringing more rapid expansion of demand than of supply and out of lines of production where new supply is increasing faster than new demand. If established industrial areas react adaptively in this way, their own business opportunities, employment, and standards of living are likely to be raised by the development of new regions. If they react non-adaptively, or anti-adaptively, then the net effect may be bad for them.

The balance can be made more favourable for all parties concerned, indeed, the net result can be made a very important gain for established industrial countries as well as for those undergoing development, if economic development is promoted and guided in desirable directions by co-operative international action. The use of multilateral agencies, including an international development authority which would assist member governments in harmonising their views on long-range developmental programmes and in the planning, financing, and execution of such programmes, would be an important means of fitting developmental and trade policies together in a pattern of greatest mutual benefit.

The Course of the Analysis, by Chapters

Investment, meaning the building of transportation systems, factories, houses, educational institutions, and increase of all sorts of equipment used in production, is in a very real sense the "activator" of modern advanced economies. At high levels of employment and income, business firms and individuals set aside large amounts of money as savings. Unless these savings are balanced by equivalent expenditures on investment—that is, unless a high rate of saving is matched by a high rate of investment—incomes shrink and jobs disappear. At times—and this may be true in the first several years after the war when pent-up demands are being released and emergency reconstruction is under way—there may be a tendency for too much investment in relation to savings; this would produce inflation. But, as a general proposition in the decades ahead, the chief danger in the industrially advanced countries will be too little investment to keep their economies going full blast, not too much. The investment stimulus that would be created by large-scale world development programmes participated in by capital from the advanced countries would therefore be a boon to them, helping them to solve their employment problems. If other world conditions were favourable, the investment stimulus from widespread developmental undertakings in many countries might well produce a new "long wave" of prosperity, in which periods of high economic activity would be longer and depressions shorter and less violent than otherwise.

The war has brought about an enormous expansion of the capital goods industries, including not only plant facilities but also the labour skills and the raw material sources that contribute to the making of capital goods. Large-scale programmes of economic development throughout the world would boost the demand for capital goods. This would have the advantage for the industrially advanced countries of offering additional outlets of a constructive sort for exactly those types of civilian goods to which their over-expanded war industries can most readily convert—machines, locomotives, ocean ships and river boats, electrical equipment, bridges, tractors, and the basic materials to make them, such as steel, copper, aluminium and rubber. Post-war readjustment problems cannot be avoided by world development activities, but they can be made easier. Also, a rising demand for capital goods and raw materials associated with large-scale investment in less developed areas, together with the increase in productive capacity of other sorts which economic development would bring about, should promote a better balance in international economic relations and lessen some of the weaknesses that have contributed to financial and currency troubles in the past.

Investment of private or public capital from advanced countries in projects that raise the productivity of less developed areas is not "playing Santa Claus". A reasonable return on productive loans and productive business undertakings abroad can and should be expected. The soundness of developmental investment, from the point of view of repayment, depends, however, not only on its effect in increasing the productive capacity of the borrowing areas, but also on the ability of the debtor countries to earn the currencies of the creditor countries. In an expanding world economy this problem of international payment need not cause great difficulty. It does call for forethought, however. Measures suggested to meet it include: longer-term international investment programmes to regularise the flow of capital and avoid sudden reversals in flow; arranging the programmes so that there can be long terms of amortisation and low rates of interest and so that the proportion of technical aid that goes along with the capital is high; increased emphasis on equity investment and direct investment, less on the rigid bond form of contract requiring fixed money payments; distinction in loan contracts between local currency obligations and foreign currency obligations, with more flexibility in the latter; and adjustment of the import trade policies of creditor countries to permit the inflow of goods and services by which they can be repaid.

What may be the order of magnitude of the investment that could reasonably be made in less developed areas after the war, and how substantial might the stimulus be to the economies and the capital goods industries of the advanced countries? Some rough calculations are made in Chapter IV. It is clear that investment in less developed areas can by no means remove the necessity for vigorous domestic programmes of expansion and stabilisation in the advanced countries. The magnitude of the possible development investment abroad is great enough, however, especially in relation to the levels of domestic investment that would otherwise prevail in time of depression, to offer a significant benefit to the advanced economies if rightly organised—especially if an element of counter-cyclical timing can be introduced. Immediately after the war the largest flows of international capital will probably be for the reconstruction of industrially advanced areas which have been bombed or fought over. Another large use of international capital funds may be in the progressive "young" countries which already have an advanced technology but lack capital. The less developed areas with low living standards and dense populations are limited in their capacity to use capital effectively at first because of social resistances encountered when modern methods are introduced, but their capital utilisation may rise to high figures once the process is well started.

The amount of mutual benefit for lending and borrowing areas alike will be greater if the international investment process is encouraged to go forward, not on a bi-lateral basis, but multi-nationally. In this connection an international development authority —which might be one institution or several related ones—could render important assistance. The functions which need to be performed are discussed in Chapter V.

The mutual benefit from well-planned development of less developed areas would be enhanced if the advanced countries were to offer inducements to the developing countries so that these latter would schedule part of the orders for capital equipment needed in their programmes in a manner to canalise this demand towards those particular industries and those particular localities which at any given time might be depressed and to step up the total volume of equipment orders in periods of actual or threatened general depression. For example, the Government of China (and of other developing countries) might be offered generous financing at especially low rates of interest for equipment orders to be placed with industries or industry branches in the advanced countries which might from time to time be officially listed as "underemployed". The benefit to China (and similar countries) would be much cheaper capital costs. The benefit to the advanced countries would be a higher and more evenly distributed level of employment and income. This mutuality of benefit is possible because the essence of the method is to manufacture additional real capital out of productive power that would otherwise run to waste in unemployment and under-utilisation of facilities.

The external trade of developing areas, according to past experience, may be expected to increase together with their capacity to produce and consume, unless this tendency is blocked by political insecurity or by a general economic depression leading to a breakdown of the world trading system. There is a very close connection (shown statistically in the charts of Chapter VIII) between income and imports. This is a large part of the explanation, for example, of the fact that fewer than 12 million people in Canada to the north buy almost as much from the United States as more than 120 million people in the twenty Latin American Republics to the south. The more rapidly incomes rise in the developing countries, the better will be the market situation for the advanced countries and the easier will be the trade adjustments forced upon them by economic development. Projects which merely enable a country to substitute high cost production at home for goods formerly imported not only fail to raise the real income of the country attempting to develop its economy in this manner but injure the trade prospects of the advanced countries. Development in accord with the broad principle of "comparative advantage", with encouragement of flexible, multi-lateral trading relationships, is the most favourable course for both groups of countries.

As less developed areas acquire modern industries the composition of world trade will undergo important changes. The role of the simple "traditional type of exchange" of manufactured goods against foodstuffs and raw materials will continue to decline; specialisation will become more complex. Each country will import some types of manufactured goods and some types of raw materials and will export other types of manufactured goods and other types of raw materials. The trade in "semi-manufactures", representing specialisation by stages of processing, is likely to increase. Services, not ordinarily included in import and export totals, will play an ever larger role as the world grows more wealthy and communication improves. For some time, at least, developmental programmes would increase the trade importance of capital goods as compared with consumers' goods. The advanced countries will export more of the goods requiring complex manufacturing skills, much capital, and newer research. They should import more of the cheaper and simpler manufactured goods, thus lowering the cost of living and raising the real incomes of their less wealthy people in particular.

Some kinds of goods—the "new" dynamic products and those for which the market expansion as incomes rise is especially great—will have much more favourable prospects than others in a world of rapid economic development. It happens that the country which is in the best position to help or hinder post-war development programmes of other countries (the United States) is also in a better position than any other major country to benefit in its export trade from rapidly rising world income. The advanced industrial countries in general can gain the most from world economic development and suffer the least detrimental effects from it by being adaptive—shifting, that is, from relatively unpromising to relatively promising lines of production.

Methods of encouraging industrial adaptability of this kind within the advanced countries and of lessening the burden of transition costs which might otherwise fall upon special groups of workers or investors or particular communities are discussed in some detail in Chapters X and XI. International arrangements could also be made to ease transition adjustments and to facilitate the process of adaptation. Some methods of encouraging "infant industries" in newly developing countries cause less disruption to established trading relationships than others; for example, direct subsidies to industries that give promise of being eventually well adapted to the country are less likely to be restrictive from the international point of view than are protective tariffs or quotas. They fit in better with a general programme of world economic expansion. The newly developing countries might be willing to take account of the impact on the trade of the advanced countries when planning their own development measures and to adopt various means to soften that impact, provided that the advanced countries on their part co-operate effectively by lending capital and technical aid, and by opening their own markets to at least some of the products that the developing countries can furnish them at lower cost than they could be produced at home.

One of the first effects of economic development in areas where living standards have been low and conditions of nutrition and health have been poor is to lower the death rate. More infants survive to grow up, and the toll from disease at all ages is less. The result is a rapid increase in the number of people. Later, after a time-lag, the birth rate also falls, population growth tapers off, and there may even be a tendency for the birth rate to fall below the death rate, bringing a decrease of population. At least, this has been the sequence of events in the past when countries have undergone modern economic development. The length of that time-lag between the fall in the death rate and the fall in the birth rate, especially in areas so heavily populated already as are China, India, and some other underdeveloped regions, can have very great significance both for the developing countries themselves and for the advanced countries. A great burst of population growth could absorb most of the effects of better production methods, leaving living standards still very low, and the resulting "population pressure" could also increase the likelihood of political tensions and wars. Remedies lie in the direction of the most rapid possible increase in living standards and great attention to raising the level of popular education as fast as rising incomes make this possible, for the birth rate tends to fall as levels of living and of education rise.

Economic development in areas hitherto lacking modern means of production will bring profound effects upon the distribution of political power and cultural influence in the world. This study does not undertake to speculate about the precise nature of these effects. It does seem safe to say that as the productivity of countries now on a very low economic level moves upward their influence in shaping the course of civilisation will also greatly increase. Mutual interchange between diverse cultures will grow, and one-way "imperialism"—political and cultural—will decline.

Some Standard Bugaboos

The analysis summarised above provides a basis for answers to some of the practical queries which continually find expression in the more advanced industrial countries. "If we help to raise living standards elsewhere, will that not lower our own living standards? If low-wage countries industrialise, what about the competition from the low-paid workers who will now be equipped with machines? If every country industrialises, then where shall we trade?"

1. "*If we help to raise living standards elsewhere, will that not lower our own living standards?*" This is a question which bothers many people, including many who have an ardent wish to see better living standards throughout the world. The answer is that the more advanced industrial countries can co-operate in a general raising of living standards, not only without detriment to themselves, but with positive advantage to their own living standards. How? By helping to increase production. It is inefficient production that causes low living standards. If the people of south-eastern Europe, Latin America, Africa and Asia learn better techniques, install better equipment, and produce more, their standards of consumption will also rise. Their increased consumption will be matched by increased production and will not subtract in the least from the total amount of goods available for consumption elsewhere. In fact, the improved development of their resources will make it possible for other countries, by exchanging with them, to satisfy their own needs more efficiently.

It is not true that the world's basic materials are so absolutely limited in amount that if newly developing countries use more, the others must use less. The very process of developing new countries is likely to reveal new sources of supply and to lead to the invention of new materials. The limit on the amount of goods available for man's use is not a rigid one; it depends on his capacity to produce, which is increased by installing better equipment and by general economic development.

2. "*If low-wage countries industrialise, what about the competition from the low-paid workers who will now be equipped with machines?*" Labour in the countries that lack modern industrial equipment gets low wages primarily because labour under those conditions is so unproductive. Economic development brings a rise in labour productivity, and as this happens wages should also rise. If they do not, then it is correct to speak of labour exploitation. It is desirable and justifiable that international influence be used to make sure that labour does receive improved conditions and improved wages as economic development proceeds.

It is cost of production per unit of output which counts in competition. The services of labour in undeveloped countries like China, though very cheap when measured in man-hours, are very expensive when measured in cost per unit of output, for all but a few types of production. Indeed, the cost of inefficiently utilised labour is so great per unit of output that China is unable to compete with the United States and other high-wage countries in most kinds of industrial production and has to confine its exports on the world market to raw materials of kinds relatively abundant in China and to certain finished goods requiring much hand labour, relatively small amounts of capital and relatively simple techniques. The only way that a Chinese manufacturer can compete with an American manufacturer in any line at all is by being able to hire labour at a very low price per hour. And then he can only compete in lines where unskilled labour is the main factor in production costs. This situation will change as modern techniques are adopted in China and as capital equipment becomes more available—in other words, as China develops its economy. But, by the same token, labour will become more productive, and wages should rise.

The real kernel of truth in the fears respecting the competition from the low-wage labour of newly developing countries is this: as the efficiency of these countries increases, they will acquire a cost advantage in certain lines of production where they have not had a cost advantage before. They will be able to sell more goods abroad, including new kinds of goods. But they will also buy more goods, including new kinds of goods, from abroad, since they will have rising consumption demands. The key to the problem of avoiding adverse effects from the new competition of newly developing countries is to be found in a policy of promoting adaptation—that is, encouraging the transfer of capital and labour into the lines where opportunities are expanding and out of those lines where other countries are becoming able to produce at lower cost. The competition of "low-wage labour" in the newly developing countries will not injure labour in the advanced countries if the advanced countries maintain flexibility, keep in the forefront of technical progress, and aim at expanding production in promising industries rather than at defending to the bitter end those industries for which the outlook is unpromising.

3. "*If every country industrialises, then where shall we trade?*" There have always been fears of this sort in established industrial countries as the next countries embarked upon the path of industrial development. Such fears are based on a mixture of truth and falsehood. The falsehood is the idea that if each country industrialises and diversifies it will not buy as much as before. Usually, the opposite is the case. Consumption rises hand in hand with production. Trade with developing countries will increase in total volume, unless political conditions are unfavourable.

The truth in these fears is that the character of the trade with countries undergoing development will shift. It will not be possible to sell them exactly the same kinds of things as before. There will be new competition in some lines of production, both within the markets of the developing countries themselves and in the general export market. On the other hand, the countries of advanced industrialism will find increasing opportunities opening up for other kinds of products. The new demands will be felt particularly in the field of the more highly technical goods, capital goods, and many kinds of consumers' goods that will be used in larger quantities as the increased real incomes of the people of the world enable them to buy more. But it will be necessary for the older established industries of the advanced countries to keep moving ahead, to adopt new techniques, and in some cases to shift to new kinds of products if they expect to maintain their industrial leadership. It will be necessary for the advanced countries to adopt a deliberate policy of encouraging industrial flexibility and mobility. Government, business and labour in the advanced countries will be wise to reject proposals for defending vested interests against the need for change and to adopt instead a policy of facilitating and encouraging adjustment.

Positive versus Negative Policies

There are two general attitudes which might be adopted in the advanced industrial countries by labour, management, and government as less developed countries begin to embark on programmes of fundamental economic improvement. The consequences of these two attitudes would be very different.

First, the advanced countries could adopt a negative attitude towards economic progress elsewhere. They could discourage and hold back development, either intentionally, or by indifference and by short-sighted policies. They could refuse loans and technical help, erect import barriers against new exports from the developing countries, and look with fear upon the increase in the productive and trading capacity of these countries.

The consequence of such an attitude would be to slow up the development of the newly developing countries, but not to stop it. Even countries on fairly low living standards, though they ordinarily have a low margin of saving available for productive investment, are able to develop their resources without the aid of foreign capital. This has been shown by the example of the Soviet Union. In any case, it is unlikely that a boycott of the newly developing countries would be maintained by all the advanced industrial countries. A negative attitude on the part of a few of them would not prevent economic development, although it would slow it down and make it more difficult.

What the negative attitude would do is this: it would make it impossible to achieve, in any considerable degree, a mutually beneficial co-ordination between the development of new countries and the industrial employment needs of advanced countries. It would throw away opportunities for arranging international co-operative measures that would ease transition adjustments. Politically, the consequences of such a negative attitude would be to earn the ill will of the peoples and the leaders of the newly developing countries. They would be stimulated to adopt policies of self-sufficiency and exclusive nationalism, and perhaps even to think of conquest when they should have acquired sufficient power.

On the other hand, the advanced industrial countries could adopt a positive attitude. They could encourage the development of new areas, co-operate with them by furnishing equipment, technical assistance, and capital loans on reasonable terms. They could look upon the increased production and the increased trading capacity of the newly developing countries as opportunities from which the established industrial countries could derive great advantage, and they could deliberately set about making the most of those opportunities. They could encourage the requisite adaptability, and could assist those groups in their own countries on whom special burdens of transition adjustment might happen to fall.

The consequences of such a positive policy would be to speed up the development of the new areas, and also to permit the working out of co-operative plans whereby development of these new areas might be co-ordinated with the economic needs and problems of the established industrial countries. If co-operation were active enough, the newly developing countries would probably be willing to construct their plans in such a way as to refrain from unnecessarily rapid expansion in those lines which would create especially difficult problems of adaptation for the older industrial countries, and to push other kinds of expansion instead. Needless to say, the economic effects on the older industrial countries would be far more beneficial in such an atmosphere, and the benefits to them would be offset by fewer disadvantages.

Politically, the positive policy would create a setting much more conducive to general international co-operation. There would be a greater chance that the world might stabilise itself after this war and might successfully make the many difficult adjustments that will have to be made if future decades are not to be marked by a fatal sequence of global wars. The chance that world political organisation might develop by a process of voluntary, democratic co-operation instead of by the alternative of repeated wars and perhaps ultimate totalitarian dictatorship would be much greater. All in all, there can be no reasonable doubt that the positive and co-operative attitude is by all odds the one most likely to make the process of modernisation in new areas a beneficial one, both for the older industrial areas and for the new.

PART I

EFFECTS ARISING OUT OF INTERNATIONAL INVESTMENT FOR DEVELOPMENTAL PURPOSES

A. The Nature of These Effects

CHAPTER I

ACTIVATION OF ECONOMIES: A CONTRIBUTION TOWARDS FULL EMPLOYMENT AND PROSPERITY

To repeat, the assumptions of this study are: a United Nations victory; establishment of a reasonably adequate international security system, and a fair degree of internal security within most areas of the world; willingness on the part of some countries to lend technical aid and important amounts of capital (whether through private business or public agencies or both), on reasonable terms, to assist in the economic development of other countries; and willingness of the latter countries to make use of such assistance. Under these circumstances, rapid development would no doubt take place in many parts of the world. The purpose of the analysis that follows is to show the probable effects of such development on established industrial areas. Part II will consider the repercussions on the advanced countries of shifts in production, consumption and trade as a result of economic development in other areas, while Part III will deal briefly with questions of population pressure, political power and cultural influence. The object of the present Part is to explore the significance for established industrial areas of the international investment relationships likely to be associated with new economic development.

On the assumptions stated, a considerable flow of capital could be anticipated in the post-war period from areas of high savings into areas where there is promising opportunity for industrial development but scarcity of capital. The high savings areas would naturally be the more advanced and prosperous countries and, at least in the early years, those least weakened economically by the war. These are also the areas most likely to have unemployed manpower and capacity in the heavy capital goods industries. The areas appropriate for investment of outside capital would include countries where reconstruction of plant and equipment and restoration of working capital has to be undertaken, highly and partly developed countries where the rate of economic growth still tends to outrun the availability of local capital, and less developed countries ripe for modernisation and industrialisation.

This flow of capital would undoubtedly take a variety of forms. Under favourable conditions, important amounts would go as direct investment by private firms of the lending countries for establishment of processing plants based on local raw materials, branch factories or assembly plants, distributing and technical service agencies, and the like. Other capital would be loaned to governments, to public developmental corporations, to private firms, or to mixed public-private corporations through the sale of securities in the financial markets of the advanced countries. Producers of equipment or their bankers would advance some capital to the newly developing countries in the form of long-credit terms, a process which might be facilitated by government funds, nationally or through an international agency. Other capital, which might be a large fraction of the total in the unsettled years of transition following the war and in the initial stages of opening a new region to comprehensive development, would be advanced by national or international public agencies. Governments acting jointly might establish an international investment bank which would both finance developmental projects out of its own funds and encourage private financing by helping to make arrangements and covering part of the risk. The analysis that follows applies generally to all these methods of making capital available internationally.

Countries on the threshold of modern development can benefit very greatly from an opportunity to borrow abroad, provided that capital can be had on reasonable terms, that it can be dissociated from political and economic domination, and that proper precautions are taken against unproductive borrowing and overborrowing. More abundant capital means better transportation by road, waterway, railway and air, better equipment for mines, factories and farms, and readier availability of necessary stocks of raw materials and inventories of all kinds which facilitate commerce. All this adds up to higher productivity for the labour of the country and hence to more remunerative jobs. It also means a stronger industrial base for military action and a greater capacity to defend the country's independence and to make its desires count in the political councils of the world. Spurred on by the thought of such advantages to be gained through the increase of industrial capital, leaders of some capital-poor countries may succeed in forcing a rapid pace of industrial development even though little or no aid is available in the form of capital from abroad. But to support such a forced pace from local resources alone the people of an area where capital is scarce would have to undergo severe privations and suffering.

A political dictatorship might, in practice, be the only means by which such a programme could be imposed. A given rate of development can be carried through much more easily and with much less suffering and loss of freedom for the people concerned if outside capital is available on reasonable terms. Also, the benefits to be had in the form of higher production and living standards can be reaped much more quickly.

But what will be the effects upon the advanced industrial countries from which the capital comes? We are not here concerned with the longer-range effects resulting from the fact that capital supplied to areas under development will sooner or later increase the production and consumption of those areas and bring about shifts in trade which may mean new market opportunities or new competition for other industrial regions.[1] The question raised here concerns the effects of the process of investment itself. What will it mean to the advanced industrial countries to transfer the use of some portion of their savings to the newly developing countries?

The answer to this question depends in part upon the timing of the developmental investment in relation to the tendencies in the capital-providing countries towards inflation and deflation, boom and depression. Special account will be taken of this fact at a later stage in the discussion.

The Savings-Investment Balance in Advanced Economies

The term "investment" as used here will mean "real" as distinct from "financial" investment. It will refer, not to the buying of stocks or bonds, but to the canalisation of a part of current productive activity into the building of railways, highways and ports, the construction of factories and school buildings, the installation of machinery, the equipment of laboratories and educational institutions and the maintenance of their personnel.

Investment performs two functions in the modern economy. In the first place, it serves to increase the productive power of society. In the second place, it offsets acts of saving (that is, non-use of current income for consumption) which individuals, business firms, governments or other public or private agencies wish to perform. In so far as investment balances these acts of saving it prevents them from leading to a decline in the level of employment and income. Investment, therefore, is not merely a means of improving productive equipment. It is also, in a very real sense, the *activator* of the economy.

The central conclusion emerging from the modern analysis of the problem of depression and unemployment may be stated thus:

[1] This topic is reserved for detailed treatment in Part II.

modern highly productive industrial societies do not desire to consume currently all the goods and services that they are capable of turning out at full employment. That is, their members desire to save. What the individual or institutional saver does is to refrain from spending for consumption some part of the income he receives. If the part of income not used for consumption (*i.e.*, saved) is spent on real investment in some form or other—for example, if the saver uses it himself to buy a machine, or lends it to someone else who buys capital goods with it—or if it is offset by an expansion of investment somewhere else in the economy, then the size of the community's total income stream is maintained. If, on the other hand, the saving is not balanced by an equivalent amount of investment the total amount spent in the economy will fall. This means that the receipts of business firms and their payments to employees will fall. Income and employment will decline. Therefore, full employment and economic prosperity can be attained and maintained only by having an adequately high rate of investment to offset the saving that members of the community wish to make when they are employed and prosperous.

The rate of saving in a modern industrial economy is highest in periods of prosperity when people are at work and when total income is large. Families with annual incomes of $2,000 a year ordinarily save a higher percentage of their incomes than those with $1,000 a year and those with $3,000 save a still higher percentage. The percentage of income saved tends to rise as income rises. Also, countries with high average income levels tend to save a larger percentage of income than those on lower income levels. The higher the productivity and living standards achieved by a given country and the more nearly it attains full employment and maximum prosperity, the higher its rate of saving is likely to be. This means that the economic well-being associated with high incomes and full employment can be maintained only if there are outlets or offsets for a large amount of saving—that is, only if the rate of investment is high. The rate of investment must be high enough to balance not merely the shrunken savings which people make at *low* levels of income in time of depression, but large enough to balance what people will want to save at the *high* levels of income attained in times of prosperity and full employment.

May we expect that the countries of advanced industrialism will continue to try to save substantial parts of their income when they are prosperous? If so, what about the outlook for investments to balance these savings and to keep their economies regularly employed and prosperous rather than more or less chronically depressed? These questions will be examined briefly.

THE OUTLOOK FOR SAVINGS

First, what indications are there of the prospective rate of saving in the industrially advanced countries? In the United States, the wealthiest of them all, this problem was subjected to study by the Temporary National Economic Committee in a series of monographs and hearings. The evidence indicated that individual savings in the United States in prosperous times are likely to be high. Persons who have adequate income want to protect their families by buying life insurance and life insurance premiums have a large savings element. People want to save for a rainy day, for a new house, or for old age, and this results in building up the funds of insurance companies, postal and mutual savings banks, pensions and trust funds, building and loan associations and the like.

From 1923 to 1929 individual savings flowing to savings institutions resulted in a growth of their assets and funds at the rate of $4,000,000,000 per year. During the depression the flow of savings declined sharply, but since 1935 has returned almost to the old level. The present flow of savings to these institutions is greater relative to national income than before the depression. If to this flow of savings is added the flow of savings through idle demand deposits in commercial banks and through trustees, foundations and investment trusts, it would appear that investment outlets (over and above replacements and refundings) running into large figures, perhaps five or six billion dollars per year, must be found for these reservoirs every year if the savings of individuals are to be put to work.[1]

At the same time, business corporations in the United States contribute heavily to the national savings by setting aside very substantial reserves for depreciation and depletion and by retaining a portion of their earnings to be used in business expansion. In fact, the bulk of the "venture capital" needed by American industry for its own further development appears to be coming from business savings.

From 1923 through 1929 business enterprises invested on the average $8,700,000,000 each year in plant and equipment and of this, $6,400,000,000, or 74 per cent., came from funds accumulated from internal sources: retained earnings, plus allowances for depreciation and depletion. During the five years 1935-39 average outlays for plant and equipment were $5,800,000,000 and of this $4,800,000,000, or 83 per cent., came from internal sources.[2]

In the United States, as well as in many other advanced industrial countries, new social security systems are in process of

---

[1] Oscar L. ALTMAN: *Saving, Investment, and National Income*, Monograph No. 37 of the T.N.E.C., 76th Congress, 3rd Session, Senate Committee Print, Investigation of Concentration of Economic Power (Washington, Government Printing Office, 1941), p. 49.

[2] *Ibid.*, p. 56. See also the *Hearings* before the Temporary National Economic Committee, Part 9, pp. 3684, 3692.

installation. If these are managed to any considerable extent on the private insurance principle of accumulation of reserves this will mean a further temporary increase in aggregate savings. Of course, eventually annual out-payments from private and public insurance funds will balance or overbalance receipts. But that stage may not be reached for some time. If post-war economic policies are successful in providing millions of individuals in many countries with steady employment at rising incomes, many people will become able, perhaps for the first time, to make substantial savings. Unless there are very drastic changes in habits and social institutions in the meantime, they will want to save, through insurance funds, banks and in other ways.

There are also counteracting forces at work, that is, forces tending to decrease the rate of saving and to increase the proportion of income spent on current consumption. High income and inheritance taxes and other measures tending to even out the distribution of incomes have this effect, for most of the individual saving is done by those who have large incomes. The extension of social security benefits, increasing public expenditures for community health, for parks and playgrounds and entertainments, and other forms of collective expenditure on consumption, have the effect of decreasing the amount of saving for which investment outlets must be found. There is evidence from Great Britain, for example, of a long-term downward trend in the proportion of the national income saved. But these changes work slowly. Furthermore, their effects may be more than offset for a time by the effects of rising income levels, if post-war economic policies do actually succeed in bringing prosperity. As new groups of people move upward into the brackets where substantial savings can be made, the aggregate amount of saving that the community desires to make is likely to rise, despite some counteracting influences.

The probabilities are, therefore, that industrial communities after the war will still want to save very substantial amounts in prosperous times, when employment and income are high. These savings will have to be put to work, that is, they will have to be balanced by investment. Otherwise the high level of employment and income cannot be maintained and prosperity will fade into depression. It is most likely that the following observation will continue to be true not only for the United States but for the advanced industrial countries generally and for the world economy as a whole after the war: "No high level of employment and income has ever been achieved without a large outlay on plant equipment and new construction".[1]

[1] Alvin H. Hansen: Testimony before the Temporary National Economic Committee, *Hearings*, Part 9, p. 3498.

The Outlook for Investment

What, then, are the prospects for post-war investment? Will the investment opportunities available to the advanced industrial countries be large enough and attractive enough for the flow of investment funds to offset the volume of savings that individuals and institutions want to make when they have high incomes? In other words, will the flow of investment be sufficient to maintain prosperity?

Some of the factors that were very important in stimulating investment and promoting economic expansion during the nineteenth and early twentieth centuries will not be present to the same degree in the decades following this war. A great stimulus to economic activity in the past has been rapid population growth, combined with the settlement of new territory. England's population increased by three and one half times during the nineteenth century, and that of Europe as a whole more than doubled. The population of the United States increased fifteen-fold. All this meant an insistent demand for new houses, new public utilities, new schools, new transport facilities and new equipment of all kinds for a constantly growing community. There was also an automatic pressure for more and more investment as the result of the opening up of the American continent, the building of all the great cities of the United States and of other new areas of settlement, enlargement of the cities of Europe and the increasing demand for all sorts of necessities and comforts as cheap foodstuffs from the newly developed areas and increasingly efficient methods of production permitted the living standards of European peoples to rise. There were ups and downs of economic activity, to be sure, but the general trend was a rising one. This enhanced the chances of success for the business man willing to invest boldly. If he overestimated the demands of his immediate market and built too big a plant there was a good chance that the growth of population and rising income would make the venture turn out right after all.

The situation in the advanced industrial countries will be quite different in the future. New land for settlement on a mass scale is no longer available. Population growth is slowing down.[1] Markets in the western world will cease to feel the expansive influence of growth in population and territory. This disappearance of certain automatic growth factors that have been very important in the past removes much of the stimulus to what may be called "extensive" investment—that is, investment required merely to equip more

[1] In eastern Europe and in Asia there continues to be rapid population growth, but the trend has definitely turned downward in the countries of more advanced industrial development. See Chapter XIII.

people and more land. This does not mean that the advanced industrial countries must necessarily suffer economic stagnation and chronic depression. But it does mean that they will suffer these evils unless new investment outlets are developed or drastic adjustments are made in the rate of saving, or both.

Another great source of investment stimulus in the past has been the industrial application of new inventions and the growth of new industries. In fact, the familiar "long waves" which economists have observed in economic activity—characterised in the upward phase by prolonged periods of high prosperity and relatively short depressions, and in the downward phase by long, intense depressions and relatively short periods of prosperity—have been accounted for by some writers largely in terms of the growth and the cessation of growth of great new industries. Thus, the buoyant period from the 1840's to the 1870's was the era of rapid investment in railroads and in the many new enterprises that the railroads made possible. Then the rate of increase in railroad investment slackened off and during the last quarter of the nineteenth century, from 1873 to 1897, bad times prevailed. Thereafter, according to this interpretation, the electrification and motorisation of the western world, bringing with it investment in street railways, telephones, electric power, automobiles, service and repair stations, and highways, produced another upward surge which lasted into the 1920's.[1]

What are the prospects for similar great new industrial innovations requiring large investments of capital in the future? There is no doubt about the innovations. Aviation, plastics, pre-fabricated housing and dozens of new products loom on the industrial horizon. Some students of these problems have expressed doubts, however, whether any of the new industries now visible will require as much capital investment as railways, automobile roads and the other great developments of the last century. One of the effects of modern technological progress, especially technological progress based on the recent great expansion of industrial research, is to increase the effectiveness of a given quantity of capital equipment. A smaller but more efficient machine, a less expensive but improved process, increases output while actually *reducing* the amount of invested capital. Many industrial corporations, merely by spending their depreciation allowances on equipment which incorporates the latest technological improvements—that is, without any additional net investment—are able to make enormous increases in output and even to branch out into new products. Therefore, even though new products and new industries will assuredly arise, some

[1] Alvin H. Hansen: *Fiscal Policy and Business Cycles* (New York, Norton, 1941), pp. 27-41, and other writers there cited.

have questioned whether they will provide as great an outlet for savings and as much stimulus to economic activity as did the railway and the motor car. Perhaps they will. Perhaps the future opportunities for *intensive* investment will be as great as or greater than ever.

But even supposing that *intensive* investment in new industrial techniques and equipment will be as important a stimulus to economic activity in the future as it has been in the past, what is to take the place of the great *extensive* expansion of investment that was formerly associated with population growth and the opening of new territory?

> Technological innovations making for *intensive* expansion may be expected to continue with unabated vigour. Our society can, therefore, remain as highly dynamic as that of the nineteenth century only if we can find a substitute for the *extensive* stimulus to investment springing from territorial expansion and population growth. Geared as our economy is to a high net investment level by a deeply rooted pattern of consumption with respect to income, we shall be compelled to seek full employment of our resources by deliberately injecting a new stimulus to investment. It is just because we have developed, in our highly dynamic society, firmly fixed institutions and habits affecting the income elasticity of saving that we cannot rely upon autonomous increases in consumption to provide full employment once the *extensive* expansionist stimulus to investment has largely disappeared.[1]

We have seen that advanced industrial countries operating at full employment levels tend to save—that is, to refrain from spending on consumption—an important part of their income. Unless these savings, or potential savings, are balanced by an equivalent amount of investment, production falls below the full employment level and the income stream shrinks until it reaches a point at which the savings that people are able to make are no greater than the volume of investment being made. Over the past century there have been both extensive and intensive outlets for investment, the former providing equipment for new territory and a rapidly growing population, the latter developing new industries and increasing the amount of capital per worker. While the post-war opportunities for intensive investment in the advanced industrial countries may be as great as ever, the lack of new lands for settlement and the tendency of population to decline rather than to increase means that the extensive stimulus will be considerably less. On the basis of considerations such as these it has been said that "The problem of our generation is, above all, the problem of inadequate private investment outlets".[2]

---

[1] *Ibid.*, p. 306.
[2] *Ibid.*, p. 362.

Developmental Investment as a Stimulus

What can be done to bring a balance between investment and savings at a high level of employment and income? The possible lines of approach are two. One is to decrease the rate of saving (that is, to increase the proportion of income currently consumed). Too much thrift, though we regard thrift as a virtue in the individual, may lead to depression in the economy as a whole, if there is a tendency to save more than can be put to use. The other is to increase the rate of investment. It has already been indicated above how tax policy and other measures encouraging a more even distribution of income, together with collective expenditures on community consumption, may help in the first type of endeavour. It has also been observed that this process is fairly slow and may be offset in considerable measure if we actually begin to be successful in attaining new high levels of income for many people in the post-war world.

In connection with the second line of attack, many suggestions have been advanced for stimulating both private and public investment within the countries where large volumes of savings are likely to appear. The methods most frequently discussed include promotion of large-scale housing programmes; encouragement of industrial research leading to new products; redevelopment of cities, especially redesigning of transportation and terminal facilities; regional developments on the pattern of the T.V.A., involving irrigation and the production of electric power, better equipment of farms and improvements of farm land, and encouragement of diversified industry. It is widely recognised, for example, that international co-ordination of domestic public works policies and other domestic measures to sustain investment and employment is desirable.[1] If many countries act together with simultaneous measures to increase their several rates of investment, the international reaction of each country's expansionist policy helps the others. The risk of failure, and especially the risk of adverse pressure on the external balance of payments as a result of internal expansionist measures, is made less for each country.

In addition to domestic measures and their international coordination, however, the investment stimulus that would be provided by a large-scale international programme of economic development might play an important role in maintaining a high level of economic activity during the decades ahead. Some of the savings of the more advanced industrial regions would be turned

---

[1] See in this connection, the Public Works (International Co-operation) Recommendation, adopted by the International Labour Conference in 1937.

into capital equipment for the rehabilitation of war-torn areas, for the hastening of economic progress in countries possessed of modern techniques but lacking capital and for the modernisation of countries lacking both techniques and capital. Such international capital development would help to sustain the rate of real investment and thus to maintain a high level of income and employment throughout the industrial world. A great international programme of development, fitting the capital needs of some areas to the abundant savings of other areas, might provide just the extra stimulus needed to put the post-war era into the upward phase of a new "long wave" in economic activity. It would help to make depressions shorter and milder and to produce more stable periods of prosperity at higher levels of employment and income. The expansionary effects of a multitude of economic development programmes over the face of the earth, especially after the outfitting of less developed countries with modern capital equipment had had time to get well under way, could conceivably compare with the great stimulus to economic life provided by population growth and the opening of new territory in the nineteenth century. It could go far, together with domestic developmental programmes, towards making the late twentieth century an era of "expanding economy"—that is, an era of high investment balancing a high level of savings, with rising incomes, progressive improvement in living standards, and relatively little unemployment.

A word should be said at this point about the *timing* of the need for additional investment stimulus which the advanced industrial countries may experience in the post-war decades. For a few years immediately after the war there is likely to be a general tendency towards boom conditions in countries which, like the United States and Great Britain, have their industrial systems intact. This would result from the release of pent-up demands for consumers' durable goods such as automobiles and refrigerators, backed by purchasing power in the form of war bonds, from the urgent industrial demands for equipment and materials with which to reconvert to peacetime production, and from the demands of war-torn areas for equipment and materials to be used in reconstruction. Providing the transition from war to peace can be managed without too much confusion, the first few post-war years may, therefore, be ones in which additional *general* demand is not immediately needed. In fact, the main problem may be to prevent the abnormally large immediate demands from producing a great price inflation and an unbalanced expansion, followed later by a disastrous collapse of demand and general depression. When this period of abnormal post-war demands has passed, however—a few years after the end of hostilities—the longer-range considerations discussed above will come into play. Even the immediate post-war boom, if it occurs, is likely to be uneven in the sense that some sectors of industry and some localities will have depression and unemployment. The practical bearing of these points upon international developmental policy designed to bring the greatest common benefits and the fewest detrimental effects to newly developing and advanced countries alike will be considered in Chapters VI and VII.

The Effects of Repayment

It is assumed in this discussion that capital funds for investment in economic development will not be free gifts, except, perhaps, for some of the funds advanced to restore production in war-damaged areas in the period immediately after the war. Rather they will be loans of capital on which a fair rate of return can and should be paid out of the increased productivity that they help to create. The problem of international payment—the transfer problem—will be discussed later. The following remarks concern the effects of repayment in the lending countries.

It has been argued above that the process of investment in international development will have a stimulating, activating effect on the economies of the advanced, capital-exporting countries. But what will happen when the time for repayment comes? Will the reverse flow create chronic economic depression in the creditor countries?

In the first place, it should be pointed out that any prospective difficulties in the advanced industrial countries when a net export of capital turns into a net import of repayment would be so far in the future, supposing a really large developmental programme to be launched, that they could be discounted rather heavily. The rate of capital outflow, given adequate measures to maintain world economic and political stability, could have a rising trend for many years. Immediately after the war there will be an urgent need for capital to be used in restarting peacetime production and in reconstructing industrialised areas damaged by the war. Soon thereafter some of the "young" countries already well started on industrial development might be absorbing sums of capital which in the aggregate would be considerable. Slowly at first, while the difficult initial adjustments are being made, but later at a rising tempo, such huge underdeveloped areas as China and India might draw upon the world's capital resources for their industrialisation. The opportunities for productive employment of capital equipment in the less developed parts of the world (and that is still most of the world, measured either in population or in territory) are so immense that a large flow of investment could continue for a generation or two. Repayments of old loans would be more than counterbalanced for many years by the making of new ones. In a fairly well-working economic world investors as a group would be quite likely to leave their *principal* invested or to reinvest it. Thus, the repayment problem would be limited to regular transfer of interest and dividends.

So many things can change while the long-term process of world economic development is going forward that fears as well as hopes projected far ahead are likely to prove baseless. In view of the urgent economic problems that will confront the world in the crucial decades after the war, the present generation could hardly be blamed for refusing to reject an otherwise promising line of policy simply on the ground that it might have adverse repercussions twenty-five or fifty years in the future—unless, indeed, those adverse repercussions could be shown to be very probable and very large. That is hardly the case, however, with the problem of receiving repayments for exports of capital used to tap the unexploited resources of the world.

It must be remembered that, assuming the developmental programme to be reasonably successful, repayment will come only after a considerable increase in productivity has taken place in the world. Not only the effects of the investments themselves, but constant improvements in technology will raise the level of productivity. Trade as well as production will expand, and there may be important increases in the purchases and sales of international services, such as tourist travel. Thus, the repayments for successful developmental investments, although larger than the original advances in absolute amounts (counting interest) will be considerably *smaller* relative to the size of national incomes and the amount of international trade and services at the time repayments are made.

Finally, in view of the fact that the net flow of capital funds need not turn back towards the original lenders for many years and that the reversal when it comes can be gradual, a world development programme would give time for economic adjustments to be made in the advanced countries. Appropriate adjustments—particularly, a substantial increase in the percentage of current income consumed (that is, a decrease in the rate of saving)—would eventually enable them to receive repayment to advantage. Such adjustments could take place over several decades. The problem of maintaining a high level of economic activity would be solved, or at least would be made much less acute, by the transformation of "high savings economies" into "high consumption economies". Under these circumstances a net inflow of capital in repayment of past investments need not depress the economies of the original lenders. Instead, it would add to their wealth.

CHAPTER II

EASING POST-WAR READJUSTMENTS

One of the great problems at the end of the war will be the re-transfer of workers and equipment from industries directed towards war demand into industries directed towards civilian consumption. This is a problem of "structural" readjustment, by which is meant a shift in the relative importance of different industries or occupations or locations in the economy as a whole.

The structural change involved in "economic demobilisation" will be much more serious after the present war than after the First World War. Conversion to the war effort has gone much further than it did in 1914-18. Enormous expansion has taken place in certain lines—for example, in shipbuilding and machine tool making—which will leave a legacy of installed capacity and trained management and men sufficient to produce an output many times larger than any conceivable regular peacetime demands. New mines have been opened, new sources of rubber and other raw materials have been tapped and impetus has been given to a great variety of technological developments, all leading to larger outputs of raw materials and foodstuffs.

We need not fear production as such. Until men are much better fed and clothed and housed there is no danger of too much production in the aggregate. The danger from this war-stimulated capacity to produce is an *unbalanced* production. The new installations of plant and equipment, the training and upgrading of labour and management, have all been directed towards the needs of war. The demand structure after the war will be markedly different from what it is now. That is, the things for which individuals, business firms and governments will be prepared to place orders will not be the same things for which production capacity has been enlarged. The production structure will not fit the demand structure. Great readjustments will have to take place.

The preceding chapter has made the point that a world programme of economic development would provide a constructive use for large savings and thus would help to maintain over-all effective demand for the products of labour. This is essential if we are to have a high level of employment and prosperity. General "activation" of the economy by large amounts of investment is also the first step, indeed a *sine qua non*, for successfully meeting the problem of structural readjustment after the war. But even with a high general level of demand there will be serious problems of excess capacity in certain lines and in certain broad fields of industry. At best, the problem of retransfer in order to make the post-war structure of production conform to the structure of demand will be troublesome and lasting.

There is another very important relation between a post-war programme of industrial development and the problem of readjusting the production structure. The character of post-war demand if we *do* have large-scale international development will be considerably different from the character of post-war demand if we *do not* have it. In the first case there would be considerably greater demand for the materials and tools of construction, for heavy goods, electrical and transportation equipment and the raw materials that go into them. Such a demand structure would be closer to the production structure that the world will inherit from the war than a peacetime demand structure *without* large-scale economic development would be. The necessary readjustments in production, difficult enough in any case, would be correspondingly less serious.

In particular, a world-wide programme of economic development would mean that there would be less need than otherwise for the advanced industrial countries to contract their war-expanded capital goods industries. There would also be less pressure for cutting output, eliminating installed capacity, and reducing the number of workers in the raw material industries that feed into capital goods production, and in other related lines of activity. The contraction of a branch of industry that for some reason has acquired a capacity to produce beyond its present demand is generally a difficult and painful operation. It will not be possible to avoid a very great amount of that sort of readjustment in the first decade or two after the war. (The greatest problems of structural readjustment may not by any means come at the time of military demobilisation; many of them may come to a head only several years later.) Therefore, if there is a chance to stimulate a constructive, continuing peacetime demand for products to which war-expanded industries can most readily convert, and to absorb in constructive ways part of the stepped-up output of raw material producers, the advantage in doing so is obvious. A post-war programme of economic development would offer such an opportunity. Developmental investment would thereby help to forestall knotty problems of excess capacity and to lessen the severity of price collapses when they come in various industries. In doing so it would lessen the pressures for restrictions on output and for nationalistic protectionism.

CAPITAL GOODS INDUSTRIES

In all the advanced industrial countries the wartime production effort has meant a decrease in the output of many civilian consumers' goods and a vast shift of workers and capital into the production of metals, machinery, motors, vehicles, aircraft, chemicals, ships, and communication devices. The new plants which have been hurriedly erected, the reconversion of old plants, the wartime training of workers in new skills, have especially increased the capacity to produce heavy, durable goods or the materials from which they are made. In other words, the war has especially expanded the capital goods industries.[1]

We may take a few examples almost at random. Machine tool output in the United States in 1942 was about 15 times as large as in the average year of the pre-war decade.[2] A machine tool industry has developed in Australia since 1939 and now supplies many of the needs of Australian factories, which are today making a wide variety of mechanisms, such as arms, motor vehicles, ships, aircraft and tanks.[3] The metal-working industries in Great Britain, the United States, Russia, Germany, Japan, Canada and in many other countries have been greatly expanded in order to increase the output of engines and bodies for aircraft, tanks and trucks.[4] Steel-making capacity is greater. Aluminium capacity all over the world has shot upward. Everywhere the metal and metal working

---

[1] In the United States, the war production effort has involved the greatest increases in manufacturing plant and equipment ever undertaken in that country. During the three years from the latter part of 1940 to the end of 1943 the amount expended on war plant expansion was roughly half the investment made in manufacturing facilities during the 'twenties and 'thirties; in other words, it equalled the industrial growth that might have taken a decade. But the greatest expansions have not taken place in the same industries that might have expanded most in peacetime. The amount of wartime investment from June 1940 to November 1942 was largest in the following fields, in order of rank: aircraft, engines, and parts; explosives; shipbuilding; iron and steel and products; chemicals, non-ferrous metals and products; ammunition, shells, bombs, etc.; guns; machinery and electrical equipment; petroleum and coal products; combat and motorised vehicles; machine tools. Not directly indicated by investment figures, but of equal if not greater importance in the shift of industrial structure brought about for war production purposes, are changes affecting workers, transportation facilities, power and other utilities, and community arrangements. "Some war plants have required the development of entire new communities, which face the danger of becoming ghost towns unless some substitute activity can be provided." (Glenn E. McLaughlin: "Wartime Expansion in Industrial Capacities", in *American Economic Review Supplement*, Mar. 1943, pp. 108-18.)

[2] *Victory*, 22 Dec. 1942, Vol. 3, No. 51, p. 9.

[3] *Economist* (London), 7 Feb. 1942, pp. 177-8.

[4] For specific items of information on this expansion, see League of Nations: *World Economic Survey, 1941/42*, pp. 34, 46, 86, 87.

industries, in particular, have been stimulated by the war, and these are the industries that in peacetime are most closely related to the production of capital goods.

It appears, therefore, that some of the most acute problems of excess capacity are likely to manifest themselves in just those industries that could turn out machines, ships, locomotives, electrical equipment, bridges and tractors, and in the industries that feed basic materials to them, such as steel, copper, aluminium, rubber and fibres. If the demand for such goods can be kept high, that will directly lessen the severity of the readjustments necessary in the advanced industrial countries. Furthermore, if the demand for capital goods can be sustained by methods that at the same time expand the world's future capacity to produce and consume all kinds of goods, that is an additional permanent gain. Such favourable effects are precisely the ones that might be expected from a vigorous world development policy which would equip the countries in greatest need of development by means of capital goods that the advanced industrial countries will be prepared to produce in huge quantities after the war.[1]

It hardly needs to be added that, of course, no programme of world development can remove the necessity for drastic readjustment in war-oriented industries after the war. What a vigorous effort of world development could do is to help make the problem somewhat easier and its solution more constructive.

One objection needs to be examined. It might be argued that from the long-term point of view it is undesirable to sustain the demand for capital goods in the advanced industrial countries, because a high proportion of capital goods production to total production makes an economy less stable. The capital goods industries are, of course, just the ones that are most subject to violent booms

---

[1] A recent report of the Delegation on Economic Depressions set up by the League of Nations observes that:

> Excess plant, whether due to expansion to meet military needs or to expansion to meet civilian needs in areas cut off from their former sources of supply, constituted one of the gravest causes of economic instability after the last war; it will without doubt constitute a grave difficulty after this war.
>
> But the expansion of the metal, the engineering and above all the machine tool industries during the war should in general prove an asset and not a liability, an asset the value of which will be determined by the policies pursued. These industries form the foundation of the whole industrial system and the strengthening of that basis should greatly facilitate the development of industry throughout the world, just as the acquisition of new skills by labour should facilitate the growth of new industries and the development of old. The war has enormously increased the number of persons in all parts of the world who have acquired a mechanical sense. For this new basic plant, as for these new skills to render the services of which they are capable, however, a high rate of investment will have to be maintained. There will lie the crucial problem. (*The Transition from War to Peace Economy* (1943. II. A. 3), pp. 33-34.)

and depressions. A high proportion of capital goods exports in a country's foreign commerce—the stage in industrial evolution reached by Germany and England, for example, in the inter-war period—has been characterised as "a form of trade peculiarly sensitive to economic fluctuations".[1] Might it not be better, therefore, to suffer the pains of more violent contraction in these industries after the war and get back to a lower ratio of capital goods output as soon as possible?

The most compelling answer to this argument is the practical fact that there will be trouble enough in achieving industrial balance after the war, even if demand for capital goods is sustained to a considerable extent by world development projects. In view of the institutional rigidities of modern capitalistic economies, great structural readjustments are very costly and disturbing at best. The future gain in stability would be quite uncertain. Deliberately to forego a constructive programme that would ease the problems of the advanced industrial countries and also help the countries in need of development, would hardly be wise. Indeed, to undertake greater adjustments than are necessary, all for the sake of being better prepared to ride out future economic storms, might be to help bring on the very storms that are feared. If fewer men and less capital have to be forced out of the capital goods industries there will, in this important sector of the advanced economies, be less pressure on wages and profits, fewer strikes and lockouts, and more time in which to make inevitable readjustments gradually. Economic stability is a very important objective, but there are other and better ways of seeking it than by deliberately sacrificing the trained men and the installed equipment which could make railroad cars, river boats, trucks, telephone cables, laboratory apparatus and electric motors for an expanding world economy.

There are other considerations. Suppose that Great Britain, for example, were to try to find alternative employment outside the capital goods industries for a large proportion of the workers and funds that will be in these industries after the war. Where would the excess metal workers, among others, look for new jobs?

Under a world development programme some of them might be making bridges and rolling stock and communication systems and printing presses for the Balkans, China and South America. If such employment were not available there would certainly be a temptation to restrict imports into Great Britain and to subsidise British agriculture more heavily. This would lessen the opportunities of overseas producers to sell their agricultural and other

[1] LEAGUE OF NATIONS: *Commercial Policy in the Inter-War Period* (1942. II. A. 6), p. 125.

products in Great Britain. If this policy put the metal workers into jobs it would do so at the sacrifice of their most productive skills. On the other hand, a high volume of international trade in capital goods, stimulated by great investment activity in the countries where equipment is needed, would permit the British metal workers to continue where they can earn the highest wages and would permit the overseas producers of agricultural products and other British imports to enjoy a better market. The same applies to other advanced industrial countries.

Agricultural and Raw Material Industries

There will also be severe structural maladjustments after the war in some of the industries producing primary materials. To meet war demands, new facilities for the production of copper, aluminium, sugar and many other raw commodities have been opened, or old facilities have been expanded. Japanese conquest of the world's rubber-producing region has led to feverish activity in the United States for the building of synthetic rubber plants. In Latin America the output of natural rubber is being stepped up. Germany and the other Axis countries are trying to compensate for the effects of the blockade by developing domestic raw material sources to the maximum and by finding substitutes for commodities that they would import in peacetime. Furthermore, the war has stimulated numerous technological innovations, which will have the effect, now or later, of increasing the output of many raw materials and lowering their cost. When hostilities have ceased, agricultural and mining areas that are temporarily out of production because of military operations and regions now more or less isolated by blockade or lack of shipping will begin to offer their supplies once more. In the meantime, war expansion in other countries may have duplicated their facilities.

The improvements in production techniques and lowering of costs in many raw material industries in the two decades following the First World War have been described by one writer under the title "A Raw Commodity Revolution".[1] These factors provoked a downward trend in most raw material prices after the middle 'twenties. Unmarketable surpluses and excess capacity plagued many agricultural and other raw material industries even before the onset of the great depression. Some of the results of these developments were "to demoralise prices for raw commodities and the manifold products into which they enter, to upset national finances

[1] Melvin T. Copeland: *A Raw Commodity Revolution*, Business Research Studies, No. 19 (Harvard University, Graduate School of Business Administration, Mar. 1938).

and foreign exchange, and to cause social and political unrest throughout the world". When a general economic crisis arrived, "the raw commodity situation helped to turn the business slump which began in 1929 into a major economic disaster".[1]

There is every reason to expect that similar problems of structural unbalance will confront the world's agricultural and raw material producers after this world war. Some of the surplus situations will be no less acute. Experience shows that the demand for most raw materials is relatively insensitive to decreases in price, so that an output larger than the market will take at current prices may lead to a drop in price so violent as to prove ruinous even for efficient producers. Nevertheless, experience shows that it is very difficult to induce agricultural and raw material producers to leave their industries. Often they have not the capital or the knowledge to try alternative products, or they are held by social ties or habit. Thus, structural maladjustments in the form of overcapacity, in particular raw material industries, may persist for years.

An expansion in the world demand for industrial equipment, such as might be brought about by a series of international development programmes on a considerable scale, would undoubtedly help to ease the post-war difficulties of structural readjustment in the raw material industries by taking up some of the slack between peacetime raw material demand and available supply. It would do this indirectly by helping to maintain a high level of employment and consumption and directly by promoting the use of more capital equipment to be used in new installations.

There is a further effect of economic development, especially industrial development in areas hitherto devoted almost exclusively to agriculture and extractive industries, which would help raw material producers in the long run. Industrial development in such areas as south-eastern Europe, China, India and Latin America would mean greater diversification of industry over the world. As areas that had been exclusively sellers of raw materials on the world market became areas of diversified industry the supply of industrial goods and the demand for raw materials would increase. Raw material prices, in general, would tend to rise relatively to the prices of other goods and services. That is, the "terms of trade" of raw materials in relation to other things would improve.

This shift would also help, in the long run, to prevent chronic tendencies to disequilibrium in international trade and balances of payments like those which have arisen out of the weak and precarious earning power in international markets of the "raw material countries". Greater diversification of their economies as a result

[1] *Ibid.*, pp. 1, 25.

of the measure of industrialisation which is normally a part of economic development would likewise make their international financial positions more stable. Expansion of industries in the less developed countries, accompanied by large-scale investment from abroad, has frequently been referred to in recent economic discussion as an important means of promoting better equilibrium in international economic relationships generally, making success more probable in efforts to stabilise currencies and to forestall pressures that lead to restrictive trade and exchange controls.

CHAPTER III

INCOME DISTRIBUTION AND THE TRANSFER PROBLEM IN REPAYMENT

THE INTERNATIONAL DISTRIBUTION OF WEALTH AND INCOME

In the countries that would be the natural sources of investment funds, proposals for international economic development are sometimes stigmatised as "share the wealth" ideas or invitations to "play Santa Claus". The implication is that a country providing capital funds is depriving itself of wealth for the benefit of the area where the capital is used.

This discussion does not concern gifts advanced to rescue victims of war from starvation and disease and to set them once more on the road to self-support. It may be assumed that such gifts will be made in considerable amounts by the more fortunate peoples who have suffered less from the war. The benefits expected by the givers will not be repayment in coin or goods, but they will nevertheless be real benefits—the satisfaction which comes from an act of sympathy and kindness, the military advantage of stabilising the civilian situation in the rear of our advancing armies, the political advantage of laying the basis for a more durable peace and the economic advantage of reviving production and trade as rapidly as possible and—at a certain stage and for certain commodities—finding a good use for surplus stocks or surplus production capacity.

A long-range programme of national and international economic development is a different matter. There is no reason why capital advanced for sound developmental purposes should be a free gift and no one expects that it will be. The capital, if wisely invested, increases the capacity of the borrower to produce. Out of this increased productivity the loan and interest charges can be repaid.[1] It is a case of mutual benefit. An investment banker who lends $10,000 to a manufacturer for improvement of his plant on the understanding that the $10,000 will be repaid out of earnings over

[1] Assuming that the creditor countries follow reasonable trade policies. It must be possible for debtor countries to sell goods abroad; otherwise they cannot earn the foreign exchange with which to pay external debts.

a period of years, with reasonable interest, is not a Santa Claus. Nor is he an exploiter. The same is true of an advanced industrial economy with a high rate of savings which makes funds available out of those savings for sound developmental loans in areas where capital is scarce, on the understanding that these loans will be repaid with reasonable interest. In fact the same banker may actually make the loan in the latter case as in the former, the difference being that the borrower is in a neighbouring country.

The indirect benefits of a sound world developmental programme, some of which have been mentioned already, are so important to the advanced industrial countries as to make it distinctly in their interest to foster such a programme. In doing so they would be well advised to make capital available on very good terms and even at quite low rates of interest. But there is no reason why they should forego the usual contract for repayment of funds advanced or why the contracts for repayment should not be fulfilled without an excessive percentage of default. Indeed, the advanced industrial countries which would be the lenders have largely within their own power the means of controlling many of the factors which determine whether international loans will be generally defaulted or generally repaid—namely, those relating to security against recurring war, to world economic stability, which depends very directly not only on the international policies of the great advanced countries but on the extent to which they stabilise their own business and employment, and to the functioning of the world trading system in a manner which permits debtor countries to earn the means of repayment by selling their export goods.

A programme of international investment for the development of some areas of the world with capital supplied from other areas is not, therefore, a programme for "sharing the wealth" by unrequited generosity on the part of peoples with more abundant savings. If the economy of the lending country gives up some of its wealth temporarily as a loan to the economy of the borrowing country, it does so on a sound promise and prospect of receiving it back, plus interest, later.

But there is a further point which puts a still more favourable aspect on developmental loan transactions from the point of view of the lending countries. So far we have been speaking as though the lending economy would actually sacrifice some of its wealth, temporarily, receiving it back with interest at a future date. In the static type of analysis which economists used to apply to such problems it was always assumed that Economy A in supplying capital to Economy B must temporarily reduce its own supply of capital, or its consumption, by that much. Capital, in this connection as in others, represented "abstinence" on the part of somebody. We now know that this is not necessarily true. The more dynamic modern type of economic analysis explains why. The capital export is additional investment for the lending country. Under certain conditions—namely, when there is less than full employment in the lending country—the stimulus thereby provided to its economy may result in an increased total production. Indeed, production may well increase by more than the amount of the investment. This would be particularly likely if the capital goods industries and their raw material suppliers had a great amount of unemployment.[1] Under such conditions, and if the domestic means actually available for stimulating investment and employment were for one reason or another somewhat less than adequate[2], then the lending country might invest abroad without even temporarily reducing the amount of real wealth and income available at home. In fact, by the very act of investing abroad the lending country might have more at home, so that there would be a net gain even if no repayment were ever received. This is a real possibility under the conditions that may exist a few years after the end of the war. On the other hand, if a considerable capital export took place under boom conditions, where substantially full employment already existed, the cost to the lending country might be more than the amount of the loans—for the additional investment stimulus might contribute to a dangerous inflationary tendency. We shall return to these points in later chapters, where methods will be suggested for influencing the timing of international developmental investment in order to maximise advantages and minimise disadvantages for all parties concerned.

The effect of international developmental investment on the countries where the development takes place is, quite obviously, favourable to the increase of their wealth and income. These countries will be those in which economic progress is held back, not by a tendency of savings to exceed investment, but by a scarcity of savings available for improving the capital equipment of the area.

[1] It is assumed that the export of capital for developmental purposes would probably be accompanied or followed by increased orders for capital goods. Of course, capital exports can take place without the export of capital goods. The "real transfer" from the lending country may take place through an increase in the export of consumers' goods and services or a decrease in imports of goods and services. But loans of capital made in connection with a world development programme would surely increase the world demand for capital goods and a large part of that increased demand would be felt by the capital goods industries of the advanced industrial countries, which would also be the lending countries. In some cases, the capital loans might be tied directly to capital goods exports, as when specific capital goods are sent abroad on credit.

[2] On the practical merits of the argument sometimes made that it is always possible to get full employment by increasing purely domestic investment, see Chapter VII.

The importation of capital from abroad will make possible better equipment, more production per worker and higher living standards.

One more point, and a very important one, remains to be added to this analysis of the effects on the international distribution of wealth and income which may be expected to result from developmental investment. Capital for developmental purposes will be accompanied on its migration to the underdeveloped areas of the world by technical knowledge, in the form of engineering advice, instruction of local people in the operation of machinery, and—if the developmental authorities are wise—assistance in carrying through broad, basic educational programmes among the people of the area. The knowledge which flows with capital into less developed areas in this way is itself a very powerful means of raising productivity and living standards. The "export of knowledge", however, costs the countries which supply it very little, for knowledge is something of which the stock is not depleted in the process of passing it on to others.

> Knowledge, together with the productive arts based upon it, is the most potent of all the resources we have for improving the economic lot of mankind. It is a most peculiar resource in that it can be added to one region without subtracting it from any other. Indeed, the more widespread knowledge becomes the more rapidly it grows, for it increases by being applied under new circumstances and by the cross-fertilisation of ideas passing back and forth among many minds. Its potency as a factor of production has been shown in recent decades by the phenomenal rise in the industrial output first of Japan and then of the Soviet Union. Both countries have imported some capital, to be sure, but in relatively small quantities. Their main borrowings from abroad have been imports of knowledge and technique.[1]

To sum up, it is clear that sound developmental loans subsequently repaid out of increased productivity increase the wealth and income of the countries under development. As for the effect on the capital-providing areas, the developmental investment certainly does not cause a permanent decrease in the wealth and income of the lending countries and may not even decrease temporarily the amount available for their use at home. In fact, under the conditions of the post-war epoch, the process of sending capital abroad for investment may even increase the amount of income left over for domestic consumption or investment, because of the stimulus that investment gives to employment and production.[2] The chief disadvantage which might arise for the lending countries

[1] Eugene STALEY: *World Economy in Transition* (New York, Council on Foreign Relations, 1939), p. 279.

[2] Again, reference is made to Chapter VII for consideration of the argument that full employment can be assured just as well by purely domestic investment.

would appear in case of boom conditions at home. An inflationary situation might be aggravated by large-scale developmental investment abroad undertaken just at that time. This suggests that attention to the timing of developmental operations is important.

Thus, it may truly be said that developmental investment in the less well-equipped regions of the world is a method of levelling income and living standards *upward*. By supplying capital and technical knowledge, under soundly conceived plans, the countries of advanced industrialism can help to raise the incomes and living standards of other countries, not only without lowering their own incomes and living standards, but with positive economic benefit to themselves. There is no taking from one and giving to another, but mutual gain.

The Internal Distribution of Wealth and Income

What effects will international developmental investment have on the distribution of wealth and income *within* the countries that supply capital? The older static type of economic analysis would have said that export of capital temporarily reduces the amount of capital per worker in the lending countries. Therefore, runs the argument, it must tend to make interest rates higher and wages lower. Even when the capital is eventually repaid with interest, the interest earnings go to the investor class. The national income is thereby increased (because the capital presumably earns a higher rate of return abroad than it could have earned at home) but the distribution of that income is altered in a way unfavourable to labour.

Even in terms of the static type of analysis this is not the whole story. Indirect gains resulting from the productive investment of capital abroad, such as cheaper food supply for working people, will diffuse themselves among all classes in the capital-exporting countries. This is quite likely to counterbalance any tendency of capital export to lessen labour's share in total income.[1]

However, there is an even more important reason why the static type of analysis outlined above leads to wrong conclusions. It ignores the relation of investment to the volume of employment. Reasoning which assumes that because investment is made abroad there will be less capital for use at home is very likely to be incorrect. Increased investment, as has been explained earlier in this study, raises the level of output, if there is less than full employment. The interest of workers in maintaining employment, production and income at high levels far overshadows their interest in the

[1] Cf. Jacob Viner: "Political Aspects of International Finance", in *Journal of Business of the University of Chicago*, Vol. I, Apr. 1928, pp. 148-9.

theoretical share of income going to labour as compared with that going to capital. If the size of the income "pie" to be divided can be made large, there will be adequate methods available to labour in these days of collective bargaining by which to make sure that it gets a fair slice. Developmental investments help to raise employment and income, and that is the main consideration.

The Transfer Problem in Repayment

Developmental investment is ordinarily rated successful if it increases production by an amount more than ample to pay interest and amortisation on the original capital. However, international loans as a rule have to be repaid, not in the currency of the country where the increased production takes place, but in some foreign currency, usually the currency of the lender. It may thus occur that a borrower who has ample earnings in, for example, pesos, may still find great difficulty in transferring those pesos into a foreign currency, such as dollars, in order to meet service charges on a loan. The debtor area may not succeed in marketing enough of its products abroad to earn the amount of foreign money needed to pay for its current imports and to provide extra cash for meeting payments on its external debt. When this happens, the so-called "transfer problem" arises. The soundness of international developmental investment, from the point of view of repayment, therefore depends not only on an increase of productivity in the borrowing area but also on its ability to earn the currency of the creditor country, or some other foreign exchange acceptable to the creditor.

The ability of a borrowing area to acquire the foreign exchange needed for external debt payment depends partly upon circumstances outside its own control, including the economic situation in the world at large and the trade policies of the creditor countries. Dollars may be scarce because of a depression in the United States, which sharply reduces American purchases abroad and thus reduces the supply of dollars available for the purchase of goods and the payment of dollar debts. Or the United States and other creditor countries may use tariffs and quotas to hold their imports at a low level, thus preventing debtor areas from earning an adequate supply of foreign currency with which to make repayments. The world is all too familiar with the serious strain imposed on international financial and trading mechanisms when creditor regions refuse for one reason or another to accept the goods and services which are the real means of repaying loans. Supposing, however, that economic conditions abroad are stable and that creditor countries adopt trade policies appropriate to their creditor positions, the solvency of the debtor country on international account will also depend on the wisdom of its own policy and on the wisdom with which the developmental operations are planned.[1]

An international loan is financially sound if it (1) increases the fundamental capacity of the debtor country to produce goods and services, and to produce them in excess of its current consumption; and (2) increases the capacity of the debtor country to earn more foreign exchange by export than it needs to balance current imports, thus creating a margin for debt payment. With reasonable foresight and co-operation, large-scale developmental investment which meets these tests should be possible, assuming always that the leading countries will pursue policies appropriate to preventing disastrous depressions and wars. As was pointed out in Chapter I, the rising trend of world productivity fostered by developmental undertakings and by the progress of technology has a tendency to lessen the size of repayments in relation to current levels of income and of international trade. Expansion of international tourist traffic, under the influence of aviation and the advertising of far places provided by this global war, may also make it easier for some developing countries to earn the currencies of the creditor nations.

Acute transfer problems are most likely to arise in connection with large financial indemnities, in connection with debts between allies resulting from war supply transactions or in connection with debts imposed as a charge for relief and rehabilitation aid which rescues people and repairs damage but does not increase the fundamental capacity of the recipient area to produce goods beyond current needs and to earn foreign exchange by export. Where an investment does raise productivity and does increase the international earning capacity of the debtor country, the transfer problem is manageable, given a tolerably sound economic environment.

If economic and political stability can be maintained, the principal of many individual investments and of international investment in the aggregate in the developing countries is likely to be left intact for an indefinitely long period. The investors as a group will not wish to withdraw the capital sum, preferring a regular flow of interest and dividends. Particular loans will, of course, mature and be paid off, and particular owners of securities

[1] See on this point J. J. Polak: "Balance of Payments Problems of Countries Reconstructing with the Help of Foreign Loans", in *Quarterly Journal of Economics*, Feb. 1943, pp. 208-40.

will liquidate their holdings by selling them to others. But the creditor countries will continue to maintain or increase their total international investments.

Under such conditions there would be no sudden reversal in the direction of capital movements to impose a brusque strain on the exchanges. The return flow of interest and dividends would build up eventually to a size which would overbalance the new capital flowing out, but that would be a gradual turning and would give time for trading relationships to shift so that the creditor countries, now becoming "rentiers", could gradually become net importers of goods and services, as would be appropriate to their position. This is the evolution that Great Britain went through earlier, until all "normal" economic adjustments were swamped in cataclysmic wars and depression. If we contrive to get a stable world, such a course of evolution would again be normal. If, on the other hand, there are more cataclysmic wars and depressions, the transfer problem will go unsolved (as will most other economic problems) and the pressure of large-scale international debts on the exchanges will again contribute to currency disorders, trade restrictions, and defaults.

Even if the world avoids wars and depressions of the cataclysmic sort its success in attaining political and economic stability is likely to be much less complete than we should wish. There will be shocks to meet, and there will be inflexibilities in the functioning of economic systems. Difficulties arising out of a rigid mass of international obligations for debt payment could help to turn a setback into a major catastrophe. Therefore, it is important that conscious thought be given at the time investment programmes are put on foot to the means by which sudden strains on the balances of payments of the developing countries can be prevented and to the means by which the building up of timely export balances on their part (and import balances on the part of the creditors) can be assisted. The following points deserve consideration.

First, sudden suspensions or reversals in the flow of capital funds after economies have become adjusted to a certain direction of flow can be disastrous. Investments that might have been sound become unsound. The outward movement of capital from the United States in the 1920's was erratic, rising to a peak in the first half of 1928 and falling abruptly thereafter in response to the counter-attraction of the stock-market boom. This is one of the factors generally cited in explaining the world economic crisis that followed. As the depression got under way, capital flowed back towards the United States in the form of contractual repayments, while new loans ceased. A fundamental requirement of sound lending policy is, as a recent American official study points out, "that investment programmes be formulated on a comprehensive and long-range basis and be executed at a reasonably regular rate . . . "[1] Under post-war conditions, governments will undoubtedly have to assume some of the responsibility if this requirement is to be met; the guiding and stabilising influence that could be exerted by a strong international public agency for assisting in the planning and execution of developmental programmes would be invaluable in this connection.

Second, developmental programmes will be more sound if attention is given in their planning to keeping the total foreign exchange obligations of a developing country low in relation to the reasonably expected increase in its capacity to meet such obligations. Long periods of amortisation and low interest rates for developmental capital are desirable. Projects need to be examined for their probable effects on the long-term export possibilities of the borrowing country as well as on its productive capacity in general. Both for the purpose of minimising difficulties of external payment and also for other reasons, good developmental policies will require that the largest possible amount of technical and educational assistance should accompany any capital funds that go into newly developing regions. The increase of capacity to produce, and to earn means of international payment, will thereby be kept at a maximum in relation to the size of the debt.

Third, it is desirable to have some flexibility in contracts of repayment. The rigid, bond type of contract which sets fixed obligations in foreign currency may lead to forcing of exports on falling markets, restriction of imports, currency disorders, and defaults. "The danger of saddling a large burden of inflexible obligations on an unstable international economic and financial system was amply demonstrated in the depression period."[2] Equity investments, in which the investor's return varies with the earnings of the enterprise, or some form of income debenture in which payment obligations likewise fluctuate with business results, would avoid these difficulties to a considerable extent. In time of sharp depression affecting the export possibilities of a newly developing country the earnings of foreign equity investments would probably also drop, and the demand for foreign exchange with which to pay out dividends would decline, thus preventing some of the conse-

[1] Hal B. LARY and associates: *The United States in the World Economy: The International Transactions of the United States during the Interwar Period* (International Economics and Statistics Unit, Bureau of Foreign and Domestic Commerce, U. S. Department of Commerce, Washington, 1943), pp. 19, 99-100.

[2] *Ibid.*, p. 106. See also, on advantages of direct investments and investments of an equity character, pp. 20, 104.

quences associated with the bond form of external debt. Some of the most promising future opportunities for private international capital investment, and among the soundest from the point of view of lending and borrowing countries alike, would appear to be in the field of direct investment—that is, investment involving participation in management, perhaps in association with nationals of the borrowing country. The returns on direct investment are ordinarily of an equity character and would fluctuate with business conditions instead of being an inflexible burden on the foreign exchange earning capacity of a developing country.

Another method of providing flexibility in repayment contracts would be to distinguish two parts of the payment obligation: (1) payment in the borrower's local currency; and (2) transfer of this payment abroad, making it available in the currency of the creditor or some other foreign currency. The first part of the obligation might be expressed as a fixed annual amount, or the amount might be made to depend on the profitability of the undertaking, or on some index of general economic conditions in the borrowing country. The second part of the obligation—transfer abroad—could be made to vary in some agreed manner in accordance with changes in the international economic situation confronting the paying country. Thus, the foreign payment obligation might be linked to an export price index (if the country's foreign earnings depend largely on the market for a few raw materials), or to an index of world trade activity, or to an index of the "supply of dollars" and the supply of other important creditor currencies (that is, the amount spent by the creditor countries for imports of goods and services or lent abroad, and thereby made available to other countries for purchase of their goods and for debt payments to them). The last would perhaps be preferable, for it would bring home to the creditor countries the fact that in the final analysis they cannot collect on debts except by taking goods and services.

Fourth, the external payment problem will be immensely eased if the principal creditor countries take measures within their own economies: (1) to keep business activity and employment at a high, sustained level, and (2) to facilitate mobility in the transfer of workers and capital from less promising lines of production, and especially those lines in which imports can be had more cheaply than domestic production, into the more promising lines that ought to expand as income rises at home and abroad.[1] Both would help to bring an increased flow of imports of goods and services, which is normally associated with repayment. A long-term rise in

[1] This second point will be dealt with in considerable detail in Part II, especially Chapters X and XI.

the rate of consumption (a fall in the rate of saving) within the industrially advanced, creditor countries would make it easier for them to maintain a high level of employment while increasing the size of their imports in relation to their exports. They would thereby adjust themselves to consume the increased incomes to which they would be entitled by reason of past investments. That adjustment seems quite feasible over a long period of years, by the time a net return flow of interest and repayment on developmental capital begins, even though, as was indicated earlier, it can hardly be counted on to sustain employment in the first decade or two after the war.

CHAPTER IV

THE ORDER OF MAGNITUDE OF THESE EFFECTS

So much for the nature of the investment effects connected with international economic development. But what may be the magnitude of these effects? Will they be important to the economies of the more advanced industrial countries, or will they be insignificant in relation to the huge normal volumes of savings and investment in these countries? No-one can answer such questions with any assurance. Proper answers will depend upon unknown factors that the future alone can reveal, including the amount of vigour, determination and co-operation that may be forthcoming in support of a world development programme. Hence the following observations are tentative.

Rate of Capital Development in Asia

A recent study by a Chinese economist reached the conclusion that *modern industrial capital* in China in 1937 could be valued at roughly 3,800 million Chinese dollars.[1] Converted into American dollars this represents not much more than a thousand million dollars' worth of modern capital equipment in a country inhabited by 400-450 million people, or less than $2.50 per capita. The same study points out that in the United States $430 per capita was invested in manufacturing industries alone in 1930, not including the large amounts of capital equipment in transportation, mining, agriculture and commerce, almost all of which is "modern".

There is no reason to doubt that the people of China and of other underdeveloped areas are ultimately capable of operating the machines of industrial society as effectively as the peoples that are today more industrialised. If this view is accepted and for the moment the time factor is disregarded, the total amount of capital which would have to be put to work by the peoples of Asia and of other underdeveloped regions in order to raise their economies to the productive level already attained in the better equipped parts

[1] Tso-fan Koh: "Capital Stock in China", in *Problems of Economic Reconstruction in China* (China Council Paper No. 2, Institute of Pacific Relations, Eighth Conference at Mont Tremblant, Dec. 1942, mimeographed).

of the world is staggering to the imagination. In Asia alone there are more than one thousand million people, half the population of the world. For the most part they work with pre-industrial tools and possess almost negligible quantities of agricultural machinery, railroads and motor roads, locomotives and trucks, manufacturing and mining installations.

Yet the rate at which modern equipment can be installed and therefore the amount of capital funds which can be productively employed in economic development during any one year is limited. The main limiting factor may not be the ability and willingness of other areas to lend funds and to supply capital goods. It may be instead the social resistance of the people to changes in habits, for the coming of industrialism imposes drastic changes. The time required to train managerial personnel and to create an industrial labour force out of an agricultural population unacquainted with mechanical techniques will retard the utilisation of modern capital. Industrial development is a *social* process, involving much more than the mere installation of machines. Fundamentally it is educational, organisational and political rather than mechanical. The rate at which new capital can be absorbed is closely related to the rate at which new ideas can be adopted.

The modernisation of Japan has commonly been regarded as a remarkable example of speedy economic development in a country previously quite without any of the methods and appliances of advanced industrial communities. An interesting question, therefore, is how rapidly capital investment might take place in Asia if other Asiatic areas, in the years after this war, were to develop at the same rate in proportion to population and area as Japan developed earlier. Such a calculation is, of course, fraught with all sorts of difficulties, both technical and logical. The results may have a wide margin of error, even within the framework of the assumptions implied by the question. There is no assurance whatever that the rest of Asia will develop at a rate anything like that of Japan, for developments there may be more slow or more rapid. Nevertheless, such a speculation, if too much is not read into it, may have some value as a starting point from which to exercise judgment.

What period of Japan's economic development should be selected as a standard of comparison? There is no satisfactory way of answering that question. It has been assumed below, more or less arbitrarily, that the rest of Asia will stand at the same point in reference to future economic development at the end of this war as Japan stood in 1900. There is some basis for this assumption in the fact that living standards in the late 1930's in the rest of eastern Asia seem to have been roughly comparable with those of Japan in 1900. If it is argued that an earlier period in Japan's economic development should be taken as the starting point, the answer is that usable data on capital investment in Japan are not available for years earlier than 1896. If, on the other hand, it is thought that development in other areas of Asia (for example, in India) has already gone further than Japanese economic development had gone by 1900 and will progress more rapidly henceforth, the necessary adjustments can be made in the figures below. Second decade figures can be regarded as first decade figures, and so on. In general, the rest of Asia will face many of the same problems in economic development that Japan faced. Most of the Asiatic area is densely populated. Complex and rigid social customs established in a pre-industrial era offer resistance to modern techniques. Although there are great local variations, Asia outside Japan is, on the whole, probably not markedly richer or poorer than was Japan in the resources required by modern industry.

Data which can be used to project Japan's rate of development upon other parts of eastern Asia have been assembled by Mr. Robert W. Tufts and are presented in the Appendix to this chapter. The basic figures, as indicated there, are imperfect and incomplete. They undoubtedly err on the low side. It is with due reservations, therefore, that the results are summarised as follows.

In the first decade after 1900 Japan was adding to its capital equipment at a rate which is roughly indicated by the figure of 78 million dollars a year (at 1936 prices). This should probably be regarded as net rather than gross investment and as an understatement rather than an overstatement. While persons accustomed to thinking in terms of the annual income and investment of such countries as the United States will regard this as an extremely small sum, it was nevertheless equal to 12 per cent. of the Japanese national income at the time. In other words, 78 million dollars worth of new capital investment each year represented a very substantial rate of advance, considering the extremely low level of wealth and income on which the Japanese people existed in 1900. In the decade after 1910 investment in Japan averaged 166 million dollars a year, or 17 per cent. of the annual income at the time. The investment figures rise to 313 million dollars and 354 million dollars in the third and fourth decades after 1900, representing 12 per cent. and 10 per cent. of the current national income in these years. (See table 1.)

On what basis can this rough indication of Japanese capital development be translated into terms that would suggest the investment magnitude of a comparable rate of development in other parts of Asia? China, for example, obviously offers a much larger base

TABLE 1. CAPITAL INVESTMENT IN JAPAN 1900-1936[1]

(in millions of 1936 dollars)

Period	Total	Average per year	Investment as per cent. of national income
1900-09	783	78	12
1910-19	1,658	166	17
1920-29	3,128	313	12
1930-36	2,476	354	10

[1] For sources and methods, see Appendix to this chapter.

for economic development. Should we multiply the Japanese capital investment figures for a given decade by the ratio between Chinese and Japanese population or by the ratio between Chinese and Japanese land area? A combination of the two has been used. As explained in the Appendix, the Japanese investment figures were split into two parts. One part, representing investment in industry and commerce and in local public works, was assumed to be more closely related to population. The other part, representing investment in undertakings connected with agriculture and transportation, was assumed to be more closely related to land area. The figures for the first type of investment for a given decade were multiplied by the ratio between the present population of China and the 1900 population of Japan, and the figures for the second type of investment were multiplied by the ratio between the two land areas. These results were then added to give a weighted average based on both population and area. A similar procedure was used for other parts of Asia. The results are summarised in table 2.

TABLE 2. CAPITAL INVESTMENT IN VARIOUS ASIATIC AREAS, BY DECADES, SUPPOSING RATE OF INVESTMENT WERE TO BE THE SAME IN RELATION TO POPULATION AND AREA AS IN JAPAN IN THE DECADES 1900-1936[1]

(in thousand millions of 1936 dollars)

Country	Population (millions)	Area (thousands of sq. kms.)	Investment by decades			
			1st	2nd	3rd	4th
China	450.0	11,103	13.6	23.1	44.9	51.6
India	365.9	4,079	7.2	14.6	27.7	31.4
Neth.E.Indies	68.4	1,904	2.2	3.7	7.2	8.3
Other areas[2]	116.5	2,865	3.5	6.0	11.7	13.3
Total	1,000.8	19,951	26.5	47.4	91.5	104.6
Average investment per year.			2.7	4.7	9.1	10.5

[1] For sources and methods, see Appendix to this chapter.
[2] Burma and British colonies, Thailand, Philippines, French colonies, Japanese colonies.

These calculations would indicate that if the parts of Asia shown in the table develop their economies after the war at a rate which leads them to add capital equipment as rapidly as Japan did after 1900 they might absorb each year during the first decade about 2,700 million dollars' worth of new capital (from all sources, internal and external). In the second decade the rate of new investment might rise to 4,700 million dollars a year, in the third decade to 9,100 million dollars a year and in the fourth decade to 10,500 million dollars a year.

These figures are based on 1936 prices. If allowance were to be made for the higher money costs of capital goods that prevail now and may prevail after the war they would have to be increased.

In the case of Japan, according to a very rough calculation, imports of capital goods amounted to some 40 per cent. of Japanese total investment during the first decade after 1900. The ratio fell in succeeding decades to 36 and 34 per cent.[1] If we assume that foreign countries will supply capital goods to these other areas of Asia in the same proportion to total investment as in Japan, then the economic development of the major part of Asia at the rates indicated would give rise to capital goods orders abroad at the rate of about one thousand million dollars yearly during the first decade (about 550 million dollars' worth from China, and nearly the same from the other areas together). In the second decade of development 1,700 million dollars' worth of foreign capital goods would be purchased yearly, 3,300 million yearly in the third decade, and 3,500 million yearly in the fourth decade. All amounts are in terms of 1936 prices.

It must be emphasised that these figures and the figures in the preceding table are not forecasts, but rather an indication of what would happen under certain assumptions. The reader must use his own judgment in deciding whether China, India, the Netherlands East Indies and other parts of Asia are likely to absorb capital and import capital goods at rates resembling those of Japan after 1900, or at higher rates, or lower ones.[2] Below are some points to be weighed on each side.

---

[1] See Appendix to this chapter.

[2] The only other remotely comparable calculation that has been discovered is one made by Colin CLARK in the course of some long-range forecasts on the future of the economic world (*The Economics of 1960*, London, Macmillan, 1942, Chapter VI). He indulges in some statistical speculation about the present size of the capital stock of India, China and the rest of Asia and Oceania (as of 1935-38), and projects ahead to 1960. The difference between the two figures would indicate an increase of 253,000 million "international units" (that is, dollars of 1925-34 purchasing power) in the capital stock of this area by 1960. Such an increase would be about 3 ½ times as great for the first two decades after the war as the total new investment indicated by the comparison with Japan on the

The thesis that actual capital investment in these countries is unlikely to approach the figures indicated within the near future could be supported by arguments based on the social resistances already referred to above. Also, the speed of economic development might be greatly retarded by internal political conflicts in China, by conflicts among religious groups, difficulties of caste and unsettled questions of self-government in India, and by colonial status or the transition form it has reached in some of the other countries.

On the other hand, it is quite possible to argue that the figures in table 2 are not nearly high enough to give a reasonable forecast of economic development in at least some of these countries. It would be interesting to attempt a similar projection based on the rate of development in the Soviet Union, which was probably higher, in the decade following 1928, than the highest rate attained by Japan, even though capital was not drawn from abroad. The economic development of Japan appeared rapid, as viewed by contemporaries, in comparison with the earlier growth of industrialism in Great Britain and in the countries of western Europe and the United States.[1] There would seem to be good reasons for expecting that, other things being equal, the *later* a country embarks on an intensive process of "catching up" to modern industry, the more quickly it would run through the process, once the main political

assumptions stated in the text above. The figures used by Mr. Clark as a basis for his calculations are open to criticism on the ground that they exaggerate the present capital stock of the area in question, and hence the probable amount of its growth.

Mr. Clark's projections into the future are based on a curve showing the relationship between real income per occupied person and real capital per occupied person in a number of countries at a number of different dates. The lower portion of the curve rests on very few cases and the evidence supporting the estimates in these cases is often quite scanty. The character of the very lowest part of the curve depends to a great extent upon the lowest point, namely, that for China in 1930. Mr. Clark's estimate of the real capital per occupied Chinese is 185 I.U. (that is, 185 dollars of 1925-34 purchasing power). The basis of the estimate is explained as follows: "Professor Buck's *Land Utilisation in China* indicates the average value of farm buildings and homesteads as 584 yuan per farm, or $240 at the rate of exchange then current. This may be put at $100 per occupied person, or allowing for the purchasing power of money in China, about 133 I.U. Making 40 per cent. allowance for capital other than farm buildings (probably generous) this becomes 185 I.U. per occupied person" (*ibid.*, pp. 86-7). This estimate would seem to be very high. It may also be asked whether it is appropriate to fit estimates of farm capital in China to the same curve as estimates of modern industrial capital in countries using very different techniques and social systems, and then to use the curve so constructed as a predictive device.

[1] Whether it was really more rapid than the economic development in these other countries at comparable stages is a question not readily answered. There is room for some very interesting statistical analysis of various aspects of the development of modern industrial economies, with comparisons among them related to different stages of advancement.

The results would be helpful as a starting basis from which to form judgments in quantitative terms when it comes to appraising programmes for future developmental undertakings.

and social obstacles have begun to crumble. The installation of up-to-date railroads, airlines, agricultural machines and factories should take much less time after the step-by-step pioneering has been done elsewhere.

This is especially true if circumstances permit the great international reservoir of technical knowledge, capital equipment and savings to be drawn upon with real effectiveness. The engineering methods of the 1940's, which China can adapt to its purposes, are enormously more productive and flexible than those of 1900 which Japan took over; modern travel makes it easier to send young people to technical schools abroad or to interest foreign specialists in helping to install their techniques and equipment in China; America and Europe turn out a much greater annual product now, and, particularly in view of the wartime expansion of their capital goods industries, can provide better equipment more quickly and in much larger quantities than would have been possible in 1900. Perhaps an effective international institution will be established to mobilise and guide capital and technical aid in promoting the development of new areas; if so, the results could well exceed past experience, for the world has never tried systematic co-operation to this end.

It can also be argued that a number of new social techniques applied to economic development in recent years make it possible to speed up the process considerably. The work of the Tennessee Valley Authority is a pattern for one type of regional programme. A much more highly centralised and completely State-managed type is expressed in the successive five-year plans of the Soviet Union.

For China, in particular, Government leaders are proposing comprehensive plans of development to be begun in the immediate post-war years and to be directed towards long-range goals that will require a generation or more to achieve. A programme put forward by Dr. Wong Wen-hao, Minister of Economic Affairs, calls for the investment of the equivalent of ten thousand million dollars (U.S.) for China's first five-year reconstruction plan after the war. This amount should be invested during the first four years, roughly one third on development of agriculture and water conservancy, one third on industry and mining, and one third on communications. It is estimated that one third of this capital could be raised in China, and that two thirds would have to be obtained abroad. This would indicate a considerably faster rate of capital development for the first decade than the hypothetical figure calculated on the assumptions of table 2 (say 25,000 million in ten years, as compared with 14 on the assumptions represented in the table). It is approximately equal to the rate shown in the table for the second decade of China's development.

The net result of the foregoing considerations, some of which might justify higher and others lower estimates of the probable rate of capital expansion in Asia, would seem to be a demonstration that the real answer will depend on circumstances best labelled "political" in the broadest sense.

The determining factors in the speed of Asia's post-war development are likely to be the *will* to adopt modern methods on the part of the peoples directly concerned and their ability to make that will effective in strong and competent government, together with the *willingness* of the more advanced countries to co-operate effectively, not only in lending capital and technical assistance but also in organising a sound foundation of international security and steady trade.

Other Areas

Well informed authorities on the amount of investment that could reasonably be made in the Latin American countries in the first post-war decade suggest the figure of 5,000 to 6,000 million dollars, of which 3,000 to 3,500 million dollars would come from foreign sources, mainly the United States, and the rest from internal savings. This would be devoted to transportation, communication, power, agriculture and food production, mining and metallurgy, manufacturing industry, colonisation and resettlement. About half the total would go for public service projects, such as transportation, other public utilities, health and sanitation. Of the rest, private capital might be counted on to undertake perhaps a third or a half. Thus as much as three quarters of the total would probably require governmental or intergovernmental auspices or encouragement of some sort.

In an article on "Problems of Industrialisation of Eastern and South-eastern Europe" by P. N. Rosenstein-Rodan[1] the author discusses the amount of capital that would be needed to make effective use, through development of industries, of the agrarian excess population in this area, which he estimates at 20 to 25 million people out of a total population of 100 to 110 million. He advocates a system of large-scale planned industrialisation under which at least half the capital would be supplied internally. Outside lenders and the borrowing area would each acquire 50 per cent. of the shares of a development trust formed of all the industries to be created in the region, and an average dividend of 3 per cent. would be guaranteed by governments on the shares subscribed in their

[1] *Economic Journal*, June-Sept. 1943, pp. 202-11.

countries. Shares could also be acquired by contributions in kind, for example, by the establishment of branch factories.

The total investment is estimated on the basis of an average figure of £300 to £350 per head required (including housing, communications and public utilities) to employ one worker in industries of the "light" category combined with a few industries of the "heavier" variety. This, assuming 12 million active workers to be employed, indicates a net investment of £3,000 million, which, allowing for maintenance of old and new capital over a period of ten years, is increased to a gross investment of £4,800 million. Another £1,200 million of capital is said to be necessary for the improvement of agriculture, but it is assumed that the bulk of this will have to be provided internally. After examination of the income of the area and the rate of investment in relation to income which would be necessary if this programme were to be achieved in ten years, the author concludes that at best some 70 to 80 per cent. of it could be financed on the basis suggested. Suppose we take, therefore, 75 per cent. of his figure of £4,800 million. This gives an investment for industrial development of £3,600 million, or, at today's exchange rate of $4.03 to the pound sterling, about 14,500 million dollars, of which half, or about 7,250 million dollars, is assumed to be provided by outside capital. This would represent an investment of foreign capital in the area amounting to 725 million dollars annually, on the average, over a decade.

In some notes on a long-run plan of economic development for Poland, Dr. Leon Baranski has attempted to work out the amount of investment required.[1] His calculations relate to an estimated population of 32 million, about a third of the population in the larger area just discussed. Assuming as an objective that the income of Poland is to be doubled over a period of nineteen years (starting after a three-year period of reconstruction), his figures show total investment rising from 450 million dollars in the first year to 640 million in the tenth year, and to 900 million in the nineteenth year.[2] Internal savings provide half of this in the first year, 70 per cent. in the tenth year, and nearly the entire amount in the nineteenth. This assumes that 7.5, 10.5, and 14.8 per cent., respectively, of the increasing national income in these years is saved and made available for the programme. Starting in the twentieth year, internal savings are assumed to rise above new investment,

---

[1] "Notes on Poland's Long-Run Economic Plan" (mimeographed). A more general discussion by the same author on economic development over a broader area is contained in his pamphlet *East and Central Europe* (published by New Europe, New York, 1943).

[2] The original zloty figures have been converted at the exchange rate of June 1939, namely, 18.81 cents, and rounded.

providing for gradual repayment of the foreign capital borrowed previously. For the first decade as a whole, Dr. Baranski's figures put total investment at approximately 5,400 million dollars, internal savings at 3,300 million dollars, and capital from abroad at 2,100 million dollars. In other words, he suggests that capital might be loaned from abroad at an average rate of slightly more than 200 million dollars yearly.

The Advanced Countries, and Past International Capital Flows

Are the figures mentioned above in connection with the economic development of various less developed areas large or insignificant by comparison with the amount of annual investment in the advanced industrial countries? Could developmental investments of this order of magnitude have any appreciable effect on the savings-investment balance and hence on the level of economic activity and employment? How do these amounts appear in comparison with past rates of international investment?

Table 3 brings together a number of statistical series which give an idea of the order of magnitude of investment and the use of savings in the United States from 1925 through 1938. A striking feature are the great fluctuations. In the year of peak business

TABLE 3. ESTIMATES OF CAPITAL FORMATION AND EXPENDITURES FOR NEW DURABLE GOODS IN THE UNITED STATES, 1925-1938[1]

(*in thousand million dollars*)

Year	Net capital formation (Kuznets)	Gross capital formation (Kuznets)	Private gross capital formation (Dept. of Commerce)	Expenditures for new durable goods (Fed. Res. Bd.)
1925	9.3	17.5	—	23.8
1926	9.2	18.0	—	25.3
1927	8.2	17.0	—	24.6
1928	7.4	16.5	—	24.9
1929	10.0	19.6	17.6	25.5
1930	4.2	13.4	12.1	20.4
1931	0.1	8.4	6.4	14.8
1932	−4.2	3.0	2.3	8.7
1933	−3.6	3.3	3.3	7.6
1934	−2.6	4.9	4.4	10.4
1935	0.7	8.3	6.7	13.0
1936	5.4	13.3	10.0	17.9
1937	6.4	15.3	11.6	20.2
1938	2.9	12.0	7.7	16.6

[1] Adapted from table 5 in "Estimates of National Output, Distributed Income, Consumer Spending, Saving, and Capital Formation", by Marvin Hoffenberg, in *Review of Economic Statistics*, May 1943, p. 143. Detailed explanations of the various series and references to the original sources are given in Mr. Hoffenberg's study.

activity, 1929, gross capital formation approached 20,000 million dollars, and in the fairly prosperous year 1937 it was 15,000 million dollars. At the bottom of the depression, however, there was only 3,000 million dollars of gross investment annually. Net investment, after allowance for depreciation and maintenance of existing capital, was 10,000 million dollars in 1929, fell below zero in the depression years and reached 6,400 million in 1937. Expenditures on durable goods ranged from a high of 25,000 million to a low of 7,600 million.

It is evident that the volume of investment necessary to maintain a high level of employment in the United States runs into very large sums. After the war the necessary volume will be still larger. The amount of net savings in the United States for which investment outlets (internal and external) will have to be found year after year if there is to be approximately full employment has been put at upwards of 12,000 million dollars at 1942 prices.[1]

The London *Economist* estimates that in Great Britain in 1938 about 900 million pounds (4,400 million dollars at 1938 exchange rates) were spent on capital goods. This includes money from savings and from depreciation allowances. But in order to provide full employment in that year an additional expenditure of about 300 million pounds (1,500 million dollars) on capital goods would have been needed, making 1,200 million pounds (5,900 million dollars) in all.[2]

Obviously, any investment figures that can reasonably be talked about in connection with post-war development of the areas of the world where living standards are lowest will not approach in magnitude the huge amounts of savings, investment, and capital goods production in the large advanced countries. The basic reason is that amounts of capital which are huge in relation to the existing capital and the annual incomes of less developed countries—and which would therefore strain their capacity to assimilate—are comparatively small in relation to the much greater capital stock and incomes of the highly developed countries.

Suppose that some years after the war there is an average annual new capital investment to the amount of 3,000 million dollars in Asia, with 2,000 million of it supplied from outside; that the eastern and south-eastern area of Europe is developing to the tune of 1,500 million dollars of new annual investment, 750 million of it from outside; and that Latin America is building up its productive equipment at the rate of 600 million dollars a year, aided by 350 million dollars of capital from abroad. This would mean a total "investment outlet" of slightly more than 3,000 million dollars annually

[1] Alvin H. Hansen and Guy Greer: "Toward Full Use of Our Resources", in *Fortune*, Nov. 1942, p. 158.

[2] "Full Employment" in *Economist* (London), Jan. 1943, p. 4.

in these three areas for the savings of the advanced countries. That would not come near to equalling the internal domestic savings and the domestic outlets for them of such a great industrial economy as the United States (when it is operating at high capacity). The conclusion is obvious that foreign investment for the equipment of undeveloped areas cannot take the place in the advanced countries of *domestic* measures designed to sustain a high level of economic activity and employment.

However, 3,000 million dollars of net investment stimulus and the sustained demand for capital goods associated with it, or even 1,000 or 2,000 million, would have been felt most gratefully in the United States economy, large as it is, when business was going downhill or struggling to advance out of depression. (Compare the figures on U. S. capital formation in the years 1931-35, table 3.)

Also, the three less developed areas mentioned above are not the only areas in which an international programme to assist economic development would stimulate investment outlets. In fact, for the first decade or so they are not likely to be the greatest fields for investment (see below). In later decades, after their development is well started, they will bulk much larger than they do today in the world's capital calculations.

The effect on the level of business activity and employment in advanced countries would be most appreciable (and appreciated) if international development programmes were planned many years ahead, with the financing also assured for a considerable period in advance, so that the investment stimulus—and the capital goods demand associated with it would be regular, year after year. In years of downward fluctuation at home such a foreign stimulus would be by no means negligible. The effect would be doubly good if the rate at which capital and capital goods flow to the newly developing areas could be *stepped up* in times of threatened or actual depression in the advanced countries.[1]

Some basis for comparing the figures discussed in this chapter with the flow of international investment capital in the past may be useful. Just before the First World War, British investments abroad were increasing at a rate equivalent to some 900 million dollars a year, French at 240-350 million dollars, German at 125 million dollars, and American at 125-150 million dollars. This totals to 1,400 or 1,500 million dollars a year, neglecting capital provided by some of the smaller countries.[2]

---

[1] This will be discussed much more fully in Chapters VI and VII.

[2] For sources of estimates and other data, see Eugene STALEY: *War and the Private Investor: A Study in the Relations of International Politics and International Private Investment* (Garden City, N. Y., Doubleday, Doran and Company, 1935), pp. 4ff., and Appendix A.

The United States was the largest exporter of capital following the First World War. At the peak of its lending activity in 1928 it sent abroad for long-term investment 1,300 million dollars and nearly 1,000 million in several other years.[1]

Investment capital supplied by other countries was in considerably smaller amounts during this period. International investment almost ceased in the depression, except for erratic movements of capital seeking security. It seems safe to say that the total amount of new international investment has never in the past exceeded 2,000 million dollars a year.

However, it should be remembered that the world's capacity to produce and to invest will be, potentially at least, much greater in the 1940's and 1950's than before the First World War or in the 1920's. Economic operations of many sorts are on an increasing scale. Compare, for example, war expenditures in the Second World War with those in the First. Furthermore, there has never been a comprehensive and systematic programme of international development promoted by governments, in peacetime, so that past standards of comparison might not be applicable if a concerted developmental effort were undertaken.

The Phases of Post-War Investment

It appears likely, for reasons advanced above, that the sums of capital used in developing China, India, south-eastern Europe, and other largely pre-industrial countries will not dominate the world investment scene in the first decade or two after the war. Domestic investments, first of all, in the advanced industrial countries will be on a much larger scale. Even in the field of international investment it is likely that other areas will, at least at first, absorb greater sums of foreign capital.

Probably the largest demands for capital from abroad in the years immediately after the end of hostilities will be from the industrially advanced areas that have been damaged by war action. Capital will be needed for emergency repair and then for reconstruction of cities and their utilities, for putting factories and transportation systems into running order, and to replenish working capital (raw materials and merchandise for factory storerooms and for the shelves of wholesalers and retailers).

[1] The amount was 821 million in 1926 and 987 million in 1927. From 1928 to 1929 came a drop of more than 50 per cent. to 636 million, and a further fall in 1930 to 364 million. There was a net withdrawal of 128 million in 1931 and 251 million in 1932 (Hal B. Lary and associates: *The United States in the World Economy: The International Transactions of the United States during the Interwar Period*, International Economics and Statistics Unit, Bureau of Foreign and Domestic Commerce, U.S. Department of Commerce, Washington, 1943, table I).

The rate of capital investment in this process will not be retarded by the social resistances which are present in areas lacking previous experience of industrialism. The limiting factors will probably be the availability of raw materials and shipping space, the willingness of the countries that have not suffered so severely from the war to extend loans, credits or gifts and the degree of success that can be attained in organising the efforts of rehabilitation and reconstruction so that they may proceed rapidly. A recent study by the Economic, Financial and Transit Department of the League of Nations estimates that continental Europe in the two post-war years 1919 and 1920 imported foodstuffs to the value of 6,300 million dollars, raw materials to the value of 7,200 million and finished goods to the value of 3,900 million, making total imports of 17,400 million dollars. This was paid for currently to the extent of 5,000 million in merchandise exports, 100 million in net gold exports and 5,600 million in "invisible" exports such as shipping services, goods and services sold to foreign armies and to tourists, emigrants' remittances, earnings on investments abroad, etc. The balance of 6,700 million dollars is estimated to have been financed by long-term loans (4,000 million) and short-term credits (2,700 million).[1] It is pure speculation what the corresponding amounts may prove to be after this war, but it can hardly be doubted that the need will be larger.

Apart from rehabilitation and reconstruction in war damaged industrial areas, the largest international flow of capital in the first decade or two after the war may well turn out to be for the further development of countries that already have fairly advanced techniques and are prepared for rapid expansion. This would include the progressive countries of young industrialism, such as Canada, Australia, New Zealand, some of the countries of Latin America and some of the countries of Europe that are industrially less developed. If the Soviet Union desires to borrow outside capital to facilitate rapid repair of war damage and further development of its resources, the amounts involved might be large.

It is also possible that non-developed and sparsely populated areas of the sort found in Africa and the Near East or even in the Arctic and Antarctic may become important investment outlets. The social and political obstacles to rapid development of their natural resources by means of airlines and roads, hydroelectric power stations and mining installations, irrigation systems and railways might be considerably less than those to be expected in densely populated countries at a similarly early stage of development.

[1] League of Nations: *Europe's Overseas Needs, 1919-1920, and How They Were Met* (1943), pp. 23, 24, 30.

Given international political and economic stability and a reasonable amount of organised co-operation, the further development of those countries which stand at an intermediate stage of advancement is capable of providing a very considerable investment outlet for the savings of the more advanced areas in the period which follows post-war rehabilitation and reconstruction. At a somewhat later stage, rising gradually to attain very large volume a few decades after the war, the capital demands of the countries of Asia and of other countries where industrial development is still at a low level may very well surpass those of the intermediate countries. Ultimately they could provide much larger investment opportunities. But for some years to come we may find that investment outlets in countries of intermediate industrial advancement (and, of course, domestic investment in advanced industrial countries) will be *quantitatively* more important. At the same time, developmental investment in such areas as China and India is likely to be particularly rich in significance for the future.

B. What Policies Will Yield Greatest Mutual Benefit?

CHAPTER V

THE ADVANTAGES OF MULTILATERAL CO-OPERATION: AN INTERNATIONAL DEVELOPMENT AUTHORITY

The previous chapters have discussed the effects which may be felt in the advanced industrial countries as a result of international investment for economic development. The problem to be considered now is how to get the greatest amount of benefit out of the investment process, both for the countries undergoing development and for the advanced industrial countries.

Capital may move from one region to another or one nation to another in many different forms. Business firms may establish branch factories abroad, install and manage public utility enterprises, operate mines and smelters and otherwise engage in what is called "direct" or "entrepreneurial" investment. Examples are the foreign operations of the great oil companies, of Pan-American Airways and of the International Telephone and Telegraph Company. Governments or business corporations in areas under development may borrow capital by floating securities in one or more of the great capital markets of the world. Before the depression of the 'thirties the stock exchanges in London, New York, Paris and Amsterdam were important channels for this type of international capital flow. Intergovernmental loans may be made for developmental purposes. For example, the Export-Import Bank and various subsidiaries of the Reconstruction Finance Corporation have been used by the United States to supply funds to South American Governments in order to increase the output of strategic materials and to assist in economic development. Government agencies or commercial banks may finance the sale of exports on credit. One of the reasons for creating the Export Credits Guarantee Department of the Board of Trade in Great Britain and the Export-Import Bank in the United States was the desire to assist in providing "medium-term" credits such as are needed in the financing of exports of machines and other capital equipment.

Intergovernmental loans, or borrowing by a government in capital markets abroad, may be organised, facilitated or even guaranteed by an international body. The League of Nations sponsored reconstruction loans of this kind in several instances.

It is not proposed to attempt any detailed analysis here of the various advantages and disadvantages of each of the particular forms of international capital investment mentioned above. All will doubtless have a place to some extent in post-war developmental activities, if conditions of reasonable political and economic stability are established in the world. The point to be stressed in this chapter is that all the various forms of international investment may be expected to yield a higher total of mutual benefits to the capital-lending and capital-borrowing areas if they take place in an environment of continuity and consistency for which the backing of an international organisation devoted to this purpose is needed.

This does not mean that all international investment must be controlled or directed. But basic programmes extending over a considerable number of years would be an enormous advantage in helping to avoid the evils of hit-or-miss fluctuations in international capital flow. Agreements are needed on broad lines of development and on the amount of international support that can be expected, not just for a year or two, but for a longer period. There is need for advance consideration of the general nature of the trade developments required in order to give the international investments a sound prospect of repayment. For these and other purposes it is essential to have an international organisation of some kind on which both the newly developing countries, which are receiving capital and technical knowledge, and the older industrial countries that supply capital and technical knowledge are represented. There are a number of reasons why the establishment of such an agency is important and why it would be best to give it a multilateral, supra-national character. These reasons may be grouped under five general headings.

First, the multilateral form of developmental operation would encourage a more efficient use of world resources. An international development authority would presumably work out arrangements which encouraged multilateral rather than merely bilateral exchange. It would be in a position to join together many scattered sources of savings and many scattered investment outlets. It would provide a disinterested means by which countries in need of technical assistance could arrange for the best such services available, regardless of nationality. By assisting in the planning of co-ordinated regional development programmes it could stimulate private initiative and private investment to supplement governmental activities in new areas.

A country which has surplus savings available for investment abroad is not necessarily the best source of the particular capital goods needed for a project where its savings might be used. Nor will countries capable of providing excellent equipment necessarily be able to finance the development projects for which this equipment is needed. A multi-national development programme has the great advantage over bi-national arrangements that it is more likely to take account of such factors. To make the best use of world resources it should be possible to arrange for the construction of a railroad in China, by using, for example, British rails, American locomotives, and German signal equipment, with the financing coming from savings collected in varying proportions in many different countries. If French capital is available only to finance purchases of French equipment, American capital only to finance purchases in America, etc., as is likely to be the tendency if bi-national arrangements are the rule, wasteful restrictions are forced upon the world economy analogous to those imposed by bilateralism in trade. Governmental organisation of international economic relations promises to be more important than ever before when this war is over. That makes it all the more essential, from the point of view of efficient use of world resources, that governments should co-operate in regular, multilateral ways, rather than in merely *ad hoc*, bilateral ways.

A multilateral agency would be in a position to combine many sources of saving into a pool of developmental funds. It could spread the risk of developmental undertakings by offering, or encouraging the offering, of securities based on many different projects, and perhaps, in addition, guaranteed by governments. Thus it could open a channel by which otherwise idle savings might be put to work in improving the productive equipment of the world. In particular, such arrangements might make it possible to utilise the savings of life insurance and other savings institutions which are limited by law to the purchase of securities of the highest grade.

An internationally co-ordinated programme would encourage private enterprises from all over the world to participate with confidence in particular projects for development. It would also enable technical assistance to be drawn from every country rather than from only a few countries which happen to be able to provide large sums of capital.

In the second place, an international development authority would provide a means for taking a longer look ahead and for studying the broad problems that would emerge in connection with widespread economic development. It would be in a position to propose co-ordinated programmes of action on the important problems of economic adjustment which developmental activities would raise.[1] Even if such problems were apparent to private investors or individual governments, they would in many cases be capable of solution only by concerted effort.

In order to guard against unbalanced development it is very important that there should be some international agency capable of watching the world economic situation as a whole, and regional sections of it. Such an agency should be in a position to point out that current plans of various countries will lead to a world overcapacity in some industries and under-capacity in others, to discourage the making of excessive loans where transfer difficulties might arise in repayment, to call attention to dangers that might threaten as a result of excessive loans on short term, and the like. The agency could also help to co-ordinate the timing and direction of orders for equipment to be used in developmental work (as explained in the next chapter), in order to assist in business cycle stabilisation and to lessen the amount of unemployment in special localities and depressed industries. There is need for co-ordination of investment programmes in order to avoid violent shocks from sudden rises and falls in the international flow of capital. Also, collaboration of a type which can best be organised through a multilateral agency would encourage the sort of mutual arrangements which, as will become more apparent in Part II of this study, are needed in order to lessen the difficulties of transitional adjustment in the economies of the advanced countries as new industries develop elsewhere. For all these reasons an international public body able to take a long view of world economic development and to initiate joint programmes of action is highly desirable.

In the third place, an international authority is needed in order to keep the political conflicts which often develop around international investments at a minimum. A multilateral agency supervising international investment should undertake to assure newly developing countries that outside capital will not become an instrument of "peaceful penetration" and political domination. It should provide an impartial means of regulating the relations between lenders and borrowers when they are citizens of different countries and of adjusting the conflicts that inevitably arise between lending and borrowing areas. It should endeavour to prevent developmental investments from being used by individual States as tools in the game of power politics and balance of power.

---

[1] This is discussed in more detail in Part II.

Investment in the less industrialised countries has in the past been more prolific than any other type of investment in providing occasions for political friction, military intervention and rivalry between the great Powers. The investment of capital in such areas is full of conflict-producing potentialities. Many of the potential conflicts derive directly from the economic relationship itself: the debtor-creditor conflict, the labour-capital conflict, the tenant-landlord conflict, the conflict between business competitors, between rival investors seeking the most attractive investment opportunities and between joint creditors of the same doubtfully solvent debtor. Other potential conflicts arise from the clash of pre-industrial with industrial cultures, as when hand workers are displaced by machines or the graves of ancestors are exposed to unfavourable magic by the building of railway lines. Local desires for social reform may clash with vested property interests held abroad. Resentment may arise against real or alleged "foreign domination" connected with the import of capital.

As experience has shown, conflicts between great Powers may come over the control of strategically located railway lines, canals and airways and over the appointment of financial advisers and customs controllers in countries of weak government. Foreign investments have been used by the great Powers in political manoeuvre to prepare the way for annexation, to mark out regional spheres of dominance and to maintain them, to cement alliances and to disguise military penetration. The increasing tendency of national governments to regulate and supervise the foreign investments of their citizens or to undertake direct governmental investment operations abroad increases rather than decreases the danger that capital invested abroad may become a pawn in political conflict. Denationalisation or multi-nationalisation of developmental investment can be accomplished through the medium of a supranational agency equipped to provide capital without political strings attached. The assistance of an international authority for impartial adjustment of the conflicts bound to trouble the relations of lending and borrowing areas is also needed. The establishment of an organisation to perform such tasks would help to remove political obstacles that might otherwise block any really great expansion of developmental investment in the post-war world.[1]

In the fourth place, a permanent international agency for the supervision of investment relationships is desirable in order to

[1] For much fuller discussion of all these points, see Eugene STALEY: *War and the Private Investor: A Study in the Relations of International Politics and International Private Investment* (Garden City, New York, Doubleday, Doran and Company, 1935). The portion of the book particularly relevant in the present connection is Chapters XIII–XVI, supplemented by some of the case studies.

protect the investors on the one hand and the peoples of the borrowing areas on the other against such abuses as repudiation of contracts, exploitation of local labour and political domination by foreign capital.

Among the obstacles to sound international investment in the years before this war was fear on the part of lenders that their ventures abroad might be exposed to expropriation and default. The system of national diplomatic protection of the rights of investors abroad had broken down. The peoples and governments of areas in need of foreign capital feared, on their side, the political consequences of letting capital in. They also feared that their independence might be jeopardised by economic pressure. Because of these fears, they often imposed restrictive regulations, many of which were highly exasperating to foreign business men and investors. This further discouraged the fruitful use of capital.

There can be little hope of removing these difficulties until the antiquated and unsuitable system of "national diplomatic protection of citizens abroad" is superseded, in so far as investment relationships are concerned, by some system of international supervision and adjustment. Under the system of national diplomatic protection a citizen who feels that his property interests in a foreign country have been unjustly treated may first attempt to get redress from the government of the country. Thereafter, his only recourse is to turn to his own government, which, through its department of foreign affairs, may or may not intercede on his behalf. Experience shows that this is a very ineffective way of adjusting conflicts that arise out of international investment and a still less effective way of preventing them. The defects in the method are many and grave: (1) it places reliance upon a process of judgment and execution by interested parties, a thoroughly discredited method of rendering justice and a method calculated to intensify disputes; (2) it has the effect of enlarging rather than restricting conflicts by enlisting national feelings and staking national prestige on disputes that may originate in private grievances; (3) it entangles justice with *haute politique* and makes it subordinate to political expediency; (4) it induces investors who want vigorous protection from their governments to engage in propaganda of a sort likely to heighten international friction; (5) it provides a readily available pretext for aggressive political penetration by strong States; (6) it induces some countries to adopt policies hostile to foreign capital through fear of penetration: these policies provoke more conflicts and generate more friction; (7) it encourages a legalistic approach to the problems raised by international investment conflicts, an approach which tends to inhibit the conscious and intelligent consideration of social issues that may be at stake.[1] Progress towards the improvement of this very unsatisfactory system of regulating international investment relationships could best be made under the auspices of an authority multi-national in character.

A fifth group of reasons of a more general character might be added. The maintenance of a durable peace, and hence the attainment of reasonably stable economic welfare, will ultimately depend in large part upon the success of the peoples of the world in building up some sort of responsible and effective world government. Perhaps one of the best ways to hasten the development of world government is to set up a variety of international agencies (ultimately under some general co-ordinating organisation) to perform concrete and important tasks, giving them adequate means to do their jobs. Let the performance of important functions, such as economic development, serve at the same time to build up habits of united action and the prestige and power of quasi-governmental organs at the world level.

All these reasons support the proposals, now advanced in many quarters, for some type of international development authority. One of the most concrete suggestions, covering particularly the financial aspects of the functions for which international organisation is needed in this field, is that published by the United States Treasury.[2]

The Bank is intended to co-operate with private financial agencies in making long-term capital available for reconstruction and development and to supplement such investment where private agencies are unable to meet fully the legitimate needs for capital for productive purposes. "The Bank would make no loans or investments that could be secured from private investors on reasonable terms. The principal function of the Bank would be to guarantee and to participate in loans made by private investment agencies and to lend directly from its own resources whatever additional capital may be needed." The authorised capital would be about 10,000 million dollars, and shares would be subscribed by member governments according to a formula that would take account of national income and international trade. A substantial part of the subscribed capital would be reserved in the form of unpaid subscriptions as a surety fund for the securities guaranteed by the Bank or issued by it.

The Bank would be empowered to guarantee, participate in, or make loans to any member country and through the government of the country to any of its political subdivisions or to business or industrial enterprises, on certain conditions. The payment of interest and principal must be fully guaranteed by the national government. The borrower must be unable to secure the funds from other sources on reasonable terms. A competent committee must study the project or pro-

---

[1] For a fuller discussion, see *ibid.*, Chapter XVI.

[2] *Preliminary Draft Outline of a Proposal for a Bank for Reconstruction and Development of the United and Associated Nations* (Washington, U. S. Treasury, 24 Nov. 1943).

gramme and report in writing that the loan would raise the productivity of the borrowing country and that the prospects are favourable to the servicing of the loan. The Bank may also encourage and facilitate international investment in equity securities by obtaining the guarantee of governments for conversion into foreign exchange of the current earnings of such foreign-held investments, and it may participate in such equity investment to a limited extent.

The Bank is to impose no condition upon a loan as to the particular member country in which the proceeds must be spent. In making loans, the Bank shall provide that the local expenses of a project be largely financed from local sources. A member country failing to meet its financial obligations to the Bank may be suspended from membership, and member governments and their agencies agree not to extend financial assistance to that country without approval of the Bank until it has been restored to membership. In the event of an acute exchange stringency the Bank may accept local currency in payment of interest and principal for periods not exceeding three years. The Bank shall scrupulously avoid interference in the political affairs of any country and may operate in any country only with the approval of the government.

This chapter has stated some of the reasons for establishing an international development authority of a broad, multi-national character. This is not the place to discuss details of organisational structure. Perhaps what is needed is not just one such authority, but a number of related institutions: one to regulate and supervise international private and public investment, from the point of view of ensuring reasonableness of contracts and adequate performance; another to provide capital for public developmental authorities, which might be organised on a national or regional basis, and to co-ordinate their activities; and a third to provide some form of pooling of risks in developmental undertakings of private enterprise. Whatever the organisational structure of the international development authority, whether it consists of one organisation with branches and subsidiaries or of several separate but co-operating organisations, it ought to be in a position to carry on the following activities:

1. To launch a world survey of resources and needs, co-ordinating for this purpose similar surveys undertaken on a national basis by national governments.

2. To raise capital by selling its own securities to governments, to savings institutions and to the general public.

3. To advance capital for use on approved projects under conditions to be stipulated by the authority.

4. To act as an intermediary in arranging for supplies of capital from other sources, for technical assistance and for purchases of equipment needed in development.

5. To work out plans for guaranteeing a minimum rate of return on approved developmental projects undertaken by private enterprise, in order to spread risks and encourage venture capital.

6. To propose standard forms of contracts and concessions, and to get the assistance of the International Labour Organisation in preparing codes relating to labour standards which might be incorporated in agreements or be made a condition for obtaining loans.

7. To act as mediator when disputes arise in connection with international developmental investments and to propose preventive measures designed to forestall defaults, violations of contract and the like.

Under the auspices of the international development authority and its related institutions there would be room for many types of developmental enterprise. These would include projects carried through entirely by private enterprise, or by private firms with the aid of financing through some branch of the international authority. They would also include purely governmental operations, undertaken, for example, by the Government of China or of Peru, and financed by the international development authority itself, by credits from another government or group of governments under supervision of the authority, by private capital under supervision of the authority, or by a combination of these. There would also be a place for "mixed" enterprises, in which stock might be owned partly by public bodies and partly by private persons or firms. The important thing is that all these types of investment activity should be guided and encouraged by an agency which is accountable not just to one government or a few governments, but to representatives of all the various peoples concerned.

CHAPTER VI

ANTI-DEPRESSION TIMING AND DIRECTION OF EQUIPMENT ORDERS

The programme proposed in this chapter is designed to enhance still further the mutual benefits that might be expected from properly managed development of the less industrialised areas with the aid of capital funds and equipment supplied by the more advanced industrial countries. In brief, the method suggested would at the same time help to: (1) outfit with modern capital equipment underdeveloped countries like China, India, the Balkans and parts of Latin America, or countries devastated by war; and (2) maintain stable post-war employment in industrially advanced countries. The basic principle is co-ordinated *timing* and *direction* of the equipment orders connected with international developmental investment, in a manner designed to counteract general or local depression conditions and, conversely, to avoid intensifying general or local boom conditions in the advanced countries. This programme might be particularly helpful in maintaining stability of employment during the immediate post-war era, from a year or two after the war through the decade or more of convalescence and difficult adjustments that must follow.

For the sake of clarity of exposition, attention will be concentrated below on China, as an example of a newly developing country, and on the United States, as an example of an advanced industrial country capable of supplying capital and equipment. This should not be taken to imply, however, that bilateral arrangements are advocated, except as they might be appropriate as part of a general world development programme under the auspices of an international development authority which would be particularly concerned with encouraging multilateral rather than merely bilateral exchange.

In brief, inducements would be offered to China (and other countries) to schedule the orders for equipment to be used in their development programmes so as to canalise this demand towards those particular industries and those particular localities which at any given time might be depressed and so as to step up the total volume of equipment orders in periods of actual or threatened general depression. For example, certain industries or industry branches or localities might from time to time be officially listed, perhaps by agreement between the international development authority and governments concerned, as "underemployed". China would then be privileged to place orders with these industries, either directly or through an international development corporation, on special terms, including generous financing at very low rates of interest.

The benefit to China (and similar countries) would be much cheaper capital costs and probably also better delivery schedules and other advantages. The benefit to the United States (and similar countries) would be a higher and more evenly distributed level of post-war employment and income. This mutuality of benefit is possible because the essence of the method is to manufacture additional real capital out of productive power that would otherwise run to waste in unemployment and under-utilisation of capacity.

The success of the method depends upon having large-scale international development programmes laid out in advance and in considerable detail. It also depends upon thorough awareness of particular industry capacities and potential post-war surpluses and shortages, in the advanced countries. Finally, quite detailed negotiations, both of a political and business sort, would have to be carried on at various stages. All this means that to wait for the end of the war to set up the necessary organisation for planning and supervising such operations would perhaps make the whole attempt "too little and too late".

Early establishment of an international development authority, charged with working out plans for operations of the type outlined, is therefore important. It is desirable that the operations be on a broadly multilateral basis, under the aegis of an international authority representing both the industrially less developed and the industrially advanced countries. This is preferable to bilateral arrangements between, for example, the United States and China, or combinations of capital-supplying countries only, as in the old China "consortiums".

The Importance of the Timing and Direction of Equipment Demands

In times of actual or threatened unemployment in the United States, when industrial capacity is going unused and incomes are depressed, an additional demand for equipment arising out of large-scale foreign development programmes would be an unmitigated benefit. It would serve to sustain incomes and employment in the industries immediately affected (equipment-supplying industries in the narrower sense, and also suppliers of materials, technical services, etc.). Beyond that, the "multiplier" effect of additional expenditures by these industries and persons deriving income from them would help to raise the level of income and employment generally. In such a situation—that is, starting from under-utilisation of capacity and under-employment of labour—the effect of supplying equipment for installation in other areas might be to raise American production and American income by considerably more than the amount actually sent abroad. In that case, the transaction would be of net benefit to the American economy quite apart from any repayment in the future, and even if no payment of any kind were ever received for the equipment.

However, exactly the same orders for equipment might not be beneficial at all to the American economy if they came at another time. In a situation of general boom, when inflationary tendencies should be held in check and when shortages of various kinds were creating production bottlenecks, additional equipment orders from abroad might intensify the shortages, make the problem of controlling inflation more difficult and contribute to an unhealthy speculative fever that would later be followed by a crash.

Thus, the repercussions of a large international development programme upon the advanced industrial countries will depend not only upon the size of the development projects themselves but also upon their timing and their type, in relation to the general economic situation in the equipment-supplying countries. It may be useful for the sake of illustration to speculate on the phases which China's economy and the economy of the United States are likely to go through in the post-war period.

Stages in the Post-War Economic Situation of China

For at least a year or two after the end of hostilities China's economic situation is likely to be dominated by emergency conditions: rescue and rehabilitation of underfed populations, repossession of territory, remigration, severe currency disorders and the like. Perhaps river developments, highway construction and other public projects that can be started on short notice and with relatively little equipment will be put under way at once as means of providing employment for soldiers. But it would be remarkable if any major development projects calling for large amounts of supplies from outside were actually ready for action before a year after the war, and the time needed for preliminary decisions and for working out plans and organisations might be longer than that.

Assuming political stability and international co-operation, this first stage of emergency relief and rehabilitation will gradually pass within a few years into the beginnings of a second stage, which might be labelled the stage of construction. This would be characterised by a rapidly rising volume of fundamental development projects—building of railways, airports, river works, power stations and industrial plants, improvement of agricultural techniques, introduction of machinery into agriculture, equipment of elementary and technical schools and training of personnel. There would be a large, and for many years a rising, demand in China for capital goods from abroad. (The demand for consumption goods from abroad would also rise rapidly, but it might be held in check at this stage by government policy, in order to save foreign exchange to be used in purchasing equipment and in order to encourage domestic industries.) China's capacity to absorb capital goods produced by advanced countries would be limited during the years of construction by: (1) the rapidity with which the Chinese people would be able and willing to learn how to operate new equipment and to make the profound changes in habits and social adjustments required by a transition from pre-industrial to industrial society; (2) the funds available to China for development purposes. As productivity rose in China, as the result of improved techniques and equipment and better education of labour and management, domestic savings would be more important and loans from abroad would be progressively less important in determining the rate at which new capital equipment could be installed; (3) the availability of the capital goods themselves in the outside world and in China. As China's technical knowledge and productive capacity advanced, a smaller *proportion* of new capital equipment would have to come from outside, for China would be able to supply many of its own needs. But the *absolute* amount of new equipment purchased abroad might go on increasing for many years, and then stay at a permanently high level, for the Chinese economy, as it became progressively wealthier, would demand a larger variety of equipment and an increasing volume of replacements.

There would be no clearly marked end of the second or construction stage of post-war Chinese economy, unless a great depression or internal political turmoil or a great war intervened to bring it to a close. Such events apart, the rate of new construction would at length slow down and taper off gradually. After three or four decades China might pass imperceptibly into a third stage, in which its problems would be more like those of countries already industrialised—namely, to maintain, so far as possible, a somewhat slower but steady rate of progress. Its economic relations with and repercussions on the United States and other "advanced" countries of today would no longer be those peculiar to a new area undergoing the first stages of modern development.

Stages in the Post-War Economic Situation of the United States

A first post-war stage of uneven prosperity—of boom marked by sectors of depression—may be anticipated in the United States. That is, the general level of effective demand is likely to be high—indeed, so high as to threaten rapid price rises and to necessitate temporary continuation of wartime controls if inflationary forces are to be held in check. But in the midst of general prosperity and shortages there will be depressions in particular industries and particular localities and actual or potential surpluses of some kinds of goods.

This situation will be the result of two main factors. The first is an abnormal demand, including pent-up domestic demand for consumer durable goods and for replacements of equipment in industry, and a foreign demand for supplies needed in emergency relief, rehabilitation, and reconstruction. The domestic demands will have been built up during the war and will be backed by purchasing power. Persons who have saved and bought war bonds will want to spend these savings on new automobiles, refrigerators, tires, houses, etc.; industries, faced by excellent opportunities for profit, will rush to convert back to civilian production. Needs of devastated areas in Europe and Asia for food and for the materials with which to revive production and to reconstruct their cities and industries will work in the same direction, on the assumption, as seems likely, that these needs will be financed to a fairly large extent by gifts, loans and credits, as well as by the available assets of the countries concerned.

The second factor is structural maladjustment of production. This will account for the depressed spots during the boom. There will be certain lines of production, and not merely direct munitions industries, for which the post-war civilian demand will represent only a fraction of the wartime output. This is likely to be true, for example, of shipbuilding and aircraft construction and some sections of the machinery industry (especially, in the latter case, after the immediate spurt of reconversion demands has passed). It will take time to shift surplus workers and equipment from such industries into permanently tenable lines of production, especially as direct munitions workers and demobilised soldiers will be trying to make similar shifts. Plants that made civilian goods before

1941 will in some cases find it difficult to shift back from war production. There are likely to be serious problems of under-employment in specific industry branches and specific localities, even while business in general is booming.

Following this first stage of uneven boom, there is likely to be a second stage of actual or threatened general depression. This will come when the abnormal, pent-up demands based on the postponed consumption of wartime have been largely satisfied. Whether the United States gets merely the threat or the actuality of a severe depression depends upon the vigour and timing of counteracting policies by Government and business. This is the stage at which domestic measures for stimulating the rate of real investment— redeveloping cities, encouraging house construction, etc.—will be urgently needed and also the stage at which international demands for equipment to be used in the economic development of China or other countries will be most beneficial.

How soon after the war may this second stage appear? After the First World War, the post-war boom in the United States lasted through 1919 and early 1920. The general collapse of effective demand occurred in the spring of 1920, about eighteen months after the armistice. The depression which followed was relatively short, but sharp, and recovery began in 1921. There are reasons for expecting a somewhat longer period of post-war boom this time. This depends, among other factors, on the length of the war and on the wisdom and success of governmental and business policies designed to keep the boom phase under control. But it is too much to hope that the policies actually adopted will prevent the appearance sooner or later of a stage of incipient depression. Once the most urgent civilian demands piled up during the war have been met there is likely to be a sharp drop, to which the lower average age of durable goods then in the hands of consumers will contribute. Structural maladjustments inherited from the war will make the threat of depression still more grave once the drop occurs. At a guess, we might place the beginning of this second stage in the American economic situation some two to four years after the end of the war.

The third stage in the post-war economic situation of the United States will be reached when the stage of depression has been passed through or has been successfully neutralised by vigorous and timely expansionist policies. This third stage might be characterised as one of settling down to more or less "normal" progressive development. The effects of the war would gradually cease to be dominant elements in the economic situation and the ups and downs of business would be determined by future events and forces about which, for present purposes, it would be fruitless to speculate. The major economic problem will be to keep an even keel at a rate of production which corresponds to substantially full employment, while increasing productive efficiency year by year so as to permit a steady improvement in living standards. A rate of consumption plus investment sufficient to balance a high and increasing capacity to produce is essential to avoid depression. So far as the investment side of this problem is concerned, a world development programme would provide an important and remunerative addition to domestic investment opportunities over a period of many decades—assuming that means are found for organising world political security and economic co-operation.

Managing Equipment Demands for Mutual Benefit

Given successive phases in the development of the post-war economic situation in China and in the United States somewhat resembling those outlined above, is it possible for Chinese needs for developmental equipment to be fitted into the industrial employment needs of the United States or other countries in such a way as to be of general benefit? Or will the supply of capital equipment for China make shortages worse, contribute to dangerous boom-time over-expansion, and in general necessitate sacrifices and intensify economic troubles in the United States? The thesis advanced here is that proper co-ordination and timing, to be attained by joint laying of plans in advance and joint execution of flexible programmes, can make the supply of equipment for China's post-war economic development very largely a matter of immediate common benefit. A considerable part of the capital supply needed by China might thus be achieved at a very low real cost, to a substantial extent out of productive power that would otherwise run to waste.

During the first post-war phase in the two countries (emergency conditions in China, reconversion and uneven boom in the United States) some types of commodities will have to be made available to China, as to other nations liberated from enemy occupation, despite a general shortage and even though the industries producing these things are already working at full capacity. This will be true of medical supplies, for example. The same will be true of key equipment and materials for emergency use in a Chinese rehabilitation programme. To some extent, left-over army equipment might be turned to this purpose. Providing the shipping shortage has been overcome, such items as army trucks, "jeeps" and the bulldozers used to make landing fields for the air forces could certainly be applied to good advantage in China for road building and river control works. After the First World War, American transport equipment overseas was sold at about 25 per cent. of original cost.[1] Perhaps occupying forces will need more equipment for a longer time after this war, but there will surely be a considerable surplus stock on hand. Perhaps some excess wartime plant capacity—for example, ammonium nitrate plants, convertible to fertiliser production—could be transferred bodily to China with advantage to China and to the United States. There are also good arguments for continuing production on some war orders after hostilities have ceased, in order to "taper off" more gradually and lessen the immediate problems of transfer to civilian lines. Special consideration might be given to continuing the output of types of war equipment that could best be adapted for use in China (and elsewhere) in rehabilitation, reconstruction, and development.

As China begins to pass beyond the period of emergency measures and embarks upon some longer-range developmental projects the United States may still be in the stage of uneven boom. But it should be possible, through working out joint plans in advance, to find a considerable volume of equipment in the first two or three years after the war that would satisfy both the following conditions: (1) be immediately useful to China in the initial phase of its economic development; and (2) be capable of being produced by industries or industry branches in the United States which would otherwise be underemployed owing to lack of demand for their full capacity output. For example, the plant facilities and the trained workers which have turned out quantities of small boats and landing barges during the war might produce a modern river transport fleet for China's rivers.[2] They would perhaps be more readily convertible to this purpose than to the meeting of any other immediate civilian demand. The war-swollen shipbuilding industry might operate some of its yards—perhaps those in localities where alternative employment for labour is least available—to turn out ocean-going cargo carriers for China's large coastwise traffic and for handling an anticipated increase in Chinese import and export trade.

This is not to say that China should be content to wait for "cast off" equipment from the United States or other advanced industrial countries, or that China's programme of economic development should depend on whatever happens to be convenient for other countries to produce with otherwise unemployed resources. Presumably China will have its regular programme of development and will place the orders that it thinks important in the regular

[1] Irving BERNSTEIN: *The Automobile Industry: Postwar Developments, 1918-1921* (U. S. Department of Labor, Bureau of Labor Statistics, Hist. Study No. 52, Sept. 1942, mimeographed), p. 16.
[2] Or for the Amazon.

manner on the best terms it can get. What is suggested here is that *in addition* China should be encouraged to develop a *supplementary* programme which could be speeded up or slowed down or turned in one of several alternative directions of development in order to take advantage of opportunities abroad for the acquisition of capital equipment on favourable terms. As an inducement to China to put as much of its programme as possible into the "supplementary" category and to concentrate in the immediate post-war years on those types of equipment which could be produced from otherwise unemployed capacity, the United States (and other equipment-supplying countries) might offer special terms. Certain industries or industry branches might be officially listed from time to time as "underemployed". For example, any industry where output threatened to fall below a certain percentage of efficient installed capacity might be so listed. The Chinese development orders placed with these industries might be covered by credits at extremely low rates of interest—much lower than the rates at which China would ordinarily be able to borrow. Financing might be through a United States Government agency, such as the Export-Import Bank, in co-operation with an international development authority, or directly by some subsidiary of the international development authority itself. Similar arrangements might be worked out in other advanced countries where an unemployment problem existed and, of course, other developing countries might be invited to participate on the same basis as China. As stated earlier, China and the United States are used throughout this discussion simply for purposes of illustration, and the methods suggested could be generally applicable.

Such a plan would supply China with large quantities of capital (in the form of real capital, not capital funds expendable at discretion). The real capital would arise out of productive power that would otherwise have been largely wasted. Where there would have been idleness of plant and unemployment or underemployment of labour and management, there would now be production, and the production would be going to China in return for promises to repay the capital plus a small rate of interest.

For the first few post-war years, if forecasts made above regarding the economic situation in the United States are at all correct, structural underemployment will be the only source of "listed" industries in which orders could be placed on the basis proposed. (The situation in other industrially advanced countries, particularly those devastated by war, would differ from that in the United States in many ways.) But within two or three or more years there is likely to be a general threat to effective demand for the output of the durable goods industries and, behind them, the raw material and other industries. At this point China should be invited to expand its supplementary equipment programme to cover an unlimited range of items, not just those for which surplus capacity had been discovered in the boom period. The United States could now afford to extend very generous terms on all sorts of equipment purchases, as part of a programme for maintaining a high level of employment not only in specific industries but throughout its economy.

Equipment supplied to China under these conditions as to *timing* in relation to the business cycle and *direction* in relation to structural maladjustments would represent little or no real cost to the supplying countries. In fact, the real income available to the people of the United States would very likely be greater rather than less as a result of furnishing equipment to China on such terms, even if no interest were paid and even if repayment were uncertain. This is because of the "multiplier" effect of orders to the durable goods industries in sustaining employment and income all along the line, in industries producing for local consumption as well as for export.

The benefit to China (and other countries to which such a plan might be extended) is obvious. It would mean an opportunity to equip the country with modern instruments of production at an extremely low capital cost. Orders placed in accordance with an inflexible domestic plan of development, without regard to the situation of the supplying countries, would very likely be placed to a considerable extent in boom-time, inflated markets. If so, they would have to be paid for at high prices and probably on disadvantageous terms as to credit, interest rates, delivery dates and the like. On the other hand, orders directed towards specific underemployed industries and orders timed so that the great bulk of them would fall in a period of depression or incipient depression could be placed on much better terms.

The Need for Advance Preparation

The actual putting into practice of such a programme would require a high degree of advance preparation as well as some means of co-ordinating its execution internationally. The benefits depend entirely upon *timeliness* of action, and *accuracy* of information on the changing economic situation in more than one country.

The most suitable auspices under which to organise and execute co-operative international programmes of this kind, as already suggested, would be that of an international development authority, on which countries desiring to borrow for large-scale development and countries prepared to supply equipment, materials and technical aid would all be represented. Such an international development authority would probably find it expedient to decentralise its work regionally. Assuming that the principles discussed above in terms of China and the United States were to be applied on a much wider basis, there might be subcommissions or special public corporations to co-operate in the development of China, of south-eastern Asia, south-eastern Europe, Africa, Latin America, etc.

There are many reasons why the kind of co-operation described should be organised on a world-wide basis, or at least on a widely multilateral basis, rather than on a bilateral basis which would lead to separate bargains between China and the United States, China and Great Britain, China and the Soviet Union, and so on. One reason is that bilateral arrangements would be more likely to produce a revival of the old commercial and political antagonisms connected with competitive pursuit of short-term "national interests" in investment operations. China, and other countries in China's position, would be more exposed to pressures of national diplomacies; attempts to make investment undertakings serve some purpose of foreign political "penetration" or "power politics" would be more probable.[1] Furthermore, the advance planning which is essential for proper timing and hence for the harvesting of some of the most important mutual benefits would be much less effective on a bilateral basis. Important elements in the situation—namely, what action other parties are going to take at the same time—would remain unknown. Of course, many details might best be worked out bilaterally, so long as the programme as a whole were co-ordinated and checked by a multilateral authority.

One of the first concerns of an international development authority, in preparing to execute the type of mutually beneficial programme discussed in this chapter, would be to work out concrete programmes in which the probable equipment needs of the regions under development would be fitted into the probable employment needs of specific industries in specific localities at specific times. This could be done only on the basis of extensive information, translated into forecasts. Then flexible programmes of action

[1] As for the old "consortium" device formerly used by capital-supplying countries in some of their dealings with China, it is inapplicable for the purposes under discussion here. An association of bankers working under the auspices of their governments could make loans, but they could not plan and execute the economic (as distinguished from the purely financial) programme which is needed. In any case, a group of which China itself was not a full member would not be appropriate for planning the international aspects of Chinese development, nor would it be acceptable to the new China.

would have to be prepared, and actual execution would be adjusted to the real course of events as this became known. Considerable time would be required for preparation—assembling a staff, establishing the right contacts for pertinent information, analysing the information from many countries and putting it into comparable form and working out key projects in some detail. *Now* is not too early to start. In fact, if the tasks of planning and research are not begun until the end of the war, or just before the end, the whole undertaking may not be in time for effectiveness in the early post-war years.

What economic information would an international development authority need to have in order to work out a programme of the type contemplated here? In the first place, it would need to know about each region where large-scale development is contemplated (*e.g.*, about China):

1. A rough estimate of the order of magnitude of the total amount of capital goods of all kinds which the region might be capable of absorbing and putting to good use over the next ten years.
2. The particular categories of equipment and the approximate amounts of each that could usefully be absorbed in the earlier years.
3. What portions of this equipment would be needed from abroad and what could be supplied locally.
4. What particular items it would be important to have first—in other words, a priority schedule or flexible time schedule.

In the second place, an international development authority would need to know about each advanced industrial region capable of supplying equipment (*e.g.*, about the United States):

1. The probable productive capacity of particular equipment industries at the end of the war, and the ease or difficulty with which these capacities might be increased or decreased (by transfers of workers, etc.).
2. The peacetime demands, aside from international development projects, which might be anticipated for the output of these industries immediately after the war and over the next five to ten years, assuming various levels of national and world income.
3. On the basis of (1) and (2), indicated structural maladjustments leading to particular shortages and surpluses and the indicated types and amounts of equipment that might beneficially be ordered in this region for international development use at various stages.

In other words, what is needed is an international survey of resources and needs, from the particular point of view of development opportunities, in co-operation with the governments of all the countries involved. Even if this were done, in the first instance, rapidly and rather superficially, it would still take time. Time-consuming negotiations on the political plane and on the business plane would also be necessary before large-scale projects could be put under way. The importance of starting *soon* cannot be too strongly emphasised.

CHAPTER VII

SOME OBJECTIONS TO THIS PROGRAMME CONSIDERED

The proposal advanced in the preceding chapter embraces both the *timing* of supplementary equipment orders, in a manner to counteract the ups and downs of the business cycle, and the *direction* of supplementary equipment orders (even in good times) towards special areas and industries, so as to make use of man-power and productive equipment that would otherwise be under-employed. Thus, the development of new areas would be aided by methods that would at the same time lessen the economic disturbances and the human suffering connected with "cyclical" and "structural" unemployment.

Counter-Cyclical Timing

The desirability of encouraging counter-cyclical timing of developmental investments needs little justification. This would fit in directly with the measures for the international co-ordination of public works programmes which were advocated by the International Labour Organisation as long ago as 1937. To link the timing of world-wide developmental programmes for the newly developing countries with international co-ordination of national programmes in the advanced industrial countries themselves would greatly increase the effectiveness of anti-depression measures. A comprehensive programme of really constructive development which can be speeded up or retarded in accordance with the stabilisation needs of the economy is preferable to the kinds of "make work" projects which can be adopted on the spur of the moment. But experience teaches that in the absence of special measures, such as are proposed here, newly developing countries such as China (for example) will be most likely to purchase their equipment from the advanced countries in the boom years. In years of depression, when such purchases would be most helpful to the advanced countries and least costly to China, both the world demand for China's export products and the willingness of lenders to make new loans are likely to be at a low ebb. This makes it necessary for China to slow down rather than to speed up its construction projects in times of world depression. That is why some special plan for organising and financing developmental purchases of equipment in a counter-cyclical manner is needed. Otherwise, investment in the development of new areas may help to unstabilise rather than to stabilise the economies of the world.

Some who would agree that the timing of equipment orders so as to help stabilise employment is in theory desirable may argue that it is impossible in practice to do the timing accurately enough to make the scheme practicable. Of course, there would be practical difficulties, as there are in any attempt to do things more sensibly than they have been done before. At the very least, however, an international development authority undertaking a programme of the type here suggested could make available speedier and more accurate information on the situation and the trends in the capital goods industries of the world in relation to proposed developmental projects of newly developing areas. It could stimulate all parties concerned to plan further ahead. It could arrange special inducements, of the type proposed earlier, which would make it worth while for newly developing countries to prepare special projects that could be put into execution rapidly when the appearance of unemployment in the advanced countries offered favourable purchasing and financing opportunities. It is very unlikely that efforts along these lines would actually result in *worse* timing of developmental equipment orders than would be the case in the absence of such efforts, and the results obtained might be surprisingly good.

Direction of Orders to Structurally Depressed Localities or Industries

The policy of adjusting the *direction* of equipment orders so as to assist special depressed areas or special depressed industries is more open to question than the policy of timing such orders counter-cyclically. Nevertheless, for the disturbed condition of the immediate post-war decade, with its severe structural maladjustments, the arguments in favour of a policy of this sort outweigh the arguments that can be raised against it. It is a policy that needs to be applied with caution and discretion, however, and not indiscriminately.

In general, the best public policy is to stimulate and encourage those industries which are promising candidates for permanent expansion—the sorts of industries which can produce efficiently because they are well adapted to the resources and skills of their locality and because they find their products in increasing demand. Workers, managers and investors should be encouraged to shift out of the weak, relatively ill-adapted and inefficient industries, or those experiencing declining demand, into more promising lines. The danger in directing developmental purchases so as to give aid to special underemployed industries or localities is that it may tend to shore up inefficient enterprises and to perpetuate situations that will continue to be weak spots in the economy. Which industries are over-built and should be allowed to decline is not easy to decide. Whether capacity is "redundant" or not depends in part on how rapidly world income and demand is going to increase. Measures to encourage economic development might supply not only a temporary demand for otherwise overexpanded industries, but a permanent increase in demand as well.

If one could trust market forces to indicate correctly at any moment which industries, from the point of view of long-run needs, ought to undergo expansion and which ought to contract, and if the free play of market forces could bring about these expansions in one direction and contractions in the other without much time-lag and without great transitional disturbances and human suffering, then there would be little reason for paying attention to structural underemployment. It would only be necessary to have sufficient total investment to raise the volume of effective demand to the full employment level. If only enough money were pumped into the market at any point it would spread throughout the economy and give work to everyone who wanted it. Unfortunately, just this impression is given by some of the less cautious writings on the relation of savings, investments, the "multiplier" and expansionary fiscal policies in general to full employment. Some such writings seem to carry the implicit assumption that the factors of production—workers, managers and natural resources—are almost perfectly mobile, so that they can shift painlessly and frictionlessly out of the industries and localities where they happen to be into those where expanding demand offers new opportunities. In real life this is far from the truth. Even in times of high general prosperity, for example, in the late 1920's, there were the famous "distressed areas" in England and Wales. Soft-coal miners in the United States were getting an average of only about 200 days of work a year. The textile and shoe towns of New England were suffering from a migration of their special industries to other regions, and wheat farmers in the American mid-West had insufficient incomes. Ideally, the redundant labour and capital in these places and occupations should have "flowed" into other types of production and perhaps other localities. But the flow was sluggish and in some cases hardly noticeable at all. Meanwhile, resources were wasted in persistent structural underemployment. Workers and their families suffered. Another striking illustration of the way in which unemployment from structural unbalance may persist in spite of a very high level of effective demand is the fact that as late as September 1942 there were 400,000 unemployed in New York City, out of a labour force of 3,500,000.[1] This was more than two years after the beginning of large-scale defence expenditure, and despite a critical "manpower shortage" in the United States as a whole.

The structural maladjustments in every part of the world after this war will probably be as severe as any ever known. The first requisite for dealing with them constructively is, of course, the maintenance of a general high level of effective demand (produced by a high level of consumption and investment). But beyond that there is a special case for mitigating structural underemployment by directing some of the stimulus from large-scale purchases of developmental equipment specifically to industries and localities where problems are acute.

There are three types of situation in which it would be sound policy to direct supplemental equipment demands towards particular depressed industries or localities. These situations will be widespread in the decade after the war.

The first type is one in which contraction is necessary as a long-run proposition, because an industry has become overexpanded or for other reasons, but in which a more gradual process of contraction would have an important effect in lessening the difficulties of the readjustment. A rear-guard action, in the form of an artificial stimulus to the demand for the product, might enable a more orderly retreat to be made. For example, alternative means of employment may be scarce in some shipbuilding communities at the end of the war. Many workers may have to move elsewhere to find jobs. In these circumstances, a large order for river boats to be used in improving transportation in the newly developing parts of the world might be given to firms that have been making landing barges for the armed forces. Even though the order were not to be repeated, it would temporarily reduce some of the pressure on the shipbuilding community and would permit adjustments to be spread over a year or two.

The second type of situation is one in which an industry or a community is suffering from a temporary lack of demand for its product. The workers and equipment, though underemployed at the moment, will be needed later, perhaps after an abnormally low replacement demand has become normal again or after an anticipated rise in world demand has had time to materialise. For example,

[1] *New York Times*, 6 Sept. 1942, quoting an official of the New York State Department of Labor.

portions of the machine tool industry might work at top speed to make the equipment needed by plants converting from war production to civilian production and then suddenly find the job done. There would be hardly any replacement demand for the time being, because most of their customers would have just installed new machinery. This could happen to various industries, even in the midst of general prosperity. If it happened to enough large and important industries, however, general depression might follow. Measures to sustain demand in specific industries would surely be justified in cases of this sort.

The third type of situation which would justify specially directed orders for equipment arises when a depressed industry or locality could be "converted" to a new and more promising line of production, provided that a special stimulus and opportunity in that line could be offered to it. For example, if the textile industry in a certain region is declining as a result of cheaper production elsewhere, while the electrical equipment industry shows signs of eventually providing alternative employment in the same region, then a special supplementary order of electrical equipment for developmental purposes would be justified, especially if it were combined with subsidised vocational retraining for workers. Similar reasoning might apply to the case of a particular kind of factory for which the product was no longer in strong demand but which could be converted to some other line.

A final argument for the policy of directing part of the supplemental purchases towards underemployed industries and localities is that such a policy would probably increase the willingness of advanced countries to provide capital for international developmental undertakings. It would help to line up special interest groups in favour of a programme which is in the general interest.

The Repayment Problem

It may be objected against the programme proposed in the preceding chapter that it is undesirable to encourage China and other countries in similar positions to make large supplemental purchases of equipment and thus to increase their foreign exchange or "transfer" problem when repayment comes due. This is true, so far as the undesirability of a large cash debt is concerned. The method proposed would help to minimise this. A method which encourages and assists China to *time* and *direct* its purchases of equipment so as to buy as much as possible in depressed markets and with the assistance of special loans at very low rates of interest should have the effect of giving China the maximum amount of real capital with the minimum amount of cash debt.

Expansion Abroad a Supplement to Expansion at Home

The arguments put forward by economists in recent years for "expansionist" measures in order to sustain the level of employment and income have usually been cast in national terms. When aggregate community income is spoken of, it is usually "national income", although the same reasoning would apply to world income or the income of the people of one county or province. Proposals for action are ordinarily in terms of the fiscal policies of national governments, national housing programmes, national development programmes, and the like. This is a natural consequence of the fact that the only governmental units in the world today which have the authority and the means to grapple with major economic problems are national units. The main barriers to the movement of goods and services and to the shifting of capital, labour and management are largely along national lines, so that the division of the world economy into national compartments has been more important in the past than any other sort of division.

Yet the interconnections between the economies of nations are such that expansionist policies in one are much more likely to be effective if others pursue similar policies at the same time. Hence the well recognised need for international co-ordination of domestic anti-depression policies. The programme suggested here would add a further form of action against depression, making three in all: (1) domestic or national measures; (2) mutual reinforcement of these measures through international co-ordination; and (3) concerted measures for developmental investment in the under-equipped areas of the world, with special encouragement to the counter-cyclical timing of equipment orders and to placing them in depressed industries or localities.

Partly because the argument for expansionist economic policies has usually been presented in national terms, the question is likely to be asked, "Why have an international development programme? Why not maintain employment by domestic expansion?" It should be said at once that nothing in this discussion of international development is intended to suggest that domestic measures of expansion should be neglected. Both are needed; they are not alternatives. The reasons in favour of having both domestic and international anti-depression measures are largely implicit in what has already been said. Nevertheless, it may be useful to set them down explicitly here. There are four main points.

First, the two phases of expansionist policy—domestic economic expansion and world economic expansion—are interdependent; they are organically related. The best domestic policy will not work as well as it should unless a suitable international policy is combined with it. It is often said that the most important contribution which great economies like the United States can make to world recovery is domestic recovery, and there is much truth in that statement. But it makes a great difference whether domestic recovery, is moulded in the light of what is needed for international recovery, not later but simultaneously.

Second, it is good to have more than one string to a bow. The danger in the decades ahead is not too much investment and employment, but too little, at least after the first few years of making up wartime deficiencies. By combining domestic and international measures to promote economic expansion there is more chance that some parts of the programme will achieve enough success to give the needed total stimulus.

Some of the domestic measures relied on to maintain employment at a high level may encounter serious resistance or may not be applied in time, or for one reason or another may not produce the results hoped for. The opposition to "government in business" may limit the scale of direct governmental investment for developmental purposes in some countries, or may force it into "non-competitive" projects of relatively low productivity. Unless the psychology of business men changes, public investment and deficit financing to stimulate employment may induce a lack of "confidence" which causes a further shrinkage in private investment. Measures designed to reduce the rate of saving (increase the rate of consumption), through shifts in tax burdens, increases in community expenditures on welfare and social security, and the like, may be effective in the long run, but they work only gradually and can hardly meet acute conditions. The same is true of attacks on monopoly as a means of promoting expansion and combating depression.

A programme of international investment, supplementing such domestic measures, offers advantages in many of these respects. Developmental investment in China and elsewhere would produce a stimulus in the form of increased demands for exports of equipment. This is a kind of stimulus which business men are accustomed to regard as good, whereas the same orders inspired by a public project at home might produce nervousness about government competition and government in business. Of course, international developmental investment will encounter resistances of its own. The point here is not that international investment is necessarily superior as an expansionist device to domestic investment, but that the two together are better than either one separately.

Third, the argument that domestic economic expansion in

the advanced countries can take care of their post-war employment problem seriously underrates the difficulties of structural readjustment. The advanced countries cannot use at home all the machines and other capital goods equipment that they can produce, especially with the new plants and the newly trained workers resulting from wartime expansion of these industries. As has been argued above, it would be much less difficult to get full employment after the war and also would be much more profitable for the advanced countries and beneficial for the world as a whole, if the utilisation of productive capacity of these types were maintained at a higher level through the development of new areas.

To get full employment through domestic measures alone it is necessary to convert the particular sorts of productive capacity that exist in the country into the particular sorts of productive capacity that are needed at home. The post-war employment problem will be much easier if both a wider area of demand and a wider area of supply can be taken in. A product that can be turned out by a particular plant in Belgium, Great Britain or the United States may be extremely useful in the economic development of Iran, Java or Bolivia, whereas the domestic demand for this particular kind of equipment may be small and the problem of converting the existing plant and existing skills to kinds of output for which there is an unsupplied domestic demand may be difficult.

Fourth, the expansion of employment through investment in the development of less developed areas increases the permanent productivity of the world much more than would an additional investment of like amount in the advanced countries themselves. Capital invested in areas where capital is scarce works at a higher "margin of productivity" than in areas where capital is already abundant. This is reflected in the higher interest rates, after allowance for risk, of the less developed areas. A new road or a new tool in China tends to increase output more than an additional road or an additional tool in the United States where the most urgent needs of this sort have already been met.

From the standpoint of the industrially advanced countries, an expanding world market, with a higher initial demand and a higher replacement demand for the technical goods and equipment that they are now particularly able to produce, would make it possible for them to expand (or to contract less) in just those lines in which their labour can be most productively employed. In this way a world development programme would enable labour to work at jobs which would yield higher wages than could otherwise be earned. In general, productive efficiency rises as a result of the greater specialisation which is made possible by a wider market and wider division of labour. If proper adaptations can be made (see Part II) the increased size of the world market resulting from world development will enable such improvements in the division of labour to be made. This permits living standards to rise.

PART II

LONGER-RANGE EFFECTS RESULTING FROM SHIFTS IN PRODUCTION, CONSUMPTION, AND TRADE

A. The Nature of These Effects

CHAPTER VIII

ECONOMIC DEVELOPMENT AND TRADE PROSPECTS

Economic development means the rise of new and more efficient production in agriculture, mining, manufacturing, and commerce. Some of the new production will undoubtedly be competitive with established industries of the advanced countries. On the other hand, rising income levels in the countries undergoing development will enable their people to consume more than ever before, including perhaps many products from the advanced countries for which they were not good markets in the past. Is the net result likely to be favourable or detrimental to the more highly developed economies? What problems will be raised for them, and what shifts in production may they have to undertake? What could be done to make the necessary readjustments less onerous and to encourage results that, so far as possible, will be of mutual benefit to the newly developing and advanced countries alike? These are the problems to be considered in Part II.

Economic Development and the Volume of External Trade

Does the economic development of a country lead to a shrinkage of its trade with the rest of the world? So far as past experience is concerned, the evidence is decidedly to the contrary.

The United Kingdom in 1775, at the beginning of a spectacular period of industrial pioneering and economic growth, imported goods from other countries to the value of 13 million pounds sterling and exported goods to the value of 8 million. By 1929 imports had increased more than 70-fold to 919 million pounds and exports had increased 69-fold to 554 million pounds (chart 1).[1]

The dollar value of United States imports in 1850 was 173 million. In 1929, after three quarters of a century of tremendous

[1] These import and export values and others shown in chart 1 have been adjusted to take account of price changes. The other charts that follow show current values, except that chart 5 on Japan gives both current and adjusted values. The data necessary to achieve consistency in this respect were not available.

Chart 1. The External Trade of the United Kingdom, 10-Year Averages, 1701 to 1930[1]

(*Millions of Pounds, Adjusted for Price Changes and Linked to 1913 Prices as Explained in Footnote*)

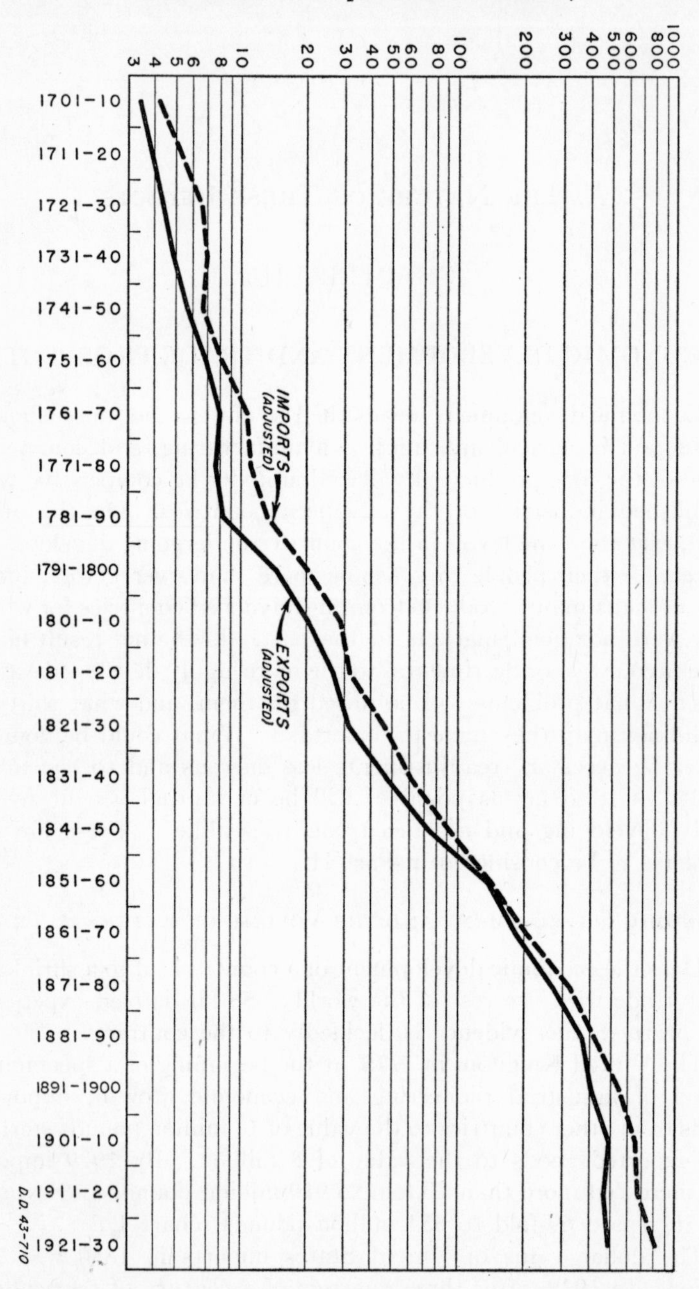

[1] Werner SCHLOTE: *Entwicklung und Strukturwandlungen des englischen Aussenhandels von 1700 bis zur Gegenwart* (*Probleme der Weltwirtschaft*, Schriften des Instituts für Weltwirtschaft an der Universität Kiel, No. 62), p. 48 and tables 7, 9, 10 in Appendix (Jena, Fischer, 1938). Schlote, as explained on pp. 29-33 of his study, has adjusted for price changes in a number of separate periods, using different weights for different periods, and has finally linked the periods to 1913 prices. The data before 1814 are based on official valuations which made use of the prices prevailing in 1694, thus providing an index of volume rather than of current value.

Chart 2. The External Trade of the United States, 5-Year Averages, 1791-1940[1]

(*Millions of Current Dollars*)

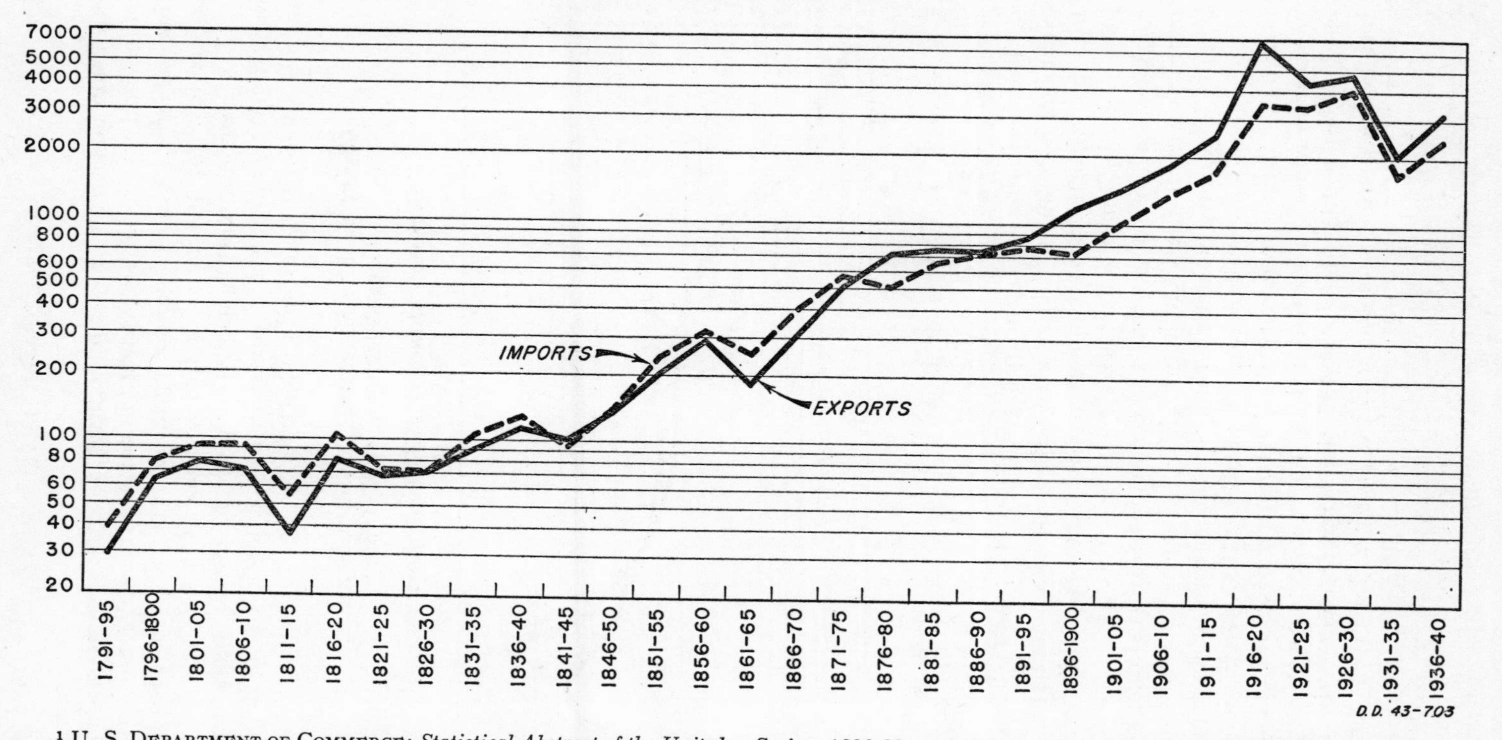

[1] U. S. Department of Commerce: *Statistical Abstract of the United States, 1941*, pp. 523, 525, and 526; and Treasury Department: *Monthly Summary of Commerce and Finance of the United States*, No. 5 Series, 1898-99, p. 1446, and No. 5 Series, 1899-1900, p. 1889. Data cover general trade. Fiscal years ended 30 Sept. through 1842; fiscal years ended 30 June through 1870; calendar years thereafter.

expansion in agriculture, manufacturing, internal commerce, population, and territorial settlement, imports were 4,399 million, 25 times as large. Exports rose during the same period from 144 million dollars to 5,241 million, a 36-fold increase (chart 2).

The economic development of France in the 75 years before the First World War was accompanied by a 10-fold rise in imports and an 8-fold rise in exports (chart 3).

Chart 3. The External Trade of France, 5-Year Averages, 1831-1935[1]

(Billions of Francs of 1928 Gold Parity)

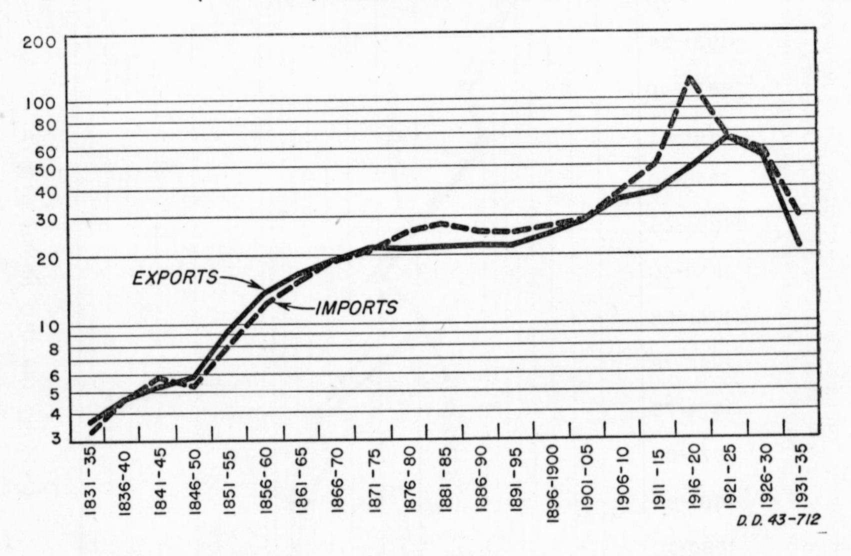

The beginning of Germany's great period of rapid industrial expansion is commonly dated about 1880. Between that year and 1913 its imports showed almost a 4-fold increase, while exports increased 3½ times (chart 4).

After centuries of political and economic isolation, Japan began to trade with the rest of the world in the 1860's. In subsequent decades it adopted western techniques and entered upon a period of rapid economic development. From 1873 to 1937, Japan's imports increased 42 times in quantity and its exports 65 times. (The increases measured in current values, not taking account of price changes, were 135 times and 147 times respectively.) In 1929, just before the onset of the world depression, Japan was buying

[1] MINISTÈRE DU TRAVAIL ET DE LA PRÉVOYANCE SOCIALE: *Annuaire statistique, 1910* (Paris, Imprimerie nationale, 1911), "Résumé rétrospectif", pp. 78*-79*; also *Annuaire statistique, 1938*, "Résumé rétrospectif", p. 125*. Data cover special trade, converted to francs of 1928 gold parity.

11 times as much from the rest of the world (after allowance for price changes) and selling 10 times as much as in 1889, forty years earlier (chart 5).

Chart 4. The External Trade of Germany, 1872-1938[1]

(*In Billions of Reichsmarks*)

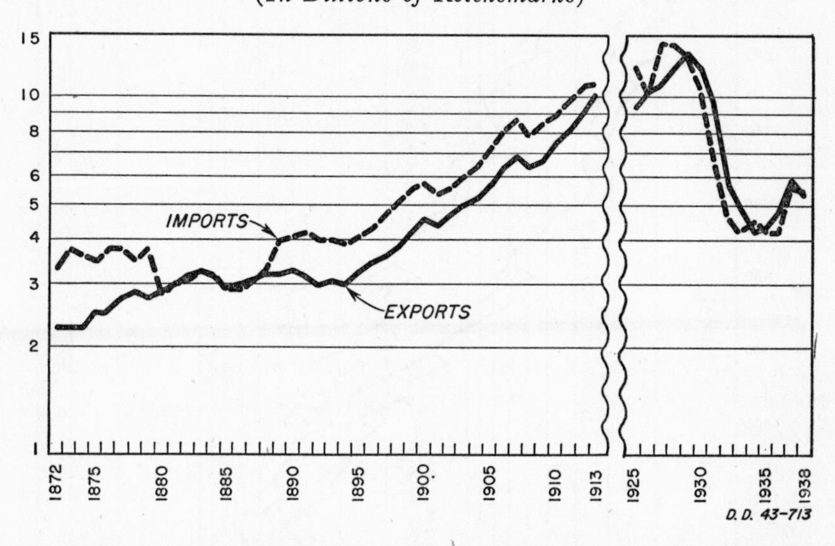

The one great exception to the general rule that the external trade of a country increases as its resources are developed is the Soviet Union (chart 6). Before the First World War there was a strong upward trend in the imports and exports of Russia. After the disturbances of war and revolution this trend was resumed under the Soviet Union in the middle 1920's, although at a lower level (so far as comparisons can be made between the pre-war and post-war values). From 1930, however, in the case of exports, and from 1931 in the case of imports, there was a sharp decline, with only a slight recovery in the latter part of the 1930's. At the same time, the Soviet economy was in a period of rapid expansion. Industries grew at a phenomenal rate under successive five-year plans. In this instance, clearly, rapid economic development and industrialisation were accompanied by a decrease in external trade.

The explanation is to be sought in a number of special circumstances. The depression in the capitalist world in the 1930's not only reversed the upward trend in the trade of most other countries (see charts 1-6) but also affected Soviet trade. It brought a severe

[1] STATISTISCHES REICHSAMT: *Statistisches Jahrbuch für das Deutsche Reich*, Annual volumes for 1885, 1937, and 1939. Data cover special trade.

Chart 5. The External Trade of Japan, 1868-1939[1]

(Millions of Current Yen, Left-Hand Scale; Index Numbers, 1913=100, Right-hand Scale)

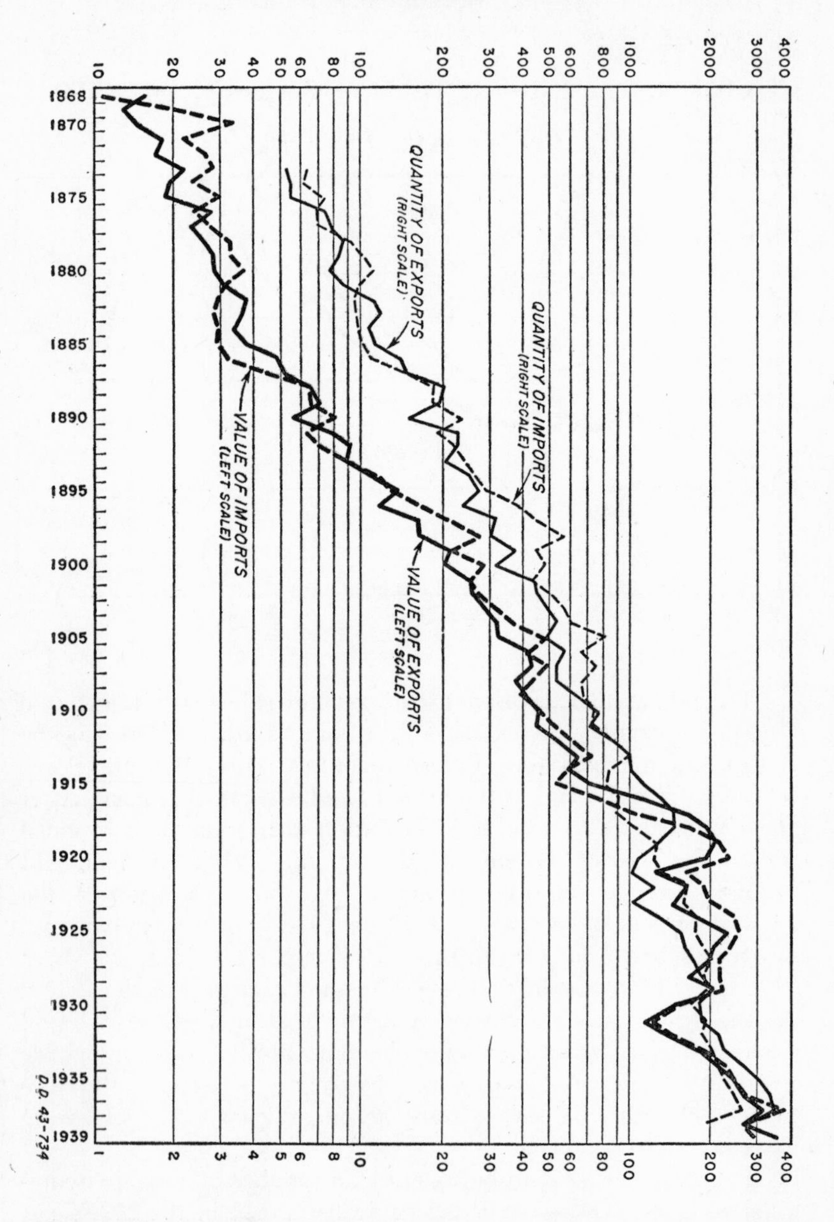

[1] TANZAN ISHIBASHI (ed.): *The Foreign Trade of Japan: A Statistical Survey* (Tokyo: issued by the *Oriental Economist*, 1935), Figure III (frontispiece), pp. 697-698, for years through 1934. For later years value figures are from U. S. DEPARTMENT OF COMMERCE: *Foreign Commerce Yearbook*, 1939, p. 266, and quantity figures are from *Oriental Economist*, Nov. 1940, p. 734 (attributed to the Yokohama Specie Bank).

Chart. 6 The External Trade of Russia and the Soviet Union, 1890-1938[1]

(*Millions of Old Gold Rubles, Left-Hand Scale; Millions of Foreign Trade Rubles, as Adopted in 1936, Right-Hand Scale*)

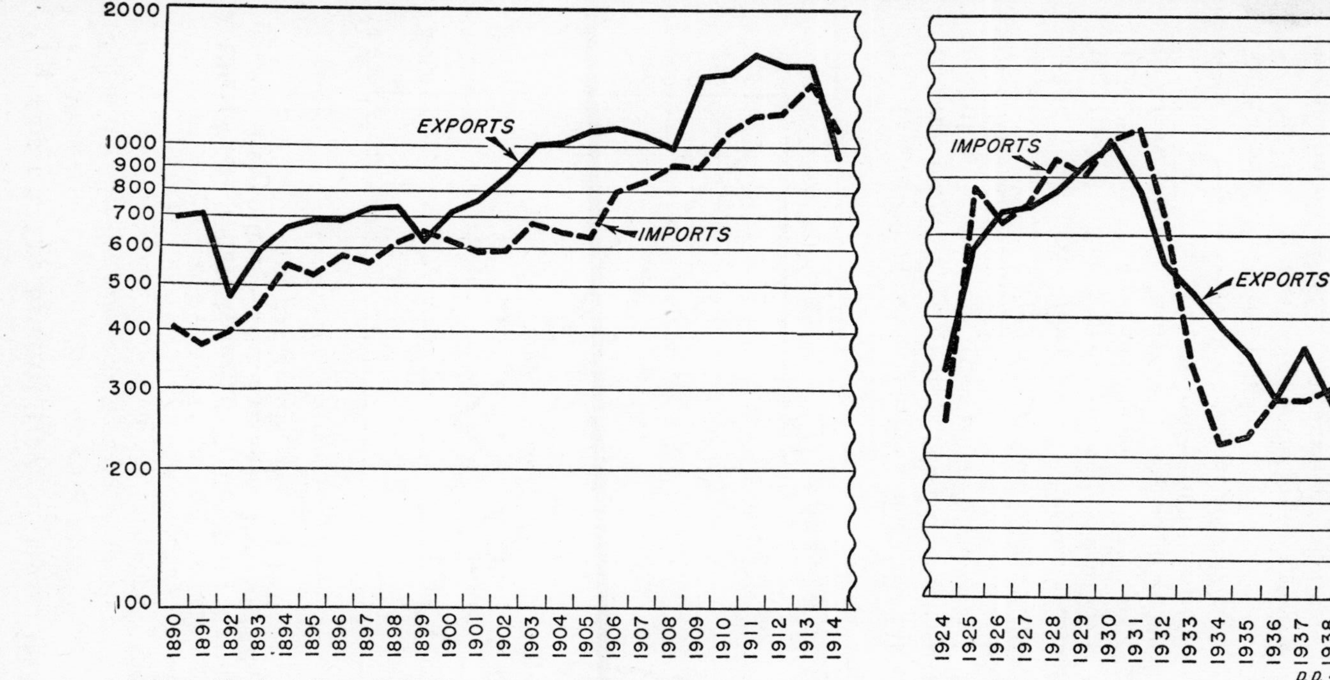

[1] *Obzor Vneshnei Torgovli Rossii*, 1890-1903 (St. Petersburg, Department of Customs Collection, 1892-1905). *Vneshnyaya Torgovlya Soyuza Sovietskikh Sotsialisticheskikh Respublik*, 1904-1928. *Statisticheskii Obzor* (Moscow, Cooperative Supply Publishing Co., 1931). *Statistika Vneshnei Torgovli*, 1929-1938 (Central Customs Directorate, Foreign Trade Publishing Bureau, 1930-1939).

(An approximate translation of the above is as follows: *Survey of the Foreign Trade of Russia*, 1890-1903. Publication of the Department of Customs Collection, St. Petersburg. Published annually 1892-1905. *Foreign Trade of the Union of Soviet Socialist Republics*, 1904-1928. *Statistical Survey*, Office of the Cooperative Supply Publishing Co., Moscow, 1931. *Statistics of Foreign Trade*, 1929-1938. Central Customs Directorate. Foreign Trade Publishing Bureau. Published annually 1930-1939.)

Data not adjusted for territorial changes. Figures for 1924 through 1935 converted, at ratio of 1 to 4.6, to foreign trade rubles as adopted in 1936. Values on left-hand scale and those on right-hand scale are also related at this ratio.

fall in the price of Soviet export products, which were at that time mainly raw materials, and made the exchange of these products for imports of machinery and other needed items less advantageous.

Chart 7. Aggregate External Trade of the Latin American Republics, 5-Year Averages, 1901 to 1920, Annual Totals, 1921 to 1939[1]

(Millions of Current U.S. Dollars)

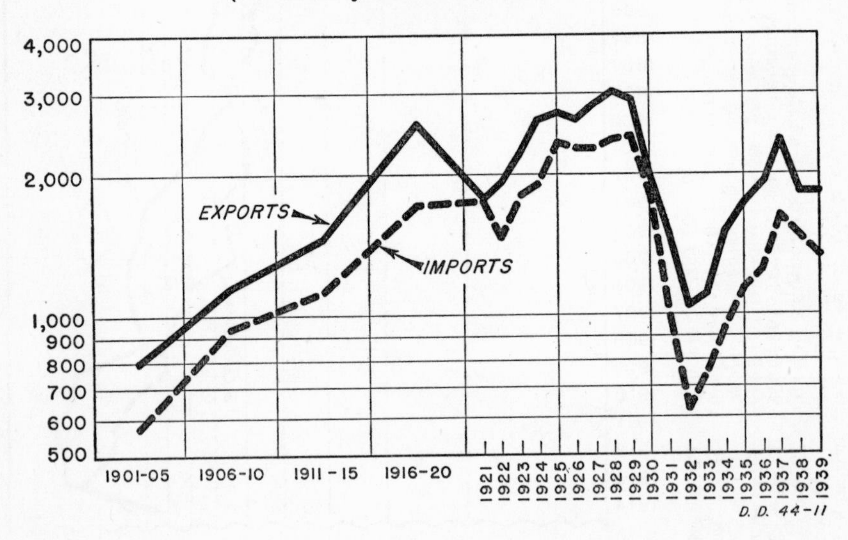

The vast extent of Soviet territory and the great variety of its resources made a largely self-contained development more feasible for it than for any other region still to be developed. Strict control over income distribution and trade prevented consumers from having or making use of increased purchasing power to buy "standard of living goods" from abroad. Finally, and most important, the strained intensity and speed with which Soviet industrialisation was pushed ahead and the course that it took (concentration on heavy industries first) was prompted in no small measure by anticipation of war. Under the circumstances there were strong strategic reasons for a deliberate policy of maximum self-sufficiency and a minimum of foreign trade.

This historical case of the Soviet Union's trade serves to underline the importance of the fundamental assumptions, stated earlier, on which this study is made. If the world after the present war is one in which the countries that are prepared for rapid economic development feel no greater sense of security against attack than

[1] U. S. DEPARTMENT OF COMMERCE: *Foreign Commerce Yearbook* for 1926, 1930, 1935, and 1939 (data assembled from summary tables covering each country).

did the Soviet Union in the last two decades, then, indeed, there will be other attempts to promote an isolated internal development without expansion of imports and exports. Other countries, lacking the enormous extent and variety of resources to be had within the boundaries of the U.S.S.R. would not be in as good a position to succeed, but they might nevertheless be impelled to try.

Chart 8. National Income and Imports of Sweden, 1861-1939[1]

(Millions of Kronor)

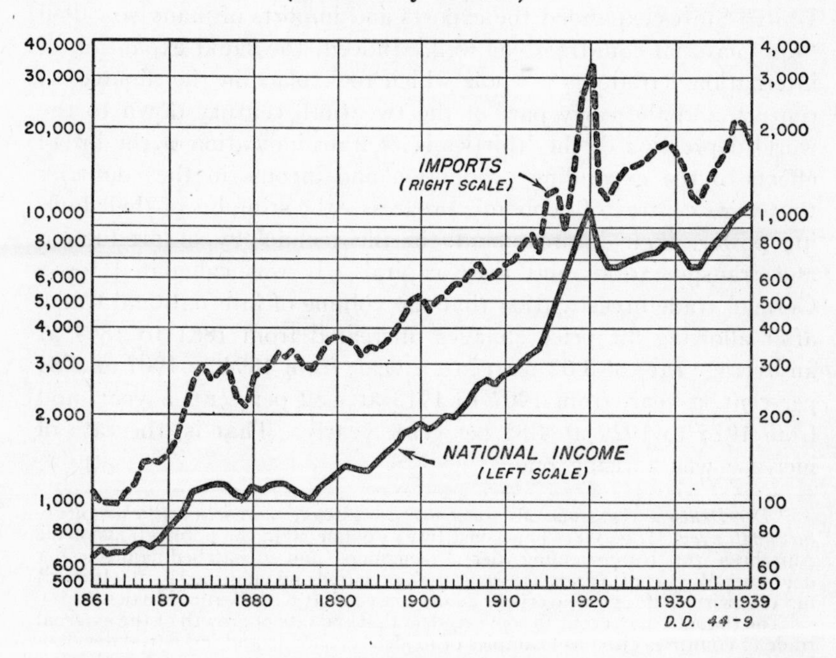

The analysis in this study, however, assumes a reasonably effective system of political security. It also assumes a willingness on the part of the advanced industrial countries to co-operate in the process of economic development in other parts of the world, and willingness of the countries undergoing development to accept that co-operation. Past experience indicates quite clearly that where there is a sense of reasonable security against attack and

[1] Income data are from: THE INSTITUTE FOR SOCIAL SCIENCES OF THE UNIVERSITY OF STOCKHOLM: *The National Income of Sweden, 1861-1930*, Part I, pp. 234-235. The series after 1930 has been extrapolated on the basis of estimates for the years 1930 to 1936 in an article by Erik LINDAHL in the *Skandinaviska Kreditaktiebolaget*, Quarterly Circular, July 1937, pp. 45-50, and on the basis of estimates for the years 1936 to 1939 by the Swedish Department of Finance. Import data are from: THE INSTITUTE FOR SOCIAL SCIENCES OF THE UNIVERSITY OF STOCKHOLM: *The National Income of Sweden, 1861-1930*, Part II, pp. 234-235, and the U. S. DEPARTMENT OF COMMERCE: *Foreign Commerce Yearbook*, 1939, p. 110.

readiness to co-operate in encouraging mutually beneficial trade, the countries undergoing development show large increases in their total purchases from other countries and in their total sales to other countries.

It would be a mistake, however, to suppose that the effect of a particular country's economic development upon world trade is reflected only in its own imports and exports or is adequately measured by the changes which they undergo. The development of modern industry in Great Britain, in western Europe and in the United States expanded the exports and imports of many so-called "raw material countries" as well. Indeed, the rapid expansion of international trade as a whole which took place in the nineteenth century and the early part of the twentieth century down to the world depression of the 'thirties is itself an indication of the broad effects of the expanding production and income in the countries that were centres of economic progress—the stimulus of their buying power, their improvements in the technology of production and transportation, and their capital. It was calculated for a German trade investigation that the volume of international trade, after allowing for price changes, increased from 1881 to 1894 at an average rate of 3.02 per cent. a year, from 1894 to 1907 at 3.84 per cent. a year, from 1907 to 1913 at 4.52 per cent. a year, and from 1925 to 1929 at 4.85 per cent. yearly. That is, the rate of increase was a rising one.[1]

---

[1] *Der Deutsche Aussenhandel unter der Einwirkung weltwirtschaftlicher Strukturwandlungen*, II, p. 348. This constitutes Volume 20 in the publications of the Ausschuss zur Untersuchung der Erzeugungs- und Absatzbedingungen der deutschen Wirtschaft (*Enquête-Ausschuss*), prepared and published by the Institut für Weltwirtschaft und Seeverkehr an der Universität Kiel (Berlin, Mittler, 1932).

The thought may occur to some readers that the rate of growth of the external trade of countries cited as examples of modern economic and industrial development should be compared with the rate of growth of international trade as a whole, or that the growth of the trade of "industrial" countries should be compared with that of "agrarian" or raw material supplying countries. Such comparisons would not be particularly significant for the present problem, however, for the growth of the international trade of a country will reflect not only its own economic development but also that of other countries. If the United States, by reason of an internal increase in productivity and income, expands its imports from Costa Rica by $100 million yearly, the Costa Ricans are soon able to buy more themselves. The percentage increase in Costa Rica's trade may be larger than the percentage increase in the foreign trade of the United States because of the difference in size of the two economies. Also, if multilateral trade channels are reasonably free to expand, the original developmental stimulus may pass on from the second to a third and fourth and other countries. The activating centres of trade expansion could have a slower rate of increase at some stages than the rate of increase shown by other, especially smaller, economies which are riding the waves thus put in motion.

Some interest may attach, none the less, to the following indications of average rates of growth of imports derived from the charts in this chapter by a rough-and-ready graphic measurement of trend lines which were laid out by eye. The period represented by each such trend line is indicated below. Rates of increase are expressed in terms of the number of years required to bring about a doubling.

Income and Imports

People who produce more can afford to buy more. As the total income of a country rises it becomes able to purchase a larger volume of imports from abroad, paying for them by correspondingly increased exports, and it is likely to do so unless special circumstances stand in the way. There is a straightforward connection between economic development and larger imports, for economic development increases production and income (both by improving the fundamental capacity to produce and by helping to maintain a high level of economic activity).

The economically less developed countries contribute to the trade of the world much less than in proportion to their size and population. For example, Albania, Bulgaria, Greece, Hungary, Poland, Rumania, Turkey and Yugoslavia together represented in 1935 a third of Europe's area, over a fourth of its population, but only 6 per cent. of its imports and 8 per cent. of its exports.[1] India and China account together for 39 per cent. of the world's population but for only 5 per cent. of its international trade.[2]

Chart 7 shows the aggregate international trade of the 20 Latin American republics, including trade among themselves. Their imports run only about twice as high in value as the imports of Canada (chart 12), not counting, of course, any of the interprovincial trade within Canada. The United States sells almost as much to less than 12 million Canadians on the north as to more than 120 million people in the Latin American republics on the

---

The following rates of increase apply to *volume* of imports (that is, after allowance for price changes):

Japan (chart 5), 1880-1913: doubling in 10 years; 1873–1929, doubling in 11 years.
United Kingdom (chart 1), 19th century: doubling in 19 years.
World (obtained by plotting the volume data of the German *Enquête-Ausschuss*, cited above, based on 33 countries), 1881-1913: doubling in 20 years.

The following rates of increase apply to imports at *current values:*

United States (chart 2), 1840's to 1913, also 1840's to 1929: doubling in 16-17 years.
France (chart 3), 1830's through 1880's: doubling in 16 years; 1830's to First World War: doubling in 22 years.
Germany (chart 4), 1880-1913: doubling in 16 years.
Japan (chart 5), 1880-1913 and also 1868-1929: doubling in somewhat more than 8 years.
Russia (chart 6), 1890-1913: doubling in 13 years.
Sweden (chart 8), 1861-1929: doubling in 17 years.
Australia (chart 11), 1901-1929: doubling in 12 years.
Canada (chart 12), 1911-1929: doubling in 12 years.
Latin America (chart 7), 1901-5 to 1916-20: doubling in 11 years; 1901-5 to 1929, doubling in 13 years.

[1] League of Nations: *Europe's Trade*, 1941, p. 49 and table 4 on p. 16.
[2] League of Nations: *The Network of World Trade*, 1942, p. 19.

Chart 9. National Income and Imports of the United States, 1799 to 1940[1]

(*Millions of Current Dollars*)

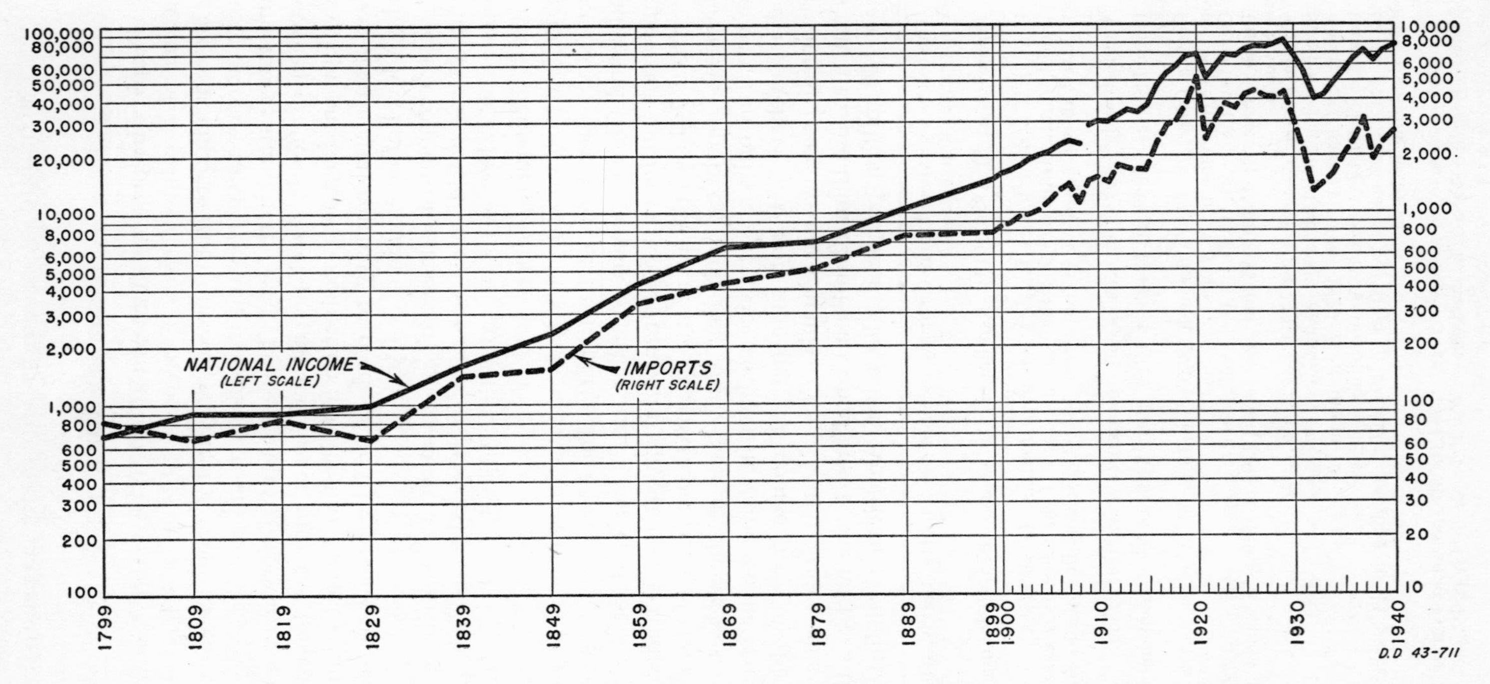

[1] Income data are from: Robert F. MARTIN: *National Income of the United States, 1799-1938*, New York, 1939, p. 6 (decennial data, 1799 to 1889, and annual data, 1900 to 1908); U. S. DEPARTMENT OF COMMERCE: *Survey of Current Business*, Mar. 1943, p. 22 (covering period 1929 to 1940); extrapolated back to 1919 on basis of Simon KUZNETS: *National Income and Capital Formation, 1919-35*, New York, 1937, and to 1909 on basis of W. I. KING: *The National Income and Its Purchasing Power*, New York, 1930. Import data are from: U. S. DEPARTMENT OF COMMERCE: *Statistical Abstract of the United States, 1941*, Washington, pp. 523, 525, and 526; and TREASURY DEPARTMENT: *Monthly Summary of Commerce and Finance of the United States*, No. 5 Series, 1898-99, p. 1446, and No. 5 Series, 1899-1900, p. 1889. Fiscal years ended 30 Sept. through 1842; fiscal yearsended 30 June through 1870; calendar years thereafter.

south.[1] This is true notwithstanding the fact that Canada has a thoroughly modern technology, like that of the United States, and a similar climate, while Latin America might be regarded by some as a more logical area for the marketing of United States products

Chart 10. National Income and Imports of Japan, 1887-1939[2]

(*Millions of Yen*)

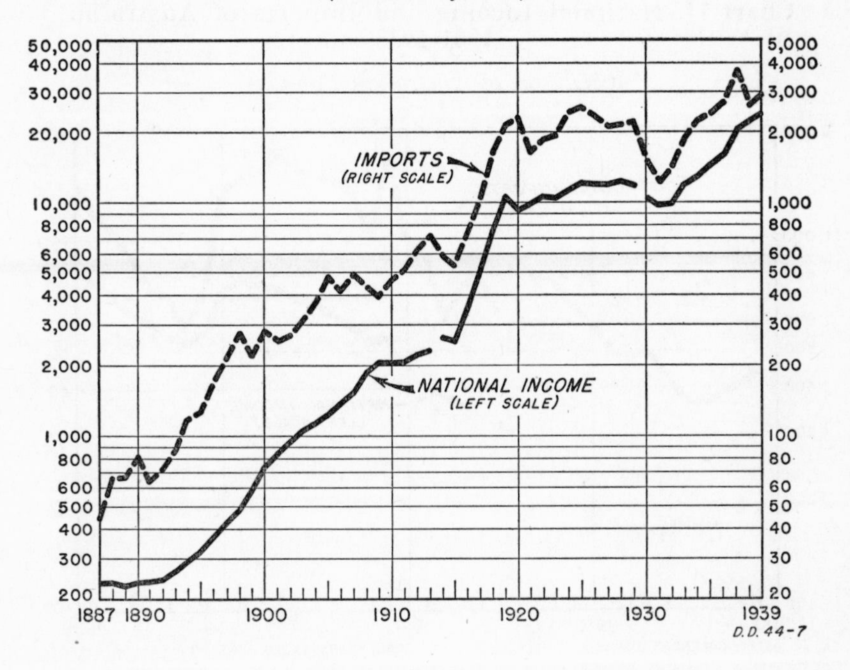

because its climate, resources, and stage of industrial development are so markedly different. A major part of the explanation is that Canada's economic development is so much further advanced that its productivity, income level, and living standards are much higher than those of Latin America. Canadians can afford to buy, and do

[1] In 1939 United States exports to Canada were 489 million dollars and to the 20 republics of Latin America were 549 million dollars. The proportion was not greatly different in other recent years. (*Statistical Abstract of the United States, 1942*, p. 562.)

[2] Income data are from: K. MORI: "The Estimate of the National Income of Japan Proper", in *Bulletin de L'Institut Internationale de Statistique* (Tokyo, 1931), Tome XXV, 2ème livraison, pp. 203-4 (for years 1887 to 1913); MITSUBISHI ECONOMIC RESEARCH BUREAU: *Monthly Circular*, Mar. 1934, p. 10, compiled by Prof. S. HIJKATA (for period 1914 to 1929); JAPAN ECONOMIC FEDERATION: *National Income of Japan* (Tokyo, 1939), pp. 48, 101 (for period 1930 to 1939). Import data are from: TANZAN ISHIBASHI (ed.): *The Foreign Trade of Japan: A Statistical Survey* (Tokyo: issued by the *Oriental Economist*, 1935), p. 697, for years through 1934, and for later years from U. S. DEPARTMENT OF COMMERCE: *Foreign Commerce Yearbook*, 1939, p. 266.

buy, for themselves and for the use of their industries, the advanced, dynamic, "high income" goods which the United States is particularly effective in producing.

The closeness of the connection between a country's income and the amount of its imports may be judged from the next series of charts. Chart 8 compares the national income and imports of Sweden from 1861 through 1939. Sweden offers the longest avail-

Chart 11. National Income and Imports of Australia, 1901-1938[1]

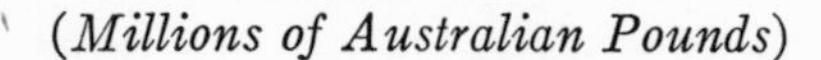

(*Millions of Australian Pounds*)

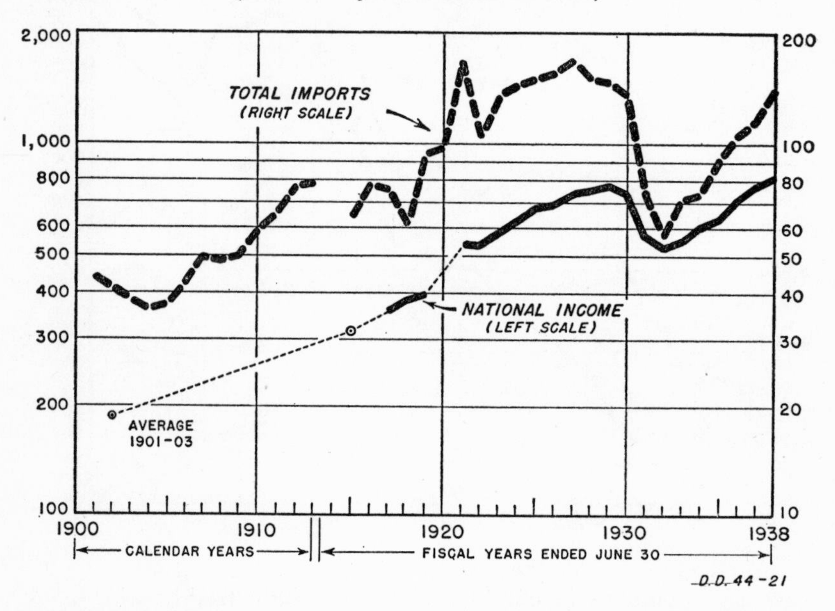

able series of annual income data. In chart 9 the national income and imports of the United States are compared at ten-year intervals from 1799 to 1900 and annually thereafter. Chart 10 shows Japan's national income and imports from 1887 to 1939. It has also been possible to get data for Australia (chart 11) and Canada (chart 12) over shorter periods. Four significant conclusions may be drawn from these comparisons.

First, there is in every case a strong parallelism between the long-run growth of national income and the growth of imports.

[1] Income data are from: Colin CLARK and J. G. CRAWFORD: *The National Income of Australia*, 1938, pp. 59-60. Import data are from: AUSTRALIAN COMMONWEALTH BUREAU OF CENSUS AND STATISTICS: *Trade and Customs and Excise Revenue, 1911*, and earlier issues; U. S. DEPARTMENT OF COMMERCE: *Foreign Commerce Yearbook*, 1939, p. 190 (original figures in pounds sterling for 1931 to 1938 adjusted for discount on Australian pound).

Second, there is also in every case where annual data are available a clear conformity between the major up-and-down fluctuations of income as affected by prosperity and depression and the fluctuations of imports. The co-variation between income and imports even extends to minor fluctuations, especially when, as in the Swedish data and in the later years on some of the other charts, the income data are good.

Chart 12. National Income and Imports of Canada, 1911 to 1939[1]

(*Millions of Canadian Dollars*)

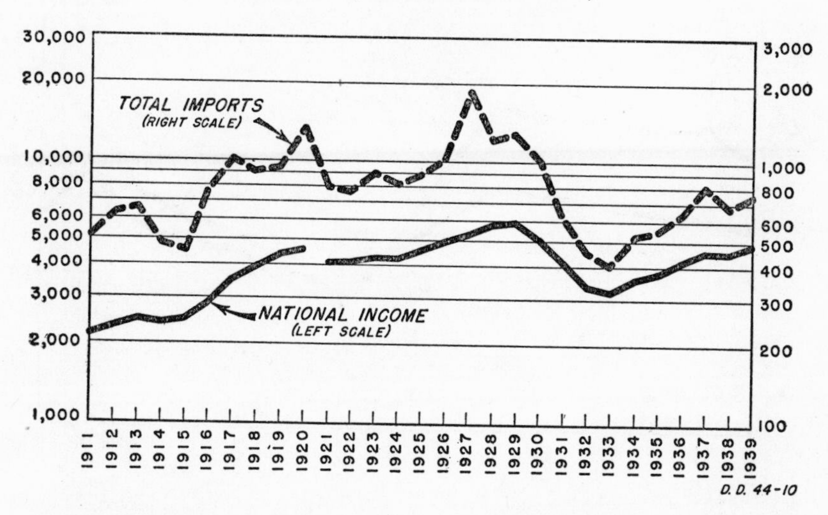

Third, although imports rise in close conformity with growth in income they do not rise quite in proportion to income. That is, over a long period there is a tendency, exhibited in all of the charts which cover enough years, for the ratio between imports and income to decline as income rises.[2] This in in accord with the expectation,

[1] Income data are from: J. J. Deutsch: "War Finance and the Canadian Economy, 1914-20", in *Canadian Journal of Economics and Political Science* (Toronto, Nov. 1940), p. 538; and the *Monthly Review of the Bank of Nova Scotia* for May 1937, Sept. 1940, and June 1941. Import data are from: Dominion Bureau of Statistics: *Monthly Report of Trade of Canada*, Dec. 1920, p. xii, and *The Canada Year Book, 1941*, p. 401.

[2] The charts are drawn on a ratio scale, so that two curves moving upward at the same rate (that is, an equal percentage of increase year by year) will have the same slope. When the import line lies above the income line the tendency of income to grow slightly faster than imports is shown by a narrowing of the space between the two curves as they move to the right along the time-axis. When the import line lies below the income line the same tendency is shown, of course, by a widening of this space. The import scales on these charts are exactly one tenth of the corresponding income scales. Therefore, the import line lies above the income line when the country's imports are more than 10 per cent. of its national income and below when they are less.

suggested in the Introduction and Summary (I), that as a community grows more wealthy it will spend a somewhat smaller percentage of its total income on commodities and a somewhat larger percentage on so-called "tertiary" production, especially services of various kinds, such as merchandising services, travel and trans-

Chart 13. Aggregate National Income and Imports of Twelve Countries, 1924 to 1938[1]

(Billions of Current Dollars)

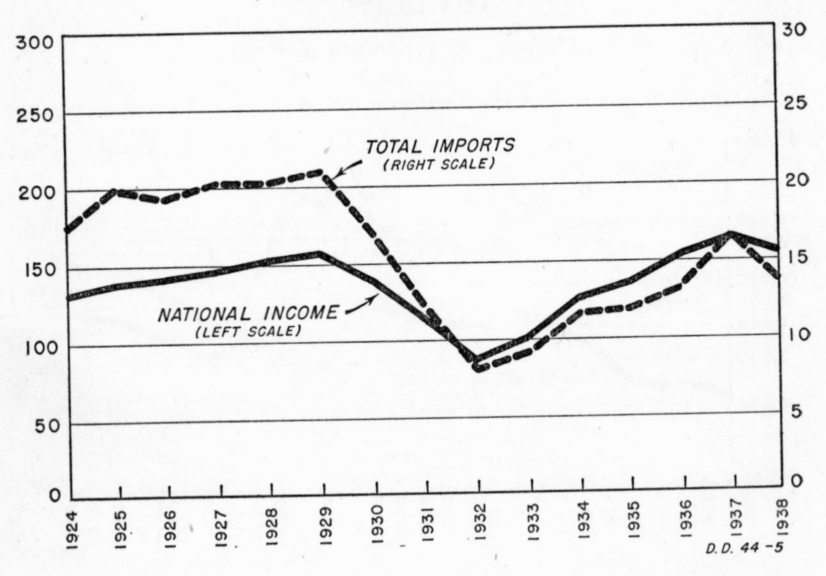

port, education, medical services, government administration, and professional services of many sorts.[2] If such services are "imported" (for example, insurance, tourist travel, education, etc., purchased abroad) they do not show up in the trade statistics.[3] Furthermore,

---

[1] Countries included are: United States, United Kingdom, Germany, France, Japan, Sweden, Norway, Denmark, Netherlands, Canada, Australia, and New Zealand. Import data are from the official foreign trade statistics of these countries. Income data are from sources cited in *The United States in the World Economy* (U. S. Department of Commerce, Washington, 1943), pp. 201 and 203. Both series are in current dollars, conversion from other currencies being made at annual average rates of exchange on the dollar.

[2] In the United States, for example, the proportion of the working population engaged in the production of services described as "domestic, personal, professional" increased from 12.4 per cent. in 1860 to 23.7 per cent. in 1935, while the percentage engaged in trade, transportation and communication activities rose from 7.4 per cent. to 22.1 per cent. Colin CLARK: *The Conditions of Economic Progress* (London, Macmillan, 1940), p. 185.

[3] The more wealthy the world becomes the more important will such "invisible" items be in international exchange, and the more urgent will it be to perfect balance of payments statistics so that they, instead of the increasingly misleading commodity trade statistics, may be used as a matter of course in discussing the broad flow of international commerce.

many of them are of a sort which by their nature have to be performed locally. Thus, merchandise imports (and probably domestic merchandise transactions also, if we were to measure them) increase with income but not quite as fast as income.

Chart 14. World Industrial Production and International Trade[1]

(Index Numbers, 1929=100)

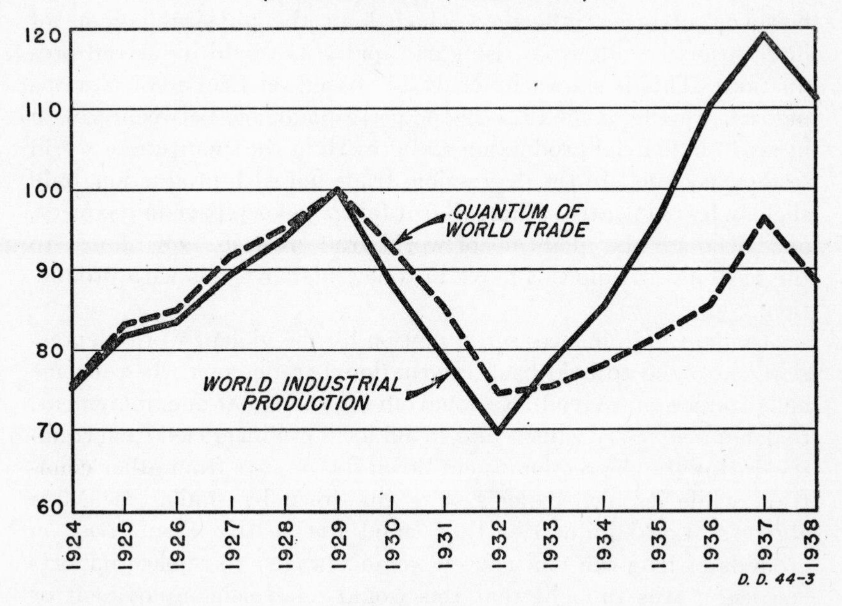

Fourth, the general economic breakdown of the early 1930's so disintegrated the world's trading system that imports fell more sharply than national incomes during the period of crisis and depression and then failed to recover their former percentage relation to national incomes in the subsequent recovery. This is evident in each of the charts for the separate countries, but it comes out much more clearly in chart 13, which shows the combined national incomes of twelve countries during the years 1924-38 and the combined value of their imports. From 1924 to 1929 imports and income rose at approximately the same rate. From 1929 to the bottom of the depression in 1932 income fell about 45 per cent., but imports fell more than 60 per cent. The subsequent recovery was no more rapid for imports than for income, with the result that

[1] LEAGUE OF NATIONS: *World Production and Prices, 1935-36,* 1936, p. 141; *World Production and Prices, 1938-39,* 1939, p. 103; *Review of World Trade, 1938,* 1939, p. 60. Industrial production for 1924 extrapolated on basis of series by Norman J. WALL: *Monthly Index of World Industrial Production, 1920-1935* (U. S. Department of Agriculture, 1936) (revised supplementary tables).

the aggregate imports of the twelve countries in 1937 represented only 10 per cent. of their aggregate incomes, whereas imports had been more than 13 per cent. of income in 1929.

Of course, the drastic fall in the total value of imports from 1929 to 1932 was partly the result of particularly severe drops in the prices of raw materials which make up an important part of international trade. That the aggregate value of imports had not recovered its old proportion to national incomes by 1937 is a reflection, however, of other influences which kept the physical volume of international trade from rising as rapidly as would industrial production. This is shown in chart 14, based on League of Nations indices. In the years 1924-29 the correspondence between growth in world industrial production and growth in the quantum of world trade was close. In the depression, trade fell with production, only slightly less violently. Thereafter it failed to keep pace in recovery, and although the quantum of world trade rose in 1937 almost to the 1929 level it did not move into new higher levels with production.

Evidently, influences set in motion by the violent world depression continued to hold back international trade even when income and production revived. National efforts to combat unemployment, to defend currency values, and to aid local producers led each country during the depression to cut down its imports from other countries, while its own exports were cut down by similar measures abroad. Local industries that could not justify themselves on grounds of long-run efficiency were encouraged to replace imports because it was thought that this would relieve unemployment or lessen the pressure on local income. Thus, a great wave of agricultural protectionism in the industrialised countries of Europe sharply reduced imports of wheat, pork, etc., from overseas in favour of a much more costly home product, while overseas countries, finding their agricultural markets shrinking, sought to aid their own people by shutting out various kinds of manufactured goods in order to replace them by a more costly home product. The result was to stop movements of trade both ways and to lower the general efficiency of production. The quotas, higher tariffs, exchange controls, discriminatory trade bargains, and other discouragements to international commerce used in this process, and the ill-adapted industries requiring protection, did not pass away automatically with recovery.

Past experience clearly supports the view that economic development designed to increase the efficiency of production and to raise income and living standards will make for expanding markets and more trade. Of course, the economic tendency for imports to rise when income rises may be checked by political decisions. A desire for strategic self-sufficiency in anticipation of war, or the influence of internal pressure groups seeking to shield themselves from competition, or the desperation born of a general depression in industry and trade, can create strong counteracting tendencies. World development programmes on a large scale would themselves, however, help in some measure to prevent these negative influences from becoming strong enough to dominate the situation.

Shifts in the Character of Trade

As the total external trade of developing areas moves upward with mounting production and income, the *character* of that trade is likely to undergo considerable change. Corresponding shifts in the trade of established industrial areas become necessary, together with changes in their types of production.

In general, the experience has been that as countries develop and modernise their production they import more of all the major categories of commodities. They import more foodstuffs, more raw materials, more semi-manufactured and more finished goods, but, as might be expected, imports of raw materials and partly finished goods increase in *relative* importance. On the export side, experience shows that a country moving from a less developed to a more advanced stage of economic development begins to sell more finished and partly finished goods abroad. It is likely, however, also to increase its exports of crude materials and foodstuffs, though these may decrease in *relative* importance.[1]

It is a common error to suppose that economic development which increases the ability of a country to carry on modern industrial processes must ordinarily lead to a decrease in its imports of manufactured goods. By and large, just the opposite has been true in the past. Of course, the new and more diversified production of the developing country does replace some goods formerly imported. But rising incomes make it possible to buy other kinds of goods abroad that the people of the developing country have not been able to enjoy before. Also, imports of equipment and repairs for industrial plants and utilities are likely to increase, while the development of a new type of international specialisation by stages of manufacture is likely to show itself in rising imports of partly finished "semi-manufactures".

[1] This refers to countries in a relatively early stage of economic development. At a later stage the proportions among the various classes of imports and exports may shift in other and more complex ways. See below on recent increases in British imports of manufactures and on exchanges of semi-manufactures. Obviously, all countries at once cannot increase the ratio of raw to manufactured products on the import side and decrease it on the export side.

Chart 15. United States Exports and Imports by Economic Classes, Annual Averages for Selected Periods from 1851 to 1940[1]

(*Millions of Current Dollars and Percentage Distribution*)

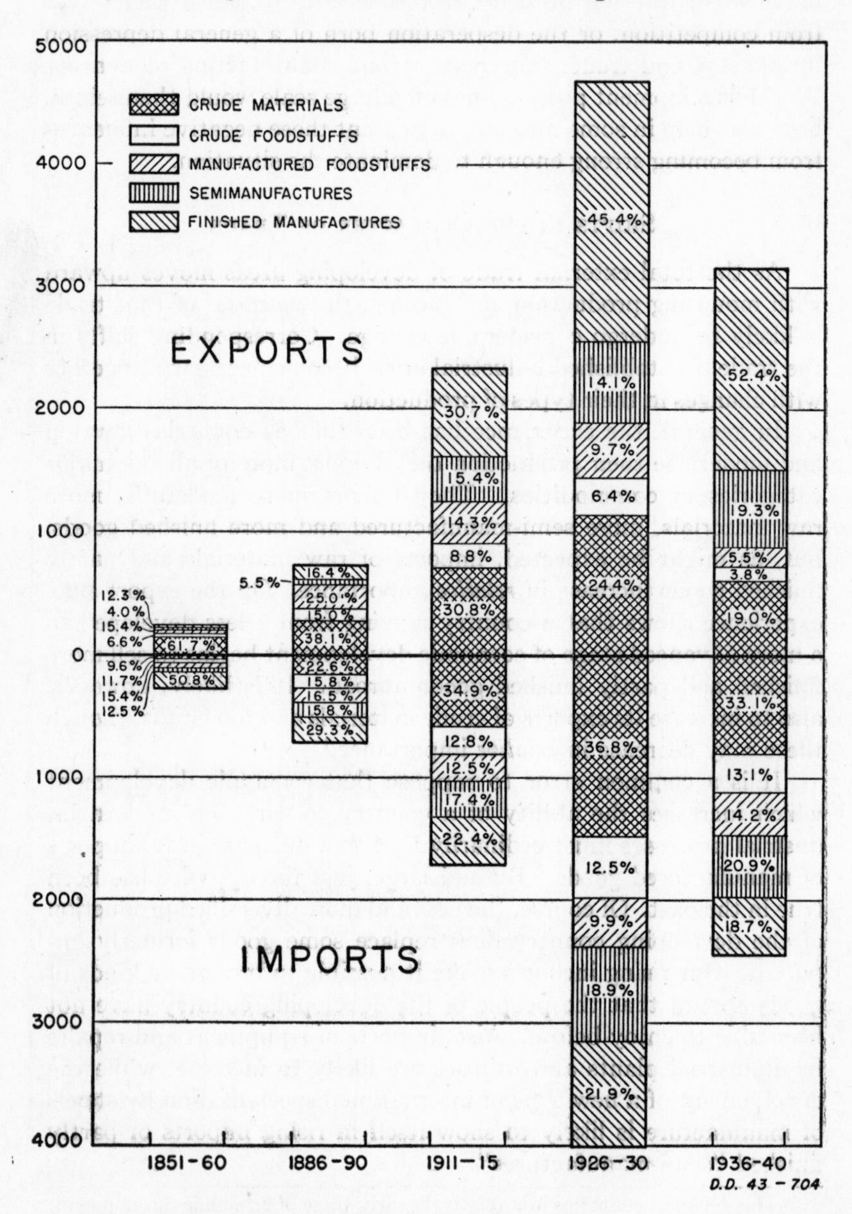

[1] U. S. Department of Commerce: *Statistical Abstract of the United States, 1941* (Washington, 1942), pp. 534 and 535.

Chart 16. Exports and Imports of Japan by Economic Classes, Annual Averages for Selected Periods from 1893 to 1938[1]

(*Millions of Current Yen and Percentage Distribution*)

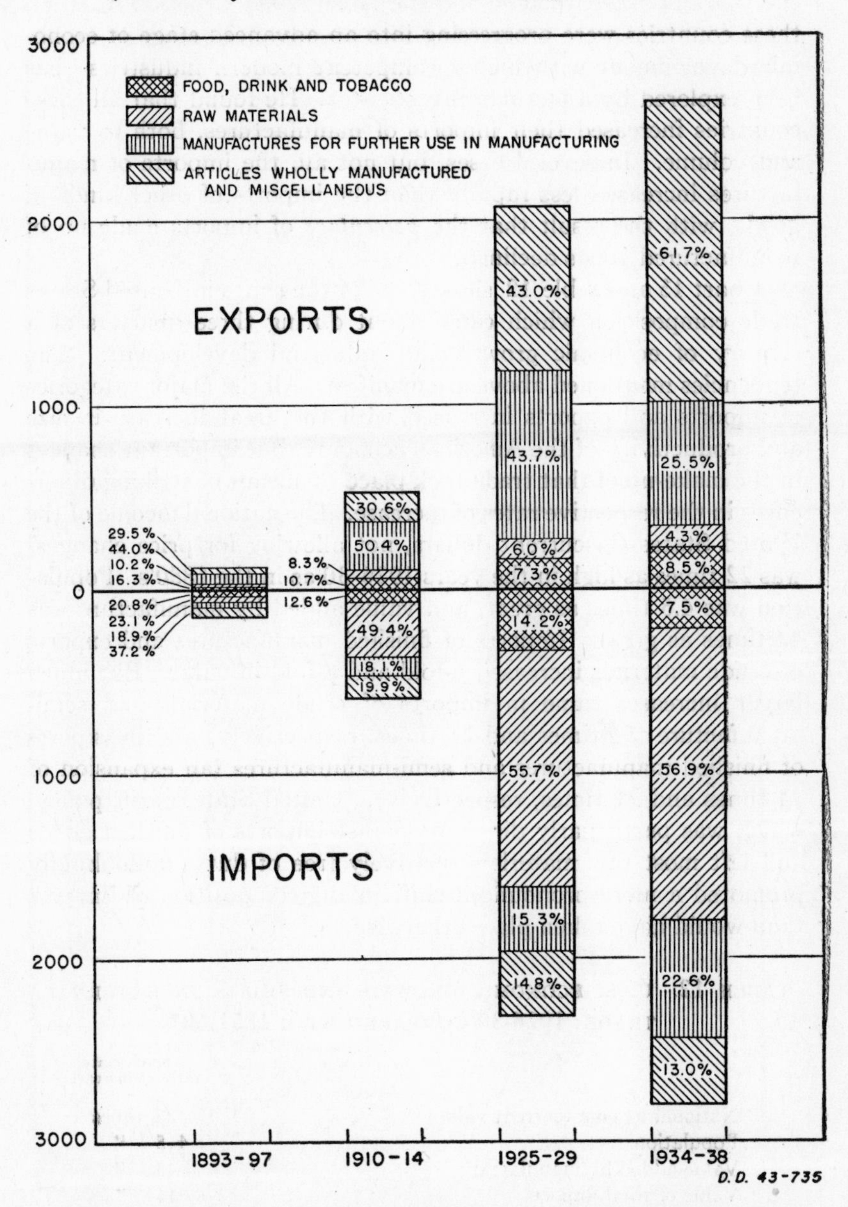

[1] TANZAN ISHIBASHI (ed.): *The Foreign Trade of Japan: A Statistical Survey* (Tokyo: issued by the *Oriental Economist,* 1935), pp. 450 and 451, and DEPARTMENT OF FINANCE: *The Annual Return of the Foreign Trade of Japan,* (Tokyo), various years.

The trend in imports of manufactured goods into the United States, Germany, France, Great Britain, and Switzerland during the years between 1860-80 and the late 1920's—a period when all these countries were progressing into an advanced stage of economic development with highly competent modern industries—has been explored by a German investigator. He found that all these countries increased their imports of manufactures, both in value and volume. In several cases, but not all, the imports of manufactures increased less rapidly than the imports of other kinds of goods, with the result that the *percentage* of imports made up of manufactured goods declined.[1]

Chart 15 and table 12 show some of the shifts in United States trade composition which came about during three quarters of a century of economic growth and industrial development. The tendencies mentioned above are manifest. All the major categories of imports and exports increased with the great increase in size and productivity of the American economy, but important changes in the make-up of that trade took place by means of striking differences in the respective rates of increase. The national income of the United States (in current dollars, not allowing for price changes) was 22 times as high in the years 1926-30 as in the 1850's. Population was 4.5 times as great, and value added by manufacture was 43 times as great. Imports of finished manufactures and exports of crude materials increased 6-fold and 8-fold in value. But much larger increases came in imports of crude materials and semi-manufacture (54 times and 21 times, respectively) and in exports of finished manufactures and semi-manufactures (an expansion of 74 times and 71 times, respectively). United States tariff policy, which was particularly directed against imports of finished goods and left most raw materials relatively free of duty, undoubtedly promoted a more pronounced shift in the composition of imports than would have taken place otherwise.

TABLE 12. U. S. ECONOMIC GROWTH AND SHIFTS IN EXTERNAL TRADE: 1926-30 COMPARED WITH 1851-60*

	Approximate ratio of increase
National income (current value)	22 times
Population	4.5 "
Value added by manufacture	43 "
Value of total imports	14 "
Value of imports of crude materials	54 "
Value of imports of semi-manufactures	21 "

[1] Walther HOFFMANN: *Stadien und Typen der Industrialisierung* (Jena, Fischer, 1931), pp. 168-70.

	Approximate ratio of increase
Value of imports of crude foodstuffs	15 times
Value of imports of manufactured foodstuffs	9 "
Value of imports of finished manufactures	6 "
Value of total exports	19 "
Value of exports of finished manufactures	74 "
Value of exports of semi-manufactures	71 "
Value of exports of crude foodstuffs	20 "
Value of exports of manufactured foodstuffs	13 "
Value of exports of crude materials	8 "

* Calculated from data as to income in Robert F. MARTIN, *op. cit.*, pp. 6-7 (average of two years, 1849 and 1859, compared with average of 1926-30 inclusive) and as to other quantities in U. S. DEPARTMENT OF COMMERCE: *Statistical Abstract of the United States, 1941.*

The changes in the make-up of Japanese imports and exports which accompanied the rapid growth of Japan's economy in the four decades from the 1890's to the 1930's are shown in chart 16 and table 13. Imports of manufactured articles and of food, drink and tobacco increased in total value, but not nearly in proportion to other increases. In fact, if one makes allowance for price changes, the trend of physical volume of imported manufactures is doubtful, varying at different times during the period. Imports of raw materials and of partly manufactured goods increased most noticeably. On the export side, the value of raw materials increased, but relatively slowly, while the great increases were registered by exports of manufactured and partly manufactured articles. Some writers have suggested that characteristics of the Japanese social system may have prevented the rise of production and income from exer-

TABLE 13. JAPANESE ECONOMIC GROWTH AND SHIFTS IN EXTERNAL TRADE: 1934-38 COMPARED WITH 1893-97[1]

	Approximate ratios of increase
National income (current value)	52 times
Population	1.7 "
Value of total imports	19 "
Value of imports of raw materials	47 "
Value of imports of manufactures for further use in manufacturing	23 "
Value of imports of food, drink, and tobacco	6.9 "
Value of imports of articles wholly manufactured and miscellaneous	6.7 "
Value of total exports	21 "
Value of exports of articles wholly manufactured and miscellaneous	45 "
Value of exports of manufactures for further use in manufacture	12 "
Value of exports of food, drink, and tobacco	11 "
Value of exports of raw materials	9 "

[1] For source of income and trade data, see footnotes to charts 9 and 16. Population data: *Japan Year Book*, 1940-41, p. 36 and E. B. SCHUMPETER (ed.): *The Industrialization of Japan and Manchukuo, 1930-1940* (New York, Macmillan, 1940), p. 72.

cising its full potential effect on the living standards of the masses of the people, and that this may have prevented expansion of certain kinds of finished imports for popular consumption.

Japan's development illustrates a complex trade specialisation, both by commodities within the broad groups and geographically. Japan imported manufactured goods from western Europe and the United States, particularly the higher-priced, better-quality goods, including capital goods. It also imported large quantities of raw materials, such as cotton. Japan exported manufactured goods to Asia and Africa, especially lower-priced goods of the sorts demanded by the low-income groups of those areas. Had there been a high, sustained level of effective demand in the world—that is, had the rate of investment not fallen so disastrously at the end of the 1920's—this particular characteristic of Japanese trade development might have widened the market for western goods, directly through new Japanese demands, and also indirectly through the tapping of new African and Asiatic and other low-income demands.

It is instructive to notice some aspects of the trade between Japan and the United States as both countries developed economically and as Japan rose from an industrially backward to an industrially advanced country. The trade increased enormously in total value, American exports to Japan rising from 8 million dollars in 1873 to 58 million in 1913 and to 289 million in 1937, while American imports from Japan rose from 9 million dollars in 1873 to 92 million in 1913 and to 204 million in 1937.[1] The principal item of trade in each direction was, interestingly enough, a raw material—raw silk from Japan to the United States, raw cotton from the United States to Japan. The fact that silk came to the United States in an unmanufactured state despite Japan's industrial advancement may have been due, however, to the American tariff. Cotton exports from the United States to Japan rose steadily throughout most of the period, reflecting the growth of the Japanese textile industry. Wheat exports from the United States to Japan rose and later fell. Exports of machinery rose steadily until 1924, after which they fell rapidly as Japan became able to build more of her own machinery. Automobile exports, however, which represented a more complex, mass-produced product in which America was preeminent, continued to increase in value. Thus, as Japanese production and income grew, some American products were more in demand than before and others less, although the total of Japanese purchases from the United States increased very greatly.

[1] Annual Report of the Bureau of Statistics, U. S. Treasury Dept., on the Commerce and Navigation of the United States, for fiscal year ending 30 June 1873; U.S. DEPARTMENT OF COMMERCE: *Foreign Commerce and Navigation of the United States*, 1940 and 1941.

The new competition in the textile industry offered by Japan's combination of abundant and cheap labour with rapidly modernising production methods raised severe problems of adjustment for

Chart 17. United Kingdom's Exports to and Imports from the United States, Annual Averages for Selected Periods from 1854 to 1929[1]

(*Millions of Current Pounds and Percentage Distribution*)

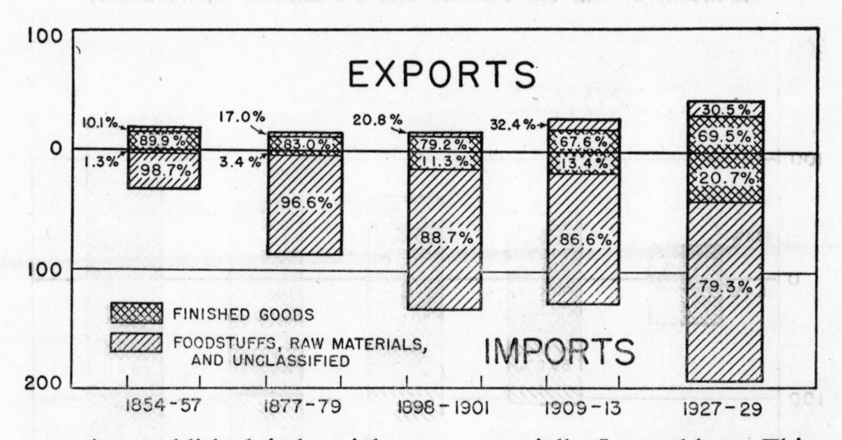

certain established industrial areas, especially Lancashire. This long established centre of the British textile industry proved to be particularly vulnerable and relatively inflexible. Similar problems were raised, but not by Japanese competition, in the textile towns of New England as new textile centres developed in the southern part of the United States. Both in Lancashire and in New England the problem was the same, though one was caused by a foreign and the other by a domestic development. In both cases industrial adaptation leading to more diversified production became necessary. Also, in both cases the problem of finding new lines of production in which to expand was greatly intensified by the onset of the world depression of the 'thirties. Here is the most difficult economic problem created by the general process of economic development. How can workers and capital shift out of certain established lines of production that are not fully able to stand up to the new competition, and how can new lines of production be found that will take advantage of the expanding opportunities and bigger markets that should be offered by the same process of development? This will be discussed in later chapters.

The make-up of Canada's external trade in 1929 and 1900 is compared in table 14. Here again, economic development bring-

[1] SCHLOTE, *op. cit.*, tables 33 and 34, pp. 91 and 92.

ing an increase in production and income was accompanied by expansion in all the principal classes of imports and exports. While the value of Canada's own manufacturing production was increas-

Chart 18. United Kingdom's Exports to and Imports from "Industrial Europe", Annual Averages for Selected Periods from 1854 to 1929[1]

(*Millions of Current Pounds and Percentage Distribution*)

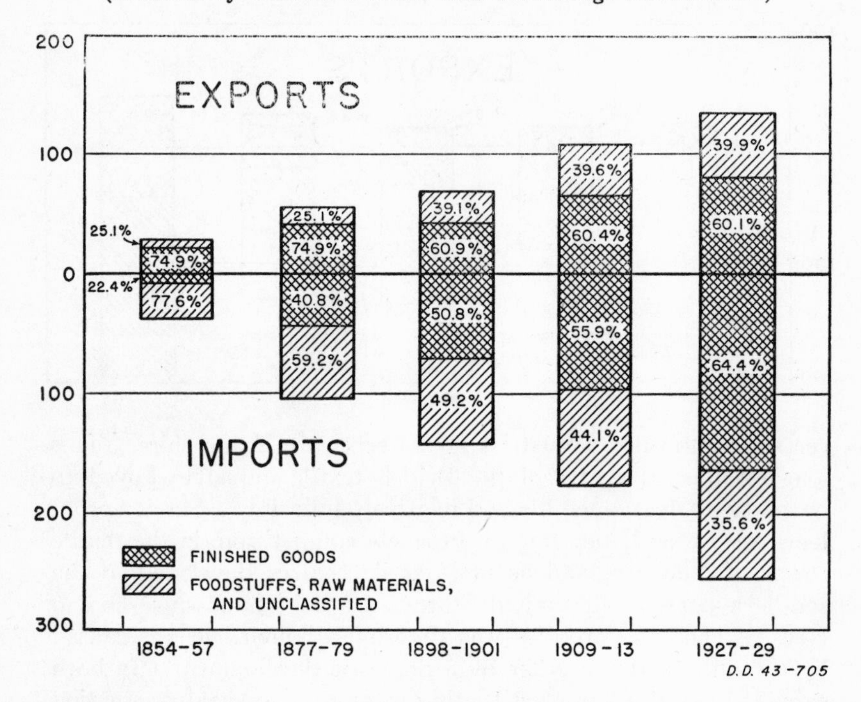

ing about 8 times, the value of its imports of fully and chiefly manufactured goods was also rising nearly 8 times, in this case faster than the value of other classes of imports. The rise was accounted for by a wide variety of commodities, no individual item or small group being chiefly responsible. Among the manufactured articles importation of which increased markedly were machinery, automobiles and parts, petroleum products, farm implements, electrical apparatus, alcoholic beverages, books, cotton and woollen goods, and paper.

In a recent pamphlet on *Industrialization and Trade*, Mr. A. J.

[1] SCHLOTE, *op. cit.*, tables 33 and 34, pp. 91 and 92. "Industrial Europe" as used in this chart comprises the following countries: Germany, Netherlands, Belgium-Luxemburg, Switzerland, Austria-Hungary and the Succession States, France and Italy.

Chart 19. United Kingdom's Exports and Imports of Finished Goods in Trade with "Industrial Europe", the United States, and All Other Countries, Annual Averages for Selected Periods from 1854 to 1929[1]

(*Millions of Current Pounds and Percentage Distribution*)

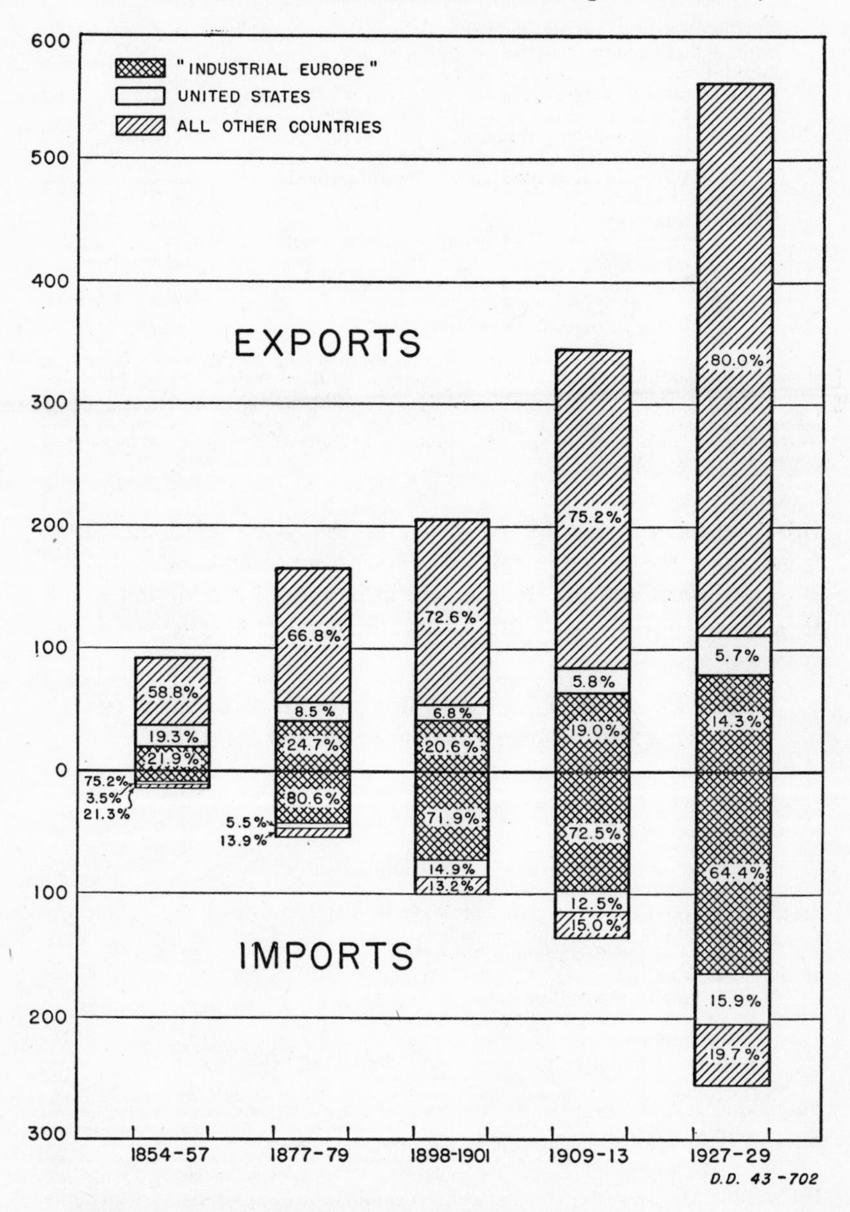

[1] Schlote, *op. cit.*, tables 33 and 34, pp. 91 and 92.

TABLE 14. CANADIAN ECONOMIC GROWTH AND SHIFTS IN EXTERNAL TRADE: 1929 COMPARED WITH 1900*

	Approximate ratio of increase
National income (current value)	at least 4.5 times
Gross value of manufactured products	8.1 "
Population	1.9 "
Value of total imports	7.2 "
Value of imported fully- and chiefly-manufactured goods	7.8 "
Value of imported raw materials	6.7 "
Value of imported partly-manufactured goods	5.2 "
Value of total exports	6.6 "
Value of exported partly-manufactured goods	7.1 "
Value of exported fully- and chiefly-manufactured goods	6.9 "
Value of exported raw materials	6.1 "

* National income figure for 1903 (1900 not available) from Sir Robert GIFFEN: "The Wealth of the Empire, and How it Should be Used", in *Journal of the Royal Statistical Society*, September 1903, p. 583. Figure for 1929 from *Monthly Review of the Bank of Nova Scotia*, Sept. 1940. Gross value of manufactured products from *Canada Year Book. 1940*, p. 397. Figure for 1900 not exactly comparable with that for 1929. Population from *Canada Year Book 1940*, p. 103. Values of imports and exports from *Trade of Canada Fiscal Year Ended March 31, 1939*, p. 856. Data for fiscal years ended 31 Mar. 1900 and 1930.

Brown presents data for Australia in the years 1915 and 1938 from which the ratios in table 15 have been calculated.[1] In Mr. Brown's interpretation, the increase in industrial production which took place (outside agriculture and the food industries) did not

TABLE 15. AUSTRALIAN ECONOMIC GROWTH AND SHIFTS IN EXTERNAL TRADE: 1938 COMPARED WITH 1915*

	Approximate ratios of increase
National income produced (current values)	2.6 times
Population	1.4 "
Value of industrial production (excluding double counting and the food industries)	3.2 "
Home-produced materials, etc., included	3.0 "
Imported materials, etc., included	2.8 "
Imports of finished manufactures	1.5 "
of which:	
plant and equipment	2.2 "
consumable goods	1.2 "
Total imports	2.2 "
Exports of manufactures and semi-manufactures	3.0 "

* Calculated from data (except population) given in A. J. BROWN: *Industrialization and Trade: The Changing World Pattern and the Position of Britain* (London, Royal Institute of International Affairs, Oxford University Press, 1943), p. 50. Population source: Australia, Commonwealth Bureau of Census and Statistics, Bulletin No. 58: *Demography 1940*.

[1] See full citation in footnote to table 15. The sources of the data are not indicated in Mr. Brown's pamphlet and an examination of Australian official publications has not revealed them. If the figures were available, it might be better to take some other base year than 1915, for imports in that year appear to have been smaller than usual owing to the effects of the First World War.

contribute greatly to the growth of Australian national income. In other words, "industrialisation", as distinguished from broad economic development of the lines of production offering the most promise in terms of income, was being pushed further in Australia than the income results justified. "Output per head (measured at world prices) in the non-food industries was almost certainly not greater than that in agriculture . . . " The shift towards non-agricultural, non-food industries therefore had a more pronounced tendency to replace certain kinds of imports than to increase income. This helps to explain why Australian importation of manufactured consumable goods rose only slightly in value and probably fell in volume over the period concerned. The author continues:

In considering what effect industrialisation is likely to have on the external trade of any given country, therefore, the first question one must ask is whether—and how much—industrialisation is likely to affect national income. It is not merely a matter of the relative *per capita* productivities of industrial and non-industrial occupations in the country . . . in the longer run one is bound also to take into account the very great effects which it may have, in favourable cases, on the possible expansion of population and the possibility, with total population at any given level, of maintaining and improving the average *per capita* output in primary occupations. One must ask, in fact, whether the country is an Australia or an Argentina, in which industrialisation would not greatly expand income, or, on the other hand, an overpopulated Japan, or an industrially well-endowed United States where it would strongly promote the expansion of purchasing-power. One has then to ask whether the country concerned is one of limited resources which, like Japan, would have to draw the materials for its new industry largely from abroad, or a United States or a Soviet Union with a great variety of resources to draw upon at home. Finally, one must ask how far the products of the new industry are likely, on the one hand, to find their way abroad, like Japanese cotton goods and American cars, or, on the other, to be designed purely to supply the home market with goods which would otherwise be imported.[1]

BRITAIN'S TRADE ADJUSTMENTS AS OTHER COUNTRIES DEVELOPED

Great Britain, which led other countries in the adoption of modern industrial methods, has seen many nations take over and even improve the production techniques in which it pioneered. Its trade experience has interest, therefore, as one instance of trade shifts in an older industrial country under the influence of the changing character of trade with developing regions.

Certain aspects of the changing geographical distribution of British trade between 1854 and 1929 are shown in charts 17, 18 and 19. Direct exports to the United States, as chart 17 shows, were not large, and they declined in the latter part of the nineteenth century, although they somewhat more than doubled in

[1] A. J. BROWN, *op. cit.*, pp. 53-4.

value over the period as a whole. The import demands created by American economic development were not, for the most part, demands for the types of products exported by Great Britain, and the American tariff further held direct exports from Great Britain to the United States in check. The great increase in United States purchases of industrial raw materials and other supplies from other parts of the world was probably more important for Great Britain. This provided purchasing power that indirectly, through multilateral trade channels, helped to enlarge the market for British goods in the "all other countries" of chart 19.

The movement of goods in the other direction—that is, British imports from the United States—was larger. It increased 6-fold between 1854-57 and 1927-29. Economic development in the United States was instrumental in raising British real income by providing large quantities of cheap foodstuffs (wheat, etc.) and raw materials (cotton, etc.). The development of manufacturing industry in the United States led to a more complex type of specialisation in which finished goods moved from the United States to Great Britain as well as in the opposite direction: British imports of finished goods rose from less than 2 per cent. of total imports from the United States to more than 20 per cent.

British exports to the countries of western Europe that were advancing rapidly towards a high level of economic and industrial development increased 5-fold over the period, from 27 million pounds in 1854-57 to 133 million in 1927-29. These European countries bought larger amounts of finished goods from Great Britain as they developed (80 million pounds sterling against 20 million), but they expanded their purchases of British raw materials and foodstuffs relatively more (53 million pounds sterling against 7 million). In the other direction, there was an increase in the sale of finished goods and of raw materials and foodstuffs by the same European countries to Great Britain, but the expansion was greatest in finished goods.

Chart 19 shows that exports of British finished goods, although they increased over the period to the countries of "Industrial Europe" and to the United States, increased much more rapidly to the other countries of the world. These other countries were also undergoing economic expansion, stimulated to a considerable extent by the growing demands of the industrialising countries, but their development was more in the field of primary production than in manufacturing. Thus, their rising income offered a more direct market for the traditional British manufactured goods, with less tendency for new local production to be substituted for imports either by straight competition or by protectionist measures in favour of local undertakings. But the economic development of these "all other countries" was not remaining exclusively in the field of agrarian or primary production, as is shown by the fact that at the end of the period they were providing Great Britain with finished goods to the value of 50 million pounds sterling, as compared with 2.4 million pounds at the beginning.

This shifting pattern of British trade in the 75 years before the world depression illustrates the importance of flexible *multilateral* adjustment in the flow of trade as economic developments of different types and at different rates of speed take place in various regions.

In its commodity make-up, British export trade in the 1860's was largely concentrated on textiles and apparel (more than 60 per cent. in value). Much of the remainder consisted of other consumable goods. During the next twenty years exports grew rapidly without changing much in composition. The outside world was experiencing a great development of agricultural production and of railways, but industrialisation, in the sense of the development of manufacturing, had not progressed very far. From the late 'seventies or early 'eighties, however, the pressure of industrialisation elsewhere was more apparent.

> Textile and apparel exports had continued to increase absolutely, but, by about 1910, were only some three eighths of the total; coal, chemical, machinery, and vehicle exports had made enormous advances; but exports of manufactured and semi-manufactured iron and steel had begun to feel the keen competition of Germany and the United States, and, though still increasing in volume, had fallen relatively to total exports.[1]

The shifts in the composition of British export trade became even more clear after the First World War. Between 1910 and 1938 the proportion of textiles and apparel fell from about 38 per cent. to less than 25 per cent. The proportions of iron and steel manufactures and semi-manufactures and of coal declined somewhat, under the influence of new competition and technological changes such as the rise of fuel oil. "On the other hand, exports of machinery, electrical appliances, vehicles, and (to a smaller extent) chemicals increased in relation to total exports—altogether, they rose from less than 14 per cent. of them in 1910 to not far short of 40 per cent. in 1938.[2]

These shifts are of the sort that one would expect on the part of an advanced industrial country in a world where other countries are developing rapidly and where the arts of manufacture, especially the older and simpler forms of manufacture, are being widely

[1] A. J. Brown, *op. cit.*, pp. 59-60.
[2] *Ibid.*

mastered. Some evidence presented later in this chapter suggests that it would have been better for Great Britain if there had been an even more rapid shift towards the newer lines of production and export. Indeed, there was considerable urging from expert opinion in Great Britain itself in endeavours to hasten the adaptive process. Thus, a leading economist, Alfred Marshall, said in 1908:

England will not be able to hold her own against other nations by the mere sedulous practice of familiar processes. These are being reduced to such mechanical routine by her own, and still more by American, ingenuity that an Englishman's labour in them will not continue long to count for very much more than that of an equally energetic man of a more backward race. Of course, the Englishman has access to relatively larger and cheaper stores of capital than any one else. But his advantage in this respect has diminished, is diminishing, and must continue to diminish; and it is not to be reckoned on as a very important element in the future. England's place among the nations in the future must depend on the extent to which she retains industrial leadership. She cannot be the leader, but she may be a leader.

The economic significance of industrial leadership generally is most clearly illustrated just now by the leadership which France, or rather Paris, has in many commodities which are on the border-line between art and luxury. New Parisian goods are sold at high prices in London and Berlin for a short time, and then good imitations of them are made in large quantities and sold at relatively low prices. But by that time Paris, which had earned high wages and profits by making them to sell at scarcity prices, is already at work on other things which will soon be imitated in a like way. Sixty years ago England had this leadership in most branches of industry. The finished commodities and, still more, the implements of production, to which her manufacturers were giving their chief attention in any one year, were those which would be occupying the attention of the more progressive of western nations two or three years later, and of the rest from five to twenty years later. It was inevitable that she should cede much of that leadership. . . It was not inevitable that she should lose so much of it as she has done.[1]

The Committee on Industry and Trade (Balfour Committee), which reported in 1929 after an extensive investigation into "the means of restoring the competitive power of British industry and trade without impairing the standard of living", laid great stress on what it called "industrial mobility". There was urgent need, the Committee said, among employers and managers and also among workers and their organisations,

for a more vivid and intelligent appreciation of the importance of "mobility" using the term in its widest sense, to cover, not only actual transference whether industrial or geographical, but also a mental attitude towards changes of environment, and the power and will to react promptly to such changes.[2]

---

[1] *Fiscal Policy of International Trade*, House of Commons Paper, No. 321, 1908, quoted in H. FRANKEL: "The Industrialization of Agricultural Countries", in *Economic Journal*, June-Sept. 1943, p. 193.

[2] *Final Report of the Committee on Industry and Trade*, Cmd. 3282, 1929, pp. 297, 300.

Mr. Alexander Loveday, in a discussion of "Britain and World Trade" argued that rigidity of production severely handicapped Great Britain in trade competition during the 1920's. As an economic mission to Argentina had observed, Great Britain had retained its position in some departments—particularly old established businesses—but in new departments of trade, such as aviation, road construction, and motor transport, it had been outdistanced. The same was true in such fields as radio and photographic apparatus, scientific instruments and agricultural machinery.

> It is on these new and growing industries that the prosperity of highly industrialised countries must ultimately depend. It is inevitable in a world that is growing richer . . . that the industries satisfying secondary needs—the smaller and the new industries—those which exist today and those which will come into being tomorrow—should constitute an ever-increasing proportion. The future lies with the countries whose whole economic organisation is the most mobile, with those which have the imagination to foresee future needs, the courage to scrap obsolete plant and the skill to adopt and adapt new inventions.[1]

Dynamic "New" Goods and the Position of the United States

Some of the older kinds of manufactured goods which the advanced industrial countries have been accustomed to offering on the international market will meet a relatively static or even falling demand as development proceeds elsewhere, while other kinds of manufactured goods will find a growing demand and expanding outlets. If an advanced trading country produces these more dynamic goods or can turn to their production its export prospects will be relatively favourable. If it insists, out of necessity or for lack of ingenuity and initiative, on trying to sell the old types of goods despite the new circumstances its trade is likely to fall behind.

An elaborate investigation of international trade trends in manufactured goods made by the German *Enquête-Ausschuss* provides the material for table 16.[2] This shows that the value of international trade in certain products, such as laces and embroideries and pianos, increased very little or actually decreased between 1913 and 1929. At the same time, international trade in cotton yarn, wearing apparel, toys, cotton cloth, and shoes increased moderately. Very high rates of increase (over 150 per cent.) were shown by such products as automobiles and tires, rayon, electrical communication devices, and phonographs.

---

[1] A. Loveday: *Britain and World Trade* (London, Longmans, Green, 1931), pp. 170, 177.

[2] The co-operation of Dr. Gerhard Colm in connection with the materials that follow is gratefully acknowledged.

The exports of the advanced industrial countries differed considerably with respect to the proportions of rapidly increasing and slowly increasing goods that they contained. As table 17 shows, the bulk of British manufactured exports consisted in 1929 of merchandise which had experienced a relatively unfavourable development in international trade. 42 per cent. were of types which had increased by less than 75 per cent. since 1913, and only some 4 per cent. of British manufactured exports were in the class which had experienced an expansion of more than 150 per cent. The bulk of the manufactured exports of the United States, on the contrary, consisted of products having a most favourable development in international trade. 29 per cent. were in the class which had expanded more than 150 per cent. since 1913, and 39 per cent. more were types that had experienced at least a 75 per cent. increase in international trade. United States producers had a weaker comparative position in the traditional products and a stronger position in the new products. The United States had a "dynamic" advantage while Great Britain had a "traditional" advantage. Germany, as the table shows, held a middle position.

TABLE 16. EXPANSION OF INTERNATIONAL TRADE IN SELECTED MANUFACTURED PRODUCTS, 1913 TO 1929, WITH PERCENTAGE OF TRADE IN EACH SUPPLIED BY GREAT BRITAIN, GERMANY AND UNITED STATES[1]

Products (examples)	Total trade of the most important countries		Nominal increase of total trade from 1913 to 1929 percentage	Percent. of exports of these products in 1929 from Great Britain, Germany, U.S.A.		
	1913	1929				
	in million marks					
Laces and embroideries	480	239	Under 25	14.6	19.4	0.5
Pianos	65	56		5.9	65.1	14.2
Cotton yarn	745	903	26–50	46.9	3.8	7.2
Wearing apparel	658	906		20.4	20.3	6.2
Toys	140	192		8.0	63.6	7.1
Cotton cloth	3039	4713	51–75	42.9	3.2	7.1
Shoes*	235	370		27.9	7.2	13.0
Steel mill products	2012	3268	76–100	25.8	29.0	20.0
Optical and mechanical products*	132	239		12.3	48.7	24.6
Machines	2537	5655	101-150	19.5	25.3	35.8
Clay and china products	250	523		22.1	60.3	4.6
Tires	200	510	More than 150	13.7	4.9	31.7
Automobiles	554	3109		5.7	2.2	74.0
Rayon and rayon yarn	35	390		13.9	21.4	0.0
Electrical communication devices	56	813		10.4	26.2	31.3
Phonographs*	25	125		40.8	10.2	26.9

[1] *Der deutsche Aussenhandel unter der Einwirkung Weltwirtschaftlicher Strukturwandlungen*, cited earlier, II, 157.

* 1928 figures used.

TABLE 17. MANUFACTURED EXPORTS OF GREAT BRITAIN, GERMANY AND UNITED STATES IN 1929: PERCENTAGE DISTRIBUTION INTO COMMODITY GROUPS BY RATE OF TRADE EXPANSION[1]

	Percentage in commodity groups expanding in international trade, 1913-1929, by:				
	Less than 25 per cent.	26-75 per cent.	76–150 per cent.	More than 150 per cent.	Total for which data available
Great Britain	2.1	**40.0**	*33.5*	4.3	79.9
Germany	2.1	*25.2*	**55.3**	4.5	87.1
U. S. A.	1.0	16.1	**38.8**	*28.6*	84.5

[1] *Der deutsche Aussenhandel unter der Einwirkung Weltwirtschaftlicher Strukturwandlungen*, II, 156.

The export of the traditional products, especially of textiles, was most subject to replacement by home production in newly developing countries which were acquiring manufacturing industries. In this way, for example, British textile exporters lost business in the Far East. The newer export products of the United States, on the other hand, were subject to great expansion in a time of prosperity in which many countries were undertaking to improve their equipment for production and communication (machinery, automobiles, communication devices, etc.) and in which rising standards of living were multiplying the demand for "high income" goods (radios, phonographs, etc.).

The United States was well equipped to supply the newer products because of its rapid strides in the quick application of a rapidly advancing technology to industry and because its highly developed mass market for such products encouraged methods of quantity production at low cost per unit. A "dynamic" advantage of this kind in any particular product, however, lapses as other countries gradually succeed in catching up, unless it is constantly renewed by new progress. The dynamic advantage consists in "keeping one jump ahead of the game".

The United States is probably in a better position than any other trading nation to gain expanded markets as a result of rapid economic development and rising income levels in the rest of the world. American goods are typically high-income goods and goods needed in developing production. Its "comparative advantage" in world trade is a dynamic advantage, which has its greatest scope when the trend of development and income is sharply upward. Thus, it so happens that the country whose governmental policies and business decisions will have the greatest influence in determining whether or not there will be vigorous economic development throughout the world is also the country whose export trade prospects are most favourably influenced by world economic development. This coincidence might play an important part in shaping the future.

Development and International Trade: More Complex Specialisation

What will happen as more and more countries acquire modern capital and modern techniques? Will international trade dry up? A substantial part of international trade in the past has involved, directly or through multilateral channels, countries that were in an agrarian or primary-producing stage of development. When these countries also become able to produce finished goods will there still be trade, and if so what kind?

Part of the answer is that if economic development goes forward along lines of "comparative advantage"—that is, if the lines of production developed in new countries are those which will yield them the greatest increase of income—not all countries will engage heavily in manufacturing, and certainly not all regions even within the highly industrialised countries will cease to specialise in agricultural or other non-manufacturing types of production.

Another part of the answer is that there is room for an enormous amount of specialisation of a complex but highly beneficial and profitable type *within* and *cutting across* broad classifications like "manufactures", "raw materials" and "foodstuffs". Unless there are such restrictive trade barriers as to prevent product specialisation, a lively trade in manufactured goods will move both ways, between different countries, as will raw materials, articles in various stages of semi-manufacture, and services. The higher income levels achieved by economic development will provide the increased volume and variety of demand to support this more sophisticated type of specialisation and exchange.

A certain type of superficial reasoning leads to the conclusion that two countries both possessed of modern capital and modern techniques have little or no economic reason to exchange products. If this were sound, then why should some of the most heavily used railway lines in the world connect Ohio, Illinois and New York? All are "industrial" States and might be regarded as "non-complementary" in climate and in similarity of industrial advancement. It is evident that in the absence of artificial barriers to trade they have developed a highly complex "complementarity"—not one based on anything so simple as exchange of raw materials and foodstuffs for manufactured goods, but of specialised individual plants which cater to the needs of more than one area, of specialisation by stage of manufacture with partly processed goods flowing from one region to the other, and of specialisation along agricultural and mineral lines indicated by variations in natural resources. The example of trade between Canada and the United States has been mentioned above.

Economic advancement means not only higher productivity and income and hence higher real purchasing power for domestic and foreign goods, but it also means a tremendous *diversification* of consumption and of commodities needed in the process of production. In every subdivision of those misleadingly simple categories "manufactured goods" and "raw materials" there are lumped together literally thousands of varieties and qualities to fit specialised purposes and tastes and different purses. Countries frequently import and export what appear to be the same commodities:

Germany, for example, exports watches and clocks of inferior quality and imports the same of higher quality (the total value of exports exceeding that of imports), whereas the United Kingdom exports watches of high quality and imports watches of low quality (the value of exports being negligible as compared with the value of imports). Germany exports electrical machinery of a superior quality and imports the same of an inferior quality, and the U. K. exports electrical machinery of an inferior quality and imports the same of a superior quality.[1]

The stage of technological advance will not be the same in all countries. For a long time the countries where modern industrial processes are new will be learning to produce manufactured products one by one; there will be many goods that they will not be equipped to produce efficiently. For example, while Japan was learning to produce cotton textiles it had to import the complicated high-quality machinery, and other countries will be in this same position with regard to many types of products consumed by a population with rising income. In many specialities they will still lag behind those countries which now are in the lead, if the latter continue to progress and to develop newer and better capital and consumption goods.[2]

The tendency to classify countries into "agricultural" or "raw material" countries on the one hand and "industrial" countries on the other will become still less justifiable than it is today as more and more countries acquire a modern and diversified pattern of production. Is the United States a raw material or an industrial country? As a matter of fact, it is both. Industrialisation has been a feature of its economic development and has brought a *relative* shift in production and exports away from raw materials and towards manufactured goods, but the production and export of

[1] H. Frankel, *op. cit.*, p. 195.
[2] *Ibid.*, p. 196.

raw materials has also expanded very greatly in *absolute* terms. Raw materials and foodstuffs, as well as manufactured goods, are produced by efficient, modern, "industrial" methods.

The economic tendency of the highly developed countries to import as well as to export many kinds of manufactured goods as other countries become able to supply manufactured specialities (a tendency likely to be resisted by protectionist influences) is illustrated in Great Britain. In the period 1921-25 Great Britain's imports of articles wholely or mainly manufactured were 45 per cent. as large as its exports of articles in the same classification. In 1926-30 this percentage was 59, in 1931-35 it was 62, and in 1936-39 it was 66. Thus, Great Britain "becomes also a heavy importer of manufactured goods, mainly of different quality and design, and the nineteenth century's conception of international trade based on the exchange of manufactured goods for raw materials and foodstuffs is retreating more and more into the background".[1]

A recent study divides the import and export trade of many countries into foodstuffs and raw materials on the one hand and manufactured products on the other.[2] A determination is then made of the extent to which, in the case of each country, manufactured exports may be considered to provide the external purchasing power with which to pay for manufactured imports. Imports of raw materials and foodstuffs are similarly matched against exports of raw materials and foodstuffs. To the extent that imports as a whole exceed exports as a whole, or *vice versa*, the difference represents an exchange of commodities against services and other so-called "invisible" items in the balance of payment. After these matchings have been made, the amounts which it has not been possible to offset are put down as representing the amount of trade that can only be accounted for by exchange of manufactures against raw materials and foodstuffs.

This method was applied to the trade of 47 countries, including approximately 92 per cent. of all international trade, for the years 1925 to 1937. The results for three selected years are shown in table 18. As the table shows, two thirds of all international trade can be accounted for in other ways than by the "traditional type of exchange" of manufactured goods against raw materials and foodstuffs.

The same study comments on the device sometimes adopted of dividing international trade into statistical categories based on

---

[1] H. Frankel, *op. cit.*, p. 201.

[2] Albert O. Hirschman: "The Commodity Structure of World Trade", in *Quarterly Journal of Economics*, Aug. 1943, pp. 565-595.

TABLE 18. ESTIMATED PERCENTAGE DISTRIBUTION OF INTERNATIONAL TRADE ACCORDING TO THE VARIOUS TYPES OF INTERCHANGE[1]

	1925	1929	1937
Commodities against services (so-called "invisible" items)	13.9	9.3	14.7
Foodstuffs and raw materials against foodstuffs and raw materials	39.6	38.3	34.8
Manufactures against manufactures	17.2	19.4	17.2
Manufactures against raw materials and foodstuffs	29.3	33.0	33.3
Total	100.0	100.0	100.0

[1] HIRSCHMAN, *op. cit.*, p. 574.

distinctions between "agricultural" countries and "industrial" countries. See, for example, table 19. Presentations of this sort can be very misleading if one assumes that trade between "industrial" countries is the same thing as exchange of manufactures against manufactures, or that trade between the two different types

TABLE 19. INTERNATIONAL TRADE DIVIDED INTO TRADE BETWEEN VARIOUS TYPES OF COUNTRIES[1]

	Foreign trade between:		
Year	"Agricultural" countries	"Industrial" countries	"Agricultural" and "industrial" countries
1913	10.7	29.2	58.8
1925	11.5	25.0	62.2
1929	12.0	23.9	62.6

[1] HIRSCHMAN, *op. cit.*, p. 577, quoting from Institut für Weltwirtschaft und Seeverkehr, "Die Aussenhandelsentwicklung und das Problem der deutschen Ausfuhrpolitik", in *Weltwirtschaftliches Archiv*, Vol. XXXVI (July 1932), p. 34.

of countries is merely another expression for exchange of foodstuffs and raw materials against manufactured goods. Historical comparisons on this basis are particularly dangerous if one forgets that "agricultural" and "industrial" countries are such in different degree in different years as economic development proceeds.

With respect to British foreign trade, the calculations according to the methods described above were carried back to 1854, with the interesting result shown in table 20. The outstanding fact, as the author points out, is the decrease of the "traditional type of exchange"—the exchange of manufactures against foodstuffs and

TABLE 20. PERCENTAGE DISTRIBUTION OF BRITISH FOREIGN TRADE ACCORDING TO THE VARIOUS TYPES OF INTERCHANGE*

	1854 -63	1864 -73	1874 -83	1884 -93	1894 -1903	1904 -13	1925 -29
Commodities against services (so-called "invisible" items)	14.2	12.1	20.3	18.2	23.9	15.1	23.1
Foodstuffs and raw materials against foodstuffs and raw materials	11.1	10.9	12.1	14.3	16.3	20.0	15.8
Manufactures against manufactures	8.8	13.2	17.2	20.1	25.3	22.7	25.7
Manufactures against raw materials and foodstuffs	65.9	63.8	50.4	47.4	34.5	42.2	35.4
Total per cent.	100.0	100.0	100.0	100.0	100.0	100.0	100.0
Total, mill. pounds	2,820	4,553	5,486	5,675	6,723	9,620	8,880

* HIRSCHMAN, *op. cit.*, p. 590, based on figures of SCHLOTE, *op. cit.*

raw materials—from two thirds of Britain's total foreign commerce to a proportion varying between one third and two fifths. This result was brought about by increases in all the other types of interchange.

The growing importance of exchange of manufactures against manufactures probably reflects in part the rise in income levels in Great Britain which let the British buy a larger variety and volume of manufactured goods from abroad at the same time that they were making and buying more at home. It also reflects the increasing importance of other countries, such as Germany and the United States, in the finished goods field. The increase in the exchange of raw materials and foodstuffs against raw materials and foodstuffs results mainly from the increase of coal exports and exports of processed foodstuffs such as canned goods and beverages. This trend, too, may be viewed as an effect upon Great Britain of the economic development of other countries, which brought increased demands for coal and foodstuffs.

Thus, British trade, under the influence of developments abroad and at home, became larger and very much more complex. "The statement that British trade consisted mainly in exports of manufactures against imports of foodstuffs and raw materials was still a valid generalisation for the period 1854-63. For the decade 1894-1903 (and for subsequent decades) it had become a distortion of the facts."[1]

[1] *Ibid.*, p. 591.

CONCLUSIONS

The analysis of this chapter leads to a number of important conclusions.

An increase in external trade ordinarily accompanies the rise in production and income brought about by economic development. However, this tendency may be offset or completely blocked by politico-economic factors hostile to a lively international exchange, such as the existence of political insecurity which demands that military considerations be dominant or a world depression which disintegrates the whole system of trade.

The *income* results of economic development are very important from the point of view of the trade prospects of advanced industrial countries. The more rapidly incomes rise in the developing countries, the easier are the necesary trade adjustments likely to be for the advanced countries and the more favourable the effects on them. Projects which merely enable a country to dispense with imports without raising the real income of the country tend to contract trade rather than to expand it and to have unfavourable effects on the trade prospects of the advanced industrial countries.

It follows that economic development in accordance with the broad principle of comparative advantage (which does not, of course, preclude establishment of industries that may have to be subsidised during an educative period, if they are likely to prove well adapted to the country in the long run) is the most favourable course both for the developing countries and the advanced countries.

A flexible, multilateral system of international trading relationships is an important favourable factor for the advanced countries in a developing world, for it makes possible the most rapid general rise of income (through more efficient and complex types of specialisation) and it offers more alternative opportunities for adjustment as shifts in the world market situation make adjustments necessary.

As the economically less developed areas "grow up" to modern industry the composition of world trade may be expected to undergo important changes (*relative* shifts within an increasing total). The importance of the simple "traditional type of exchange" of manufactured goods against foodstuffs and raw materials will continue to decline. Specialisation will become more complex, with each country importing some types of manufactured goods and exporting others, importing some types of raw materials and exporting others, and with a considerable increase in the trade in partly finished goods ("semi-manufactures") representing specialisation by stage of processing. Services (not included among the commodity imports and exports of the ordinary trade returns) will play a more important role in international exchange as the world grows more wealthy and as communication improves.

For some time, at least, the relative importance of capital goods as compared with directly consumable goods is likely to rise. The advanced industrial countries will export increasing quantities of some kinds of consumable goods, especially those involving higher technical skills, newer research, and those adapted to mass production for relatively high income groups such as are found in their own domestic markets. The cheaper and simpler manufactured goods adapted for use in poorer communities will increasingly be supplied by the newly developing countries. They can be imported by the relatively advanced countries with benefit, especially to their less wealthy inhabitants.

Some kinds of goods—the "new" dynamic products—will have much more favourable prospects than others as economic development proceeds and world income rises. The more adaptive the industries of the advanced industrial countries can be in shifting from relatively unpromising to relatively promising lines of production the less serious will be the problems raised by new competition and the more the advanced countries will gain from world economic development.

The chapters that follow will attempt to explore somewhat more fully the problems of adaptive adjustment to world economic development. For, if these conclusions are reasonably sound, the key to policies on the part of the advanced industrial countries which will serve them best in a world of progressive change is industrial adaptability.

CHAPTER IX

NEW OPPORTUNITIES AND NEW COMPETITION

Increased consumption and increased production go hand in hand in newly developing areas. From the point of view of the effects on advanced industrial countries this is a point that must never be lost from sight. Yet it is frequently overlooked in popular analyses of these effects.

As a long-run proposition, the amount by which consumption increases in countries undergoing economic development will be no more than and no less than the amounts by which production increases in the same countries. The amount which the Chinese, for example, are able to produce will determine, except for relatively temporary and minor qualifications, the amount that they will be able to eat, to wear and otherwise to use (including the use of goods and services for real investment in roads, factories, airports, schoolhouses, etc., which for present purposes may be regarded as consumption that is spread over a longer time). A community cannot indefinitely consume more than it produces. Conversely it cannot indefinitely produce more than it consumes (and invests).

Temporarily, of course, a country such as China can borrow abroad. Thus for a time the total of its consumption and investment can exceed its current production. Later, when payments of interest and principal on old loans begin to exceed the amount of new loans, the total production of goods and services in the country must run ahead of the total use of goods and services in the country. In the long run, however, the two are equal, if international gifts or unpaid international debts are excepted. But such gifts or defaulted debts are very unlikely to amount to more than a tiny fraction of the total production and consumption over a period of years.

Hence, the analysis of trade effects resulting from economic development of new areas may start with the proposition that over a long period the increased production of these areas will be balanced in value by their increased consumption, but that for the first few decades, while they are borrowing abroad for developmental purposes, their consumption (and investment) will run ahead of their production. Later, while they are repaying, the reverse will be true.

While this is true of *total* production and consumption, it is decidedly not true of the production and consumption of *specific kinds* of goods and services. The increased production of certain commodities will run far ahead of the increased consumption of those same commodities in the newly developing countries. In the case of other commodities, the opposite will be true. In other words, while the increase in imports of newly developing countries will be equal to the increase in their exports (with the qualifications noted above as to timing) this over-all result will represent a net increase in the *imports* of some goods and services and a net increase in the *exports* of others. Thus, the effect of economic development in new areas, from the point of view of older industrial areas, will be to cause important shifts in the world trading situation. For some goods and services the effect will be to increase world demand relative to world supply. For other goods and services the effect will be the opposite. Still other goods and services will not be affected importantly one way or the other, either because they are not readily exchanged between countries or because the effect of economic development on their demand just balances the effect on their supply.

Probable Nature of the Shift in Trade

It will be convenient in subsequent analysis to classify the goods and services of the world into three groups as follows:

A-products: items in increased demand as a result of economic development of new areas. The market for these items would be expanding. Their prices, unless counteracted by changes in production methods and costs, would tend to rise. Employment of labour and capital connected with their production would tend to increase.

These would, for the most part, be net import items for the newly developing countries as a group. However, some export items of these countries might become A-products by reason of greatly increased home consumption.

B-products: items in increased world supply as a result of the economic development of new areas. From the point of view of established industries in the advanced countries the market for these items would be a contracting one. Their prices would tend to fall, and the amount of employment available for labour and capital in the advanced countries by reason of their production would tend to decrease.

These would, in general, be the characteristic export products of the newly developing countries. However, some items might become B-products through a decrease in imports into these countries as some kinds of foreign goods are replaced partially or wholly by home production.

C-products: items not directly affected one way or the other on the world market by the economic development of new areas. The increased consumption of these items in the newly developing countries would be exactly balanced by the increased production in the same group of countries. This would necessarily be true, for example, of non-transportable goods and services, such as buildings and the heavier building materials, retail merchandising services, automobile repair services, local newspapers, etc. It might also happen to be true of other items in which industrial development would increase consumption and production by the same amounts.

As economic development proceeds, what specific kinds of commodities or groups of commodities will fall into these categories A, B, and C? What quantities will be involved? These are questions that cannot be answered exactly. In part, the answers will depend upon what policies are adopted by the governments of areas undergoing economic development, especially in the subsidy of certain lines of production. The answers will also depend in part upon the policies of other governments, including the amount and kind of developmental assistance that may be made available, and the extent to which tariffs, quotas, exchange controls and the like are applied to international trade. Nevertheless, it is possible to make some useful general statements.

First of all, it is worth pointing out that economic development is likely to increase domestic production and consumption even more than it increases international trade, even though the increase in international trade is large. The output of C-products will absorb a very important part of the increased productivity of a country with such vast and varied resources and needs as China, for example. Houses, bridges, streetcars, highways, telephone services, merchandising, amusements, hairdressing, automobile services and repairing, will represent a great enlargement of domestic production and consumption and a great increase in living standards, but will not enter directly into international trade.

In the case of A-products and B-products and the shifts in trade, it will be useful to look separately at two time-periods. The first may be called the stage of development, and the second the stage of approaching industrial maturity.

In the stage of development the total imports of the newly developing countries will presumably increase faster than their total exports, since they will be receiving loans of capital from abroad. This means that the increase in demand for A-products will bulk larger in the world market situation than the increase in supply of B-products, and the net effect will be an increase in the effective demand for the products of the advanced industrial countries. This is only another way of describing the stimulating effect on the economies of the advanced industrial countries which, as shown in Part I, may be expected to arise out of the investment phase of the process of economic development.

What will the A-products be? That is, on what internationally traded commodities will the newly developing countries spend their increased income? Conversely, what kinds of goods will they be able to produce in large quantities at low cost, and to offer for export during the early years of development (B-products)?

Let us look at the particular case of China. In so far as China's import and export trade follows the course of greatest immediate economic advantage (whether it is carried on by private traders or by government organisations) those goods will be imported which require in their production large proportions of the factors of production that are relatively scarcer in relation to demand, or less efficient in relation to their prices, in China than they are in other countries. Conversely, China will find its greatest immediate economic advantage in exporting those goods which embody large proportions of the factors of production that China has in greater relative abundance, after taking account of local demand, than have other countries. Factors of production that are very scarce in China include labour skilled in modern mechanical art, trained engineers, modern machinery and industrial equipment, the free capital funds necessary to install the more expensive types of equipment, and managers experienced in the more complex types of mass production. The same is true, in varying degrees, of other newly developing countries; it will be true of the newly developing countries as a group in comparison with the advanced industrial countries. On the other hand, China and most of the other newly developing countries will have a relative abundance of the less skilled grades of labour, much of which can be used, after brief training, in semi-skilled occupations. They will have a relative abundance of certain kinds of natural resources—the particular kinds of natural resources varying markedly, of course, from one area to another.

Now, in so far as China and the other newly developing countries follow the course of greatest economic advantage, they will buy abroad and import products embodying large proportions of the scarcer, and therefore more expensive, factors of production mentioned above. A-products will therefore typically include items requiring large amounts of complicated machinery or complicated technical processes in their manufacture, requiring large amounts of research and managerial skill and development, or very complex organisation and precision manufacture in order to obtain low cost through mass production. Concretely, this would mean such items as automobiles and trucks, dynamos, aircraft motors, machine tools, many types of chemical and medical equipment, refrigerators, air-conditioning apparatus, etc. The money with which the newly developing countries would pay for these A-goods (or, if the goods were bought initially on loans, the money to repay the loans), would be earned by selling abroad products that embody large proportions of their own characteristic resources. These B-products would typically include commodities that can be turned out by fairly routine methods with a relatively small amount of technical labour, relatively small amounts of capital per worker, and relatively simple organisation, or in which hand labour and craftmanship continue to play an important part, or in which labour having a fairly low level of education may be quickly trained to operate the machines that are required.

From this analysis it is obvious why the textile industry, except for types and grades of textiles dependent on highly technical processes or slowly acquired skills, has blossomed quickly in newly developing countries as a "home market" industry and, in Japan, as an export industry. One may expect that countries like China and India will quickly be able to expand production in textiles, the processing of foods, the early stages in the fabrication of their special local materials, and the simpler types of manufacture in general. They will find it less costly to import than to produce locally, on the other hand, the types of consumers' goods which require the most capital equipment and the most complex processes in their production, and they will find a great many items of capital equipment in the advanced countries which, in the early stage of development, they could hardly produce satisfactorily for themselves.

There are two reasons, however, why the newly developing countries are not likely to follow exactly the formula of immediate "comparative advantage" in determining what kinds of goods to import and what kinds to export. One is the reason of military defence. After the experiences of this war, the Chinese Government, for example, will undoubtedly want to encourage the development within China of some types of industries whose products might be obtained more cheaply abroad but which are essential for support of a modern fighting force. What would be an unduly rapid development of heavy industry from the point of view of immediate economic cost and gain may be determined upon as a necessary defence programme. The Soviet Union plunged directly into a heavy industry programme and built modern metallurgical, electrical and other installations capable of producing tanks and planes. This was done at great expense, measured in terms of the immediate sacrifice of consumer goods and retardation of the progress in living standards that could otherwise have been made. But who will say, in the light of experience, that the decision was not the right one? The extent to which China and other newly developing countries may feel it necessary to build high-cost, capital-consuming industries adapted to military production will depend in part upon the type of security system that emerges out of the war. In any case, the military preparedness viewpoint will certainly influence development programmes for some time to come.

The second reason why the newly developing countries are not likely to confine their development programmes to those lines in which they would find the greatest immediate economic gain relates to the process of industrial education. At present, technical skill, industrial organising ability and engineering talent are scarce in these countries. But this is a situation which can be changed. One of the ways to change it is to plunge rapidly into the production of modern industrial goods of many kinds, even though that production may be so expensive at first that it would be cheaper to purchase the goods abroad. At a certain cost in immediate economic welfare, in short, the process of industrial education may be speeded up. There may be a long-run gain from subsidising, over the initial period of learning by trial and error, many industries that on the basis of current cost comparisons would not be immediately advantageous.

This, of course, is an application of the familiar "infant industry" argument for State subsidies or protection. The validity of this argument, when it is honestly and correctly applied, has generally been admitted, even by ardent free traders. The problem is to select the industries to be subsidised so as to bring the maximum benefit in industrial education with the minimum of cost and sacrifice to the people of the newly developing countries and also to the people of the advanced industrial countries—for the latter have a stake in the matter. Certainly it would be a mistake for the newly developing countries to erect protective tariff barriers indiscriminately on the grounds of fostering "infant industries", just as it would be a mistake to try to prevent the fostering of industries that genuinely possess the characteristics of "infants"—that is, the capacity to grow up to self-sustaining maturity. There are also strong reasons for thinking that the traditional method of the protective tariff is much inferior to other methods of subsidy for fostering genuine infant industries. More will be said on this in a later chapter. In any case, a policy of educative subsidies, whether by tariff protection or by other and better ways, will limit the working of immediate "comparative advantage" in determining the kind of industries that will develop, and consequently the kinds of commodities that will fall into the A-products and B-products classifications.

In the second stage reached by newly developing countries, designated above as the stage of approaching industrial maturity, they will engage increasingly in the more technical and highly capitalised lines of production. If industrialisation really "takes" in such countries as China or India or Latin America, the supply of industrial capital, of technical skill, of engineering ability and of talent for complex organisation may some day be relatively as abundant there as in the United States, Great Britain, Germany or other present-day "advanced" countries. As that begins to happen—and it will be a very gradual process—new influences will affect the import and export specialisation of all these areas. Trade between the advanced industrial countries and the newly developing countries, which will by this time also be fairly "advanced", will be much more complex. It will resemble the trade between two such industrial regions as western Europe and the United States, or between New York and Michigan. Economies of large-scale production, special local resources, local conditions and labour skill, historical factors such as the accident of a particular industry having started first in a particular place, or the influence of an individual industrial genius and his associates will determine the course of trade between countries after industrialisation has spread more or less evenly over many regions. This, of course, is on the assumption of reasonable political security and reasonable freedom of international trade. As has already been shown (Chapter VIII) it is false to think that inter-regional and international trade depend in any fundamental way upon the present historical situation in which some areas happen to have much more advanced industrial techniques than others. As industrial development proceeds, and as all countries attain a considerable degree of industrial maturity, incomes will rise, transport costs will probably continue to decline, and the total value of inter-area trade will undoubtedly grow—always assuming that political conditions permit. Inter-area trade will probably not rise as rapidly as income (in other words, C-products will become more important in relation to A- and B-products) but there is every reason to expect it to increase substantially in absolute amount. There will be a larger volume of trade, substantial shifts in the particular items imported by various countries, and a much more complex division of labour among various localities.

Repercussions in the Advanced Industrial Countries

In the advanced industrial countries, workers, business men and communities whose incomes depend on A-products will find themselves favourably affected as economic development proceeds elsewhere. For these products will be in increasing demand. There will be more employment in these lines. Wages and the return to capital in the establishments producing A-products may rise relatively to wages and the return to capital elsewhere in the same country. That will depend upon how readily new labour and capital can move in to meet the increased demand, and on other factors, such as the types of collective bargaining agreements in force.

What will be the effect on consumers of A-products? Because the world demand has increased more markedly than the world supply, the first effect will be a tendency of the prices of A-products to rise. If this were the end of the matter, the repercussions on consumers of A-products would be adverse. However, the A-products as a group are likely to be of a kind for which output is readily expansible as demand increases and for which the cost per unit may even fall as world output grows. A-products, as we have seen, will characteristically embody large proportions of fixed capital and technical skill and scientific methods of production. Increased demand for goods of this sort is very likely to encourage still further improvements in production technique, which will ultimately result in lower costs per unit of output. Thus, if an enormous world demand for electrical refrigerators develops in future years, there is every reason to expect that moulded plastic bodies and continuous assembly-line manufacture of the working parts will be applied on a mass output basis so as to make better refrigerators available at a price which can be afforded even by families with quite moderate incomes. In this way, because of the dynamic effects on enterprise and technology, the reaction of the increased demand for A-products upon the consumers of A-products may be favourable.[1]

It thus appears that industries, workers and communities in

[1] It might be argued that the increasing costs per unit of the primary raw materials and labour skills from which A-products are made will more than offset any decline in manufacturing costs as the demand for A-products rises. But such a "static" analysis based on the law of diminishing returns assumes as constant some of the things that are most affected by the process of change which is under consideration here. Even in the case of primary raw materials the fact that production is being carried on by and for new groups of people and with the prospect of greatly increased markets is likely to hasten the application of existing technical knowledge and also to hasten research and the discovery of new and cheaper techniques and alternative materials. The rise of the chemical industries is particularly important in this connection. The production of plastics is not subject in anything like the same degree to the natural limitations that affect the output of minerals. Even in the case of minerals, new methods of handling and treating ores continue to make available low-grade deposits which earlier were not considered workable.

the advanced industrial countries which are engaged directly or indirectly in the production of machinery, electrical goods, transportation equipment and other A-products will find their economic opportunities expanding as economic development takes place in new areas. On the other hand, the effect on consumers of these products is not likely to be particularly adverse, if it is adverse at all. Indeed, the response of production technique to the new market opportunities offered by world development will quite possibly benefit the consumers of A-products directly.

What then will be the effects on the producers and consumers of B-products in the advanced industrial countries? The B-products are those in which the newly developing countries will offer new competition on the world market. Supply will increase relative to demand. It is in these lines—preliminary processing of some kinds of raw materials, various kinds of light manufacturing, as in the field of textiles, and production of goods not requiring large amounts of capital or the more complex technical skills—that the newly developing countries will have a cost advantage. In the case of A-products, the superior technology and industrial organisation and the larger capital supplies of the advanced countries more than offset the lower wages of the less developed countries. Therefore costs per unit of output (and that is what matters in competition) are lower for A-products in the advanced countries. In the case of B-products, which we are now discussing, the superior technology and organisation and capital equipment of the advanced countries are not effective in the same degree and costs will tend to be lower in the newly developing countries.

The new production of B-products in the countries undergoing development may mean new competition in several markets—the markets of the newly developing countries themselves (local production replacing imports of some kinds of goods), markets of third countries (export competition), and even in the advanced countries' home markets.

The tendency of increased B-product imports to enter the home markets of the advanced countries is likely to encounter barriers in the form of protective tariffs or quotas. As the new countries develop their cost advantages in such products there will probably be demands for boosting these import barriers still higher. The prevention of a substantial increase in imports of B-products into the advanced countries, however, would imperil the success of the whole process of international economic development (by depriving the newly developing countries of a means of paying returns on investments), and would represent refusal on the part of the advanced countries themselves to accept an important benefit to their own living standards. Cheaper imports of B-products, which will include many items of common necessity, would especially benefit the lower-income groups by making their wages buy more.

As a result of one or more of these forms of new competition the B-product producers in the advanced countries will find their markets shrinking. In an expanding economic world the shrinkage may be merely a *relative* one—that is, their markets may simply not grow as fast as markets for other kinds of things. In cases of that sort the adjustment is not difficult. However, the most difficult case will be discussed here, namely, one in which new competition from the developing countries actually results in a decrease in the marketing possibilities of some specific industries in advanced countries.

In such a situation the established producers of B-products will be affected adversely. They will get lower prices for their products. Business opportunities will be less bright and employment will shrink in B-product establishments, *unless they are able to adopt radical innovations and thus adjust themselves to the new situation.*

Some established enterprises in the B-product industries may adjust themselves by improving their own methods of production, instituting new and more efficient processes which lower costs per unit without lowering wages. They may improve the quality of their products and cater for a somewhat different demand from that which is satisfied by the new cheap goods. Other B-product establishments may shift to new lines of production in which the market situation is more favourable—either into A-products which are in increasing world demand, or into those C-products for which the domestic outlook is good. Some will, of course, seek subsidies from the government. The most frequent form of subsidy is the protective import duty which permits the industry to charge domestic consumers more than those consumers would otherwise have to pay for the same article from lower-cost sources abroad. Subsidies might even be sought to assist in maintaining exports. In general, however, experience shows that the attempt to keep an industry going in its old ways by means of subsidies from the community is likely to be a losing battle, accompanied by heart-breaking experiences for workers, managers, investors and the localities in which they live. B-product industries which hang on without making constructive adjustments will offer less regular employment, lower wages and lower profits than in the past. At the least, they will not be able to get the advantage of improvements in these respects to the same extent as producers of A-products and C-products.

What of the people in the advanced industrial countries who consume B-products but do not derive their employment or income from B-products? Their real incomes will be increased as a result of the new economic developments which enlarge the supply and lower the prices of these commodities. The benefit to them is likely to be quite substantial in the aggregate. The benefit is received through a fall in the cost of living brought about by the fall in prices of B-products[1], and B-products are likely to include many of the items of every-day consumption. It is well known, for example, that an important influence in raising the real income of British workers in the latter half of the nineteenth century was the cheapening of British food supplies as a result of the fall in the price of grain (a B-product in those days) which accompanied overseas development. However, if the governments in the advanced industrial countries adopt a policy of "protecting" domestic producers of B-products instead of assisting them to shift into more promising lines of production or to improve their techniques, the consumer will not be able to harvest this benefit.

Finally, consider the effect of increased production and consumption in the newly developing countries on the C-products of the advanced industrial countries. These are products not *directly* affected one way or the other by the economic development of new areas. There will, however, be important indirect effects.

In the advanced industrial countries, C-product industries make up a high percentage of the total economy. Persons engaged in such activities as the construction, supply and maintenance of houses and office buildings, automobile repair, teaching, legal practice, the practice of medicine, local merchandising, and in the service industries and professions generally, will be affected both as consumers and as producers. As consumers, they will benefit from the lower prices of B-products. As producers, they will be affected favourably to the extent that investment in world development gives a useful stimulus to the economies of the advanced countries, for their business opportunities will rise or fall with the rise or fall of general prosperity. The cheapening of B-products is very likely to benefit them by making a greater amount of domestic consumer income available for expenditure on local goods and services. Again, they will benefit by the increased expenditures of persons connected with the expanding A-industries. On the other hand, they will be affected unfavourably by the depressed condition of some B-product industries, where capital and labour try to remain in these relatively unpromising lines instead of moving into more promising ones. If, however, workers and capital in the B-product industries adapt themselves fairly quickly to the new conditions of

[1] Of course, this may be a *relative* fall. That is, B-product prices may not fall absolutely, but may rise less rapidly than other prices and incomes.

supply and demand, shifting the emphasis in the country towards the production of A-products and away from the production of B-products, then the over-all productive efficiency of the country will rise. There will be a general rise in the standard of living and the producers of C-products for the domestic market, along with everybody else, will be more prosperous.

An Example

The particular case of a town that will undoubtedly be affected in very important ways by the course of world development after this war may serve to clarify these general principles. Consider the future prospects of Lynn, Massachusetts, in the United States of America. Lynn has long been known as a textile and shoe town. More recently the General Electric Company has established a large plant there.

It is quite likely that many kinds of electrical goods will be in great demand in the newly developing countries and that for a considerable time the more advanced countries will continue to have a cost advantage in the production of at least many types of electrical equipment. This is especially likely in view of the great importance of research and new product development and improvement of old products in this field. Electrical goods, in other words, will be A-products. Textiles and shoes, on the other hand, will probably belong to the B-product classification. The balance of cost advantage in these lines will no doubt shift in the direction of the newly developing countries as they adopt modern industrial methods. Supposing that this is a correct forecast of the situation, Lynn will have both A-product and B-product industries. Of course, it will also have C-product industries, represented by its retail trade establishments, its automobile service stations, its banks, its physicians, teachers, hairdressers and other producers for the local market.

Thus, industrial progress in China, India, Latin America and elsewhere will affect Lynn in two ways. It will mean increased opportunities and it will also mean increased competition. Because of economic development elsewhere, the plants producing electrical goods will be able to offer steadier employment at higher wages and will probably expand, taking on more workers. The textile and shoe industries, after a few years, will probably begin to feel the pressure of increasing competition from the newly developing countries, both in the export market and (unless they get a very high subsidy from domestic consumers by means of protective import duties or quotas) even in the domestic market. The people in Lynn who make

C-products will be more prosperous, on the one hand, because of the prosperity of the local electrical industry, and on the other hand, less prosperous because of the depression in the local textile and shoe industries. The balance of gain or loss to them will depend in the first instance on the size of these two opposite effects.

However, business men and workers would presumably make some attempt to adapt themselves to the new situation. The rational adjustment would be for capital and workers to shift from the textile and shoe industries into the electrical and other expanding industries. The retraining of workers which would be necessary might even up-grade the skill and raise the earning power of some of them; this would be analogous to the result sometimes observed in personal affairs where misfortune turns out to have been a blessing in disguise by stimulating an individual to make a courageous readjustment that was long overdue.

Additional plants for producing electrical goods, and also plants for producing other sorts of A-products—perhaps new types of products just out of the laboratory—could also be established in Lynn to take advantage of the improved opportunities in these lines created by economic development of new areas and to absorb the workers and capital which would otherwise be less well employed in the declining textile and shoe trades. If these adjustments are brought about the people of Lynn will be better off. Adjustments of this kind represent a shift of productive efforts out of less efficient or less well adapted industries into more efficient or better adapted kinds of work. Methods of facilitating and encouraging such shifts will be discussed in a later chapter. If such adaptations are not possible, on the other hand, many of the people of Lynn would, for a considerable time, be worse off, as a result of economic development elsewhere.

Now suppose that Lynn had only an electrical industry, and no textile or shoe industry. In that case, the economic development of the new areas would bring to Lynn a great upsurge of prosperity. The real income of the people of Lynn, including the people in local trades, would increase both by reason of the rising demands for their products and by reason of the fall in the prices of the B-products which they consume.

Again, suppose that Lynn had only textile and shoe industries and no electrical industry. In that case, its people would be very badly hit by the economic development of new areas. The best remedy would be to encourage new industries of the A-products type or to shift to C-products which (if it were a period of general prosperity) would be in increasing demand elsewhere in the country. Failing this, the community would be chronically depressed and would have to exist on subsidies from the rest of the country, or people would have to move to other more fortunate or more adaptable communities where they could find jobs in expanding A-product and C-product industries.

ADAPTABILITY: THE MAIN FACTOR DETERMINING THE BALANCE OF ADVANTAGE AND DISADVANTAGE

The case of Lynn points clearly to the conclusion that the balance of advantage or disadvantage to established industrial areas from shifts in the world trading situation will depend above all upon *adaptability*. The same conclusion holds for countries.

To generalise further, the following formula may be laid down. The net balance of advantage and disadvantage to a particular country from the new opportunities and the new competition created in the world market by development of other countries will depend upon:

1. The country's specific stake in A-products
   (*a*) as producer, and
   (*b*) as consumer.
2. The country's specific stake in B-products
   (*a*) as producer, and
   (*b*) as consumer.
3. Its adaptability
   (*a*) in shifting its production out of B-products into A-products and C-products, and
   (*b*) in shifting its consumption towards imported B-products, which can now be had more cheaply.

The first two of the factors mentioned above depend upon past circumstances—the original endowment of the country in natural resources, the kinds of industries that have grown up, the skill and aptitudes and habits and tastes acquired by the people. They are "history". History may have placed one country in a very favourable situation, from which it can gain directly and easily by world economic development. In the case of another country, its past history and resources may have left it in a vulnerable position where it has important industries which suffer from the competition of new areas but has few industries in a position to take advantage of the new market opportunities created by the rise of income abroad.

The only controllable factor is the third, that is, adaptability.

Nothing can be done to change history. But, starting with the situation of the present, whether relatively good or relatively bad, a vast difference in a country's future prospects can be made by the willingness or unwillingness of its people to be energetically adaptable. The third factor, therefore, is the one on which practical leaders of business and labour and practical economic statesmen will need to focus their attention. *Adaptability* is the key to the situation of the advanced industrial countries as economic development proceeds in other areas. Proper adaptation will enable them to take maximum advantage of new opportunities offered by rising world standards of consumption. Proper adaptation will make it possible for them to reap great benefit from access to new and cheaper sources of supply of those goods which newly developing countries can best produce. And proper adaptation will keep to a minimum the troubles and losses occasioned by new competition from low-cost industries in other parts of the world.

Denmark's history in the late nineteenth century offers a concrete instance. A problem of direct, specific competition brought about by new economic development abroad was successfully solved by an adaptive adjustment which benefited all parties and enabled the adaptive country to raise its living standards very markedly.

The economic development of the United States, particularly the building of new railway lines and the rise of large-scale mechanised farming, together with the effect of steamships in cheapening ocean transport, raised serious new competition for the cereal-growing regions of Europe in the latter part of the nineteenth century. Denmark, for example, was accustomed to rely on grain as its principal export product (mainly to England). Now the cheap new grain supplies from overseas appeared to threaten Denmark's economy with ruin. The cheap grain, however, together with general advances in industry and trade which increased the productivity of the average worker, contributed to a rise in real income and living standards in Great Britain. The British were able to spend more on a greater variety of foods, including butter, cheese, ham and bacon. These more expensive animal products can be purchased in large quantities only by people on a relatively high standard of living. At the same time, improvements were made in the techniques of producing these products (the cream separator, etc.). Thus new and profitable opportunities for Danish agriculture appeared in the field of animal products, at the same time as, and partly as a result of, the new competition in grains. Gradually, over the latter part of the nineteenth century, and very consciously in the 1880's and 1890's, the Danes shifted the emphasis of their agricultural production towards animal products. Denmark actually became a cereal importer instead of a cereal exporter, although —and this is an interesting feature of the adaptation—as much grain continued to be produced within Denmark as before. Indeed, home production of grain increased slightly. What the Danes did was to feed the cheap grain to animals, "processing" it into higher quality foods which were then sold in the English market. By this method, and with the help of producers' co-operative societies and educational movements that contributed to intelligent farming and marketing, the Danish farmer acquired one of the highest living standards in the world. An overseas development which had offered menacing new competition was thus converted by successful economic adaptation into a tremendous new opportunity.[1]

---

[1] Summarised from the manuscript (now available in hectographed form) by Carl Major WRIGHT: "The Adaptation of Danish Agriculture from Grain Production to Dairy Products and Meats".

CHAPTER X

THE IMPORTANCE OF INDUSTRIAL ADAPTATION IN THE ADVANCED COUNTRIES

The key to the problems which will face advanced countries as economic development proceeds in other areas is *industrial adaptation*—that is, shifting of the uses of labour and capital. Progress elsewhere in the world will mean economic gain for established industrial regions if there is sufficient mobility of labour and capital into those industries where the new conditions create better prospects, and out of the industries where opportunities are less good.

The position of established industrial countries in a progressive economic world may be likened to that of a private firm in a growing country. New customers and new rivals are appearing all the time. The market is constantly changing. Costs and prices are shifting. If the firm's managers are alert and imaginative they will study the trends in consumer demand and the trends in their own production costs. They will use that information in deciding from time to time to expand one department and contract another, to push certain old products that are profitable and to withdraw others that have ceased to be profitable, and to add appropriate new products at the right time.

Employers, workers and governments in the advanced industrial countries will have a choice of three basic policies when the increases in production and consumption in newly developing areas begin to affect the economic outlook. The first might be called a policy of restriction, the second a policy of *laissez faire* and the third a policy of adaptation.

The restrictive policy, if it is adopted, is most likely to be entered upon piecemeal in response to pressure from particular groups of workers or employers. Producers attached to industries which feel the new competition of developing areas will regard that competition as a threat to their livelihood. The public will be susceptible to the fallacious argument that low-priced imports, especially if they come from countries with low standards of living, must undermine the living standards of a country that receives them. Such fears can easily lead to non-co-operation or hostility towards economic development abroad, and to high protective tariffs or restrictive import quotas on goods coming from newly developing regions. The effect of such policies, designed to "protect" certain industries where new competition makes the future outlook relatively unpromising, is, of course, to throttle opportunities for expansion in other and more promising industries. For trade is a two-way street and refusal to import cuts down opportunities for export. In this case exports as well as imports from the newly developing areas would be sacrificed. The workers of Lynn would be "protected" in their attachment to a weak and struggling textile or shoe industry, but they would be hampered in their search for better employment in the electric equipment field. They would have to live precariously on a subsidy from the consumers of the country, instead of being assisted to transfer into an expanding industry able to meet competition both at home and abroad by keeping in the forefront of technical progress. Workers and owners of capital would be spared, temporarily, some part of the necessity of making difficult changes. But probably their own direct interests would be badly served by this policy over a period of years, and certainly the community and the country as a whole would be poorer.

The *laissez faire* policy may be a theoretical alternative to a policy of restriction, but it is hardly a practical one. "As a surgeon *laissez faire* cuts unnecessarily deep, while its treatment in general is so rough that the whole constitution of the patient will be undermined, and in any case it is so brutal that the patient will run away."[1] In the difficult and disorderly economic conditions that are sure to characterise the post-war situation, industrial groups in every country will demand help from governments in meeting their problems. Especially where the difficulty arises out of new competition from new industries abroad, the advice to let the automatic market system take its course and ruthlessly eliminate the "unfit" will hardly be accepted. Furthermore, it is not certain that the free play of competition would actually bring about a successful readjustment. The free market is sometimes very ineffective in ironing out pockets of unemployment, such as might tend to develop as a result of new competition from abroad, and workers and business men are often not mobile enough, in the absence of special public measures of encouragement and assistance, to transfer quickly to new lines in search of better opportunities. The history of "distressed areas" shows that market forces may operate in vicious circles at times, so that the very pressure of economic adversity which is supposed to compel readjustment makes it next

[1] J. W. F. Rowe: *International Control in the Non-Ferrous Metals* (W. Y. Elliot, ed., New York, Macmillan, 1937), p. 74.

to impossible to launch new types of production or to attract new industries to a depressed locality.

The traditional theory of international trade has usually ignored the transition problem of adjustment to new conditions of trade. By the device of looking only at long-run results it assumed away the whole problem of changes in production structure. In the "long run" of theory the factors of production are treated as highly mobile, but in meeting a practical need for immediate adjustment they may be very immobile. The enlargement of international trade by development of hitherto less developed areas will lead to beneficial results *if we successfully come to grips with the transition problems that arise in the process of changing the structure of production.* Protectionists, on the other hand, usually argue as though all changes which cause a shrinkage in any established industry are bad. They propose to "protect" workers and capital by insulating them from the necessity for change. Most national legislation, under the protectionist influence, hampers industrial adjustment and perpetuates weaknesses and vulnerabilities instead of helping in a positive way to remedy them.

The most feasible and also the most constructive alternative to restrictive intervention by the State is not non-intervention (*laissez-faire*), but intervention of a more constructive kind—namely, a positive programme of industrial adaptation. Such a programme would be designed to assist industry and labour in reorienting themselves, so that they can take maximum advantage of new opportunities. In this way the enterprise and initiative of citizens will be preserved and will be exerted in the most promising directions. The results of such a programme, assuming that it is successful, might well be in many (but not all) respects similar to that which the automatic market system would accomplish if it were able to function with the theoretical perfection assumed in older text books. But the process of adjustment ought to go forward with more attention to the human problems of the individuals directly involved and with less infliction of suffering on particular groups.

An adaptive policy, recognising transition costs and difficulties but also recognising that there is greater gain to be had by making adjustments than by not making them. would include two types of measures: (1) measures designed to stimulate industrial mobility, so that transfers of capital and labour would take place more rapidly out of the industries that ought to contract and into the industries that ought to expand; and (2) measures designed to soften the impact of changing economic conditions on the groups of people most directly affected, thus distributing more equitably the burden of transition costs which would otherwise fall unjustly on certain particular groups and individuals. The effects of future economic development on advanced industrial countries will depend very much upon the degree to which these countries are willing and able to adopt positive policies of encouraging industrial adaptation.

Problems of Changing the Production Structure

"The production structure" might be defined as a set of going arrangements under which workers, capital and natural resources are employed in certain ways and at certain places. A change in the production structure means that workers must shift their occupations or the location of their employment, or that capital must take new forms and be applied in new ways or new places, or that new patterns of land use must be adopted.

"Adaptation" or "adaptive changes in the structure of production" may be defined as changes that, once accomplished, result in *better* use of resources—better in the sense that more real income is produced (or steadier real income, or better distributed real income). All changes in the structure of production, however, including adaptive changes, take place at a certain cost to society and to individuals or groups in society. Such changes often encounter serious resistance. In fact, some of the most dangerous economic problems of our day arise out of blockages of one kind or another which serve to rigidify the production structure and to prevent adaptive readjustment. The economic troubles caused by rigid production structures in turn contribute to other social and political difficulties.

This is one of the key problems of the modern world. Methods of facilitating adaptive changes in production structure, if they could be successfully devised and applied, would help not only to maximise advantages and minimise disadvantages from economic development of new areas, but would also contribute in no small measure to the solution of a host of other international and national problems. For example, positive adaptation policy—that is, deliberate encouragement of the mobility of the factors of production in a manner calculated to bring about desirable changes in production structure—is a key to the successful handling of the following post-war problems of economic adjustment, among others:

1. *The immediate problem of economic demobilisation and the reconversion of war industries to civilian uses.* The necessity for facilitating the shift of workers and investment from one occupation to another and from one place to another as a part of this process is so well appreciated these days that no comment is required.

2. *The problem of actually harvesting the benefits of expansionist policy designed to provide full employment.* Economists discussing the problem of full employment in recent years have put most emphasis on maintaining aggregate "effective demand" for the products of industry. For this purpose it is essential to maintain the flow of money income at a high level by stimulating investment and consumption. But a high effective demand is not enough to maintain full employment. It is also necessary to achieve a harmonious structure of production. Where immobilities of labour and capital (resulting, for example, from monopolistic practices) prevent the expansion of certain kinds of production, the remedy is not further expansion of over-all money income and demand. That might merely result in inflating prices. In order to have full employment there must be a harmonious relationship between the structure of production and the demands of purchasers. If it is possible to find means for redirecting labour, plants and natural resources into new uses more quickly and smoothly when the need arises, the task of promoting full employment by expansionist policies is made easier and the danger that expansionist policies will merely result in price inflations in certain sectors of the economy without achieving full employment is reduced.

3. *The problem of adjusting to technological development.* Technological change will doubtless be rapid in the post-war world. New applications of science and of machine technology give rise to changes in the production structure and create transition problems—that is, problems of shifting resources out of old uses into new uses. The cotton economy of the southern United States had to undergo profound readjustments because of the coming of tractors and other mechanical equipment. The advent of electrical refrigeration threatened the livelihood of icemen, but increased the opportunities for electricians and machinists. If adaptation can be made smoothly and successfully, improved technology is an obvious economic gain to society. The evils of the machine—technological unemployment and kindred problems—represent unsolved problems in industrial adaptation.

4. *The problem of bringing about effective international access to raw materials and markets.* This problem, so far as its peacetime aspects are concerned, is largely one of lowering restrictive barriers to trade. The real ground for complaint about "access to raw materials" is not refusal of the possessors to sell (in time of peace). Rather, it is that many important countries maintain high protective import barriers around their domestic markets, thereby making it less possible for other countries to obtain international

purchasing power through the sale of their own export products. Lacking this purchasing power, the latter countries are hampered in obtaining the foreign raw materials that they need.

Thus, the real means of increasing the access of all peoples to the raw material resources of the world, thereby increasing the chances of maintaining future peace, is through a lowering of trade barriers, coupled with general economic expansion. Yet trade barriers cannot be lowered without causing some shifts in the employment of labour and capital in the countries that agree to lower barriers. One of the effects of protective import duties and of protective quotas and exchange controls is to distort the production structure of the countries applying them. That is, types of production are encouraged which are not able to stand alone in unsubsidised competition with the rest of the world, while other types of production that could compete in the world export market are handicapped by the restrictive influence which import barriers also exercise on exports. If the protective barriers are then removed, a second shift in production structure must take place before the benefits of freer trade can really be had. The hitherto protected industries must shrink to some extent, and new opportunities for expansion appear in the export industries. An adaptation policy which facilitates the shifting of resources out of the former and into the latter groups of industries will assist, therefore, in removing the obstacles to international trade expansion, and in meeting the complaints about inequality of access to the world's resources.

5. *The problem of working out a harmonious relationship between governmental economic intervention and private enterprise.* The great practical problem in the relation between government and private enterprise is one of making a "mixed system" work satisfactorily. By "mixed system" is meant a system in which both conscious control of economic affairs through government and "automatic" responses of private enterprise to market forces play a role. Much discussion of the proper limits of governmental intervention in economic affairs has run in terms of the *amount* of intervention. An equally important issue is the *kind* of intervention. Adaptive intervention, on the whole, works with the price system rather than against it. Therefore, its tendency is self-liquidating rather than cumulative. Anti-adaptive intervention, meaning all policies which protect vested interests in old ways of doing things, helps to establish situations that cannot be maintained except by further government aid. Measures of this sort tend to perpetuate themselves and to produce new necessities for government control. For ex-

ample, a policy of protecting the New England textile industry against foreign competition by means of tariffs or against competition from the South by means of differential freight rates would be anti-adaptive. Stimulation of new industries in the area and subsidisation of vocational retraining for textile workers would be adaptive. The first policy works against the forces of the free price system to which private enterprise responds; the latter works with these forces. In general, adaptive types of economic intervention by government help to make it possible for private enterprise and governmental economic activity to work together harmoniously. Anti-adaptive policies, however, have the effect of demanding more and more intervention. Hence, positive adaptation policy is the key to making a "mixed" system work satisfactorily.

6. *The problem of commodity stabilisation without uneconomic restriction.* Most commodity control schemes adopted thus far have performed only restrictive operations. They have limited output or exports in order to raise prices (or to prevent them from falling) and to prevent the accumulation of surpluses. An important part of the pressure which has led to such restrictions in the past will be removed if the world can maintain a reasonably high level of general employment, and hence a high level of demand for raw materials, after the war. However, changes in the technology of production, discovery of new resources, the development of substitutes and other factors are bound to create situations in which from time to time there is over-capacity in some raw material industries and under-capacity in others. Positive adaptation measures designed to facilitate a movement out of those industries which are overdeveloped and into other lines of production (for example, industrial diversification in areas hitherto dependent on a single crop or a single product) will be a necessary part of any successful commodity stabilisation programme.

This list of post-war economic problems which depend upon industrial adaptation for their successful solution is not exhaustive. Yet it is enough to indicate that the development of new industries in less developed areas, or the further expansion of industries in other areas, is not the only type of economic change which imposes on workers and capital a necessity for shifting from one employment to another. It is clear that an adaptive programme which would meet the conditions created by new economic development would also have beneficial effects in many other connections.

Costs of Transition and their Distribution

It was noted above that all changes in the structure of production, including adaptive changes that result in better utilisation of resources and hence lead to higher incomes, take place at some cost. These "transition costs", as they will be called, may be looked at from the point of view of particular individuals and groups or from the point of view of society as a whole. From the first point of view, transition costs include earnings foregone between the loss of one job and the finding of another, the expenses of moving a family to a new location, the decline in the earning power of a particular skill, the decrease of earnings and dividends from an established business, and the like. From the point of view of society as a whole, transition costs include production loss in shifting workers or equipment from one job to another, the cost of transporting workers or equipment to a new location, the cost of retraining workers and refitting capital equipment for new uses, etc. If the decline of some industries and the expansion of others cannot take place without dislocations which spread outward from the sectors of industry immediately affected and cause a general depression, then that, too, adds to the social and individual costs of transition.

The necessity of shifting out of some lines of production into others also carries with it certain "non-economic" transition costs which cannot well be expressed in pecuniary terms. Established habits of working and living may be disrupted, family or neighbourhood groups may be broken up. Some trade unions may lose membership. The prestige and position of some business leaders may fall. Even where the individuals concerned succeed in making adjustments which turn out to be to their distinct advantage, the initial necessity of making a change and facing an uncertain future is likely to be felt as a psychic cost.

The existence of these various costs, together with ordinary human inertia and ignorance of opportunities elsewhere, plus the obstructions placed in the path of adaptive adjustments by monopolistic groups or restrictive legislation, act as barriers to adaptive changes in the production structure. Experience has shown that coal miners, wheat farmers, and textile workers will remain attached to a depressed industry with surprising tenacity, even in times of general prosperity when there are opportunities elsewhere. Attachments of this sort, reinforced by uncertainty and by the fear of economic losses, also go far to explain the often passionate insistence of industrial groups upon getting or maintaining "protection". They quite naturally wish to insulate themselves from competition and especially from new competition which might force them to undergo change and readjustment.

Where an adaptation in the production structure is successfully achieved, the costs of transition are of course offset by gains. The gains in such a case are likely to be considerably greater than the costs, at least from the social point of view, and not infrequently from the individual viewpoints of the workers and others directly involved. Transition costs are characteristically temporary, while the gains from an adaptive change in production structure make themselves felt in increased earnings year after year. One can, in fact, lay down the following general principle: it is worth while to undertake a given adaptation in the production structure if, regarding the transition costs as an investment, the increase in annual real income likely to result from the proposed adjustment is large enough to represent a satisfactory rate of return on the investment. A "satisfactory" rate of return in this connection is, of course, impossible to define precisely. But it is clear, for example, that a change in production structure which might represent an initial "investment" of one million dollars in the form of various costs and losses would be eminently worth while if it had the effect of increasing annual income by 500,000 dollars a year for a considerable time in the future. The return would in this case be 50 per cent. annually. On the other hand, transition costs valued at one million dollars would hardly be justified for the sake of increasing annual income by 40,000 dollars (4 per cent.) in view of the uncertainty of the future and in view of the likelihood that non-pecuniary costs of transition would also be involved.

It must be pointed out, however, that the *gains* from successful transition adjustment are often more widely shared than the *costs*. The costs may be concentrated upon relatively few people. For example, it has been estimated that American consumers, in 1937, were paying about $290,000,000 more per year for their sugar than they would have had to pay if sugar could have been imported, free of duty and restrictions, from the cheaper sources abroad.[1] In all likelihood this considerable sum could have been saved to consumers annually by free import of sugar and the consumers would have been able to increase their purchase of other kinds of goods. For the individuals dependent on the protected sugar industry, however, and for some communities in sugar-growing States such as Colorado, the immediate losses involved in a drastic shrinkage of the domestic beet sugar industry would have overshadowed the immediate gains, even though the country as a whole would have been richer. The persons directly affected might sooner or later transfer into alternative lines of production in which they would earn as much or even more than they earned in the sugar industry. But if the adjustment required several years they would suffer considerable losses in the meantime. This is obviously a case in

[1] J. P. Cavin: "The Sugar Quota System of the United States, 1933–37", manuscript doctoral dissertation in Library of Harvard University, 1938, p. 313.

which an adaptive change in production structure would bring permanent net gain of considerable magnitude to society as a whole, but would do so at the cost of sharp, even though temporary, losses to particular individuals and groups.

It is cases of this sort which suggest that it might be equitable and useful to compensate private interests for transition costs made necessary by industrial adjustments undertaken in the general social interest. If the people of the United States, for example, could after the war somehow arrange to "buy out" the beet sugar interests at a price not exceeding, say, two or three times the annual amount which consumers would save by free imports of sugar, the bargain would be a good one. The "buying out" might consist partly of compensation in money, and partly in free vocational retraining, subsidisation of developmental projects, and research directed to discovery of new products which would lead to industrial expansion in the regions affected.

The bearing of all this on the way in which economic development of new areas will affect established industrial areas is clear. The advanced countries will be able to get maximum advantage from world economic development if they are able to adapt their production structures as development proceeds, shifting the use of their particular resources and skills into those lines where new opportunities are opening up and out of those lines where new low-cost competition makes it more advantageous to become a buyer. In a changing world situation, the countries that benefit most will be those that can make continual adaptations of this sort. The "best" production structure for a particular country—that is, the one likely to give the most in terms of real income, stability of real income and employment, and equitable distribution of income—is not a static thing. It changes gradually—in the form of better prospects for some industries and worse for others—from decade to decade and even from year to year. It depends not only on the country's resources and skills and preferences, but also on the state of technology and on what other countries are producing and consuming. Flexibility and adaptability in the production structures of the advanced industrial countries are extremely important in order that they may (1) adapt and readapt their industrial output so as to make the most of new opportunities, and (2) achieve these adaptations with a minimum of transition costs.

B. What Policies Will Yield Greatest Mutual Benefit?

CHAPTER XI

MEASURES TO ENCOURAGE INDUSTRIAL ADAPTATION WITHIN EACH COUNTRY

It is clear from all that has gone before that internal adjustments in the production structure of each country are a necessary counterpart of world economic progress. Each country will need a positive policy of industrial adaptation, not merely to meet the new situations created by economic development elsewhere, but also to deal with the serious structural problems that will be left by the war and to meet constructively the many other economic changes that require expansion of some industries and contraction of others. Under the conditions of modern economic life the problem of changing the production structure cannot be met successfully simply by leaving it to time and to the working of so-called automatic market forces. Methods of deliberately encouraging mobility and promoting the process of adaptation will be essential. What methods are appropriate?

The basic principles of a positive adaptation policy such as each advanced industrial country might apply to advantage within its own borders are:

1. To encourage expansion of the stronger and more promising industries and contraction of the weaker and less promising ones (unless the latter, by drastic enough changes in techniques, can so increase their efficiency and improve their prospects as to cease to be "weaker" industries).

2. To assist the transfer of workers and capital from less promising to more promising lines. In other words, the object would be to encourage mobility of the factors of production in adaptive directions.[1] This would be in some respects a revolutionary reorientation in the policy of most governments, for a large part of governmental intervention in the past has been motivated by the pressure of groups that wanted to be shielded

[1] Cf. Report III to the 26th Session of the International Labour Conference: *The Organisation of Employment in the Transition from War to Peace* (Montreal, 1944).

from competition—that is from the necessity of making adjustments. The new policy would be to help them make the adjustments.

3. To protect persons and communities against serious loss of income and employment arising out of a reorientation of production which is in the social interest, but not to protect them against the necessity of making adjustments. This means, in addition to a basic social security system, special methods of distributing transition costs more equitably. The object of an adaptive policy would not be to protect industries or occupations as such, but to protect people. The best way to protect the people connected with a weak industry or with an unpromising occupation might be to help them to shift into a type of production that offers better prospects.

The practical application of a policy like that outlined above would consist of many different kinds of action by governments and by business and labour groups. Devices used in wartime to assist transfers in accordance with the unprecedented need for industrial mobility that war creates, while not applicable without modification in peacetime, would undoubtedly repay careful study. Some tentative suggestions for peacetime adaptation policy are offered below.

Measures to Create an Economic Environment Favourable to Adaptation

Labour and capital shift to new occupations more easily in times of general prosperity than in times of general depression. In fact, little can be done about cleaning up pockets of unemployment caused by structural maladjustment if there is not a high level of activity and employment in the better adapted industries towards which labour and capital should move. The first requisite of successful adaptation policy, therefore, is "full employment" policy, meaning stimulation of private and public investment, stimulation of consumption, fiscal and public works measures designed to counteract a cyclical down-swing, and the like. As was seen in Part I, a world development programme can itself help to provide part of the volume of real investment needed to keep the economies of the advanced countries operating at a high level.

General economic expansion, both domestic and world-wide, such as is connected with increasing production, rising incomes, growing populations, new markets, and advancing living standards, makes structural readjustments easier. It softens the impact of forces necessitating readjustment by maintaining a demand higher than would otherwise be the case for the products of industries that must decline, and it makes new alternative lines of employment readily available. Thus, absolute contraction of particular industries is less frequently necessary, and the reorientation of production can more often take place by diversion of the stream of new labour and capital to new uses, rather than by actual shifts of labour and capital already employed.

It must be insisted once more, however, that measures to promote full employment and economic expansion, while *essential* in meeting the problems of transition to a new production structure, are not *sufficient*. In fact, "full employment policy" is a misnomer if it is taken to mean merely pumping up the circulation of money income. By monetary expansion it is possible to boost aggregate effective demand to a level that might soak up all unemployment if there were no important obstacles to the mobility of the factors of production. But under the conditions of today there are such obstacles and though they are less when aggregate demand is high they do not disappear. The distinction is sometimes made between "general" unemployment and "special" unemployment. It is the second, "special" unemployment, resulting from imperfect adjustment of the production structure, with which adaptation policy is primarily concerned. The point being made here is that effective measures to combat general unemployment would also go a long way, but not all the way, towards easing the problems of special unemployment. Economic expansion and deliberate encouragement of adaptation need to go hand in hand.

The interest of business men in higher profits, if it is not allowed to take the form of monopolistic restrictions of output, is the most important adaptive force of all in a private enterprise system. Adaptation is facilitated by every improvement in the information at the disposal of business managers who decide what lines of production it would pay to expand and what lines ought to be liquidated. In addition to the ordinary business outlook services, governments co-operating with an international development authority might publish estimates of the volume of new expenditures to be expected on particular kinds of goods and services and within given regions under various assumptions as to the trends of national and world income. When interpreted into terms usable by business men such studies of income elasticity of demand should help industrial managers to make more accurate plans. On the supply side, corresponding estimates might be published on anticipated new production, especially in the newly developing areas of the world. The two types of information taken together would help to indicate to the managers of economic enterprises, and also to investors of capital, to labour organisations, to vocational counsellors, to trade schools, placement services, etc. what lines of production are most promising and what are relatively unpromising.

Better organisation of the capital market, better organisation of the labour market, and other measures to improve the functioning of the market mechanism in general serve to make adaptation easier by increasing the likelihood that personal initiative will be applied in directions where the economic outlook is best. Of great importance are effective employment exchanges, so set up as to be able to offer intelligent guidance to workers who might be able to improve their opportunities by shifting occupation or locality. They might well be supplemented by increased efforts through professional associations and otherwise to facilitate correct placement of scientific, engineering and managerial workers as well as the types of workers more usually listed in employment exchanges. Wartime devices such as the National Roster of Scientific and Specialized Personnel set up in the United States are suggestive. Better vocational outlook services and application of their findings through vocational guidance in school systems would promote occupational adaptation by directing new recruits to the most promising lines of industry.

Experience has shown that high specialisation of an industrial area on a few products and a few types of skill makes it more difficult for the area to adapt itself when its specialised industry encounters depressing influences.[1] In order to make future adaptations easier, therefore, it is desirable to encourage industrial diversification wherever this can be done without great sacrifice of economic productivity.

Measures to Encourage Industries that should Expand

A cardinal point of adaptation policy is to encourage the expansion of the promising industries rather than to prevent the contraction of the weak ones. But various obstacles may stand in the way of expansion in the very lines of production that should increase most as income rises. Monopolistic practices and agreements of business firms which restrict the use of new technology, divide up markets, and limit output are such obstacles. Prices are maintained at unjustifiably high levels, consumption of the products affected does not increase as much as it should, the expanded production and new employment which should arise is throttled back.

[1] (British) *Royal Commission on the Distribution of the Industrial Population* (Cmd. 6153, London, 1940), pp. 87-8, 199-200.

Restrictions on entry to an industry or occupation enforced by trade union practices, cartels, professional associations, or by special legislation have similar effects. So do outmoded jurisdictional rules and requirements designed to make work for traditional skills by preventing the use of semi-skilled men and new processes. In all these ways the expansion of output is prevented in lines that should absorb increasing numbers of workers.

One of the C-product industries, for example, which is widely considered to offer great promise for useful post-war expansion in the advanced countries is the housing industry. The redevelopment of terminal and other facilities in cities is also an enormous potential field for highly beneficial construction. Yet there are many obstacles to be overcome. These include unsuitable conditions of land ownership, monopolistic practices in the building industries, high interest rates, inflexible trade union rules, and fears of property owners that new developments will lower the value of existing investments.[1]

In a progressive society with high and rising standards of living one of the most important fields for expansion is professional services. For example, the services of physicians, dentists and the various technicians associated with medical care are increasingly demanded as incomes increase. If entrance to these and other professions is unduly limited by high educational costs, by quota limitations of professional schools or by restrictive entrance requirements of professional groups, the effect is to create group monopolies, to raise prices to levels that prevent many people from buying such services and to restrict expansion of employment in a highly useful direction.

On the positive side, measures to encourage the development and marketing of new products and processes are important. The advanced industrial countries will get the greatest benefit and least detriment from economic advancement elsewhere if leaders of industry, labour, and government adopt the progressive attitude that their job is to continue pioneering work. Industrial research by private institutions, encouragement of research and development and scientific and vocational training by governments, are all methods of facilitating adaptation. Private financial institutions as well as governments could promote industrial adaptation by finding means to make capital more readily available for launching new products. Tax adjustments might be made in favour of enterprises which use their own funds for pioneering research and development.

---

[1] Cf. Miles L. COLEAN: *The Role of the Housebuilding Industry* (pamphlet of the U. S. National Resources Planning Board, July 1942).

Comprehensive regional development programmes, such as that sponsored by the Tennessee Valley Authority in the United States, encourage the expansion of well adapted industries in many different ways. The T.V.A., for instance, has provided expert analysis of the resources of its region and has added to those resources by flood control, irrigation, power development and education of the population in improved techniques and skills. It has also carried the overhead costs of pioneer work in starting new lines of production. For example, T.V.A. engineers developed machinery which could be used by small mills in pressing oil from cottonseed. They then found several manufacturers who were willing to make the new machinery according to these designs and to market it at low cost.

Some interesting suggestions have been advanced for governmental encouragement of a concerted private industrial expansion. Agreed programmes would be worked out by government and industry for a considerable number of important products, based on the increased sales of each product that could be expected according to past experience, if national production as a whole were to expand by a certain amount. The government would then guarantee to purchase itself any amount by which total sales might happen to fall short of the expanded amount in the plans.[1]

Expansion of industry into desirable lines can also be encouraged by educational and other measures to promote better nutrition and to raise mass consumption standards generally. Better nutrition makes it possible to divert labour and capital from some of the agricultural staples in which overproduction threatens into increasing the output of the so-called "protective foods"—dairy products, vegetables, meat, etc.

In this connection one "new product" that is no longer quite new but which still offers great opportunities for expansion should not be forgotten. That is leisure time. Shorter hours and a shorter work week should absorb a considerable part of the labour time saved by improvements in production and exchange. Leisure for workers, in other words, can be one of the most important "expanding industries" for the advanced industrial countries as world development proceeds and as the necessities of life become available at lower cost through import and through general improvement in productivity.

---

[1] Cf. Mordecai EZEKIEL: *Twenty-Five Hundred Dollars A Year* (New York, Harcourt, Brace, 1936), and *Jobs for All through Industrial Expansion* (New York, Knopf, 1939). The Industrial Expansion Bill based on this principle and introduced into the U.S. Congress by a group of representatives in 1937 is described in an article by Herbert HARRIS: "This Bill Bears Watching", in *Survey Graphic*, Apr. 1938, pp. 227ff.

Measures to Encourage Transfers out of Industries that should Contract and to Relieve Distress Connected with Contracting Industries

It is not industries as such that we want to protect, but people. In order to protect people, it is sometimes desirable to liquidate a particular industry in a particular locality. The problem then is to facilitate transfers into other industries, and to see that the burden of transition does not fall unfairly upon particular groups. That is a requirement not only of social justice but of practical economic statesmanship as well, for much of the resistance to adaptive adjustments ordinarily comes from the imposition of disproportionate transition costs upon an industry that should contract.

If an industry shows symptoms of weakness, such as inability to meet competition, the first adaptive possibility to be explored is in the industry's own technical and business efficiency. New methods and improvement in management may be enough to change the outlook entirely. Managers, industrial associations, trade unions and government agencies all have an interest in seeing that possibilities of this sort are explored.

When it is clear, however, that changes in the demand and supply situation make it advisable that a particular kind of production should be, if not completely abandoned, at least curtailed to the capacity of the more efficient plants, then what measures are available for assisting in an orderly withdrawal? One such measure is the provision of special vocational retraining facilities for workers on very favourable terms or even gratis. Job aptitude studies may help workers to find new fields in which their talents could best be used. The war has forced the development of many new methods of assisting workers to convert their skills from one occupation to another or to learn new skills quickly. Careful study of wartime experiments in this field should yield many suggestions for peacetime adaptation policy. The linking of vocational retraining to the system of unemployment benefits would be an important step in the direction of assisting adaptation.[1] The Unemployment Provi-

[1] Sir William Beveridge wrote in his social insurance report:

> Men and women who have been unemployed for a certain period should be required as a condition of continued benefit to attend a work or training centre, such attendance being designed both as a means of preventing habituation to idleness and as a means of improving capacity for earning. Incidentally, though this is an altogether minor reason for the proposal, such a condition is the most effective way of unmasking the relatively few persons who may be suspected of malingering, who have perhaps some concealed means of earning which they are combining with an appearance of unemployment. The period after which attendance should be required need not be the same at all times for all persons. It might be extended in times of high unemployment and reduced in times of good employment: six months for adults would

(Footnote continued overleaf)

sion Convention of 1934 provides that the right to receive benefit or an allowance may be made conditional on attendance at a course of vocational or other instruction.[1] Of course such a system would not be very successful if frequently at the end of a period of training there were no job to go to. Like most other measures to promote adaptive shifts in occupation, it needs as a basis a generally expanding economy.

An adequate and widely inclusive programme of social security, guaranteeing a certain minimum of well-being to all workers and their families in spite of industrial hazards, including hazards of occupational readjustments, is a basic fundamental for assuring more equitable distribution of transition costs among individuals. Also, it should increase the willingness of most workers to take risks in order to improve their position and therefore to adapt themselves more quickly to changing situations. Of course, certain methods of administering social security provisions might have exactly the opposite effect. For example, provisions that penalise workers who shift out of a particular industry or occupation or move from a particular locality are definitely anti-adaptive.

The social security systems of the advanced countries might well be re-examined from the point of view of their adequacy in protecting workers who are willing to be "mobile" and their influence on economic adaptation in other ways.

A suggestion has been advanced in England whereby special contracts of employment would be offered to workers prepared to make shifts more readily. Thus, *The Times* of London asks:

> Would it not be possible to introduce a scheme by which the State itself should become a third party to contracts of employment and carry the employee through intervals of unemployment at full pay? He would then be in the position of a soldier waiting at his depot for posting orders, and not of one flung out of the service. . . Such a scheme would have to be optional, since as a *quid pro quo* for the security provided, the worker would have to be willing, as occasion required, to change his trade and even the place of his home. . . The worker clamours for more security; the community needs more mobility. Both needs can be met if we give one as the price of the other.[2]

> perhaps be a reasonable average period of benefit without conditions. But for young persons who have not yet the habit of continuous work the period should be shorter; for boys and girls there should ideally be no unconditional benefit at all; their enforced abstention from work should be made an occasion of further training.

It is also proposed that authors, shopkeepers, business managers, housewives and others whose circumstances may have changed unfavourably should be entitled to a training benefit. (Sir William BEVERIDGE: *Social Insurance and Allied Services*, American ed., New York, Macmillan Company, 1942, paragraphs 122, 131, 328, 346, 349, 353, 383.)

[1] INTERNATIONAL LABOUR OFFICE: *The International Labour Code* (Montreal, 1941), p. 30.

[2] *The Times* (London), 2 Mar. 1943, quoted in a forthcoming book on economic adaptation problems by Mr. A. G. B. Fisher, kindly made available in preliminary manuscript by the author.

Loans or subsidies to cover the costs incurred by a person who takes work or training at some distance from his present home would also assist in the transfer of labour.[1]

Provision of adequate housing facilities for workers in localities where jobs are available is also important.

A dismissal wage or severance pay for workers dropped from regular employment is a measure that has been proposed as a means of equalising the burden of transition costs.

The traditions of business and the rules of the taxing authorities regarding capital obsolescence may be important influences in some countries in preventing adaptive changes in industry. Unwillingness to write obsolete equipment off the books sometimes makes managers hesitate to introduce new methods that, once adopted, would actually increase the profitability of the enterprise. Remedies must be sought with care, for the subject is a complicated one, but at least it would be worth while to explore whether accounting processes as applied to amortisation of capital equipment are suited to modern industrial undertakings, and whether it would be feasible to encourage adaptive adjustments by allowing shorter terms of amortisation for tax purposes in certain industries on condition that funds accumulated in this way be put periodically into complete plant modernisation or into the launching of new products.

Experiments have been made in Great Britain and elsewhere with various methods of encouraging new industries to establish themselves in "distressed areas". One method that has given encouraging results is the setting up of private or mixed corporations to provide desirable factory accommodations, well equipped with modern facilities for power, transportation and other industrial needs. So called "trading estates" in Great Britain have developed centres of diversified industry in this way. Other methods include public developmental projects designed to improve the physical facilities for industry and also the education and labour skills of the people in distressed areas; tax reforms to lessen the burden of local taxation and thus to break a vicious circle by which high taxes in areas of declining industry discourage new industries from entering; and a policy of locating defence industries or public works in areas otherwise likely to be centres of chronic depression.

Educational Measures

An outstanding example of successful economic adaptation is Denmark's agricultural shift in the late nineteenth century from export of grain to export of high quality bacon, eggs,

---

[1] This is also included in Sir William BEVERIDGE's recommendations.

cheese and butter. The adjustment, in retrospect, seems simple and obvious enough. Cheap grain was coming by improved transport methods from great new farmlands overseas. Great Britain with its rapidly rising standard of living made still higher by the cheaper grain imports, was now in a position to pay good prices for higher qualities of food. The invention of the cream separator and other changes in agricultural technique made a combination of dairying and hog-feeding more profitable. Yet to take full advantage of the new opportunities, the Danish farmers had to adopt a host of new ways of doing things. Also, they had to show political wisdom. The idea of "protecting" the domestic grain-growers from the new overseas competition was considered and rejected. The shift to new products was carried through to the accompaniment of much discussion. There was conscious adaptation, not merely blind response to changes in prices and costs, although the adaptation was firmly rooted in a new price-cost situation. After the adjustment was under way, the Danes organised their famous agricultural co-operatives, which helped still further to adapt the farmers' techniques, especially their marketing methods, to the new conditions. In political adversity (defeat at the hands of Prussia and loss of Schleswig, in 1864) influential Danish leaders had responded with the slogan "outward loss, inward gain". The Folk High School movement which rose out of that epoch did much to infuse into Danish rural people an attitude, a broadening of the spirit and of the understanding and a capacity to co-operate, which enabled them to meet economic problems adaptively. Confronted by new competition the Danish farmers might have clung fast to their old patterns of production, seeking "protection" against change. The course they adopted needed intelligence, flexibility and courage.

What can be done to encourage these human qualities of adaptability in all the countries where economic change will demand readjustments? No doubt something can be hoped for as a result of progressive increase in the general level of education. Perhaps it is even more important to place a new emphasis in education on the idea that ours is a world of constant change. Ability to understand new situations, courage and initiative in adjusting to them, both individually and in co-operation with others, are the qualities that will be needed more than ever in the world of tomorrow.

Mr. A. G. B. Fisher, in a suggestive discussion of factors which help and hinder economic adjustments, mentions a number of false beliefs and dogmas.[1] If these could be dispelled from the minds of

[1] *The Clash of Progress and Security* (London, Macmillan, 1935), pp. 46-7, 54-6, 62.

business men, workers, politicians and the general public the prospect for successful adaptations would be much improved. One of them is the notion that there exists some natural ratio which ought to be maintained between the numbers of people in primary production, such as agriculture, and the numbers in other kinds of work. Another is the notion that all resources should be fully employed—including land or equipment incapable of producing things in demand as cheaply as these things could be produced in other places by other means. The common notion of "aid to depressed industries" is also an obstacle to readjustment. Men who would not advise their sons to prepare themselves for a career in a depressed industry will nevertheless think it right that the government should keep depressed industries going or even help them to expand. It would be much better to encourage more rapid expansion of the promising industries which are doing relatively well, while aiding individuals in the depressed industries to make transfers. The whole notion of protectionism, which in its cruder forms denies, in effect, that any type of production once in existence should ever be expected to contract or that it is possible to shift to new types of production, likewise delays and impedes transfers that would benefit the community. Another impediment to adaptation is the widespread habit of thinking in terms of static economy with a fixed volume of demand. In a progressive economy, demand increases as productivity and living standards move upward. The educational function of bodies such as the International Labour Organisation is important in the effort to promote adaptive adjustments by dispelling just such false beliefs as these.

The managers of industry make many of the fundamental decisions on which industrial adaptation primarily depends. If their traditions and education fail to impart the requisite qualities of imaginative leadership, the consequences for a country's economy in a period of rapid change may be very grave. The task of the true industrial leader—the entrepreneur, the undertaker of something new—is beset with difficulties.[1] It requires a special kind of effort of will to work out new combinations and to bring oneself to look upon new ideas as real possibilities and not merely as daydreams. Outside accustomed channels of action decisions have to be based on incomplete information. The social environment may react against those who attempt to do something new, even to the extent of social ostracism or, in extreme cases, physical violence. To surmount such opposition and to overcome the obstacles which stand between the invention of a new idea

[1] Cf. Joseph A. SCHUMPETER: *The Theory of Economic Development* (Cambridge, Harvard University Press, 1934), pp. 84-90.

and its practical application in industry is a pioneering task. Some kinds of temperament and some kinds of education fail to encourage that pioneering impulse which is essential to economic adjustment in a changing world.

Broader education of workers may also assist in industrial adaptation. Unduly narrow skills are likely to impede transfer from one occupation to another. Perhaps still more important is lack on the part of some workers of the wider knowledge and wider outlook which makes it possible to see opportunities elsewhere and to appraise them correctly. Adaptiveness is increased by a type of education which emphasises broad, basic skills, and aims, not at preparing a man for one particular groove, but at enabling him to learn quickly the special techniques of a number of different occupations, and to adjust himself intelligently to the unforeseeable events of life. Not only a broad type of education, but also further democratisation of educational opportunities would make industrial adaptation easier. In the advanced industrial countries it is less true than in the past, but still true to some extent, that social and educational barriers prevent sons from entering occupations better than those of their fathers.[1]

A Comprehensive Programme of Expansion and Adaptation after the War

Where attention is centred exclusively on problem industries and the scope of remedial action is limited to those industries, there is very little room for adaptive policies. If the textile industry, for example, is in the doldrums because of new low-cost competition or for other reasons, an attempt to "do something for" that one industry alone almost invariably leads to restrictive proposals. The tendency is to seek measures that will exclude competition and that will restrict output in order to raise selling prices. Such measures, as we have seen, are likely to give only an illusion of security. If they are applied in any considerable number of industries they become mutually self-defeating, and everybody is worse off than before. The adaptive way to "do something for" the *people* of the

[1] Colin Clark, in *Conditions of Economic Progress*, pp. 76-7, 230, says that a surplus of unskilled workers in relation to skilled and higher professional workers has developed in some advanced industrial countries. There has been a shift in demand in favour of the higher skill, but transference from the ranks of the less skilled has not taken place rapidly enough. The resulting insufficient numbers of highly skilled workers, engineers, etc., in turn slows down the process of industrial expansion and prevents absorption of displaced workers. Clark quotes figures by Bowley to show that the most potent factor causing the surplus of unskilled labour in Great Britain is the existence of a large unskilled population in the last generation (for whom there was work at that date) coupled with the marked difficulty under the laws and customs governing apprenticeship and education for persons with low incomes to rise above the occupation of the father.

textile industry may be to expand the electrical industry, to increase the productivity and hence the buying power of consumers, including consumers in other regions, and to promote industrial diversification in textile areas. For this reason comprehensive programmes of development and adaptation covering many industries and whole regions have a better chance of success than attempts to deal with separate depressed industries one at a time.

Another reason why the comprehensive approach is important, if adaptive rather than restrictive measures are to be adopted, is that attention can be focused in this way on new opportunities, and not merely on new competition. A great contribution can be made to smoother economic adaptation after the war if governmental agencies, business groups and labour groups in each country keep realistically informed, by periodical surveys and re-surveys of trends which affect the viability and the prospects for future expansion or contraction of *all* the important phases of national industry. If one set of industry shows competitive strength and faces a rising demand for its products, while another set of industries finds its demand falling off and its products being produced more cheaply elsewhere, the constructive thing to do is evident. The situation should be made clear for the guidance of all persons concerned. In fact, an attempt to approach this problem realistically in the various advanced countries by means of periodic surveys of broad industrial trends and prospects, perhaps carried on jointly by representatives of government, business, and labour, could prove very worthwhile. An attempt might be made to group the different industries, or parts and subdivisions of industries, into broad classes, and especially to ascertain in each instance as clearly as possible why an industry fell within a particular group. The two major classes would be:

Group A industries: Well adapted to the country and having a promising outlook; likely to respond especially well to efforts at encouragement and expansion.

Group B industries: Industries in which the outlook is unpromising and from which there may be need to assist the transfer of workers and capital.

The industries of group A are those which vocational advisers and investment counsellors ought to recommend, towards which training programmes should be directed and in which public authorities should be most concerned to preserve freedom of expansion against monopolistic influences or domestic or foreign trade barriers. Public funds for assistance in industry in the form of basic research, marketing help, etc., would probably yield higher returns to the nation in stimulating the growth of this group of industries than would any other form of public aid to industry.

The group B industries, on the other hand, are the problem industries. They are the ones in which labour is most likely to be underpaid and to have poor working conditions and irregular employment. Some of them would be found to exist only by reason of State subsidies, including import restrictions which confer the right to charge consumers more than an equivalent product would cost if consumers were allowed to import it freely. Others would be receiving subsidies, in effect, from labour or investors, through inability to pay wages or returns on capital equal to the current rate for equivalent services in better situated industries. This situation might be due to a slow rate of technical progress in the industry as compared with other industries, or to bad management, or to appearance of new sources of supply for the products of the industry, or to shifts in consumers' demands which affect it adversely. Public funds spent to sustain these industries or protective measures which enable them to exist by exploiting consumers are likely to be wasteful and extravagant, unless fundamental remedial measures are taken at the same time.

It would be good policy to require annual surveys of all industries currently receiving public assistance (subsidies, or import protection which enables them to charge as much as, say, 25 per cent. more than consumers would otherwise have to pay), reporting particularly in regard to each:

1. Measures taken in the industry during the past year to improve its productive efficiency by methods which lower the unit cost of output without a lowering, or with an actual increase, of the remuneration paid to labour and to invested capital.
2. The amount spent in the industry during the past year on research into new products and new methods of production and the effectiveness with which this research is being carried on.

A mixed committee drawn from business, labour and government might be set up to stimulate research for the benefit of these industries and perhaps to undertake laboratory studies and market analyses which would be made available free to the industries concerned. Their object would be to find either improved and more efficient methods of turning out the old products, or new products to which the industry's personnel and capital could be assisted to transfer.

The International Interest in National Adaptiveness

If internal production structures within the different countries are flexible and adaptable, then world industrial development can proceed with a minimum of disadvantage and maximum of advantage to all. If, on the other hand, the expansion of some industries and the contraction of others is so beset with obstacles and difficulties that industrial adaptation takes place very slowly or not at all, then there will be chronically depressed industries, needlessly high unemployment and needlessly low wages. The bitterness arising out of these conditions may express itself in "protective" measures that throttle international trade and block international assistance in the development of new areas.

The analysis of the preceding chapters leads, therefore, to the conclusion that *internal* production structures within the various countries, and especially their flexibility, have an enormous *international* importance. Particularly in the post-war period, the willingness of peoples and governments in the various countries not merely to permit, but positively to encourage, the shrinkage of some lines of industry and the expansion of others, and to facilitate transfers of workers and capital will have a great deal to do with determining whether the world as a whole can achieve stable economic progress. It is, therefore, in the interest of all countries that each should be encouraged and assisted to take measures that will help adapt its own production structure more quickly and effectively to changing economic conditions. Here is a new field in which international consultation and agreed action might well be sought.

The "reconversion" of industry after the war from military to civilian pursuits will offer, together with great problems, some important opportunities to convert to a *better adapted* structure of production than that which existed before. It would be a sad mistake to go back to the old lines of production that required government subsidies and protection, rather than the new and more promising lines which, assuming that we can maintain a generally high level of employment and income, will be natural candidates for expansion. This is only one of many urgent reasons for immediate surveys by national agencies, public and private, and for some co-ordinating survey under the auspices of an international body in order to give indications on a world-wide basis of those lines of production that will need to be expanded after the war and those that will need to be contracted. In many, if not most, cases the costs of post-war transition from the wartime structure to a fairly rational peacetime structure—one permitting increased exchanges of goods with other countries—would be no greater than the transition costs of returning to the old irrational structure. And the long-range benefits would be immeasurably greater.

CHAPTER XII

INTERNATIONAL ARRANGEMENTS TO EASE TRANSITION ADJUSTMENTS

What can be done internationally to prevent unduly disruptive impacts upon the trade and production of established industrial areas as a consequence of progressive improvement in productivity and levels of living abroad? How can the transition adjustments set in motion by economic developments be made to result in the greatest permanent gains and the smallest detriments, both for the newly developing economies and the advanced economies?

In the first place, referring back to the analysis of Chapter VIII, it is in the interest not only of the people of the newly developing countries but of the advanced countries as well that development programmes should raise the real income and purchasing power of the masses of the population as rapidly as possible. Projects which merely substitute local production at high cost for articles previously imported, with no gain for consumers and no expansion in total use of these or other commodities, would deprive the advanced countries of export demands without opening alternative demands. And they would not benefit the local people very effectively. Such projects could justifiably be discouraged by agreement and by withholding international aid.

The kinds of projects that most deserve international assistance are those offering the greatest practicable prospects of raising the efficiency of production. These will generally include, in the less developed countries, improvements in communication and transportation; development of cheap sources of power; experimental, pioneering and demonstration work in introducing better methods and equipment into agriculture and into the production of simple goods of mass consumption; fostering of a few carefully selected new types of industry chosen for their appropriateness to the resources of the country and to the country's lines of "comparative advantage" in the world market; and—likely to pay higher dividends in increased productivity than almost any other investment—elementary mass education, training in vocational skills, and advanced technical training for those who show special talent.

It is to the advantage of both groups of countries, from the point of view of encouraging the most rapid rise in real income as well as from other points of view (such as the soundness of loans), that the best technical assistance and the maximum amount of it that can effectively be used should accompany international capital investment.

If the rising national income resulting from increased productivity in a newly developing country flows quickly to the people and is widely distributed, this will mean a faster increase in consumption than would otherwise be the case, and consequently a somewhat slower increase in the local availability of capital funds for further productive development. The period during which the country draws on foreign capital will thus be longer. That, however, would not be an important disadvantage in an expanding world economy, if we may assume a reasonable degree of success in preventing great wars and great depressions. It is economically sound that the savings which go into world development should come in large part from the areas where saving is least costly, namely, from the advanced countries, and should be repaid later out of the increased productivity of the developing countries when it has become easier for them to save. As was indicated in Part I, additional outlets for savings are likely to be of direct benefit to some of the more advanced countries for some time to come. They can well afford to finance the expansion of the newly developing countries over a longer time in order that, without slowing down the rate of development, the incomes of the people spendable on consumption may rise more nearly in pace with rising productivity. For the trade prospects of the advanced countries are brightest and their own problems of industrial adaptation are easiest under these circumstances.

Also, a rapid rise of popular income in a newly developing country, combined with freedom for the people to spend their incomes on imported goods if they wish, will add to the foreign exchange needs of the country. The soundest adjustment would be a rapid increase in exports, rather than a restriction of imports. But if there are obstacles which prevent such two-way trade expansion, there will be three alternatives open: (1) to slow down the rate of development of projects requiring imported supplies and technical services; (2) to restrict the importation of goods for consumption; or (3) to obtain additional foreign loans. The advanced countries, which have an interest in a rapid rise of income and trade in order to make adjustments easier for their own industries, will need to weigh factors of this sort in determining their own import policies and in deciding how extensively they should finance the new areas.

In the second place, it is important in the interest of all concerned that developmental policy be linked to appropriate trade policy, both in the newly developing and in the advanced countries. Policies that encourage flexibility and expansion are needed.

It has already been observed that in promoting economic development the newly developing countries will want to subsidise, for a time, some "infant industries" which cannot initially turn out their products as cheaply as similar goods could be imported. The choice of methods in this connection is important. It would be better, both for the newly developing and the advanced industrial countries, if the methods selected were such as not to restrict consumption. The time-honoured protective import tariff as a device for subsidising infant industries has important disadvantages in this respect. In order to permit the infant industries to operate at a profit the protective duties raise prices to domestic consumers above the prices that they would otherwise have to pay for imported goods. This causes a decline in the consumption of precisely those articles in which it is desired to expand local production. The restriction of consumption also affects the producers in countries from which the goods have hitherto been imported. They are forced to contract their output, unless alternative markets are available. Can other methods which will not have these disadvantages be used to encourage infant industries?

A preferable way would be to give direct assistance to carefully selected infant industries by paying costs of initial research and engineering out of the government treasury, by training workers in government subsidised technical and vocational schools, by offering suitably developed sites and buildings for lease at less than full cost, and by covering deficits during the first few years of operation. The industry so aided would sell at prices competitive with imported goods. The consumers would thus continue to get their goods at low cost, and, as income and consumption expanded in the country, they would presumably expand their purchases very markedly. The new industry would expand to meet the increasing domestic demand for its product.

If the infant industry had been properly selected in the first place, it would be able to take over and hold most of the growing domestic market on a competitive basis, subsidies having ceased. Foreign producers might continue to supply part of the market. Perhaps their sales would decline in absolute volume, or their sales might even increase while supplying a declining percentage of the growing total demand. In any case, they would not suddenly be confronted with a high protective tariff or a restrictive quota and would have time to make adjustments gradually.

This method would have the additional advantage over the usual protective tariff or quota that the government would be able to know at any time just how much a particular assisted industry was costing the country, and whether the infant was showing any capacity to grow up. A protective tariff or quota, on the other hand, gives a *concealed* subsidy in the form of higher prices collected directly from consumers, and the amount of such a subsidy is never clearly apparent.

The main difficulty, from the point of view of governments of less developed countries, would doubtless be the necessity of raising funds through taxes in order to finance the programme, including the direct subsidies proposed. A protective tariff or quota, in effect, levies a tax on every purchaser of the protected commodity by causing him to pay a higher price than he would otherwise need to pay. This concealed tax, which never gets into the government treasury, goes directly to the producer as a concealed subsidy. Like any other tax on consumption, it is anti-expansionary and, when applied to an article of common necessity, bears most heavily on low-income groups. From the point of view of good tax policy, it would be better to raise money by some more suitable and equitable type of taxation, such as a tax graduated according to income. However, the concealed tax-subsidy of the protective import tariff or quota does have the administrative and political advantage that it is easy to apply, and the more equitable types of direct taxation may be beyond the administrative capacity of less developed countries.

A possible compromise is as follows. An import tariff *for revenue* could be imposed on practically all goods entering the country and especially on luxury goods. Low rates which would not greatly restrict imports but would permit collection of a moderate tax on a large total volume of goods would be most effective for this purpose. Then the revenue so raised could be used to assist selected infant industries, as outlined above. This would be less restrictive on international trade and on domestic consumption than tariff protection.

In general, *sudden* readjustments of established industries are harder than *gradual* ones, and the necessity of an *absolute* contraction in output poses much more troublesome problems and involves much heavier transition costs than a *relative* contraction which represents merely a slowing down of the rate of growth as compared with other industries. Agreements might be sought with the newly developing countries in which they would undertake to take such considerations into account, as applied to established suppliers of import goods, when framing their own programmes of industrial promotion. If infant industries could be reared to meet new and growing demands as income rises, avoiding bitter and costly struggles with established industries for the right to satisfy the previously existing amount of demand, the advantage to all concerned might be considerable.

The newly developing countries might be expected to cooperate in arrangements of this sort, and also to permit their people to import the specialities of the advanced countries rather freely as their incomes rise (especially the goods called "A-products" in a previous chapter), *provided* that the advanced countries were willing to do two things: (1) make capital and technical assistance available to the newly developing countries on reasonable terms, in considerable amounts, and regularly over a period of years; and (2) permit increased imports into their home markets of goods produced in the newly developing countries. The first point has been discussed earlier. The second needs emphasis.

The newly developing countries will need to acquire foreign purchasing power with which to pay for capital goods imported from abroad and to pay the interest and amortisation on loans. If they are not able to earn adequate amounts of foreign exchange by increased exports (and a rigidly protectionist attitude on the part of the advanced countries would prevent them from doing so) they will find it necessary to restrict their own imports artificially below what their enlarged incomes would suggest, in order to protect their balance of payments positions. In other words, if China, India, Latin America and the other newly developing areas are able to market increasing amounts of their characteristic specialities in the advanced industrial countries, directly and through multilateral trade channels, they will be able, reciprocally, to admit the products of the advanced countries into their own expanding internal markets on favourable terms, but not otherwise.

A policy of welcoming considerably increased imports of low-cost goods (the "B-products" discussed earlier) is, as part of a general programme of world economic expansion, distinctly in the interest of the advanced countries. If the markets for the types of goods in which they have a comparative advantage and for locally consumed goods and services (the "A-products" and "C-products") are expanding, necessary adaptations to the inflow of new "competitive" goods can be made rather easily and with lasting gain. The opportunity to purchase the low-cost imported goods then makes the pay-check go that much further; that is, it means an effective increase in real income. The benefit is likely to be especially marked for low-income groups, because the export products of newly developing countries in so far as they are finished goods are likely to be the cheaper necessities of mass consumption.

A final means of softening the impact of new economic development upon the established industries of advanced countries is the adoption and enforcement of progressively rising labour standards, adjusted to the circumstances of each area, but moving upward as productivity moves upward. Such standards would also be in the interests of the people of the newly developing countries themselves. It would be desirable to specify, perhaps in connection with the granting of developmental loans, that modern working conditions should be introduced along with modern equipment and that wage rates should not be permitted to lag far behind the increasing efficiency of labour. A gap has often appeared in the past in areas of new industrialisation between the increased productivity of labour, made possible by new methods and machines, and the current rate of wages. The existence of this gap creates a strong inducement for capital and enterprise to enter the highly profitable new industries and expand them and the tremendous rate of profits is a means of accumulating capital. However, these same functions (expansion of new industries, accumulation of capital) can be performed otherwise. More moderate rates of profit, combined with public developmental aid (national and international) in such fields as education, vocational training, developmental research, underwriting of initial risks, etc. would encourage industrial expansion by methods more socially beneficial. If decent working conditions are introduced together with modern industrialism, and if the wages paid to labour rise in proportion to the increase in labour productivity, then the fears in the advanced industrial countries of low-priced competition based on labour exploitation will lose their basis in fact. Shifts in the production structure of the world can take place more smoothly and with greater mutual benefit to all parties.

PART III

SOME BROADER IMPLICATIONS OF ECONOMIC DEVELOPMENT IN NEW AREAS

CHAPTER XIII

POPULATION PRESSURES, POLITICAL POWER AND CULTURAL INFLUENCE

This study has been focused upon the economic repercussions in the advanced industrial countries of economic development elsewhere, especially the modernisation of areas that at present have very little up-to-date equipment. Part I considered the effects arising out of the process of investment and the various methods by which these investment effects might be made as advantageous as possible to all parties concerned. Part II offered a similar analysis respecting the shifts in production and trade which may be expected as the productive efficiency of newly developing areas increases.

No more than passing mention can be given here to influences which may impinge upon the advanced industrial countries by way of political and social channels. It is well to recognise, however, that their consequences may be very important. The political effects, for example, of the spread of modern industrial methods to that half of the world's population which still is pre-industrial may, in the long run, have more influence on the advanced industrial countries—even on their *economic* conditions, such as the employment and income of their workers—than any of the investment or trade repercussions dealt with in preceding chapters. The truth of this observation is instanced by the manner in which Japan's newly developed industrial efficiency, turned to the support of aggressive political designs, has involved other countries in devastating and costly war and has contributed to a sequence of events by which peaceful economies across the world have been converted into arsenals.

This chapter, therefore, will outline very briefly and with no pretence of adequate treatment—simply in order to call attention to these issues—three topics which certainly deserve more systematic exploration than it is possible to give them here. They relate to the effects of economic development in new areas upon: (1) population trends; (2) the distribution of political power; and (3) the flow of cultural influence.

Population Trends

On the basis of past experience it seems certain that one of the first effects of economic development in less developed areas will be to lower the death rate. Economic development brings increased wealth and increased technical knowledge, making possible more healthful living conditions. These include better food and clothing, improved housing, drainage and water supply, and better facilities for the prevention and cure of disease. In the countries where modern production methods first appeared the death rate declined drastically as industries progressed. This experience has been repeated in every country that has since built up a modern economy. For example, in England and Wales the annual number of deaths per 1,000 of the total population fell from the neighbourhood of 30 to 35 in 1740 to 27 in 1800, to 22 in 1860 and to about 12 or 13 in the 1920's.[1]

Birth rates, on the other hand, respond more slowly to the changes associated with economic development, although they, too, fall after a certain time-lag. In England and Wales, according to the best information available, the annual number of births per 1,000 total population remained fairly steady in the neighbourhood of 35 to 37 during the whole century from 1740 to 1840. Then it fell slightly and remained around 35 until 1880. Thereafter, it dropped rapidly, reaching 30 in 1900, 25 in 1920 and 15 in the 1930's.[2] This reaction of the birth rate in England and Wales is also typical of the experience, so far as it has been observed thus far, in countries that have later adopted modern industrial methods, including Japan as well as western countries.[3]

Voluntary limitation of the size of families appears to be the major means by which a fall in the birth rate comes about as economic development proceeds, though there may be other influences at work which are still imperfectly understood.[4] Each infant born has a greater chance of living to maturity, so parents have less reason to want a large number of births. Also, the urban life associated with modern industry is less conducive to large families than rural conditions, in which children become economic assets at an

---

[1] A. M. Carr-Saunders: *World Population: Past Growth and Present Trends* (Oxford, Clarendon Press, 1936), pp. 61, 72. When the crude death rate is corrected for changes in age composition of the population the fall in later years is shown to be somewhat less, from about 23 in 1860 to about 16 in the 1920's.

[2] *Ibid.*, p. 61, and League of Nations: *Statistical Year-Book, 1938-39.* On trends of birth rates and other significant points connected with populations, see the very interesting chapter in League of Nations: *World Economic Survey, 1938-39*, on "Population and Migration".

[3] Cf. E. F. Penrose: *Population Theories and Their Application, With Special Reference to Japan* (Stanford University, California, Food Research Institute, 1934), Chapter IV.

[4] A. M. Carr-Saunders, *op. cit.*, Chapters VII–IX.

early age. When living standards rise, parents wish to give each child a better start in life, including a longer and more expensive education. Therefore they limit the number of children. The emancipation of women, another consequence of more modern production methods, also leads to smaller families. Whatever the exact reasons, the universal effect of modern economic development has been to lower the birth rate as well as the death rate. But the death rate falls first, and the birth rate follows only after an interval of time.

The immediate effect of a rapid fall in the death rate while the birth rate lags behind is, of course, to produce a surplus of births over deaths. The total population of a country adopting modern industrial methods may thus increase swiftly. For example, the population of England and Wales grew from about six million in 1740 to about 16 million in 1840 and 41 million in 1940, despite the fact that vast numbers migrated out of this area in the nineteenth century to North America and other overseas destinations. The expansion of population in Europe as a whole and in the areas of European settlement overseas can be seen in the table of world population below.

TABLE 21. ESTIMATED POPULATION OF THE WORLD 1650-1933[1]
(*millions*)

Continent	1650	1750	1800	1850	1900	1933
Europe	100	140	187	266	401	519
North America	1	1.3	5.7	26	81	137
Central and South Amer.	12	11.1	18.9	33	63	125
Oceania	2	2	2	2	6	10
Africa	100	95	90	95	120	145
Asia	330	479	602	749	927	1,121
World total	545	728	906	1,171	1,608	2,057

[1] Estimates by W. F. Wilcox as revised in A. M. CARR-SAUNDERS, *op. cit.*, p. 42.

The growth in population of western Europe and countries of western European civilisation in the last century or so has been phenomenal, but the indications are clear that it is now slackening off. The best index of population trends is the "net reproduction rate". A net reproduction rate of one means that, if current rates of fertility and mortality remain unchanged, the population will just reproduce itself from generation to generation. A rate above one indicates, on the same assumption, a long-run tendency to grow, and a rate below one a long-run tendency to decline. Recent data for some countries of western Europe show: England and Wales 0.80 (1938), Sweden 0.78 (1940), Belgium 0.86 (1939), Switzerland 0.79 (1940), France 0.90 (1939), Germany 0.98

(1940). In the United States the net reproduction rate was 1.02 in 1940.[1] These are industrially advanced countries which have attained high living standards. The extent to which the net reproduction rate has fallen in some of these countries from high levels that prevailed during their earlier periods of rapid economic development and rapid population growth is shown in chart 20. On the other hand, the population is still more than reproducing itself in countries that have been exposed for a shorter time to the influence of modern production and rising living standards. This is evident in the rates shown in chart 20 for countries of eastern and southern Europe and for Japan. Even in these areas, however, it seems clear that as economic development brings better living standards the tendency of the birth rate, as well as the death rate, to fall is manifesting itself.

We are justified then in saying that a "population cycle" has been associated with economic development in the past. First, the death rate falls, followed after a time-lag by the birth rate. The immediate result is that total population increases rapidly. Later, the birth rate begins to overtake the death rate and population grows less rapidly. At a stage of advanced economic development and high living standards, the stage now attained by many of the countries which first adopted modern production methods, population growth ceases and the total population even begins to decline. There is some indication that efforts to regularise employment, social security measures and better provision for children in advanced industrial communities may arrest this downward trend, but that remains to be seen. Meanwhile, it is the first phase of the population cycle in countries of new economic development that concerns us.

Some of the countries that are potential candidates for modern development are sparsely populated. This is true of many of the Latin American countries. Others, however, and especially those in Asia, are already densely inhabited—much more so than any of the countries of Europe at the beginning of Europe's industrial revolution. Indeed the average density of population in China is about 250 per square mile, which compares with 184 in modern Europe west of Russia and 41 in the United States. The concentration in certain areas of China is much greater, estimates running as high as 600 per square mile in Shantung and 900 in Kiangsu.[2] In India the density per square mile is 250 and in Java it is 950, the highest of any country in the world. These are very high figures for countries living predominantly upon agriculture. It will be ob-

[1] League of Nations: *Statistical Year-Book, 1941-42*, pp. 50-51.
[2] A.M. Carr-Saunders, *op. cit.*, pp. 286-7.

Chart 20. Net Reproduction Rates in Selected Countries[1]

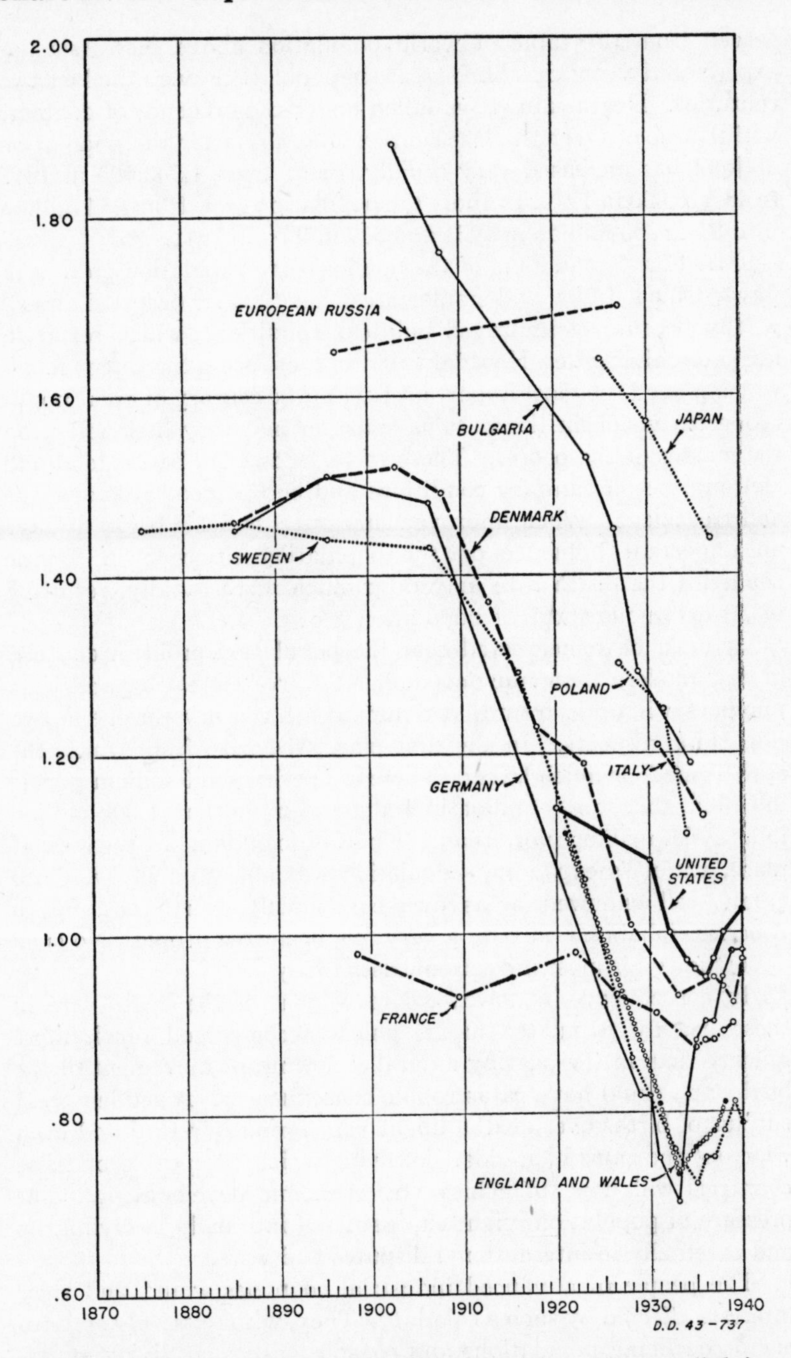

[1] A. M. CARR-SAUNDERS, *op. cit.*, table opposite p. 123, compiled from KUCZYNSKI and other sources; LEAGUE OF NATIONS: *Statistical Year-Book, 1941*, pp. 48-9, and *Monthly Bulletin of Statistics*, Nov. 1943, pp. 291-2.

served from the table of world population above that Asia has experienced a considerable increase in population over the last two centuries. Death rates have fallen under the influence of contacts with Europe. Over the last century and a quarter the population of Java has increased very rapidly, rising from 4,500,000 in 1815 to 48,416,000 in 1940. In India the population grew from 254 million in 1881 to 353 million in 1931 and 389 million in 1941.

Mr. Carr-Saunders is of the opinion that population growth in Java, China, India and similar areas has thus far occurred almost wholly because of improved political conditions, which result in less internal disorder. He thinks that in these areas the three factors which played the most important part in the century of great population expansion in the west have as yet had very little effect on the masses of the people. These three factors are better food and clothing, better sanitary conditions and better medical knowledge and treatment. When improvements in these fields begin to have an important influence on the population situation in Asiatic countries the death rate may drop much more rapidly, resulting in an enormous spurt of population growth.

It must be remembered that the population problem of Asia, in case modern economic development starts a great expansion in numbers, will differ from that of nineteenth century Europe in two important respects. In the first place, the base from which the start is made is much larger. There are perhaps 450 million people in China, almost 400 million in India and more than 1,000 million in all Asia, making more than one half of mankind. In the second place, Europe's expanding population was able to spill over into great, well endowed areas overseas, mainly North and South America, inhabited only by a very few primitive people. No such opportunity for mass migration exists today.

In consequence, a downward movement of the death rate in these heavily populated areas, unless accompanied much more quickly than in the past by a parallel downward movement of the birth rate, could have catastrophic consequences. A sudden great surplus of births over deaths might very seriously retard and even wipe out the gains in standard of living which the people of these countries will hope to achieve by economic development. The pressure of population might express itself internally in civil strife and externally in international disputes and wars.

The people of advanced industrial countries would find their interests affected by such a calamity. They, with relatively stationary or declining populations and conspicuously high living standards, would be spread relatively sparsely over some of the most fruitful and certainly the best developed land areas of the world.

Their condition would present a striking contrast to that of the Asiatic countries. The world might thus divide into two great contending groups. The one, disappointed in its hopes of progress towards popular welfare by adoption of industrial techniques, might be pathologically conscious of "population pressure". Despite its low living standards, it could nevertheless be proficient enough in modern industrial techniques to be militarily powerful. It would be tempted to blame the "haves" of the world for its economic troubles and to contemplate redistribution of territory by conquest. The "haves", on the other hand, might feel driven in self-defence to adopt a non-co-operative or even a repressive attitude which would add fuel to the conflict. This is not a prediction of what *will* happen, but simply a statement of what *might* happen if insufficient attention is given to population reactions as modern economic development takes place in these most heavily populated areas.

In an attempt to avoid the troubles outlined above three lines of thought suggest themselves. First, it is desirable that improvements in technique, equipment and skills be pushed ahead *very rapidly* in these heavily populated countries so that the rise in the efficiency of production will outstrip the tendency of the population to increase. In this way important gains in living standards may be made quickly, even if the population does increase rapidly for a time. The more speedy attainment of high living standards may be expected to hasten the tendency of the birth rate to fall, since experience shows that high living standards produce low birth rates. It is possible, of course, that revolutionary improvements in productive techniques, including new applications of science and new methods of social organisation for production, may so increase output from a given land area that population densities which today would seem fantastic will be supportable in the future with no great difficulty. But it is probably unwise to count on this as a safety valve for the population problems outlined above.

Secondly, it is desirable that economic development in heavily populated areas should bring not merely increased production and consumption of goods, not merely improved sanitation and medical care, but also rapid educational progress for the masses of the people and a rapid rise of cultural standards. If a large proportion of the increased income resulting from improved production is devoted to mass education, this would tend to reduce the lag between the fall in the death rate and the fall in the birth rate, for everywhere better educated people have manifested a desire for fewer but better reared children.

Thirdly, in view of the problem which will arise if the time-lag between the fall of the death rate and the fall of the birth rate is not reduced, peoples in the areas of potential population pressure will seek knowledge of methods of birth control at the same time that they are acquiring industrial and medical knowledge enabling more people to survive. Agencies administering international development funds and assisting with technical advice may have to concern themselves with the question whether population is pressing too heavily on the means of subsistence. Indeed, it may even be argued that an educational campagin to counteract social and religious taboos which might prevent family limitations should be recognised as an essential part of the effort to raise living standards in regions where overpopulation threatens.

Political Power

The profound effect of economic development upon political power and political leadership in world affairs is easily demonstrated. It is only necessary to point to the political position which pre-eminence in modern industry gave England in the nineteenth century, to the military and political effects of the rise of German industry, to the industrial and political rise of the United States and to the political influence which modern industry brought to Japan and the Soviet Union. Industrial development is not only a means of attaining higher standards of living but is also the basis of modern military power. Indeed, one of the main incentives to rapid economic transformations in a number of countries during recent years has been the urgent desire on their part to increase their military power. This incentive will certainly play a major role in China, for example, after the present war. The adoption of modern industrial methods in Asia, Africa, south-eastern Europe and Latin America will ultimately have the effect of enhancing the influence of the peoples of these areas in determining the political destinies of their own regions and of the world. Economic development of new areas will no doubt lead to a more even distribution of world political power.

Will the rise of all these newly developing countries in the world power scale mean a lengthy series of wars for power and position, until some new "balance" is established, or until some super-dictator subjugates most of the world and establishes a Roman peace? Will each new practitioner of modern industrialism be impelled to follow in the path of Germany or Japan? Or will the newly developing countries throw their influence on the side of a world system of security and orderly change under law, thus assuming in a more peaceful manner the new responsibilities that will come with rising power?

It is not proposed to speculate on the answers to these questions. But it is necessary to point out that the entire validity—or rather, the relevance—of the economic analysis in the preceding chapters depends upon the answers to them. Repercussions felt through the channels of investment and trade will be quite secondary in importance if political relationships are allowed to take such a course that the economies of the advanced industrial countries have to be used primarily for military production and only secondarily to raise civilian living standards. The kind of political—and hence, economic—consequences which may be expected as a result of new economic development will depend upon the reactions of the people in the newly developed areas to greater power and responsibility and upon the kind of world political system which the victorious United Nations may initiate after the present war. If it is to prove possible to turn the enormous industrial apparatus of the advanced countries and the skills of their workers towards the building of houses, improvement of cities and production of better civilian goods, instead of towards bombing planes and tanks and air-raid shelters, then the general political problem of world organisation for security and the specific political problem of bringing newly developing areas peacefully into their rightful place in the world system will have to be satisfactorily solved.

Cultural Influence

Just as profound in its long-run effects, and still more difficult to predict, may be the cultural influence of the newly developing countries as their peoples acquire greater capacity to produce and consume. Economic advancement brings with it a much more ample material basis for universal education, more leisure and means for cultural self-expression, and the enhanced prestige which is associated with greater wealth and power.

In the field of science, it seems reasonable to suppose that successful economic development in areas not now able to support scientific research would eventually provide the world with twice as many laboratories as it might otherwise have and that mankind could thereby uncover the secrets of nature so much faster. Who can say what this may mean to the intellectual and economic life of the world? Great Britain's system of parliamentary government and the common law, American skyscrapers and methods of mass production, the comprehensive five-year economic plans and the social experiments of the Soviet Union, are examples of new cultural patterns which, partly because of the industrial power of the countries that originated them, have exerted a pervasive influence in many other parts of the world. What new trends in law, ethics, morals, religion, education, music and literature will be set going as the newly developing countries rise in cultural influence? This is not the place to discuss such questions, but it does seem safe to say that the development of modern production methods in the countries hitherto on a low economic level will greatly enhance their influence, for good or ill, on the trend of civilisation in general. Mutual interchange between diverse cultures will supersede the predominantly one-way flow that was founded on the superiority of western industrial techniques, and the period of cultural "imperialism" of the West will recede into the past.

INTERNATIONAL LABOUR OFFICE

WORLD ECONOMIC DEVELOPMENT

EFFECTS ON ADVANCED INDUSTRIAL COUNTRIES

by

Eugene STALEY

APPENDIX TO CHAPTER IV

by

Robert W. TUFTS

APPENDIX TO CHAPTER IV

by

Robert W. TUFTS

Data are available in the official *Financial and Economic Annual of Japan*[1] on the paid-up capital and, after 1905, the reserves of companies (limited partnerships, ordinary partnerships, and joint stock companies). The data are grouped by branches of enterprise: agriculture, industry, commerce, and transportation. The same official publication also gives figures on national and local governmental debts outstanding, and the purposes for which the debts were incurred. Local governmental loans have been used almost exclusively for local public works. One of the classes into which national loans are divided is "economic undertakings". This class has averaged 25.3 per cent. of the total outstanding national debt throughout the years under consideration. National governmental loans incurred for economic undertakings have been devoted almost entirely to railroad development.

An estimate of yearly capital investment in Japan has been derived by adding the increment in paid-up capital and reserves of private companies, the increment in local governmental debts, and 25.3 per cent. of the increment in national governmental debts. The estimate of capital investment so obtained undoubtedly errs on the conservative side. It excludes much which has ordinarily been included by economists treating the growth of capital in various countries. It excludes, for example, investments by individuals (by the individual farmer in livestock, farm machinery, farm buildings, etc.; by professional people, such as doctors, in buildings and equipment; by individual investors in buildings, etc.) and investment by local and national governments made from current revenues. It also excludes investments by the national Government for armaments, war, and all purposes other than economic undertakings. These various classes of investment have been excluded partly because it would be almost impossible to

---

[1] Published by the Department of Finance, Tokyo, since 1900.

secure accurate estimates of the amount of such investment but also because we are chiefly concerned with the process of industrialisation, with the accumulation, that is, of capital used in production. We deliberately exclude, therefore, national loans incurred for armaments and war and similar purposes. Investments by individual farmers and professional people are not deliberately excluded, but the data are not available. These exclusions should be remembered in any comparisons of the data used here for Japan with estimates on the growth of capital in other countries by writers who do make provision for these other types of investment.

The data used here probably reflect net investment, not gross investment. Under the usual accounting methods this would probably be true of the statistics on paid-up capital and reserves of private companies. Since the data used for local investment in public works are the annual increases in local indebtedness, these figures may even understate the net investment in local public works. The figures used for national investment in economic undertakings are, as explained above, a certain percentage of the annual issue of government obligations. These figures would represent, therefore, gross investment, but since no allowance is made for investment in economic undertakings from current revenues, it is probable that these figures, too, more nearly correspond to net investment.

In order to facilitate the use of these figures in speculating about the possible rate of investment for industrial development in other parts of Asia, the estimates of annual investments in Japan are presented below in two parts. First, investment in industry and commerce and local public works is presented as "Type I Investment". It will later be assumed that investment of this sort is more closely related to population than to extent of territory. A second tabulation is then presented of "Type II Investment", consisting of investment in agriculture and transportation. It will later be assumed that investment of this kind is more closely related to land area than to population.

Capital Investment in Industry and Commerce, and Capital Investment in Local Public Works, 1896-1936 (Type I Investment)

In table I are presented the data on the annual increase in the paid-up capital and reserves of private companies engaged in industry and commerce and the annual increase in the indebtedness of local governments. Columns 1 and 2 show, respectively, the annual increase in the paid-up capital and reserves of industrial and commercial companies and the annual increase in local debt, while column 3 gives the total of the first two columns. In column 4 is presented an unweighted index (1900 equals 100) of pig iron, steel and coal prices, which has been computed from data appearing in the *Financial and Economic Annual of Japan.* This index is

TABLE I. TYPE I INVESTMENT

Capital Investment in Industrial and Commercial Companies and Capital Investment by Municipal Governments in Japan, 1896-1936

Year	1 Industrial and commer-cial	2 Local	3 Total Type I	4 Price index of invest-ment goods	5 Total at 1900 prices (million yen)	6 Total at 1936 prices (million yen)	7 Total in 1936 dollars (millions)
1896	268.8	.2	269.0	87	309	557	161
1897	70.1	5.8	75.9	99	77	139	40
1898	60.8	7.1	67.9	102	67	121	35
1899	79.1	8.4	87.5	103	85	153	44
1900	72.9	10.9	83.8	100	84	151	44
1901	48.8	8.5	57.3	99	58	104	30
1902	60.8	9.9	70.7	99	71	128	37
1903	19.0	6.0	25.0	92	27	49	14
1904	26.7	−2.1	24.6	87	28	50	14
1905	57.1	−2.4	54.7	111	49	88	25
1906	150.7	15.0	165.7	127	130	234	68
1907	214.0	11.7	225.7	122	185	333	96
1908	123.7	10.6	134.3	117	115	207	60
1909	156.7	63.1	219.8	112	196	353	102
1910	103.1	5.1	108.2	101	107	193	56
1911	131.1	16.7	147.8	91	162	292	85
1912	226.0	115.7	341.7	96	356	641	186
1913	285.1	5.0	290.1	100	290	522	151
1914	114.0	15.8	129.8	95	137	247	72
1915	167.5	13.7	181.2	98	185	333	96
1916	375.8	13.2	389.0	154	253	455	132
1917	862.9	23.7	886.6	243	365	657	190
1918	1,716.7	34.7	1,751.4	449	390	702	203
1919	1,249.9	27.4	1,277.3	418	306	551	160
1920	2,239.1	57.4	2,296.5	356	645	1,161	336
1921	1,368.9	39.4	1,408.3	261	540	972	281
1922	−111.8	133.1	21.3	206	10	18	5
1923	1,361.4	206.9	1,568.3	196	800	1,440	417
1924	783.8	136.2	920.0	196	469	844	244
1925	−167.1	194.6	27.5	178	15	27	8
1926	998.9	148.8	1,147.7	165	696	1,253	363
1927	542.0	245.3	787.3	151	521	938	272
1928	394.8	331.2	726.0	155	468	842	244
1929	605.3	205.9	811.2	155	523	941	272
1930	203.5	171.3	374.8	144	260	468	135
1931	18.5	152.7	171.2	120	143	257	74
1932	75.5	160.7	236.2	129	183	329	95
1933	524.6	193.2	717.8	164	438	788	228
1934	1,505.6	229.2	1,734.8	180	964	1,735	502
1935	1,087.9	229.5	1,317.4	174	757	1,363	395
1936	1,413.0	240.9	1,653.9	180 (estimate)	919	1,654	479

used to deflate the series presented in column 3, and the results are shown in column 5, capital investment in industrial and commercial companies and by municipal governments at 1900 prices. Column 6 merely translates column 5 into terms of 1936 prices, and column 7 translates column 6 into 1936 dollars, the exchange rate used being 1 yen equals $.28951.

In table II the data of table I are summarised by decades, all amounts being expressed in 1936 dollars. It will be seen (column 2) that the average investment of this type rose steadily, from 49 million dollars per year in the decade 1900-1909, to 273 million dollars per year in the period 1930-1936.

TABLE II. TYPE I INVESTMENT BY DECADES

Capital Investment in Industrial and Commercial Companies, and Capital Investment by Municipal Governments in Japan, 1900-36, Summarised by Decades

Period	1 Total such investment (million 1936 $)	2 Average per year (million 1936 $)	3 Average per year per head of the 1900 population (1936 $)
1900–1909	490	49	1.12
1910–1919	1,331	133	3.04
1920–1929	2,442	244	5.57
1930–1936	1,908	273	6.23

Private Capital Investment in Agriculture and Transportation, and Capital Investment by the National Government in Economic Undertakings, 1896-1936 (Type II Investment)

In table III are presented data on private investment in agriculture and transportation and data on investment by the national Government in economic undertakings in Japan during the period 1896-1936. Columns 1 and 2 show, respectively, the annual increase in the paid-up capital and reserves of agricultural and transportation companies and the annual increase in the national Government's debt incurred for economic undertakings, while column 3 gives the total of the first two columns. In column 4 is presented the price index of investment goods used in table I, by which the series in column 3 is deflated, the results being shown in column 5, capital investment in agricultural and transportation companies and in economic undertakings by the national Government at

1900 prices. Column 6 merely translates column 5 into terms of 1936 prices and column 7 translates column 6 into 1936 dollars, the exchange rate used being 1 yen equals $.28951.

TABLE III. TYPE II INVESTMENT

Private Investment in Agriculture and Transportation and Investment in Economic Undertakings by the National Government, Japan, 1896-1936

Year	1 Agri-culture and trans-portation	2 National	3 Total Type II	4 Price index of invest-ment goods	5 Total at 1900 prices (million yen)	6 Total at 1936 prices (million yen)	7 Total in 1936 dollars (millions)
1896	29.7	20.7	50.4	87	58	104	30
1897	52.3	5.9	58.2	99	59	106	31
1898	33.4	12.3	45.7	102	45	81	23
1899	1.8	.1	1.9	103	2	4	1
1900	37.6	25.3	62.9	100	63	113	33
1901	13.9	3.8	17.7	99	18	32	9
1902	23.7	6.5	30.2	99	31	56	16
1903	5.8	10.4	16.2	92	18	32	9
1904	51.1	3.2	54.3	87	62	112	32
1905	10.9	107.5	118.4	111	107	193	56
1906	−3.3	230.4	227.1	127	179	322	93
1907	−160.4	130.6	−29.8	122	−24	−43	−12
1908	17.0	76.5	93.5	117	80	144	42
1909	14.7	17.8	32.5	112	29	52	15
1910	28.2	121.2	149.4	101	148	266	77
1911	24.9	130.3	105.4	91	116	209	61
1912	36.9	1.4	38.3	96	40	72	21
1913	26.4	.3	26.7	100	27	49	14
1914	36.8	33.2	70.0	95	74	133	39
1915	9.2	7.3	16.5	98	17	31	9
1916	47.1	9.3	56.4	154	37	67	19
1917	183.4	17.5	200.9	243	83	149	43
1918	287.7	23.2	310.9	449	69	124	36
1919	38.8	27.4	66.2	418	16	29	8
1920	296.2	53.5	349.7	356	98	176	51
1921	2.3	121.2	123.5	261	47	85	25
1922	−180.4	103.9	−76.5	206	−37	−67	−19
1923	242.9	127.1	370.0	196	189	340	98
1924	113.2	189.9	303.1	196	155	279	81
1925	85.5	179.6	265.1	178	149	268	78
1926	85.5	150.2	235.7	165	143	257	74
1927	142.2	134.4	276.6	151	183	329	95
1928	108.3	185.6	293.9	155	190	342	99
1929	138.4	172.3	310.7	155	200	360	104
1930	−30.8	139.7	108.9	144	76	137	40
1931	−3.3	132.5	129.2	120	108	194	56
1932	53.6	114.4	168.0	129	130	234	68
1933	33.6	274.2	307.8	164	188	338	98
1934	44.3	276.3	320.6	180	178	320	93
1935	88.6	265.8	354.4	174	204	367	106
1936	106.4	262.8	369.2	180 (estimate)	205	369	107

In table IV these data are summarised by decades, and the amounts are expressed in 1936 dollars. It will be noted (column 2) that average investment of this type rose from 29 million dollars per year in the decade 1900-09 to 81 million dollars per year in the period 1930-36.

Total of Types I and II

In table V the data presented in tables II and IV are combined. It will be seen (column 4) that the rate of investment rose from 78 million dollars per year in the first decade, to 313 million dollars per year in the third decade and to 354 million per year in the period 1930-1936.

TABLE IV. TYPE II INVESTMENT BY DECADES

Capital Investment in Agricultural and Transportation Companies, and Capital Investment by the National Government in Economic Undertakings, Japan, 1900-36, Summarised by Decades

Period	1 Total such investment (million 1936 $)	2 Average per yr. (million 1936 $)	3 Average per yr. per sq. km. (1936 $)
1900–09	293	29	77
1910–19	327	33	85
1920–29	686	69	179
1930–36	568	81	212

TABLE V

Capital Investment in Japan, 1900-36 (in 1936 dollars)

Period	1 Type I investment (million 1936 $)	2 Type II investment (million 1936 $)	3 Total	4 Average investment per year (million 1936 $)	5 Average annual investment as per cent. of average annual national income
1900–09	490	293	783	78	12.1
1910–19	1,331	327	1,658	166	16.9
1920–29	2,442	686	3,128	313	12.3
1930–36	1,908	568	2,476	354	9.6

This annual investment may seem very low when viewed from the standpoint of countries like the United States or Great Britain. Expressed as percentages of average Japanese annual income during the same period, however, the rate is not low. Investment was 12.1 per cent. of estimated national income in the first decade, rose to 16.9 per cent. in the second, then fell to 12.3 per cent. in the third, and fell again in the 1930's to slightly less than 10 per cent.[1]

Capital Goods from Foreign Countries in the Industrialisation of Japan

Most of the capital goods required in the industrialisation of Japan were produced domestically. We are, however, especially interested in imports of capital goods, in order to see what role these played in the industrialisation process. From the *Financial and Economic Annual of Japan* data can be secured on Japanese imports of iron and steel, other metals, metal manufactures and machines. No doubt certain other imports should be included, but it is not clear from the trade statistics just what these should be. Our figures underestimate, but probably not seriously, Japanese imports of capital goods.

These data for the period 1896-1936 are presented in table VI. In column 1 the total imports of the four classes of capital goods mentioned above are shown. Column 2 translates column 1 into 1900 prices by means of the price index of investment goods used previously. Column 3 translates column 2 into 1936 prices, and column 4 translates the 1936 yen prices into 1936 dollars, the exchange rate used being 1 yen equals $.28951.

[1] The data on Japanese national income used for comparison with the estimated capital investment are those of K. Mori for the period 1900–20, of Hijkata for the period 1921-33, and of the Mitsubishi Economic Research Bureau for the years 1934-36. At the 19th Conference of the International Institute of Statistics, held in Tokyo in 1930, K. Mori presented an estimate of the National Income of Japan for each year of the period 1887-1925 (K. Mori: "The Estimate of the National Wealth and Income of Japan Proper", in *Bulletin de l'Institut International de Statistique*, Tome XXV, Vol. 2, pp. 203-4). Colin Clark has strongly criticised these estimates, saying: "It is quite impossible, however, to reconcile these figures with other sources of information, particularly statistics of agricultural production. It appears that Mori was working on a limited definition of national income not applicable for comparison with other figures." (*The Conditions of Economic Progress*, London, Macmillan, 1940, p. 113.) Mori's estimates for the year 1914 and the period 1919-1925 may be compared with separate estimates for these years prepared by Professor Hijkata "on a basis and definitions similar to those used in Europe and America . . ." (*Analytical and Statistical Survey of Economic Conditions in Japan*, Mitsubishi Economic Research Bureau, Tokyo, April 1937.) While this comparison reveals differences which would be significant for some purposes, the two series correspond fairly well. Hijkata's series have been carried through the years 1934–36 by the Mitsubishi Economic Research Bureau (same publication as cited above), using similar methods and definitions.

TABLE VI

Japanese Imports of Iron and Steel, Other Metals, Metal Manufactures and Machines, 1897-1936

Year	1 Total imports (current prices, million yen)	2 Total at 1900 prices (million yen)	3 Total at 1936 prices (million yen)	4 Total in 1936 $ (million $)
1897	42.2	43	77	22
1898	44.2	43	77	22
1899	29.0	28	50	14
1900	51.7	52	94	27
1901	42.2	43	77	22
1902	36.0	36	65	19
1903	41.0	45	81	23
1904	47.5	55	99	29
1905	85.9	77	139	40
1906	75.2	59	106	31
1907	107.8	88	158	46
1908	104.5	89	160	46
1909	69.5	62	112	32
1910	78.0	77	139	40
1911	114.3	126	227	66
1912	137.9	144	259	75
1913	139.1	139	250	72
1914	96.7	102	184	53
1915	77.5	79	142	41
1916	192.8	125	225	65
1917	329.7	136	245	71
1918	470.4	105	189	55
1919	486.7	116	209	61
1920	533.0	150	270	78
1921	394.3	151	272	79
1922	424.1	206	371	107
1923	369.6	189	340	98
1924	528.0	269	484	140
1925	348.8	196	353	102
1926	388.1	235	423	122
1927	356.1	236	425	123
1928	428.3	276	497	144
1929	456.6	295	531	154
1930	298.2	207	373	108
1931	182.4	152	274	79
1932	226.0	175	315	91
1933	346.7	211	380	110
1934	459.6	255	459	133
1935	553.5	318	572	166
1936	538.6	299	538	156

In table VII the data presented in table VI are summarised by decades. The first column gives the total imports of capital goods in 1936 dollars and prices, while the second column shows average imports of capital goods per year. Column 3 shows average imports of capital goods per year as a percentage of average total investment per year (as shown in table V). As might be expected, imports of capital goods made their greatest percentage contribution to Japanese industrialisation in the first decade after 1900, accounting for slightly more than two fifths of the total investment in that decade. In the following two decades the share of imports dropped to 36 or 37 per cent., and fell again in the 1930's to 34 per cent. In other words, imports of capital goods have accounted for between one third and two fifths of Japan's total investment in the period 1900-1936. This is a remarkably high share.

TABLE VII

Japanese Imports of Capital Goods, 1900-36, Summarised by Decades (in 1936 prices and dollars)

Period	Total (million $)	Average per year (million $)	Average per year as percentage of average total investment per year
1900–09	315	32	40.2
1910–19	599	60	36.1
1920–29	1,147	115	36.7
1930–36	843	120	34.0

The Application of Japanese Experience to Other Asiatic Areas

Eastern Asia will face many of the same problems in industrial development which Japan faced. Most of the area is highly populated; standards of living are very low; complex and rigid social customs are a more or less serious bar to the creation of modern industrial conditions; the area, with certain notable exceptions, is not particularly rich in resources required by modern industry. For these reasons the application of Japanese experience to these other areas may be suggestive.

The first thing which we need to know in order to apply Japanese experience is the present stage of industrialisation in these areas. To what period in the Japanese development does the situation in eastern Asia correspond? The question cannot, of course, be answered exactly. We have assumed, more or less arbitrarily, that the immediate post-war situation in eastern Asia will resemble, so far as progress in industrialisation is concerned, that in Japan in 1900. The only basis for this assumption is that living standards in Japan in 1900 seem to have been about what they were in the late 1930's in the rest of eastern Asia. This may underestimate the development in certain areas (*e.g.*, India), but as a general basis it is probably as satisfactory as any other year which might be chosen. If it should underestimate the level of industrialisation already attained by eastern Asia, the effect on our calculations will be to understate the future investment which would correspond to previous Japanese experiences.

Table VIII, showing possible capital absorption by various Asiatic areas in the decades after the war, on the assumption that these areas experience capital investment at the same rate, in proportion to population and area, as Japan did after 1900, is calculated as follows. The amount of "population-related" (Type I) investment previously shown in table II is multiplied for each country, decade by decade, by the ratio between that country's population today and Japan's population in 1900. Similarly, the amount of "area-related" investment (Type II) previously shown in table IV is multiplied for each country, decade by decade, by the ratio between that country's total area and the total area of Japan proper. Popula-

TABLE VIII

Capital Absorption by Various Asiatic Areas by Decades of the Post-War Period, on the Assumptions Made in the Text

Country	Population[1] (million)	Area[1] (thousand sq. km.)	Capital investment (million 1936 $)			
			First decade	Second decade	Third decade	Fourth decade
China	450.0		5,040	13,680	25,065	28,035
		11,103	8,549	9,437	19,874	23,538
			13,589	23,117	44,939	51,573
India	365.9		4,098	11,123	20,381	22,796
		4,079	3,141	3,467	7,301	8,647
			7,239	14,590	27,682	31,443
N.E.I.	68.4		766	2,079	3,810	4,261
		1,904	1,466	1,618	3,408	4,036
			2,232	3,697	7,218	8,297
Other areas[2]	116.5		1,305	3,542	6,489	7,258
		2,865	2,206	2,435	5,128	6,074
			3,511	5,977	11,617	13,332
	Total		26,571	47,381	91,456	104,645
Average investment per year			2,657	4,738	9,146	10,465

[1] *Statistical Year-book of the League of Nations.*

[2] Burma and British colonies—population, 29.6 million and area, 1,012 thousand sq. kms.; Thailand—population, 14.9 million and area, 518 thousand sq. kms.; Phillippines—population, 16.3 million and area 296 thousand sq. kms.; French colonies—population, 24.1 million and area, 742 thousand sq. kms.; Japanese colonies—population, 3.16 million and area, 297 thousand sq kms.

tion-related and area-related investment are shown separately in table VIII in the first two horizontal lines of figures corresponding to each country, and then are added in the third line to give an estimate, based on both population and area, of total investment. Thus, the table shows that if China develops after the war as rapidly as Japan developed after 1900 it would experience total capital investment to the amount of 13,589 million dollars in the first decade, 23,117 million in the second decade, 44,939 million in the third decade, and 51,573 million in the fourth decade. In other words, the rate of investment would rise from 1.3 thousand million dollars per year in the first decade to 5.1 thousand million in the fourth decade. Adding the corresponding figures for the other countries shown, the total comes to something over 2,500 million dollars as the annual investment in these Asiatic areas in the first decade after the war. The figure rises to 4.7 thousand million dollars per year in the second decade, to 9.1 thousand million dollars per year in the third decade, and to 10.5 thousand million dollars per year in the fourth decade.

Perhaps the most striking thing about these estimates is their lowness, especially in the first and second decades. Even if all the capital were to come from outside, as, of course, will not be the case, the rest of the world would be able, according to these figures, to invest only about two or three thousand million dollars per year (1936 prices) in eastern Asia in the first post-war decade. While this is not a small sum, it is not by any means as large as many people have predicted. Colin Clark, for example, has estimated that the stock of capital in India, China and the rest of Asia and Oceania, not including the U.S.S.R., will increase from 147 thousand million I.U. (that is, dollars of 1925-34 purchasing power) to 400 thousand million I.U. between 1938 and 1960, or an increase of 253 thousand million I.U. in 22 years. This would mean an annual average investment of 11.5 thousand million I.U. Other writers have suggested that the possibilities for capital investment in eastern Asia are boundless.

The process of industrialisation is a time-consuming process, involving much more than the supply of capital goods. It seems unlikely that eastern Asia will be able to industrialise much faster than Japan did. Even if it should industrialise twice as rapidly as Japan, however, the amount of capital which it would absorb in the first post-war decade would be only about five thousand million dollars a year, at 1936 prices. Much of this would be domestically produced. If we assume for the moment that foreign countries supply capital goods to these areas in the same proportion to total investment as in Japan, then the participation of other countries in the industrialisation of eastern Asia would amount to little more than one thousand million dollars per year in the first post-war decade (or to a little more than two thousand million dollars per year if we assume that eastern Asia industrialises twice as rapidly as Japan). Divided among the industrial nations of the world, these sums are not huge.
